ANNUAL REVIEW OF EARTH AND PLANETARY SCIENCES

ANNUAL REVIEW OF EARTH AND PLANETARY SCIENCES

FRED A. DONATH, *Editor*
University of Illinois—Urbana

FRANCIS G. STEHLI, *Associate Editor*
Case Western Reserve University

GEORGE W. WETHERILL, *Associate Editor*
Carnegie Institution of Washington

VOLUME 7

1979

ANNUAL REVIEWS INC. 4139 EL CAMINO WAY PALO ALTO, CALIFORNIA 94306

ANNUAL REVIEWS INC.
Palo Alto, California, USA

REPRINTS The conspicuous number aligned in the margin with the title of each article in this volume is a key for use in ordering reprints. Available reprints are priced at the uniform rate of $1.00 each postpaid. The minimum acceptable reprint order is 5 reprints and/or $5.00 prepaid. A quantity discount is available.

International Standard Serial Number: 0084-6597
International Standard Book Number: 0-8243-2007-7
Library of Congress Catalog Card Number: 72-82137

FILMSET BY TYPESETTING SERVICES LTD, GLASGOW, SCOTLAND
PRINTED AND BOUND IN THE UNITED STATES OF AMERICA

PREFACE

With the advance of science old problems are solved and new ones uncovered. Some continue to be addressed without resolution. As we attempt to define our individual and collective goals for science and engineering, for basic and applied research, an assessment of where we stand today is essential. What do we know and not know? Where do the problems lie? Which of these are most interesting and important? Which are most likely to lead to innovation in approach or concept?

We do not pretend to be able to answer these questions. However, we do attempt, as editors, to bring to the scientific community, and to other interested persons, comprehensive reviews of the state of knowledge in the Earth and Planetary Sciences. It is our hope that these reviews will summarize existing knowledge, identify important questions, lead to innovation, and provide the background that will bring new expertise and points of view to bear on the more important and challenging problems awaiting attention.

THE EDITORS

SOME RELATED ARTICLES APPEARING IN OTHER *ANNUAL REVIEWS*

From the *Annual Review of Astronomy and Astrophysics*, Volume 16 (1978)

The Asteroids, Clark R. Chapman, James G. Williams, and William K. Hartmann
Results of Venus Missions, M. Ya. Marov
Physical Properties of the Planets and Satellites from Radar Observations, G. H. Pettengill
Isotopic Structures in Solar System Materials, Frank A. Podosek
Recent Advances in Coronal Physics, Giuseppe S. Vaiana and Robert Rosner
A View of Solar Magnetic Fields, the Solar Corona, and the Solar Wind in Three Dimensions, Leif Svalgaard and John M. Wilcox

From the *Annual Review of Biochemistry*, Volume 47 (1978)

The Photochemical Electron Transfer Reactions of Photosynthetic Bacteria and Plants, Robert E. Blankenship and William W. Parson

From the *Annual Review of Ecology and Systematics*, Volume 9 (1978)

Temperature Adaptation of Enzymes: Biological Optimization Through Structure-Function Compromises, George N. Somero
Optimization Theory in Evolution, J. Maynard Smith
The Ecology of Micro- and Meiobenthos, Tom M. Fenchel
Lotka-Volterra Population Models, Peter J. Wangersky
Convergence Versus Nonconvergence in Mediterranean-Climate Ecosystems, M. L. Cody and H. A. Mooney
Predator-Prey Theory and Variability, Peter Chesson
Origin of Angiosperms, James A. Doyle
A History of Savanna Vertebrates in the New World. Part II: South America and the Great Interchange, S. David Webb
The Statistical Prediction of Population Fluctuations, Robert W. Poole
Phytoplankton and Their Dynamics in Oligotrophic and Eutrophic Lakes, J. Kalff and R. Knoechel
Speciation Patterns in the Amazonian Forest Biota, Beryl B. Simpson and Jürgen Haffer

From the *Annual Review of Fluid Mechanics*, Volume 11 (1979)

Rotating, Self-Gravitating Masses, N. R. Lebovitz
Internal Waves in the Ocean, Christopher Garrett and Walter Munk
Geostrophic Turbulence, Peter B. Rhines

From the *Annual Review of Materials Science*, Volume 8 (1978)

Data Sources for Materials Scientists and Engineers, J. H. Westbrook and J. D. Desai

From the *Annual Review of Microbiology*, Volume 32 (1978)

Growth Yield and Efficiency in Chemosynthetic Microorganisms, W. J. Payne and W. J. Wiebe

Microbial Transformations of Metals, Anne O. Summers and Simon Silver

From the *Annual Review of Nuclear and Particle Science*, Volume 28 (1978)

Isotopic Anomalies in the Early Solar System, Robert N. Clayton

From the *Annual Review of Physical Chemistry*, Volume 29 (1978)

High Resolution Electron Microscopy of Crystal Defects and Surfaces, J. M. Cowley

ANNUAL REVIEWS INC. is a nonprofit corporation established to promote the advancement of the sciences. Beginning in 1932 with the *Annual Review of Biochemistry*, the Company has pursued as its principal function the publication of high quality, reasonably priced Annual Review volumes. The volumes are organized by Editors and Editorial Committees who invite qualified authors to contribute critical articles reviewing significant developments within each major discipline.

Annual Reviews are published in the following sciences: Anthropology, Astronomy and Astrophysics, Biochemistry, Biophysics and Bioengineering, Earth and Planetary Sciences, Ecology and Systematics, Energy, Entomology, Fluid Mechanics, Genetics, Materials Science, Medicine, Microbiology, Neuroscience, Nuclear and Particle Science, Pharmacology and Toxicology, Physical Chemistry, Physiology, Phytopathology, Plant Physiology, Psychology, and Sociology. The *Annual Review of Public Health* will begin publication in 1980. In addition, four special volumes have been published by Annual Reviews Inc.: *History of Entomology* (1973), *The Excitement and Fascination of Science* (1965), *The Excitement and Fascination of Science, Volume Two* (1978), and *Annual Reviews Reprints: Cell Membranes, 1975–1977* (published 1978). For the convenience of readers, a detachable order form/envelope is bound into the back of this volume.

Annual Review of Earth and Planetary Sciences
Volume 7, 1979

CONTENTS

Francis Birch

Ann. Rev. Earth Planet. Sci. 1979. 7: 1–9

REMINISCENCES AND DIGRESSIONS

Francis Birch
Hoffman Laboratory, Harvard University, Cambridge, Massachusetts 02138

> *Geology is rapidly approaching a point where an ultimate problem which has been staring it in the face can no longer be sidestepped, namely, to determine the actual specific physical and chemical behavior of those materials which do actually constitute the earth's crust.*
>
> P. W. Bridgman, 1936

It is rash to invite reminiscences; unless firmly contained they may overflow the allotted bounds, so great is the pleasure of recalling the associations and activities of one's younger days. I mean to remain, for the most part, in the decade of the 1930s, when geophysics was beginning to find its modern form, but had not yet experienced the mushroom growth of the postwar years. Most geophysicists were then born-again physicists, engineers, geologists, as formal instruction in geophysics was to be had at only a few American universities and mining schools. Just for scale, in 1936 the American Geophysical Union had about 800 members; the Society of Exploration Geophysicists had about 230. Each of these societies now has more than 12,000 members.

The existence of a science called geophysics first came to my notice in 1930 when, a graduate student in the Harvard Physics Department, I visited the Geophysical Laboratory of the Carnegie Institution of Washington for consultation regarding my thesis. The subject, suggested by Professor P. W. Bridgman, was the electrical resistance and critical point of mercury. No satisfactory measurement had then been made of the critical (liquid-vapor) constants of a metal, but Bridgman believed that the critical region of mercury might be attainable in pressure equipment such as he had developed. Earlier work, however, had shown the need for a temperature near 1400°C and this required a furnace inside the pressure vessel. Such an arrangement had been used by several members of the Geophysical Laboratory: Leason Adams, Ralph Gibson, and Roy Goranson, whom I then met for the first time, had useful

0084-6597/79/0515-0001$01.00

suggestions. My final values, 1460±20°C and 1610±50 bars, were received with some skepticism, as the pressure was higher than expected by analogy with nonmetallic liquids. By the 1960s, however, the properties of liquid metals had become interesting to reactor designers, and groups in England, Germany, and the Soviet Union repeated the experiments with improved methods and much more data. The new numbers, I was happy to learn through the kindness of Drs. Kikoin and Senchenkov of the Kurchatov Institute, are in reasonable agreement with mine of 1932.

With more curiosity I might have learned somewhat sooner about geophysics. As an American Field Service Fellow, I had spent the years 1926–1928 at the University of Strasbourg, working in the Institut de Physique, then directed by Pierre Weiss. Weiss was one of the founders of modern magnetism, and his laboratory, entirely devoted to magnetic measurements, contained the most powerful electromagnet of the time. Wound with water-cooled copper tubing, it could maintain a steady field of 10,000 gauss in a volume of several cubic centimeters. I was permitted to make measurements with this monster and to appear as a junior author in several publications by Weiss. This was very gratifying and reinforced my intention to continue with research. So also did the kindness shown me by Weiss's associates, Paule Collet, G. Foëx, and Robert Forrer.

Near the Institut de Physique was the Institut de Physique du Globe, an important center for seismological research, first under German then under French administration. The International Seismological Association was founded there in 1903, and its journal, *Gerlands Beiträge zur Geophysik*, must be one of the earliest to use the term, geophysics. It never occurred to me to look into the activities of the Institut de Physique du Globe when it would have been very easy to do so.

The word "geophysics" seems to have made its way from German to English rather slowly, gradually displacing the older "Physics of the Earth." The famous eleventh edition of the Encyclopaedia Britannica (1911) has entries under geodesy, geography, geology, and geometry, but nothing under geophysics (or geomancy or geonomy). The International Union of Geodesy and Geophysics, and the American Geophysical Union, were organized in 1919, the Society of Exploration Geophysicists in 1930. The German usage goes back at least as far as 1884 when S. Günter published a *Lehrbuch der Geophysik*. The first professorial chair in geophysics may have been that of E. Wiechert at Göttingen in 1900.

In the United States, responsibility for various geophysical disciplines was distributed among a number of Federal departments, bureaus, and surveys. Besides the Coast and Geodetic Survey and the National Bureau of Standards (Commerce), there were the Geological Survey and Bureau of Reclamation (Interior), the Weather Bureau (Agriculture), and the

Hydrographic Office (Navy). These all had specific duties laid down by law, and after drastic budget cuts in the 1930s they had little manpower or money for research. The most notable achievements were perhaps the first-order triangulation and leveling of the United States and the pendulum surveys by the Coast and Geodetic Survey, with the reductions and discussions of isostasy by Hayford and Bowie.

University commitments to geophysics (except meteorology) remained scanty through the 1930s, though there were a few seismograph stations, especially in California and at the Jesuit colleges. A map of seismograph stations in 1934 (Macelwane 1934) shows 16 Jesuit stations of a total of 36; others were operated by the Coast and Geodetic Survey. The Carnegie Institution had founded two important centers of research in 1905: the Geophysical Laboratory and the Department of Terrestrial Magnetism, and also took part in beginning the seismograph network in southern California in 1927. James B. Macelwane, S.J., was appointed professor of geophysics at St. Louis University in 1925. Perry Byerly became assistant professor of seismology at Berkeley in 1927. Beno Gutenberg, Wiechert's student, came to the California Institute of Technology as professor of geophysics in 1930. The number of departments of geophysics or geophysical engineering increased from two in 1930 to seven in 1952 (Macelwane 1952). This does not reveal the whole story, as courses in various aspects of geophysics, mainly as applied to prospecting, were taught in 55 colleges and universities, in departments of geology and physics as well as of geophysics.

Some aspects of geophysics had been fitfully treated at Harvard according to the interests of individual professors. William Morris Davis, known chiefly for his work on the evolution of land forms, was first interested in meteorology and published a much-used text in 1894. One of his students, Robert DeCourcy Ward, became the first professor of climatology in the United States in 1910; research and instruction in meteorology have continued to the present time. Henry Lloyd Smyth published a treatise on magnetic prospecting for ore in 1899 and gave a course on this subject in the Department of Mining Engineering and Metallurgy. A. E. Kennelly, in the Department of Electrical Engineering, inferred the existence of an ionosphere in 1902, simultaneously with Heaviside; Professor Kennelly was still teaching in 1930 when I took his course. J. B. Woodworth, professor of geology, installed two Bosch-Omori seismographs in the University Museum in 1908; at that time, they were considered the best available and this station was renowned for excellent time-keeping, achieved through a connection with the Astronomical Observatory. Woodworth also gave instruction in seismology. It seems fair to conclude, however, that before 1930 geophysics played a minor part in the education of geologists and that studies now accepted as part of geophysics, broadly defined, had not yet become linked together.

Despite short rations, geophysics made impressive advances in the 1920s and 1930s; we have only to think of the work of Jeffreys, Bullen, Gutenberg, Vening Meinesz, Heiskanen, to say nothing of the rapid growth of applications in exploration. Geologists could no longer discuss their major problems without reference to internal structure as revealed by seismology and to the mechanical properties revealed by the pendulum. This was clearly recognized by Reginald A. Daly, appointed professor of geology at Harvard in 1912. He was initially interested in the origin of the igneous rocks; this led him first to volcanoes, but gradually he was drawn to the deeper interior and the indirect evidence of geophysics. A basic difficulty was, and is, the interpretation of physical measurements made on the Earth's surface in terms of materials and physical conditions in the interior. Daly compiled the available data on physical properties of rocks and minerals, but found them far from adequate for his purpose. In particular, the role of pressure was still too little known.

Aware, as a friend and colleague, of Bridgman's studies, Daly became a frequent visitor to the high-pressure laboratory and brought his problems to Bridgman's attention. Bridgman was induced to make several studies of rocks and minerals, and in two prescient papers (Bridgman 1936, 1945), he discussed pressure effects of possible significance for geology, especially pressure-induced polymorphism, of which he had found many examples for simple substances. In 1930 he joined Daly and others in the formation of an interdepartmental committee empowered to seek funds and to supervise their use for two ventures: a new program in seismology, and research on physical properties of geologically interesting materials. The original committee, besides Daly and Bridgman, included Harlow Shapley, astronomer, L. C. Graton and D. H. McLaughlin, geologists, and G. P. Baxter, chemist. With gradual replacement of personnel, this committee survives and has kept its interdepartmental character.

The new seismograph station, built on bedrock on the grounds of the Astronomical Observatory at Harvard, Massachusetts, was equipped with three Benioff units. Don Leet, appointed director in 1931, operated the station until 1955, and published 45 bulletins of observations in addition to numerous papers and books. He also developed a three-component portable seismometer, which he used for monitoring quarry blasts and other explosions, most notably the "Trinity" test in the Jornada del Muerto in 1945 when he recorded the ground motion at the Elephant Butte Dam. Several early studies of seismic velocities in quarries were done in collaboration with Maurice Ewing, then instructor in physics at Lehigh University.

At the same time, W. A. Zisman was appointed Research Fellow and began studies on elastic properties of rocks. Zisman had done his thesis in Bridgman's laboratory and was thoroughly familiar with high-pressure technique, but he found the idiosyncrasies of rocks uncongenial and resigned after completing several interesting studies. The position was offered to me just as I finished my own thesis. I was glad to accept; the date was 1932, and I valued the prospect of remaining in Cambridge and continuing my association with Bridgman. I carried a copy of *The Earth* around with me that summer, hoping for a painless osmotic infusion of geophysical knowledge.

Bridgman's laboratories, like many others of the 1920s and 1930s, were filled with homemade equipment manufactured by the students and a few technicians in the departmental shops. Bridgman's instrumentation consisted of long slide-wires, galvanometers, and dial gauges; vacuum tubes played no part, nor did recorders, except for a single employment of photographic recording. His own measurements must add up to millions of entries, written down, plotted, and smoothed graphically. Estimates of fit were based on summations with a pair of dividers. I cannot recall much use even of the rackety calculating machines of the 1930s, although I remember my pleasure when automatic division was introduced. The extraordinary sensitivity to linear displacement achieved by Bridgman with his slide-wire potentiometer, reaching a few thousand angstroms, made possible such remarkable measurements as the compressions, to 100 kb, of samples a few millimeters long. Innumerable corrections for dimensional and electrical changes under pressure were carefully applied. Later work, with more elaborate equipment, has detected some errors; for example, the freezing point of mercury at 0°C, used as a fixed point for calibration of pressure gauges, has been raised about one percent. Bridgman's free-piston gauge, still stored in the Hoffman Laboratory, has a piston only 1.6 mm in diameter; it reaches 10 kb with easily manageable weights and probably cost a few hundred dollars; the modern free-piston gauge, with its larger piston, requires a formidable machine simply to handle the weights.

For about twenty years Bridgman's experiments remained, with a few exceptions, limited to pressures of 12 kb. The reasons were, first, that there were many kinds of new measurements to be made in this range, and second, that it was economical in that failures were few and the same equipment could be used for many runs. At a time when typical research grants were a few hundred dollars, breakage was a serious matter. In the middle 1930s, however, there were improvements in steel and, more important, sintered tungsten carbide became available. Bridgman was supplied with this material, at first in tiny quantities, by Dr. Zay Jeffries of the General Electric Company, and quickly moved to much higher pressures—first to 30–50 kb and later, with his cascaded pressure systems, one inside the other, to 100 kb. Tungsten carbide was used lavishly during the war, and subsequently has become an essential component in the postwar high-pressure devices.

My own high-pressure research began in the machine shop, where I drilled a 1-inch hole down the axis of a bar of alloy steel to form my pressure chamber. The nitrogen that I used as the pressure medium had a greater potential for damage than had the liquids ordinarily used by Bridgman. This was readily measured by the time required to raise the pressure with the hand-pumps, which were perfectly suitable for compressing small volumes of liquid; after a time I coupled a geared-down electric motor to the pump handle. I was prudently installed in the former below-ground coal cellar of the Jefferson Laboratory, with the corridor above barricaded with stone slabs. After a year or so, I emerged to a room in the new Lyman Laboratory, a room later used by David T. Griggs for his experiments on the strength of rocks, and there I completed my work in the Physics Department.

The precautions in the location of my pressure system were probably occasioned by an accident in Jefferson in 1922. Atherton K. Dunbar, Fellow for Research in Cryogenic Engineering, was compressing oxygen from a storage balloon into a conventional steel cylinder when a violent explosion took place. Dunbar and William Connell, the departmental carpenter, were killed, and eleven students received injuries of various degrees. I was in college at the time but luckily nowhere near Jefferson. The explosion was eventually traced to chemical reaction between the oxygen, at about 100 bars, and oil in the cylinder. During the investigation, the oxygen cylinders around the University were stored in a concrete vault built in the garage of the ROTC unit, which then acquired the name of Dunbar.

It was in this building, with its constant reminder of the need for attention to safety, that my studies of physical properties of rocks began in 1932 and continued, except for four war years, until 1963. After the war, a small wing with shop and offices was added. In 1963, all the experimental work of the Department of Geological Sciences was concentrated in the new Hoffman Laboratory. Except in details, these early arrangements must parallel those of many projects of the 1930s, begun in sheds and makeshift quarters which often proved to be unexpectedly suitable for their purposes. Many of the wartime projects began in the same way. Dunbar is still in use, at present by the Physics Department, confirming the adage that "nothing lasts like the temporary." Many "permanent" buildings have been demolished during its sixty-year existence.

In the depression years of the 1930s, it was not uncommon for physicists who had just received their degrees to pass months in search of jobs. A floating population of well-qualified experimenters took part in the geophysics work, with great profit to the program and to my education. Half of Bridgman's Ph.D.'s between 1932 and 1954, as well as others with different specialties, participated for various periods. Harry Clark, John Ide, Dennison Bancroft, Manson Benedict, and Blaney Dane, to mention only a few, all left their marks with advancements of technique and new measurements. The surplus of physicists quickly became a shortage after 1940, with the organization of defense-oriented laboratories, and we suspended operations in Dunbar for the duration of the war.

My connection with the Department of Geological Sciences was tenuous at first, consisting mainly of frequent consultations with Professor Daly. But there could have been no better guide and tutor. His knowledge and interests were broad, and his enthusiasm was contagious. He saw to the assembling of representative rocks for our experiments, calling on his international acquaintanceships for samples from type localities, no matter where. Expanded through the kindness of other geologists, this collection has served many later investigators, as shown by the honeycombed residues. Daly was remarkably open-minded; he was even hospitable to the ideas of Taylor and Wegener at a time when most American geologists had little good to say about them, and he looked for ways to overcome the difficulties, some of which have still to be surmounted. I think that Daly, with his emphasis on the Earth's mobility, would have been pleased with the recent theories of drift, which he missed by just a few years.

It is difficult to exaggerate the changes in the scientific world between 1941 and 1945. World War I had drawn physicists into a few projects, especially the development of underwater sound for the antisubmarine campaign. But the laboratories of World War II mobilized virtually the entire profession, plus mathematicians, engineers, chemists, geologists, and geophysicists. The rapid development of technology through massive expenditures and high priorities showed the effectiveness of well-organized, well-funded groups under skillful direction, in a cause that had nearly unanimous national support. The armed services acquired the habit of subsidizing research and of cooperating with civilian scientists; after the war and before NSF, the Office of Naval Research became one of the chief suppliers of funds for unclassified research and displayed a remarkable breadth of interests.

The problems related to nuclear weapons assured a continuing interaction among governmental, academic, and industrial scientists, and after 1949, detection of underground tests fell squarely in the laps of earth scientists. Seismologists were called upon to distinguish, with a high degree of confidence, between earthquakes and explosions, and it soon became evident that they were not ready to do so. Looking back, it may seem that the importance of this problem was overemphasized, but in any case, the effort to solve it led to unprecedented investments in geophysics with fully proportionate returns. While the postwar demands revealed the effects of earlier deprivation, at the same time the forced feeding of the war years produced a hypertrophied development of new instruments and facilities. Most of the new technology soon became commercially available and found applications in geophysical research.

With the beginning of the space age in 1957, expenditures for geophysical and planetary studies rose again by orders of magnitude, even when they were incidental rather than primary objectives. There is no need to review the accomplishments of the last twenty years. The Moon landings, the close-up photography of other planets, and the automatic analytical laboratory are astonishing achievements, rightly contrasted to the muddles of earth-bound affairs. A new perspective of the Earth, and perhaps a new appreciation of its uniqueness in our solar system, has been gained. On the other hand, these immensely costly successes have encouraged visions of colonies of astronauts existing, like experimental mice, in completely artificial environments. Much of this may be the product of youthful imprinting with science fiction. Undeniably our lives are increasingly passed in the sealed spaces of cars, planes, offices, even houses. But it seems unlikely that a significant fraction of the Earth's people can escape overcrowding or the poisoning of our terrestrial environment by setting up housekeeping in space or on any of our sister planets. Noah's Ark was a family affair, but projects in space strain the resources, already perceptibly depleted, of the richest nations. A more promising task for earth scientists is to determine the acceptable limits of human misuse of our unique store of the Aristotelian "elements": air, water, fire, and earth.

Visions of communication with "advanced" civilizations elsewhere in our galaxy are also nourished. In view of the sorry record of cultural clashes on the Earth, it is hard to believe that even radio contact could be mutually beneficial. It is a humiliating thought that our strongest signal to the creatures of other worlds is our television aura (Sullivan et al 1978); what if they could tune in the programs? Or even worse, understand what they saw? I prefer to entertain a different fantasy: that other inhabited planets are ruled by pacific counterparts of dolphins or sea otters, happily indifferent to man-made radiation of all wavelengths.

Literature Cited

Bridgman, P. W. 1936. Shearing phenomena at high pressures of possible importance for geology. *J. Geol.* 44:653–69

Bridgman, P. W. 1945. Polymorphic transitions and geological phenomena. *Am. J. Sci.* 243a (Daly Volume): 90–97

Macelwane, J. B. 1934. Report of the Jesuit Seismological Association for 1933. *Trans. Am. Geophys. Union*, pp. 62–64

Macelwane, J. B., Chairman. 1952. Annual Survey of Geophysical Education 1951–1952. *Geophysics* 17:622–25

Sullivan, W. T. III, Brown, S., Wetherill, C. 1978. Eavesdropping: the radio signature of the Earth. *Science* 199:377–88

Ann. Rev. Earth Planet. Sci. 1979. 7: 11–38

GEOCHEMICAL AND COSMOCHEMICAL APPLICATIONS OF Nd ISOTOPE ANALYSIS[1]

R. K. O'Nions, S. R. Carter, N. M. Evensen, and P. J. Hamilton
Lamont-Doherty Geological Observatory, Columbia University,
Palisades, New York 10964

INTRODUCTION

The realm of the isotope geochemist consists of the 288 naturally occurring isotopes. The portions of interest to the geochemist are those isotopes that show measurable variations relatable to natural processes, and this domain has expanded as analytical refinements have permitted the measurement of progressively smaller variations in isotopic abundance. The most recent isotopic system to come within the expanding frontiers of geochemistry is that of the rare earth element neodymium (Nd). Although the analysis of Nd isotope variations is only a few years old, it has proven its worth in a variety of terrestrial and extraterrestrial studies, and is rapidly becoming a standard geochemical technique, capable of providing information not accessible by other means.

The primary cause of natural variations in the isotopic composition of Nd is the α-decay of the long-lived isotope ^{147}Sm to ^{143}Nd. The variations thus produced have been employed as a geochronologic tool as well as a tracer of geochemical provenance. In addition, variations in other isotopes of Nd have apparently resulted from incomplete mixing of the products of various nucleosynthetic regimes during formation of the solar nebula, from the α-decay of the short-lived extinct radioactivity ^{146}Sm, and from the spontaneous fission of U. The ability to discern the often minor effects resulting from these causes depends heavily upon the advances in mass spectrometry made during the last decade.

This paper discusses the natural processes capable of inducing varia-

[1] L.-D.G.O. Contribution No. 2729. This work has been supported by NSF Grant EAR-75-20891.

0084-6597/79/0515-0011$01.00

tions in the isotopic composition of Nd, the techniques used for precise measurement of the variations, and a variety of applications of Nd isotope analysis in geochemistry and cosmochemistry.

PROCESSES GOVERNING THE ISOTOPIC COMPOSITION OF Nd

Nucleosynthesis

Neodymium is one of the 14 naturally occurring rare earth elements (REE). Its even atomic number ($Z = 60$) favors nuclear stability, and it possesses seven stable isotopes, at mass numbers 142, 143, 144, 145, 146, 148, and 150. The general features of the nucleosynthesis of Nd are readily inferred from Figure 1. In the classic scheme of Burbidge et al (1957), the s-process nuclides are formed by the successive addition of neutrons until terminated by an unstable β-emitter, which carries the process to a new element having the next higher value of Z. ^{142}Nd through ^{146}Nd are s-process nuclides, while ^{147}Nd is a short-lived (11 day half-life) nuclide decaying to ^{147}Pm and thence to ^{147}Sm (Pm has no stable isotopes). The r-process consists of the much more rapid addition of neutrons, with neutron capture occurring on a time scale shorter than the β-decay half-lives of the resulting nuclides, so that very neutron-rich nuclides are formed, which subsequently β-decay to the valley of nuclear stability. ^{148}Nd and ^{150}Nd are isolated from the s-process by ^{147}Nd, and so can only be formed by the r-process. However, ^{143}Nd through ^{146}Nd can also be formed by the r-process. Neutron-rich nuclides of mass 142 will β-decay to stable ^{142}Ce before reaching ^{142}Nd, which is therefore shielded from r-process contributions. None of the isotopes of Nd are sufficiently proton-rich to require the somewhat mysterious p-process of proton addition for their formation, but there is some evidence (Lugmair et al 1978) for a p-process contribution to the lighter isotopes of Nd, either directly or through the α-decay of the short-lived p-process nuclides ^{146}Sm and ^{150}Gd to ^{142}Nd.

Each process described above is believed to occur in a different cosmic environment. Variations in the proportions of material derived from different nucleosynthetic regimes, therefore, will result in variations in the isotopic abundances of the corresponding nuclides (e.g. in the ratio of s-only ^{142}Nd to r-only ^{148}Nd). The relative abundances of nuclides derived from a *single* process (e.g. the ratio of the r-process ^{148}Nd to r-process ^{150}Nd) will not change, however. This latter property is useful in searching for Nd isotopic variations resulting from nucleosynthetic causes, since instrumental fractionation can be corrected by normalizing to the ^{148}Nd/^{150}Nd ratio, which is unaffected by variable s-process con-

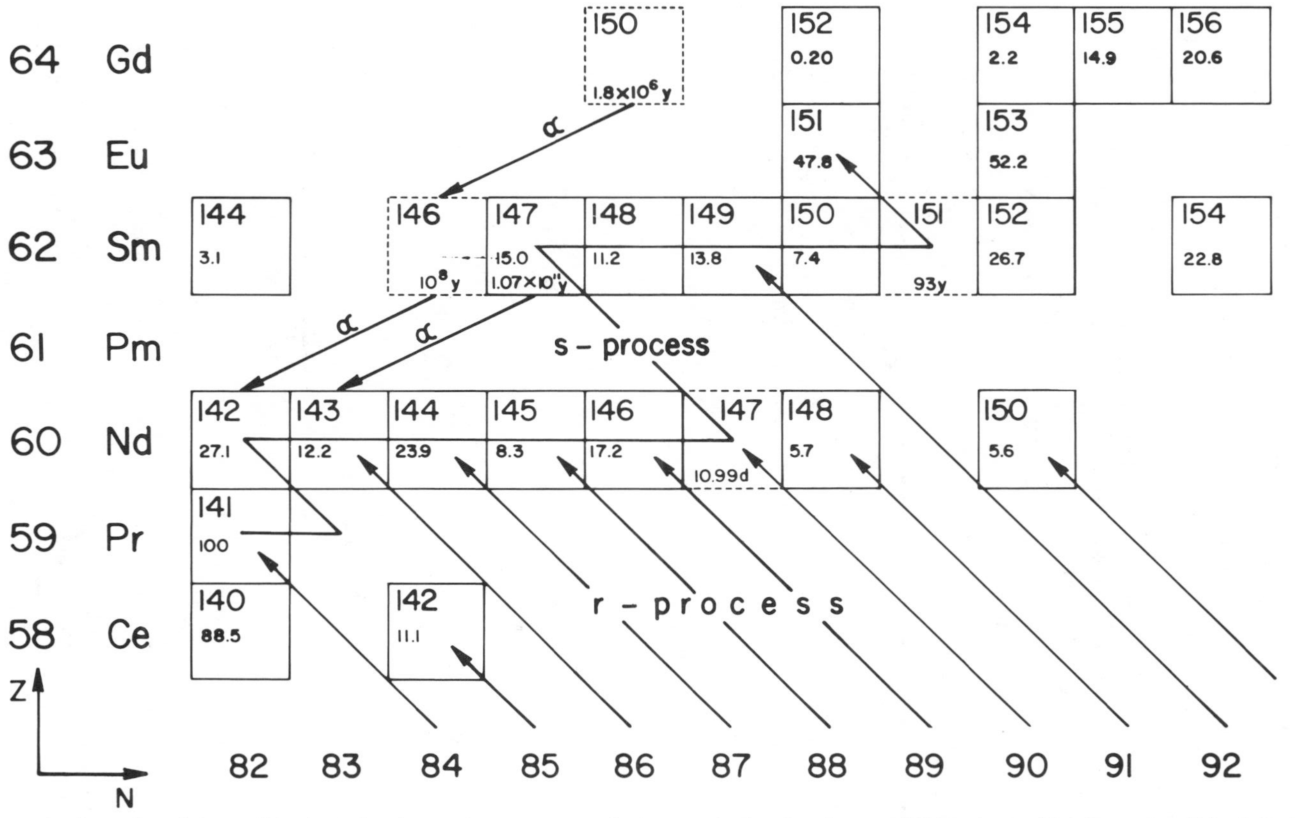

Figure 1 A portion of the nuclide chart showing nuclear processes of relevance to the abundances of Nd isotopes. Atomic number Z is plotted against neutron number N. Naturally occurring isotopes with their percent abundances are enclosed by heavy lines, unstable isotopes by dashed lines. Only those unstable isotopes involved in nuclear processes of interest are shown. A portion of the s-process path of nucleosynthesis appears as a zig-zag line from lower left to upper right; r-process paths are parallel lines from lower right, terminating at the first stable nuclide. Several α-decay paths are shown; the half-lives of the parents are given at the bottom of the corresponding square.

tributions. McCulloch & Wasserburg (1978b,c) and Lugmair et al (1978) independently observed variations in the isotopic composition of Sm and Nd in an inclusion from Allende. This inclusion has an excess of r-process nuclides, the excesses forming smooth functions of atomic mass (Figure 2). The effect of odd and even mass numbers on nuclear stability is clearly seen, and results in lower abundances of odd-mass nuclides. A small p-process contribution may also be present, but the inference of its presence is somewhat model-dependent.

The existence of the pure r-process as originally visualized by Burbidge et al has recently been questioned by Blake & Schramm (1976), who suggest that neutron fluxes lower than those required for the idealized

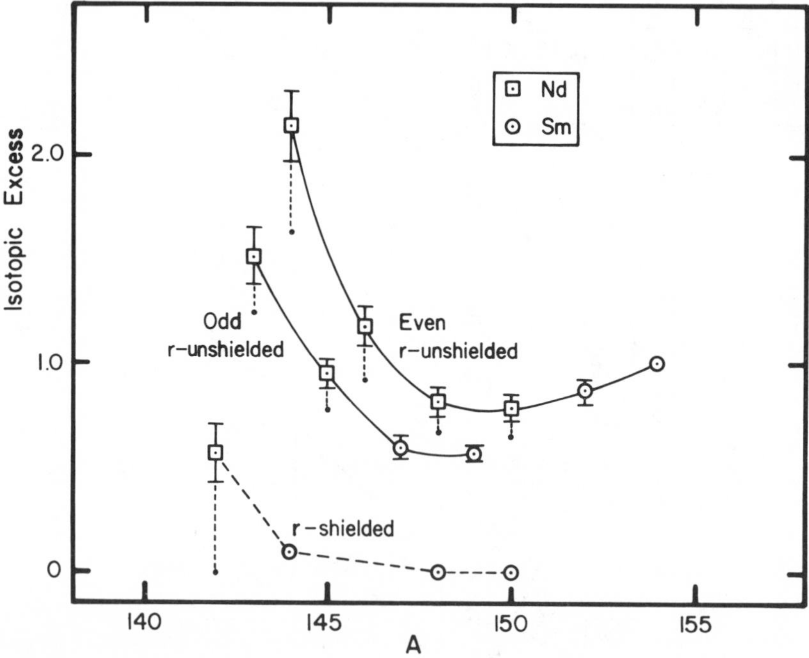

Figure 2 Excess abundances of Sm and Nd isotopes in Allende inclusion EK 1-04-1, plotted against mass number *A* (after Lugmair et al 1978). Sm points (circles) are obtained by normalizing the measured Sm isotopic composition of EK 1-04-1 to match the ratio of s-only isotopes $^{148}Sm/^{150}Sm$ determined for a terrestrial standard. Excesses of other Sm isotopes over the values in the standard are then plotted relative to ^{154}Sm excess $\equiv 1$. Nd data were normalized to the terrestrial standard ratio of the r-only isotopes $^{148}Nd/^{150}Nd$. Assuming excess of r-shielded ^{142}Nd to be zero, excesses of other Nd isotopes are shown as small dots below the squares. The squares indicate excesses obtained by fitting r-unshielded Nd isotopes to the Sm excesses; the ^{142}Nd excess obtained in this case may represent a differential p-process contribution.

r-process can reproduce the observed distribution of nuclides equally well or better. Their n- (or neutron-) process is essentially a mixture of the classical s- and r-processes, and requires less extreme and more astrophysically reasonable sites than the pure r-process.

Radioactive Decay

The abundance of a given nuclide at any point in time reflects the relative rates of synthesis and decay, integrated over all past time. Once significant nucleosynthesis has ceased, many of the resultant nuclides begin to decay at rates which vary over many orders of magnitude. For radioactive decay to produce observable variations in isotopic abundances of the daughter nuclide

1. variable parent-daughter ratios must exist over some interval while the parent nuclide is decaying,
2. the variations thus produced must be preserved, i.e. any subsequent mixing must be incomplete,
3. the analytical techniques employed must be capable of resolving these variations.

The first condition requires that some fractionation process be operative while significant amounts of the parent nuclide are still present—hence the half-life must be sufficiently long to allow this. Since any decay involving charged particles (e.g. α- and β-decay) results in a daughter of differing atomic number, parent-daughter fractionation can be induced by a variety of cosmochemical and geochemical processes. The second requirement implies that at least some of the fractionated phases be persistent to the present day. If the half-life of the nuclide involved is short, this means that the fractionations must have persisted since a similarly brief interval following nucleosynthesis. The third condition merely relates quantitatively the analytical resolution available to the size of the induced variations. The magnitude of effects produced is dependent upon the range of parent-daughter ratios, the half-life of the parent, and the interval over which the variations in parent-daughter ratio were present.

In the case of Nd, the only naturally occurring nuclides decaying to isotopes of Nd are ^{147}Sm, ^{148}Sm, and ^{149}Sm, which undergo α-decay to ^{143}Nd, ^{144}Nd, and ^{145}Nd respectively. Sm and Nd are rare earths and hence similar chemically, so that variations in the Sm/Nd ratio by more than a factor of two are not common in nature. The long-lived ($T_{1/2} \sim 10^{16}$y) ^{148}Sm and ^{149}Sm are not capable of producing measurable variations in ^{144}Nd or ^{145}Nd even over cosmological intervals ($\sim 10^{10}$y). However, the half-life of ^{147}Sm, about 10^{11}y, is sufficiently short to produce small but readily measurable differences in ^{143}Nd abundance over time spans of 10^8y or more. This provides the basis for the Sm-Nd technique of age determination, to be discussed in detail shortly.

The other decay scheme of possible cosmochemical interest involving Nd is the α-decay of ^{146}Sm to ^{142}Nd. ^{146}Sm is unstable, but its half-life of $\sim 10^8$y is relatively long. If fractionation of Sm/Nd took place within a few 10^8y after nucleosynthesis of the REE, any fractionated phases still extant should show measurable anomalies of ^{142}Nd, in addition to any variations caused by differential p-process contributions (see Figure 1).

If at time = 0, a number $^x\text{Nd}_0$ of ^{142}Nd or ^{143}Nd atoms are present, together with $^y\text{Sm}_0$ atoms of ^{146}Sm or ^{147}Sm respectively, then after a time t has elapsed a proportion $(1-e^{-\lambda_y t})$ of the parent isotope will have decayed, where $\lambda_{146} = 6.93 \times 10^{-9}\text{y}^{-1}$ and $\lambda_{147} = 6.54 \times 10^{-12}\text{y}^{-1}$. Then the total number of ^{x}Nd atoms at time t is

$$^x\text{Nd}_t = {}^x\text{Nd}_0 + {}^y\text{Sm}_0(1-e^{-\lambda_y t}).$$

The equation may be divided through by any nonradiogenic isotope of Nd, say ^{z}Nd, and since $^z\text{Nd}_0 = {}^z\text{Nd}_t$, we have:

$$\left(\frac{^x\text{Nd}}{^z\text{Nd}}\right)_t = \left(\frac{^x\text{Nd}}{^z\text{Nd}}\right)_0 + \left(\frac{^y\text{Sm}}{^z\text{Nd}}\right)_0 (1-e^{-\lambda_y t}). \tag{1}$$

In subsequent discussion, $z = 144$ will be assumed, with the understanding that any non-radiogenic Nd isotope could be substituted.

For the case of ^{147}Sm–^{143}Nd, one generally makes the substitution

$$^{147}\text{Sm}_0 = {}^{147}\text{Sm}_t e^{\lambda t},$$

so that

$$\left(\frac{^{143}\text{Nd}}{^{144}\text{Nd}}\right)_t = \left(\frac{^{143}\text{Nd}}{^{144}\text{Nd}}\right)_0 + \left(\frac{^{147}\text{Sm}}{^{144}\text{Nd}}\right)_t (e^{\lambda t}-1). \tag{2}$$

This equation is linear in the parameters $(^{143}\text{Nd}/^{144}\text{Nd})_t$ and $(^{147}\text{Sm}/^{144}\text{Nd})_t$, so that for a given $(^{143}\text{Nd}/^{144}\text{Nd})_0$ and t, a plot of ^{143}Nd/^{144}Nd vs ^{147}Sm/^{144}Nd as measured at present (the "Nd evolution diagram") is a straight line, with intercept $(^{143}\text{Nd}/^{144}\text{Nd})_0$ and slope $e^{\lambda t}-1$. This is the basis of the isochron plot familiar from the Rb-Sr system and equally applicable to the Sm-Nd system. Note however that the range of Sm/Nd in natural systems is normally less than that of Rb/Sr; in particular, very low Sm/Nd ratios are seldom found, so that the initial ^{143}Nd/^{144}Nd must be obtained from a considerable extrapolation to the ordinate, whereas it is often possible to obtain phases of very low Rb/Sr in which the measured ^{87}Sr/^{86}Sr is very close to the initial value.

In the case of the ^{146}Sm–^{142}Nd system, the time from nucleosynthesis

($t = 0$) to the present ($t \geqq 4.5$ G.y.) is many half-lives of ^{146}Sm, so the term $\exp(-\lambda t) \approx 0$ in Equation (1), and we have,

$$\left(\frac{^{142}\mathrm{Nd}}{^{144}\mathrm{Nd}}\right)_t = \left(\frac{^{142}\mathrm{Nd}}{^{144}\mathrm{Nd}}\right)_0 + \left(\frac{^{146}\mathrm{Sm}}{^{144}\mathrm{Nd}}\right)_0.$$

If we assume the ratio of ^{146}Sm to some stable, nonradiogenic isotope of Sm, say ^{144}Sm, to have been constant at $t = 0$, we may write

$$\left(\frac{^{142}\mathrm{Nd}}{^{144}\mathrm{Nd}}\right)_t = \left(\frac{^{142}\mathrm{Nd}}{^{144}\mathrm{Nd}}\right)_0 + \left(\frac{^{144}\mathrm{Sm}}{^{144}\mathrm{Nd}}\right)_t \left(\frac{^{146}\mathrm{Sm}}{^{144}\mathrm{Sm}}\right)_0. \qquad (3)$$

Again we have an equation linear in the quantities $(^{142}\mathrm{Nd}/^{144}\mathrm{Nd})_t$ and $(^{144}\mathrm{Sm}/^{144}\mathrm{Nd})_t$, assuming constant $(^{142}\mathrm{Nd}/^{144}\mathrm{Nd})_0$, in which case samples of varying Sm/Nd will define a straight line with intercept $(^{142}\mathrm{Nd}/^{144}\mathrm{Nd})_0$. However, the slope no longer has time significance, but rather defines the $^{146}\mathrm{Sm}/^{144}\mathrm{Sm}$ ratio at time zero, the time at which the $^{142}\mathrm{Nd}/^{144}\mathrm{Nd}$ was last homogeneous.

Nuclear Fission

Fission of the actinide elements results in two major fragments of unequal and somewhat variable mass. The heavier fragments of each pair generally have masses in the range of the lighter REE, so that fission products contain a significant proportion of various Nd isotopes. Spontaneous fission occurs only for certain isotopes of transuranic elements, notably the extinct radioactivity ^{244}Pu. Evidence of ^{244}Pu fission has been found in some meteorites, in the form of anomalous isotopic compositions of Xe. Because of the volatility of Xe, phases with high Pu/Xe were formed while ^{244}Pu was still extant, and thus a fairly high proportion of their Xe is fission-produced. However no such extreme Pu/Nd enrichment is likely, so that the small amount of ^{244}Pu present at solar system formation could not produce significant change in Nd isotope abundances.

Slow neutron-induced fission of natural U is a rare but not unknown process in nature. The Oklo natural reactor represents a rare combination of circumstances in which a sufficient natural concentration of U has been moderated at some past times by sufficient water to allow a chain reaction to be sustained. The resulting high concentrations of fission products have led to marked changes in the isotopic composition of the REE (Loubet & Allègre 1977).

Measurement of Nd-Isotope Compositions

Although neutron activation analysis is capable of detecting large ($\sim 1\%$) differences in the isotopic composition of an element, mass spectrometry remains the technique of choice for the precise determination of small isotopic variations. The geochemical applications of Nd isotopes have in fact been entirely dependent upon the advances in high-precision isotope ratio mass spectrometry first applied to Sr (Wasserburg et al 1969, Papanastassiou & Wasserburg 1969).

Because of the many isobars of the REE (Figure 1), good chemical separation of Nd from the adjacent REE, Ce and Sm, is required for successful mass spectrometry. ^{142}Ce, ^{144}Sm, ^{148}Sm, and ^{150}Sm are capable of causing interferences on the corresponding Nd isotopes. The REE are readily separated from other elements by standard ion-exchange techniques (e.g. Schnetzler et al 1967, Hooker et al 1975), but the separation of Nd from the other REE is more difficult. A variety of cation- and anion-exchange techniques have been developed, but in general require careful control of some parameters such as temperature or pH (e.g. Lugmair et al 1975a, Richard et al 1976, O'Nions et al 1977).

Mass spectrometry is performed on either metal or oxide species; each technique has its potential isobaric interferences, and use of the oxide species requires corrections to the data for the mutual interferences produced by the three isotopes of O. In either case correction must be made for instrumental mass fractionation during mass spectrometry. As with Sr, the usual procedure is to choose an isotope pair (not including the daughter isotope) that can be assumed constant in nature, compare its value during the analysis with a standard value, and extrapolate the observed fractionation to the ratio of interest (usually assuming a linearly mass-dependent fractionation). Unlike the case of Sr, however, there is no general agreement on a normalizing ratio or its standard value for Nd. This is a potential cause of confusion in comparing Nd measurements from various laboratories, and conversion must often be made. If two laboratories use the same normalizing ratio but assume different values for it, the conversion is straightforward, but if different normalizing ratios are used, the conversion requires knowledge of the isotopic composition of normal Nd (Table 1). In our laboratory and throughout this paper

Table 1 Nonradiogenic isotope ratios in natural neodymium[a]

142/144	1.1382602 ± 58
145/144	0.3489695 ± 50
146/144	0.7241103 ± 53
148/144	0.2430801 ± 36

[a] Weighted means and errors from DePaolo & Wasserburg (1976a). Ratios normalized to ^{150}Nd/^{142}Nd = 0.2096.

results are quoted relative to ^{144}Nd (e.g. $^{143}Nd/^{144}Nd$ ratios) and corrected to a $^{146}Nd/^{144}Nd$ ratio of 0.7219, the mean of a large number of Nd analyses made in our laboratory. Use is also commonly made of derived parameters that relate observed ratios to various simple models; these are defined when introduced. At present, Nd ratios are generally measured with 95% confidence limits of 0.005% or less. Lugmair & Marti (1978) reviewed measurements of the ^{147}Sm half-life and obtain a weighted mean value of 1.06×10^{11}y, corresponding to a decay constant of $6.54 \times 10^{-12}y^{-1}$. This is the value that has been used for all published ^{147}Sm-^{143}Nd chronometry, and appears consistent with the recently revised U decay constants (Lugmair & Marti, 1978).

COSMOCHEMICAL APPLICATIONS

Information from ^{146}Sm-^{142}Nd

The practical utility of Nd isotope studies was first demonstrated in cosmochemical applications. Following the suggestion by Kohman (1954) that ^{146}Sm might have existed as an extinct radioactivity, searches for ^{142}Nd anomalies were carried out by Murthy (1964) with negative results and by Notsu et al (1973), who claimed to have found effects in the achondrite Juvinas attributable to ^{146}Sm decay. More precise work by Lugmair et al (1975a) failed to confirm these results, and placed an upper limit of 0.013 on the $^{146}Sm/^{144}Sm$ ratio at the time of solidification. The actual $^{146}Sm/^{144}Sm$ determined by a best fit line to their data, using Equation (3), was 0.0054 ± 0.0072, indistinguishable from zero. Subsequently, Lugmair & Marti (1977) measured $^{142}Nd/^{144}Nd$ in mineral separates from the unusual achondrite Angra dos Reis, and found a barely resolvable effect from ^{146}Sm decay. Their results are plotted on a ^{146}Sm-^{142}Nd evolution diagram in Figure 3; the slope of the best fit line (Equation 3) corresponds to an initial $^{146}Sm/^{144}Sm$ of 0.0047 ± 0.0023, in agreement with their Juvinas data. Thus ^{146}Sm may be tentatively added to the small list of extinct radioactivities for which actual evidence has been obtained.

^{147}Sm-^{143}Nd Dating of the Moon and Meteorites

The Nd measurements of Notsu et al (1973) on Juvinas allowed them to calculate an isochron age of 4.3 ± 2.5 (2σ) G.y. for this meteorite. Because of the long ^{147}Sm half-life and the limited spread in Sm/Nd obtainable from mineral separates, very precise measurements of $^{143}Nd/^{144}Nd$ are required to closely define an age. The more precise Juvinas data obtained by Lugmair et al (1975a) enabled them to define an age of 4.56 ± 0.08 G.y., with a corresponding initial $^{143}Nd/^{144}Nd$ of 0.50677 ± 0.00010. This work

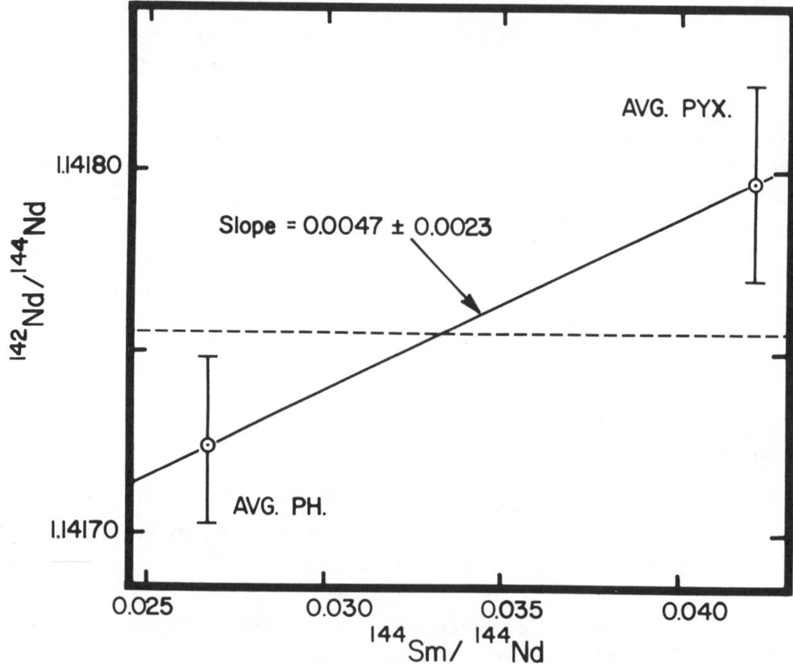

Figure 3 Plot of $^{142}Nd/^{144}Nd$ versus $^{144}Sm/^{144}Nd$ for phosphate and pyroxene components from the Angra dos Reis achondrite (after Lugmair & Marti 1977). The slope of the line corresponds to the $^{146}Sm/^{144}Nd$ ratio in Angra dos Reis at the time of its formation and is significantly greater than zero attesting to the former existence of ^{146}Sm in the solar system.

demonstrated conclusively that Sm-Nd dating was capable of precision comparable to that obtainable from Rb-Sr.

The Sm-Nd and Rb-Sr systems have proven complementary in many areas of application. In the case of the meteorites, achondrites are very difficult to date by Rb-Sr because of their low Rb-Sr ratios, but lend themselves readily to Sm-Nd dating. (Chondrites, on the other hand, have been dated by Rb-Sr, but their low REE contents make Sm-Nd dating more difficult.) So a new class of meteorites was opened to chronologic studies, and subsequent to Juvinas several other achondrites (e.g. Figure 4) have been dated by Sm-Nd internal isochrons (Lugmair & Scheinin 1975, Lugmair & Marti 1977, Unruh et al 1977, Nakamura et al 1977).

The ages obtained range from 4.55–4.60 G.y. except that of Stannern, which may be slightly disturbed and yields an age of 4.48 ± 0.07 G.y. (Lugmair & Scheinin 1975). These ages are in excellent agreement with U-Pb dates on achondrites (e.g. Unruh et al 1977, Wasserburg et al 1977) and delimit the time of achondrite formation much more narrowly than Rb-Sr studies (Papanastassiou & Wasserburg 1969, Birck & Allègre 1978). The broader significance of the Sm-Nd studies of meteorites, however, has been in establishing the initial $^{143}Nd/^{144}Nd$ ratio of solar system material, a parameter of central importance in the application of the analysis of Nd isotopes to problems of planetary evolution. The most precisely determined initial Nd for an achondrite is that for Angra dos Reis (Lugmair & Marti 1977; Figure 4), with a value of 0.50682 ± 5, but there is agreement among all initial values within their respective errors. This initial value assumes the importance for Nd isotope systematics that the achondritic initial $^{87}Sr/^{86}Sr$ (BABI; Papanastassiou & Wasserburg 1969) has for Sr. From it, a model age can be derived for any sample of known $^{147}Sm/^{144}Nd$ and $^{143}Nd/^{144}Nd$, or the time-integrated Sm/Nd ratio required to produce the present $^{143}Nd/^{144}Nd$ ratio can be calculated. Extensive use of these concepts will be made in the discussion of terrestrial Sm-Nd work.

The dating of lunar samples was another early application of Nd isotope systematics (Lugmair et al 1975b). Lunar samples have been intensively

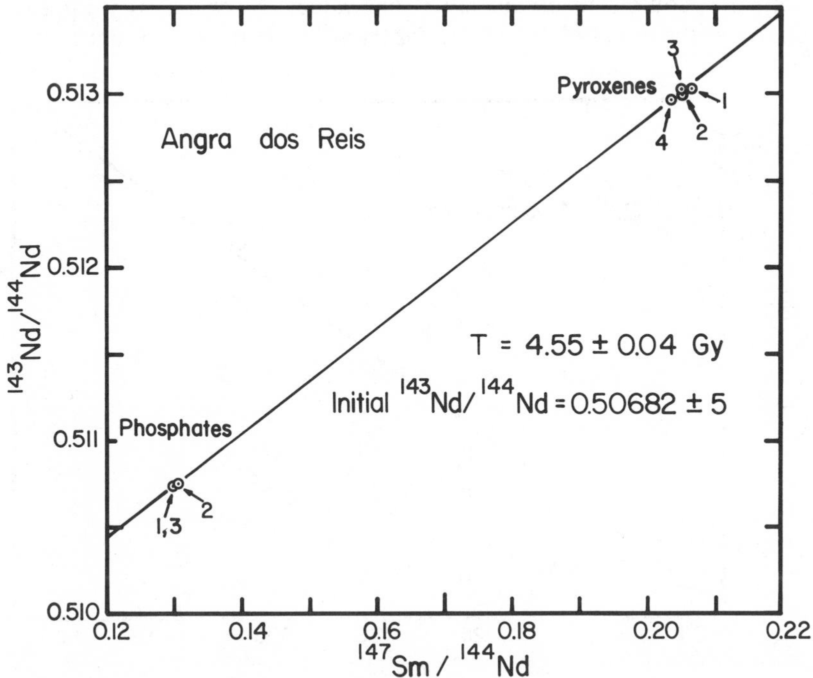

Figure 4 Sm-Nd evolution diagram for separated components of the achondrite Angra dos Reis (after Lugmair & Marti 1977).

studied, and are amenable to dating by Sm-Nd, Rb-Sr, K-Ar, and U-Th-Pb, and so offer a potential testing ground for comparison of ages and isotopic systematics in general. Only limited comparisons are available to date, however, and while Rb-Sr and Sm-Nd ages have been found to agree well in most cases (Nakamura et al 1976, Nakamura et al 1977, Papanastassiou et al 1977), a considerable discrepancy has appeared with sample 76535. This is a troctolite from the Apollo 17 site that has been dated by Rb-Sr (Papanastassiou & Wasserburg 1976) and Sm-Nd (Lugmair et al 1976). The puzzling feature is that the Rb-Sr age (4.61 ± 0.07) is older than the Sm-Nd age (4.26 ± 0.06; Figure 5), despite the fact that experience with terrestrial rocks has shown the Sm-Nd system to be much less readily disturbed by metamorphism than the Rb-Sr system (see discussion below).

Initial $^{143}Nd/^{144}Nd$ values for lunar samples have been used to understand the Sm/Nd fractionations undergone by lunar source areas prior to formation of the rocks. The difficulties of this approach are greater than in the case of Rb-Sr because of the much smaller fractionations of Sm/Nd that are encountered and because of the greater extrapolation normally required to obtain initial Nd values. Nevertheless it appears clear that the initial $^{143}Nd/^{144}Nd$ for the moon is similar to that for the

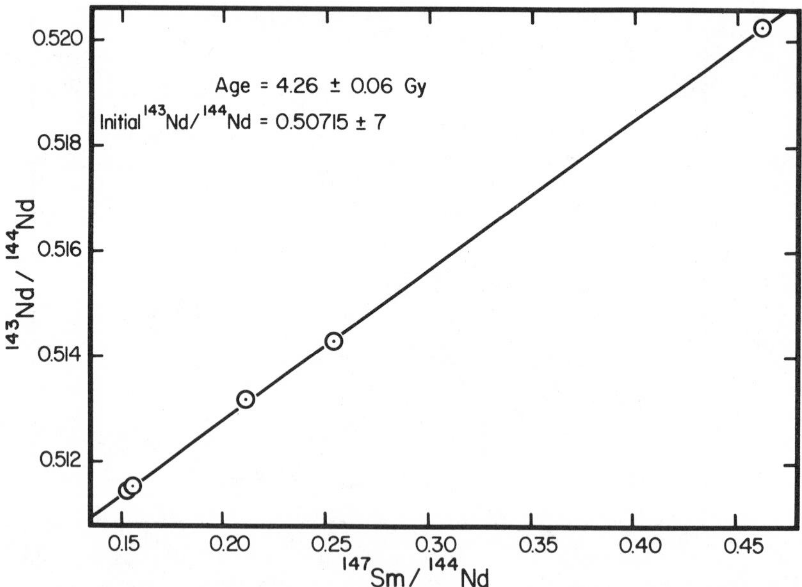

Figure 5 Sm-Nd evolution diagram for separated components of a lunar troctolite returned by the Apollo 17 mission (after Lugmair et al 1976).

achondrites, and that the bulk moon may have a Nd evolution comparable to the achondrites (i.e. a similar initial Nd and Sm/Nd ratio, the latter in turn equivalent to the chondritic ratio). There is some evidence for differential evolution of crust and mantle, with Sm/Nd ratios lower and higher respectively than chondritic since some time early in lunar history (Lugmair & Marti 1978).

Cosmochemistry has not only provided the first application of the analysis of Nd isotopes but also has provided the framework within which the extensive Nd studies that have been made on terrestrial samples can be interpreted in terms of the broader processes of planetary formation and evolution.

GEOCHRONOLOGICAL APPLICATIONS

Subsequent to the application of the ^{147}Sm-^{143}Nd decay scheme to extra-terrestrial materials by Lugmair and co-workers, others have successfully applied the scheme to the dating of Archaean (>2.5 G.y.) terrestrial rocks. Dating of such materials using alternative parent-daughter systems such as Rb-Sr has frequently been hampered by post-crystallization alteration and metamorphism leading to the differential movement of parent and daughter isotopes. Thus temporal relationships between the various components of Archaean terrains such as the granitic-gneisses and greenstone belts, which are usually equivocal on structural grounds, have not always been readily resolved using Rb-Sr and U-Pb methods.

In marked contrast to the modification of alkali and alkaline earth element abundances during post-crystallization alteration, the rare earths seem to be comparatively unaffected by such processes (Herrman et al 1974, Kay et al 1970, Smewing & Potts 1976, O'Nions & Pankhurst 1976, Frey et al 1968, O'Nions et al 1977). This point is well illustrated by Sm-Nd ages obtained thus far on Archaean rocks. At the time of writing, four Sm-Nd whole-rock isochrons have been determined on Archaean rocks at the Lamont-Doherty Geological Observatory (Hamilton et al 1977, 1978a,b,c). These studies were intended both to assess the value of the Sm-Nd method for dating Archaean basic and acid rocks and to resolve some important geochronological problems.

Specifically the following have been dated:

(*a*) Rhodesian greenstone belts
Age $= 2.64 \pm 0.14$ G.y., I $= 0.50919 \pm 18$ (Hamilton et al 1977);

(*b*) Isua supracrustals, West Greenland
Age $= 3.770 \pm 0.042$ G.y., I $= 0.507831 \pm 46$ (Hamilton et al 1978a);

(*c*) Onverwacht Volcanics, Southern Africa
Age = 3.54 ± 0.03 G.y., I = 0.50809 ± 4 (Hamilton et al 1978c);

(*d*) Lewisian gneisses, Scotland
Age = 2.92 ± 0.05 G.y., I = 0.508959 ± 49 (Hamilton et al 1978b).

A detailed discussion of these results is not entered into at this juncture; however, the considerable advantages of Sm-Nd whole-rock dating compared with other whole-rock methods such as Rb-Sr, U-Pb, and Pb-Pb

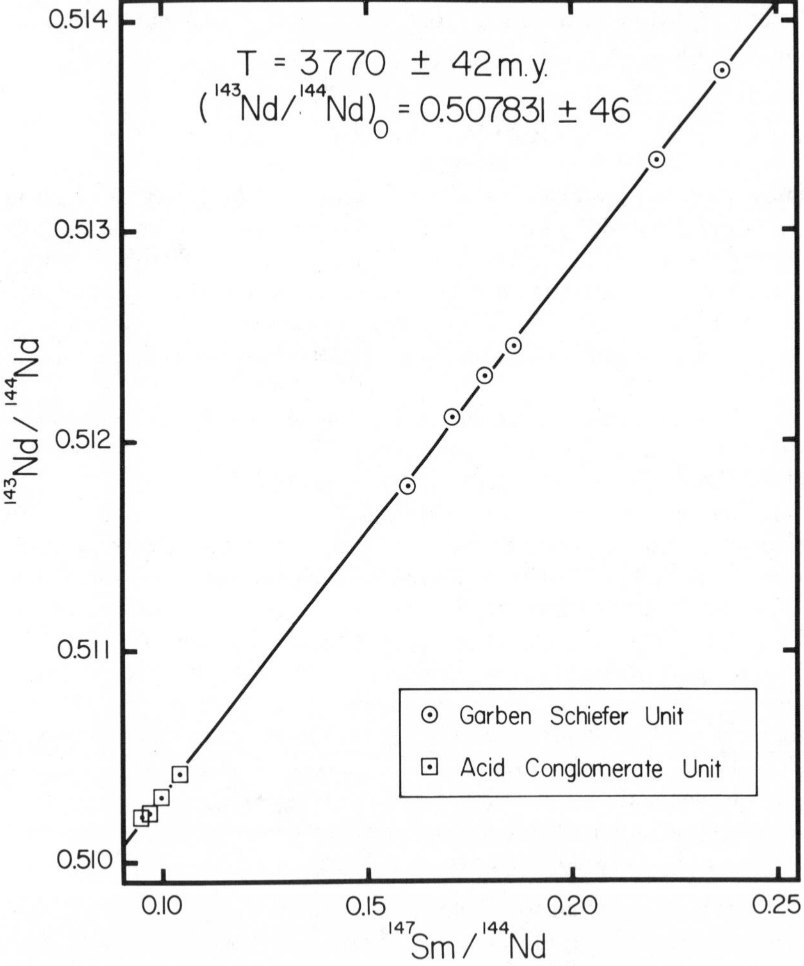

Figure 6 Sm-Nd evolution diagram for acid (conglomerate unit) and basic (garben schieffer) metavolcanics from the Isua supracrustals, West Greenland (after Hamilton et al 1978a).

is illustrated with respect to the results obtained on the Isua supracrustals and the Onverwacht Volcanics.

The Isua area of West Greenland contains the oldest known terrestrial rocks. Sm-Nd data for samples from the Isua supracrustal succession (the acid conglomerate and basic garben schiefer units) are plotted on a Sm-Nd evolution diagram in Figure 6. The age obtained (3.77±0.04 G.y.) is in agreement with the results obtained from zircons (Baadsgaard 1973, Baadsgaard, Lambert & Krupicka 1976, Michard-Vitrac et al 1977) that Amitsoq tonalitic gneisses were formed about 150 m.y. after the deposition and metamorphism of the supracrustal succession at Isua. In this respect the Sm-Nd result is less equivocal than the available Rb-Sr and Pb-Pb whole-rock ages (Moorbath et al 1973, 1975, 1977), which are less precise.

Attempts to date the Onverwacht Volcanic series of Southern Africa provide a further comparison of the application of the Rb-Sr and Sm-Nd systems. A comparatively well-preserved komatiite yielded an internal Rb-Sr isochron age of 3.5±0.2 G.y. (Jahn & Shih 1974). Highly scattered data obtained on ultrabasic and basic whole-rock samples and acidic

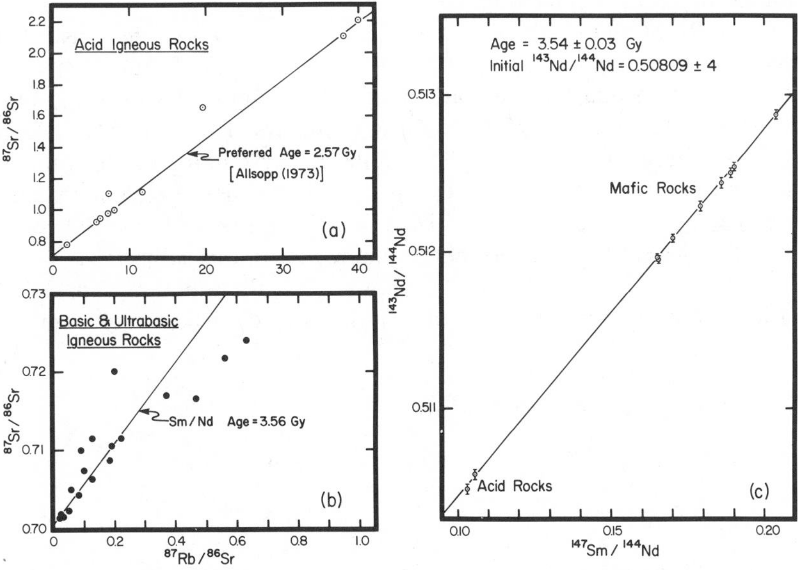

Figure 7 Comparison of Rb-Sr and Sm-Nd evolution diagrams for the Onverwacht Volcanics. (*a*) Rb-Sr evolution diagram for basic and ultrabasic samples of the Onverwacht Volcanics. Data from Allsopp et al (1973) and Jahn & Shih (1974). (*b*) Rb-Sr evolution diagram for acid samples of the Onverwacht Volcanics. Data sources as in (*a*). (*c*) Sm-Nd evolution diagram for ultrabasic, basic, and acid samples of the Onverwacht Volcanics (Hamilton 1978c).

volcanics suggest that Rb-Sr redistribution occurred about 2.6 G.y. ago (Figure 7*a* and *b*). In marked contrast Sm-Nd data for ultrabasic, basic, and acid volcanics yield a whole-rock isochron age of 3.54 ± 0.03 G.y. and an initial $^{143}Nd/^{144}Nd$ ratio of 0.50908 ± 4, with no indication of younger events (Figure 7*c*).

It can be seen, therefore, that although limited Sm/Nd fractionation and the long half-life of ^{147}Sm presently limit the use of the Sm-Nd system for dating more recent samples, it does provide a useful tool for the dating of ancient rocks, which almost inevitably have suffered alteration since their formation.

EVOLUTION OF THE EARTH'S CRUST AND MANTLE

The Earth has evolved into two distinct reservoirs of rare earth elements: the mantle (upper and lower) and the continental crust. The continental crust is enriched in light REE relative to the cosmic abundances (Table 2) whereas the suboceanic mantle has much lower overall REE abundances and is commonly depleted in light REE.

From Rb-Sr and U-Pb isotope investigations it has been demonstrated that the continental crust is a secondary geological feature that has developed in a quasi-continuous manner from about 3.8 G.y. ago (see reviews by Moorbath 1975, O'Nions & Pankhurst 1978, and McCulloch & Wasserburg 1978, for example). A fuller understanding of the complementary nature of these two reservoirs and their time-dependent changes has become possible as a result of Nd isotope studies of terrestrial rocks (see O'Nions et al 1979a,b for additional review of this subject).

DePaolo & Wasserburg (1976a) showed that the initial $^{143}Nd/^{144}Nd$ ratios computed for a number of well-dated single whole-rock samples

Table 2 Sm and Nd abundance data

	Sm (ppm)	Nd (ppm)	Sm/Nd (atomic)
CI chondrite average[a]	0.154	0.4738	0.31
Bulk Earth (Model 1)[b]	0.32	0.97	0.31
Bulk Earth (Model 2)[c]	0.74	2.2	0.31
Continental crust[d]	3.7	16	0.22

[a] Evensen et al 1978.
[b] Assumes Earth has solar REE/U ratios and Earth's heat loss equals heat production (after O'Nions et al 1978a).
[c] Assumes solar Ca/U and REE/U ratios for Earth (O'Nions et al 1979a).
[d] Assumes crust has average composition of island arcs (after Taylor 1977).

ranging in age from 3.6 G.y. to approximately 1 G.y. were derived from a reservoir having Sm/Nd $\simeq$ 0.308 (the chondrite average value of Nakamura et al 1976). Whole-rock Sm-Nd isochrons obtained on the Isua metavolcanics, West Greenland (Figure 6), the Onverwacht lavas, Southern Africa (Figure 7), and Bulawayan volcanics, Rhodesia (Hamilton et al 1977), have initial $^{143}Nd/^{144}Nd$ ratios that are consistent with DePaolo & Wasserburg's (1976a) claim (Figure 8). If it is assumed that the Earth developed from material that had a $^{143}Nd/^{144}Nd$ ratio identical to Angra dos Reis at the time of its formation 4.55 G.y. ago (Figure 4), then the initial ratios of the Isua and Onverwacht volcanics provide independent estimates of the bulk Earth Sm/Nd ratio equal to 0.306 ± 6 and 0.298 ± 4 respectively for the period between 4.55 and their formation. Because only small amounts of the Earth's total inventory of REE had been partitioned into stable continental crust by about 3.5 G.y., these Archaean metavolcanics should provide a reliable estimate of Sm/Nd in the bulk Earth.

Development of continental crust has increased the Sm/Nd ratio of the residual mantle and *in toto* it should have a $^{143}Nd/^{144}Nd$ ratio greater

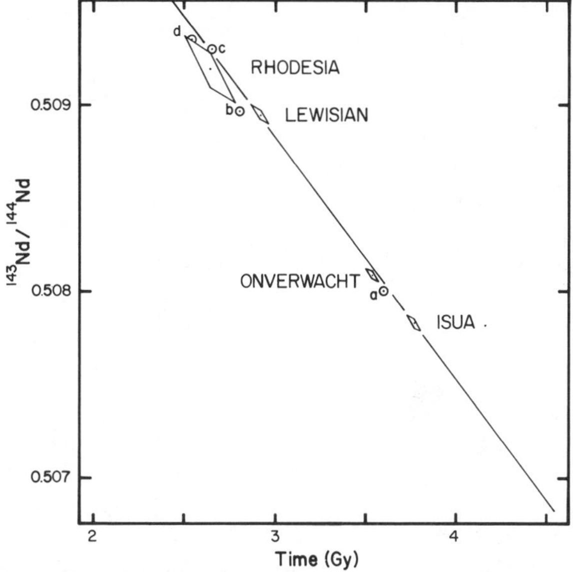

Figure 8 Initial $^{143}Nd/^{144}Nd$ ratios of Archaean rocks. The ages and initial ratios of the Isua, Onverwacht, Lewisian, and Rhodesia samples were determined from Sm-Nd whole-rock isochrons (Hamilton et al 1977, 1978a,b,c). The initial ratios of the Amitsoq gneiss, Great Dyke, and Fiskenaesset anorthosite were computed from single whole-rock analyses using ages derived by other parent-daughter systems (DePaolo & Wasserburg 1976a). Collectively these data are consistent with a chondritic Sm/Nd ratio for the Earth: the line plotted is for Sm/Nd = 0.308.

than the bulk Earth value (estimated at 0.51262). The range of $^{143}Nd/^{144}Nd$ ratios in oceanic basalts (Figure 9) indicates that the partition of REE into the crust has not had a uniform effect on the mantle; various segments have made contributions of differing magnitudes and at different times.

The Nd-isotopic evolution of the continental crust contrasts markedly with that of the suboceanic mantle, as illustrated by the 3.6 G.y. Amitsoq gneisses from West Greenland (Figure 10). The Amitsoq gneisses, like acid gneisses in general, are light REE enriched, and have average Sm/Nd and $^{143}Nd/^{144}Nd$ ratios considerably lower than the bulk Earth, at the present day (Figure 9). McCulloch & Wasserburg (1978) computed the times of separation of crustal material from a chondritic reservoir using measurements of Sm/Nd ratios and Nd-isotope compositions obtained on composite gneiss samples from Precambrian shields. The ages obtained from these model calculations are in good agreement with other geochronological estimates of the age of the shield segments. The corollary of this observation is that the large areas of shield represented by these composite samples were generated from reservoirs that evolved with chondritic Sm/Nd ratios. This point is further illustrated by the age and initial $^{143}Nd/^{144}Nd$ ratio obtained for the Lewisian gneisses of northwestern Scotland (Figure 8).

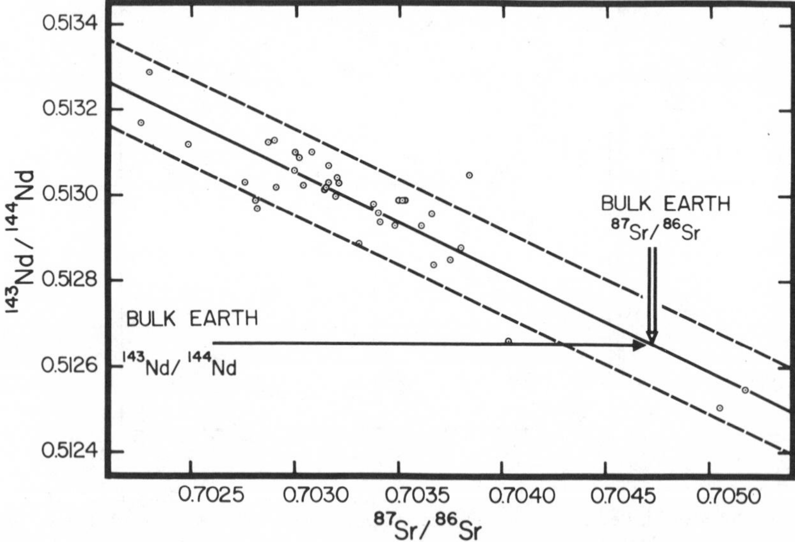

Figure 9 $^{143}Nd/^{144}Nd$ versus $^{87}Sr/^{86}Sr$ for unaltered oceanic ridge and oceanic island basalts. The dashed lines encompass 90% of the data. The present-day bulk Earth $^{87}Sr/^{86}Sr$ ratio of about 0.7047 can be deduced from the intersection of the present-day bulk Earth $^{143}Nd/^{144}Nd$ (0.51265) with the best-fit line (solid line) of the oceanic data. Data from DePaolo & Wasserburg (1976a) and O'Nions et al (1977).

The apparent uniformity of Sm/Nd in the mantle reservoir from which continental crust was extracted during the Archaean is perhaps surprising in view of the fact that the reservoir must change in response to continental

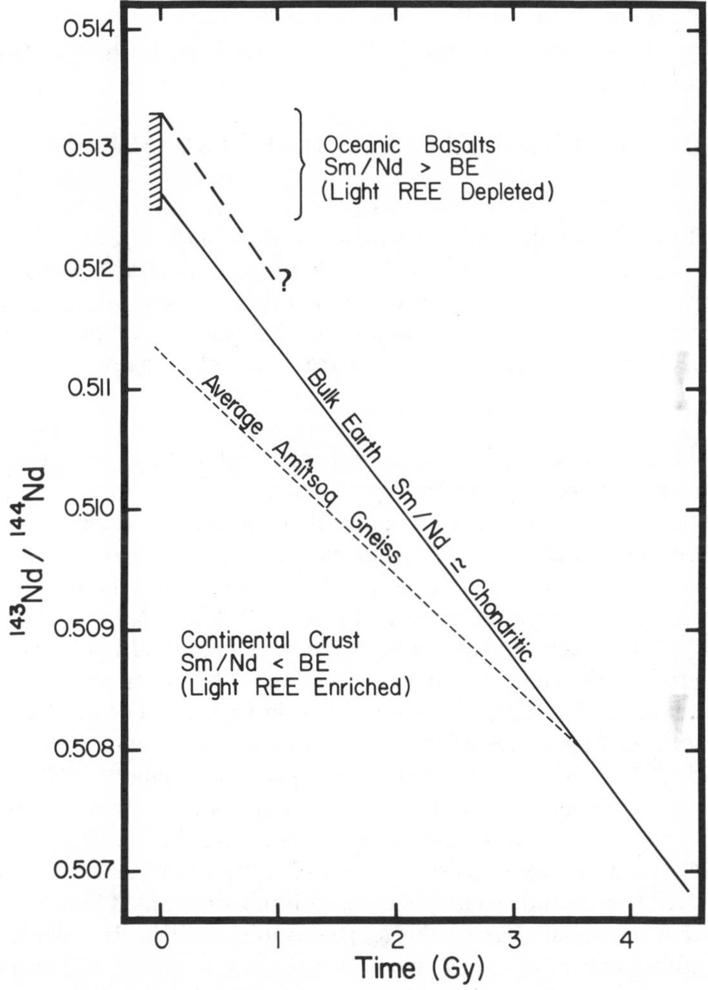

Figure 10 Evolution of $^{143}Nd/^{144}Nd$ in the Earth's crust and mantle. The bulk Earth is assumed to have a Sm/Nd ratio of 0.308 (see Figure 8) and to have developed from material with an initial $^{143}Nd/^{144}Nd = 0.50682$ identical to that of Angra dos Reis 4.55 G.y. ago (Lugmair & Marti 1977), such that its present-day $^{143}Nd/^{144}Nd$ ratio is predicted to be 0.51262. The range of Recent oceanic basalts is indicated and they are seen to be largely derived from source regions having higher Sm/Nd than the bulk Earth (≡ light REE depleted). The continental crust has evolved with a Sm/Nd ratio less than the bulk Earth (≡ light REE enriched) as illustrated for 3.7 G.y. Amitsoq gneiss from West Greenland (average Sm/Nd ratio computed from O'Nions & Pankhurst 1974).

growth. This apparent uniformity may reflect the combined effects of large reservoir size (e.g. whole mantle) and uniform extraction of REE in the Archaean such that the change in Sm/Nd in the whole mantle was insignificant. Alternatively the uniformity may reflect isolation of mantle source regions such that they escaped earlier fractionation events. Differentiating between these two possibilities presents an intriguing problem for the future.

Nd ISOTOPES AS TRACERS IN PETROGENESIS

Because the continental crust has grown by nonuniform extraction of REE from the mantle and has evolved with a markedly lower Sm/Nd ratio, there is a diversity of $^{143}Nd/^{144}Nd$ ratios in the Earth which serve as powerful tracers of provenance in various petrogenetic problems. Some instances in which the determination of Nd-isotope compositions has helped elucidate the petrogenesis of terrestrial rocks are reviewed here.

Petrogenesis of Oceanic Basalts

Nd-isotope compositions of oceanic basalts exhibit a significant range and have now been investigated quite extensively. Although Nd-isotope compositions alone can place important constraints on the petrogenetic relationships of oceanic basalts, they have been particularly useful where considered in conjunction with $^{87}Sr/^{86}Sr$ ratios. Thus in the ensuing discussion the constraints imposed by both $^{143}Nd/^{144}Nd$ and $^{87}Sr/^{86}Sr$ ratios are reviewed. Sr- and Nd-isotope data published by DePaolo & Wasserburg (1976a,b), Richard et al (1976), O'Nions et al (1977), and Hawkesworth et al (1977) are summarized in Figure 9. The most striking and noteworthy features of these data are the strong anticorrelation of $^{143}Nd/^{144}Nd$ and $^{87}Sr/^{86}Sr$ ratios, the generally higher $^{143}Nd/^{144}Nd$ ratios of submarine ocean ridge basalts and the paucity of oceanic basalts with $^{143}Nd/^{144}Nd$ ratios less than the predicted bulk Earth value of 0.5126. These features suggest that *virtually all* oceanic basalts have been generated from mantle source regions with time-integrated Sm/Nd ratios greater than the bulk Earth value (corresponding to light REE depletion). Because submarine ocean ridge basalts are more voluminous than ocean island basalts and also have higher $^{143}Nd/^{144}Nd$ ratios, only trivial amounts of the suboceanic upper mantle can be in an undifferentiated or pristine condition. The anticorrelation of $^{143}Nd/^{144}Nd$ and $^{87}Sr/^{86}Sr$ ratios requires that Rb/Sr and Sm/Nd have fractionated coherently during the differentiation and evolution of the mantle, and leads to an estimate of the Rb/Sr ratio in the bulk Earth as indicated in Figure 9. DePaolo & Wasserburg (1976b) and O'Nions et al (1977) independently estimated this ratio to be approximately 0.03.

The genetic relationships of oceanic island and ridge basalts can be conveniently made in reference to the Rb/Sr and Sm/Nd ratios of the bulk Earth utilizing the values ΔND and ΔSR. These were defined by O'Nions et al (1978a) and Carter et al (1978a) as follows:

$$\Delta \text{ND} = \frac{(^{147}\text{Sm}/^{144}\text{Nd})_{\text{SS}} - (^{147}\text{Sm}/^{144}\text{Nd})_{\text{BE}}}{(^{147}\text{Sm}/^{144}\text{Nd})_{\text{BE}}} \cdot 10^2$$

where the subscripts BE and SS refer to the bulk Earth and single stage values respectively,

$$(^{147}\text{Sm}/^{144}\text{Nd})_{\text{SS}} = \frac{(^{143}\text{Nd}/^{144}\text{Nd})_{\text{M}} - I}{(e^{\lambda t} - 1)}$$

where I is the assumed initial $^{143}Nd/^{144}Nd$ ratio of terrestrial material 4.55 G.y. ago (Lugmair & Marti 1977), and for a Recent basalt $t = 4.55$ G.y.

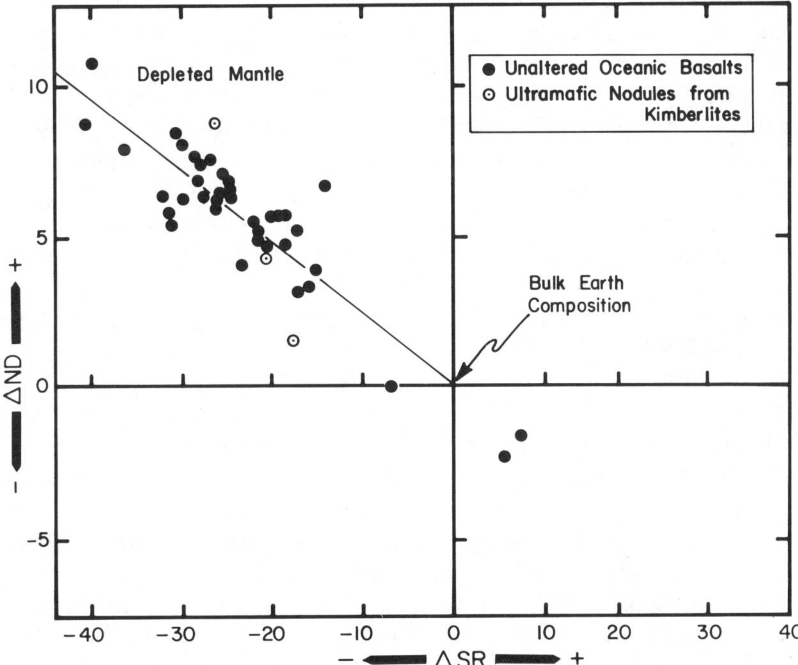

Figure 11 Plot of ΔND and ΔSR for Recent uncontaminated oceanic basalts and garnet lherzolite xenoliths from South African kimberlites. See text for definitions of ΔND and ΔSR. The ΔND and ΔSR parameters of the garnet lherzolites (Thaba Putsoa and Premier) are calculated for their initial $^{143}Nd/^{144}Nd$ and $^{87}Sr/^{86}Sr$ ratios. All samples plotted fall close to the bulk Earth point or within the −ΔSR, +ΔND quadrant, which indicates that they were derived from depleted source regions. Data from O'Nions et al (1977, and Lamont-Doherty Geological Observatory, unpublished).

ΔND and ΔSR values for Recent oceanic basalts are shown in Figure 11. A basalt generated from undifferentiated mantle will have ΔND = ΔSR = 0 and plot at the origin of Figure 11. The majority of the basalts have negative ΔSR and positive ΔND values consistent with their time-integrated or single stage Rb/Sr and Sm/Nd ratios, which are lower and higher respectively than the bulk Earth values. It is immediately clear from this diagram that very few basalts have been generated from undifferentiated mantle.

Continental Basalts and Subcontinental Mantle

Interpretations of the isotope geochemistry of continental basalts are frequently hampered by uncertainties concerning the extent of contamination by continental crust. Some recent interpretations of their Rb-Sr isotope geochemistry for example imply that contamination effects are negligible (Brooks et al 1976, Brooks & Hart 1978), whereas Carter et al (1978b) have claimed on the basis of $^{143}Nd/^{144}Nd$ and $^{87}Sr/^{86}Sr$ ratios that some Tertiary basalts from northwestern Scotland have been severely contaminated. In favorable circumstances crustal contamination can be elegantly demonstrated in continental basalts as indicated in Figure 12. In this diagram the ΔND and ΔSR values for 2.9 G.y. Lewisian granulite and amphibolite facies basement in northwestern Scotland are illustrated. It is noteworthy that the range of ΔSR values is much larger than the range of ΔND values in the Lewisian, reflecting the wide variation of Rb/Sr ratios between granulites and amphibolites and the more similar Sm/Nd ratios. Also illustrated in Figure 12 is the area occupied by oceanic basalts, and thus by inference suboceanic upper mantle. The contrasting isotopic evolution of continental crust and mantle alluded to previously in this article is well illustrated in this diagram. Some Tertiary basalts from northwestern Scotland (e.g. Mull) have ΔND and ΔSR values similar to Recent oceanic basalts, while others such as Skye, Eigg, and Muck basalts are markedly displaced from the oceanic basalt region towards the areas occupied by the Lewisian basement. Carter et al (1978b) have claimed on the basis of this evidence that these northwestern Scotland basalts have been contaminated by Lewisian basement. Volcanics from Roccamonfina, Italy (Carter et al 1978a), are shown for comparative purposes in Figure 12, and at this stage the possibility that the ΔND and ΔSR values of these volcanics reflect some form of enriched mantle and/or have been contaminated by continental crust cannot be resolved. Granitic and gabbroic intrusives from northwestern Scotland have ΔND and ΔSR values suggesting that Lewisian amphibolite facies gneisses are a significant component, which is consistent with previous petrogenetic models (Carter et al 1978b).

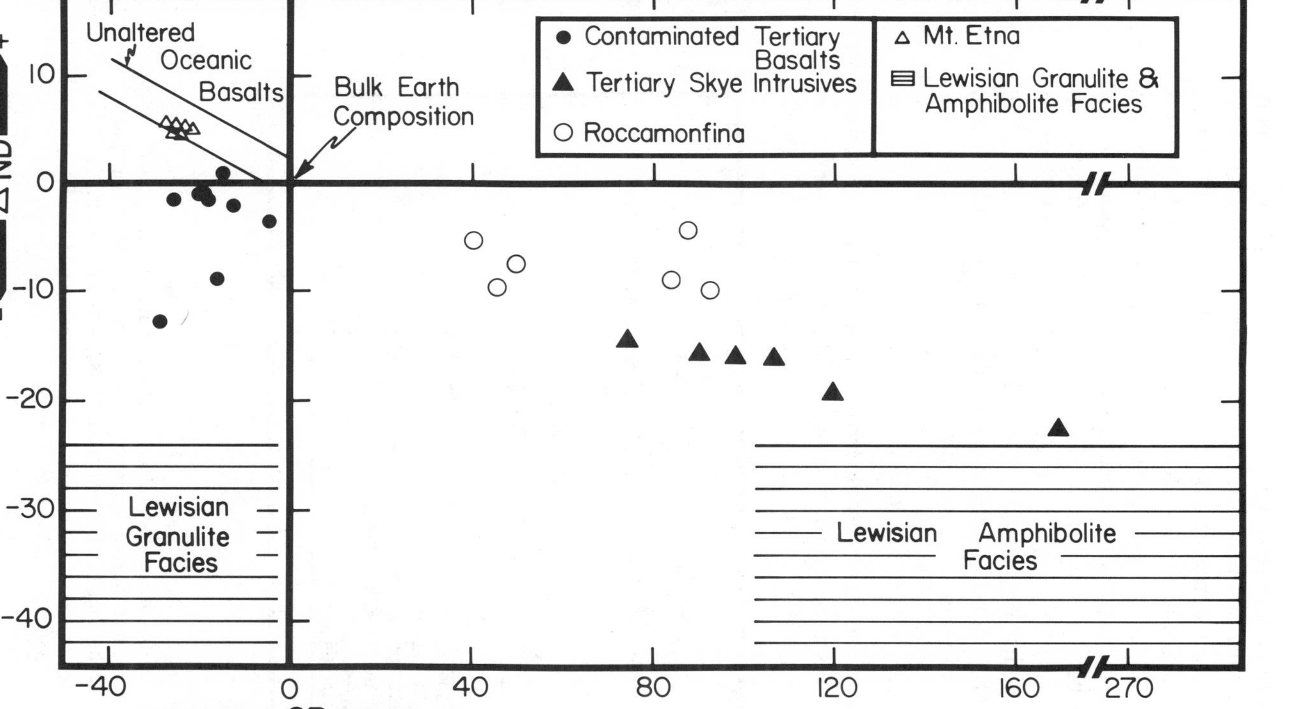

Figure 12 ΔND versus ΔSR parameters of various continental volcanics and continental basement rocks, illustrating the effects of crustal contamination of mantle-derived magmas. For the definition of the Δ parameters, see text. The Lewisian basement of northwestern Scotland has low −ΔND values and a wide range of ΔSR values. The granulite facies have −ΔSR values and the amphibolite facies have +ΔSR values. The unique character of the granulite facies (−ΔND/−ΔSR) identifies it as the crustal component that contaminated the Scottish Tertiary basalts, which also fall in the −ΔND/−ΔSR quadrant. The Skye intrusives, which fall in the −ΔND/+ΔSR quadrant, contain a high proportion of Lewisian amphibolite facies and a small component of mantle-derived magma.

The Etna (Sicily) basalts fall on the unaltered oceanic basalt trend in the +ΔND/−ΔSR quadrant and are therefore considered to be uncontaminated products of depleted mantle. The Roccamonfina (Italy) volcanics fall in the −ΔND/+ΔSR (enriched in Rb and light REE), but it is not possible to ascertain whether they have been contaminated by continental crust or are the products of enriched mantle. Data from Carter et al (1978a,b) and Hamilton et al (1978b).

The possible contamination of continental basalts, so clearly evident in the case of the northwestern Scotland basalts, confuses attempts to compare the evolutionary histories of suboceanic and subcontinental mantle. An alternative approach to the investigation of subcontinental mantle is via the study of mantle-derived lherzolites transported to the surface as xenoliths in kimberlites and other volcanics. There are comparatively few relevant Nd-isotope data, but those that are available are plotted in Figure 11. The particular samples are garnet lherzolites from Southern African kimberlites that range in age from 1.1 G.y. to 60 M.y. These all have ΔND and ΔSR values coincident with those for oceanic basalts, which indicates a similar evolutionary history.

Oceanic Ferromanganese Deposits

There are two broad classes of oceanic ferromanganese deposits, namely the Fe-rich sediments occurring at mid-ocean ridge crests and the manganese nodules more characteristic of the abyssal depths of the oceans.

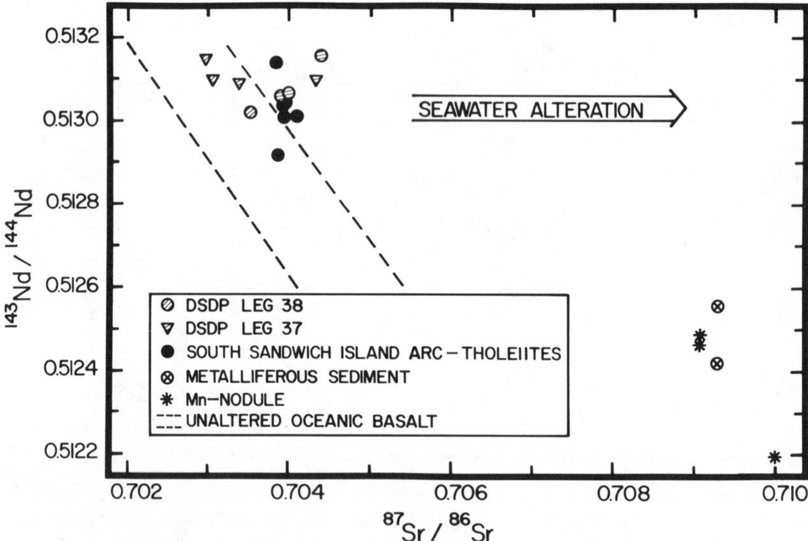

Figure 13 Comparison of $^{143}Nd/^{144}Nd$ and $^{87}Sr/^{86}Sr$ ratios in unaltered oceanic basalts (dashed lines; see Figure 10), altered ocean floor basalts, and island-arc volcanics with Mn nodules and Fe-rich metalliferous sediments. Alteration of ocean floor basalts increases the $^{87}Sr/^{86}Sr$ ratios but has little effect on $^{143}Nd/^{144}Nd$ ratios. $^{143}Nd/^{144}Nd$ and $^{87}Sr/^{86}Sr$ ratios in Mn nodules and Fe-rich sediments are similar, and require a substantial proportion of both Nd and Sr from a continental source. The isotope composition of island-arc tholeiites suggests that they contain a component of altered ocean floor basalt. Data from DePaolo & Wasserburg (1977), Hawkesworth et al (1977), and O'Nions et al (1978b).

The source of metals in these ferromanganese deposits is a subject of considerable debate; both continental weathering and the basaltic ocean crust feature as possible sources of metals in various hypotheses. Elucidation of the provenance of metals in ferromanganese deposits will require knowledge of the flux for each metal concerned from the continents (via rivers and aerosols) and from the oceanic crust, together with their residence times in the oceans. Clearly this is a long-term goal, however some progress has been made in elucidating the provenance of Nd in ferromanganese deposits from measurements of $^{143}Nd/^{144}Nd$ ratios (O'Nions et al 1978b), which adds to earlier studies of Sr and Pb isotopes. The $^{143}Nd/^{144}Nd$ and $^{87}Sr/^{86}Sr$ ratios obtained by O'Nions et al (1978b) are presented in Figure 13 and compared with the range for unaltered oceanic basalts and ocean floor basalts altered by seawater. The $^{143}Nd/^{144}Nd$ ratios of both the manganese nodules and Fe-rich sediments are similar (indicating an apparently homogeneous provenance for Nd), and have lower $^{143}Nd/^{144}Nd$ ratios than ocean floor basalt.

The effect of seawater alteration on the Nd- and Sr-isotope compositions of ocean floor basalts is also indicated in Figure 13. Whereas the $^{87}Sr/^{86}Sr$ ratios are readily modified by seawater contamination (seawater contains approximately 8×10^{-6}g.g^{-1} Sr), the $^{143}Nd/^{144}Nd$ ratios do not appear to be affected, presumably in part because of the substantially lower concentration of Nd in seawater (approximately 3×10^{-12}g.g^{-1}). Thus whereas the basaltic ocean crust behaves as a sink for marine Sr (derived largely from the continents) and ultimately cycles it into the mantle, it does not appear to be a substantial sink for Nd. The $^{143}Nd/^{144}Nd$ ratio of the manganese nodules suggests that seawater Nd is comparatively well mixed isotopically and has a large component derived from the continents, since the $^{143}Nd/^{144}Nd$ ratio of the basaltic oceanic crust is about 0.5130.

Island-Arc Tholeiites

Only a few island-arc tholeiite samples have been analyzed for $^{143}Nd/^{144}Nd$ to date (Hawkesworth et al 1977, DePaolo & Wasserburg 1977), and these are shown in Figure 13. The particular samples studied are from the South Sandwich Islands and New Britain, and both areas have samples that plot to the right of the trend defined by unaltered ocean basalts, which suggests that altered ocean floor is a component of their source region.

CONCLUSIONS

Variations in the isotopic composition of natural Nd have been identified with high-precision solid-source mass spectrometry. These variations have resulted from the α-decay of $^{147}Sm(\lambda = 6.54 \times 10^{-12}y^{-1})$, ^{146}Sm $(\lambda = 6.93 \times 10^{-9}y^{-1})$, incomplete mixing of r- and s-process Nd isotopes, and fission of natural U. Evidence for incomplete mixing of different Nd isotopes has been obtained from the Allende meteorite, and for the former existence of ^{146}Sm in the solar system from ^{142}Nd excesses in the achondrite Angra dos Reis. The Nd isotope fission products of uranium are only known from the highly unusual Oklo natural reactor.

The α-decay of ^{147}Sm to ^{143}Nd has been rapidly exploited as a cosmochronological and geochronological dating tool. It has some advantages over other systems such as Rb-Sr and U-Pb in both the coherence of Sm and Nd in natural fractionation processes, and the favorable abundances of Sm and Nd in some situations where those of Rb and Sr are unfavorable for dating.

In the Earth the isotopic composition of Nd varies both within the mantle and between the continental crust and mantle, and is a valuable indicator of provenance of rare earths in a variety of materials and thus an aid to the elucidating of their petrogenesis. The potential applications of Nd isotopes have still not been fully exploited and to date only a small number of laboratories have participated in their exploitation. Nd-isotope studies will now undoubtedly become a routine technique alongside the more established Rb-Sr and U-Pb investigations as more laboratories join the chase.

Literature Cited

Allsopp, H. L., Viljoen, M. J., Viljoen, R. P. 1973. Strontium isotopic studies of the mafic and felsic rocks of the Onverwacht Group of the Swaziland Sequence. *Geol. Rundsch.* 62:902

Baadsgaard, H. 1973. U-Th-Pb dates on zircons from the early Precambrian Amitsoq gneisses, Godthaab District, West Greenland. *Earth Planet. Sci. Lett.* 19:22–28

Baadsgaard, H., Lambert, R. St. J., Krupicka, J. 1976. Mineral isotopic age relationships in the polymetamorphic Amitsoq gneisses, Godthaab District, West Greenland. *Geochim. Cosmochim. Acta* 40:513–27

Birck, J. L., Allègre, C. J. 1978. Chronology and chemical history of the parent body of basaltic achondrites studied by the ^{87}Rb-^{87}Sr method. *Earth Planet. Sci. Lett.* 39:37–51

Blake, J. B., Schramm, D. M. 1976. A possible alternative to the r-process. *Astrophys. J.* 209:846–49

Brooks, C., Hart, S. R. 1978. Rb-Sr mantle isochrons and variations in the chemistry of Gondwanaland's lithosphere. *Nature* 271:220–23

Brooks, C., James, D. E., Hart, S. R. 1976. Ancient lithosphere: Its role in young continental volcanism. *Science* 193:1086–94

Burbidge, E. M., Burbidge, G. R., Fowler, W. A., Hoyle, F. 1957. Synthesis of the elements in stars. *Rev. Mod. Phys.* 29: 547–650

Carter, S. R., Evensen, N. M., Hamilton, P. J., O'Nions, R. K. 1978a. Continental volcanics derived from enriched and depleted source regions: Nd- and Sr-isotope evidence. *Earth Planet. Sci. Lett.* 37:401–8

Carter, S. R., Evensen, N. M., Hamilton, P. J., O'Nions, R. L. 1978b. Neodymium and strontium isotope evidence for crustal contamination of continental volcanics. *Science* 202:743–47

DePaolo, D. J., Wasserburg, G. J. 1976a. Nd-isotope variations and petrogenetic models. *Geophys. Res. Lett.* 3:249–52

DePaolo, D. J., Wasserburg, G. J. 1976b. Inferences about magma sources and mantle structure from variations of $^{143}Nd/^{144}Nd$. *Geophys. Res. Lett.* 3:743–46

DePaolo, D. J., Wasserburg, G. J. 1977. The

sources of island arcs as indicated by Nd and Sr isotopic studies. *Geophys. Res. Lett.* 41:465–68

Evensen, N. M., Hamilton, P. J., O'Nions, R. K. 1978. Rare-earth abundances in chondritic meteorites. *Geochim. Cosmochim. Acta* 42:1199–1212

Frey, F. A., Haskin, M. A., Poetz, J. A., Haskin, L. A. 1968. Rare-earth abundances in some basic rocks. *J. Geophys. Res.* 73:6085–98

Hamilton, P. J., O'Nions, R. K., Evensen, N. M. 1977. Sm-Nd dating of Archaean basic and ultrabasic volcanics. *Earth Planet. Sci. Lett.* 36:263–68

Hamilton, P. J., O'Nions, R. K., Evensen, N. M., Bridgwater, D., Allaart, J. H. 1978a. Sm-Nd isotopic investigations of the Isua supracrustals, West Greenland: implications for mantle evolution. *Nature* 272: 41–43

Hamilton, P. J., O'Nions, R. K., Evensen, N. M., Turney, J. 1978b. Sm-Nd systematics of Lewisian gneisses: Implications for the origin of granulites. *Nature.* In press

Hamilton, P. J., Evensen, N. M., O'Nions, R. K., Erlank, A. J., Smith, H. S. 1978c. Sm-Nd dating of Onverwacht Group volcanics, South Africa. *Nature.* In press

Hawkesworth, C. J., O'Nions, R. K., Pankhurst, R. J., Hamilton, P. J., Evensen, N. M. 1977. A geochemical study of island-arc and back-arc tholeiites from the Scotia Sea. *Earth Planet. Sci. Lett.* 36:253–63

Herrmann, A. G., Potts, M. J., Knake, D. 1974. Geochemistry of the rare earth elements in spilites from the oceanic and continental crust. *Contrib. Mineral. Petrol.* 44:1–16

Hooker, P., O'Nions, R. K., Pankhurst, R. J. 1975. Determination of rare-earth elements in U.S.G.S. standard rocks by mixed-solvent ion exchange and mass spectrometric isotope dilution. *Chem. Geol.* 16:189–96

Jahn, B. J., Shih, C. Y. 1974. On the age of the Onverwacht Group, Swaziland Sequence, South Africa. *Geochim. Cosmochim. Acta* 38:873–85

Kay, R., Hubbard, N. J., Gast, P. W. 1970. Chemical characteristics and origins of oceanic ridge volcanics. *J. Geophys. Res.* 75:1585–1613

Kohman, T. P. 1954. Geochronological significance of extinct natural radioactivity. *Science* 119:851–52

Loubet, M., Allègre, C. J. 1977. Behavior of the rare earth elements in the Oklo natural reactor. *Geochim. Cosmochim. Acta* 41:1539–48

Lugmair, G. W., Marti, K. 1977. Sm-Nd-Pu timepieces in the Angra dos Reis meteorite. *Earth Planet. Sci. Lett.* 35:273–84

Lugmair, G. W., Marti, K. 1978. Lunar initial ^{143}Nd/^{144}Nd: differential evolution of the lunar crust and mantle. *Earth Planet. Sci. Lett.* 39:349–57

Lugmair, G. W., Scheinin, N. B. 1975. Sm-Nd systematics of the Stannern meteorite. *Meteoritics* 10:447–48 (Abstr.)

Lugmair, G. W., Scheinin, N. B., Marti, K. 1975a. Search for extinct ^{146}Sm, 1. The isotopic abundance of ^{142}Nd in the Juvinas meteorite. *Earth Planet. Sci. Lett.* 27:79–84

Lugmair, G. W., Scheinin, N. B., Marti, K. 1975b. Sm-Nd age and history of Apollo 17 basalt 75075: evidence for early differentiation of the lunar exterior. *Proc. Lunar Sci. Conf. 6th.* 6:1419–29

Lugmair, G. W., Marti, K., Kurtz, J. P., Scheinin, N. B. 1976. History and genesis of lunar troctolite 76535 or: how old is old? *Proc. Lunar Sci. Conf. 7th* 7:2009–33

Lugmair, G. W., Marti, K., Scheinin, N. B. 1978. Incomplete mixing of products from r-, p-, and s-process nucleosynthesis: Sm-Nd systematics in Allende inclusion Ek 1-04-1. *Lunar and Planetary Science IX*, pp. 672–74. Lunar Science Institute, Houston, Texas

McCulloch, M. T., Wasserburg, G. J. 1978a. Sm-Nd and Rb-Sr chronology of continental crust formation. *Science* 200: 1003–11

McCulloch, M. T., Wasserburg, G. J. 1978b. Barium and neodymium isotopic anomalies in the Allende meteorite. *Astrophys. J.* 220:15–19

McCulloch, M. T., Wasserburg, G. J. 1978c. More anomalies from the Allende meteorite: samarium. Preprint

Michard-Vitrac, A., Lancelot, J., Allègre, C. J., Moorbath, S. 1977. U-Pb ages on single zircons from the early Precambrian rocks of West Greenland and the Minnesota River Valley. *Earth Planet. Sci. Lett.* 35:449–53

Moorbath, S. 1975. Evolution of Precambrian crust from strontium isotopic evidence. *Nature* 254:395–98

Moorbath, S., O'Nions, R. K., Pankhurst, R. J. 1973. Early Archaean age for the Isua iron formation, West Greenland. *Nature* 245:138–39

Moorbath, S., O'Nions, R. K., Pankhurst, R. J. 1975. The evolution of early Precambrian crustal rocks at Isua, West Greenland—geochemical and isotopic evidence. *Earth Planet. Sci. Lett.* 27: 229–39

Moorbath, S. Allaart, J. H., Bridgwater, D., McGregor, V. R. 1977. Rb-Sr ages of early

Archaean supercrustal rocks and Amitsoq gneisses at Isua. *Nature* 270:43

Murthy, V. R. 1964. Stable isotope studies of some heavy elements in meteorites. In *Isotopic and Cosmic Chemistry*, ed. H. Craig, S. L. Miller, G. J. Wasserburg, pp. 488–515. Amsterdam: North-Holland.

Nakamura, N., Tatsumoto, M., Nunes, P. D., Unruh, D. M., Schwab, A. P., Wildeman, R. R. 1976. 4.4-b.y.-old clast in Boulder 7, Apollo 17: a comprehensive chronological study by U-Pb, Rb-Sr and Sm-Nd methods. *Proc. Lunar Sci. Conf. 7th.* 7: 2107–29

Nakamura, N., Unruh, D. M., Gensho, R., Tatsumoto, M. 1977. Evolution history of lunar mare basalts: Apollo 15 samples revisited. *Lunar Science VIII*, pp. 712–13. Lunar Science Institute, Houston, Texas

Notsu, K., Mabuchi, H., Yoshioka, O., Matsuda, J., Ozima, M. 1973. Evidence of the extinct nuclide ^{146}Sm in "Juvinas" achondrite. *Earth Planet. Sci. Lett.* 19: 29–36

O'Nions, R. K., Pankhurst, R. J. 1974. Rare-earth element distribution in Archaean gneisses and anorthosites, Godthåb area, West Greenland. *Earth Planet. Sci. Lett.* 22:328–38

O'Nions, R. K., Pankhurst, R. J. 1976. Sr isotope and rare-earth element geochemistry of DSDP Leg 37 basalts. *Earth Planet. Sci. Lett.* 31:255–61

O'Nions, R. K., Pankhurst, R. J. 1978. Early Archaean rocks and geochemical evolution of the Earth's crust. *Earth Planet. Sci. Lett.* 38:211–36

O'Nions, R. K., Hamilton, P. J., Evensen, N. M. 1977. Variations in ^{143}Nd/^{144}Nd and ^{87}Sr/^{86}Sr ratios in oceanic basalts. *Earth Planet. Sci. Lett.* 34:13–22

O'Nions, R. K., Evensen, N. M., Hamilton, P. J., Carter, S. R. 1978a. Melting of the mantle past and present: isotope and trace element evidence. *Philos. Trans. R. Soc. Lond.* 258:547–59

O'Nions, R. K., Carter, S. R., Cohen, R. S., Evensen, N. M., Hamilton, P. J. 1978b. Pb, Nd and Sr isotopes in oceanic ferromanganese deposits and ocean floor basalts. *Nature* 273:435–38

O'Nions, R. K., Carter, S. R., Evensen, N. M., Hamilton, P. J. 1979a. Upper mantle geochemistry. In *The Sea*, ed. E. Emiliani, Vol. VII. In press

O'Nions, R. K., Carter, S. R., Evensen, N. M., Hamilton, P. J. 1979b. Isotope geochemical studies of North Atlantic Ocean basalts and their implications for mantle evolution. In *Second Maurice Ewing Volume*. Am. Geophys. Union. In press

Papanastassiou, D. A., Wasserburg, G. J. 1969. Initial strontium isotopic abundances and the resolution of small time differences in the formation of planetary objects. *Earth Planet. Sci. Lett.* 5:361–76

Papanastassiou, D. A., Wasserburg, G. J. 1976. Rb-Sr age of troctolite 76535. *Proc. Lunar Sci. Conf. 7th.* 7:2035–54

Papanastassiou, D. A., DePaolo, D. J., Tera, F., Wasserburg, G. J. 1977. An isotopic triptych on mare basalts: Rb-Sr, Sm-Nd, U-Pb. *Lunar Science VIII*, pp. 750–52. Lunar Science Institute

Richard, P., Shimizu, N., Allègre, C. J. 1976. ^{143}Nd/^{146}Nd a natural tracer, an application to oceanic basalts. *Earth Planet. Sci. Lett.* 31:269

Schnetzler, C. C., Thomas, H. H., Philpotts, J. A. 1967. Determination of rare-earth elements in rocks and minerals by mass spectrometric, stable isotope dilution technique. *Anal. Chem.* 39:1888–90

Smewing, J. D., Potts, P. J. 1976. Rare-earth abundances in basalts and metabasalts from the Troodos Massif, Cyprus. *Contrib. Mineral. Petrol.* 57:245–58

Taylor, S. R. 1977. Island-arc models and the composition of the continental crust. In *Island Arcs, Deep Sea Trenches and Back Arc Basins*, ed. M. Talwani, W. Pitman. Ewing Series, Am. Geophys. Union

Unruh, D. M., Nakamura, N., Tatsumoto, M. 1977. History of the Pasamonte achondrite: relative susceptibility of the Sm-Nd, Rb-Sr and U-Pb systems to metamorphic events. *Earth Planet. Sci. Lett.* 37:1–12

Wasserburg, G. J., Papanastassiou, D. A., Nenow, E. V., Bauman, C. A. 1969. A programmable magnetic field mass spectrometer with on-line data processing. *Rev. Sci. Instrum.* 40:288–95

Wasserburg, G. J., Tera, F., Papanastassiou, D. A., Huneke, J. C. 1977. Isotopic and chemical investigation on Angra dos Reis. *Earth Planet. Sci. Lett.* 35:294–316

Ann. Rev. Earth Planet. Sci. 1979. 7:39–62

RECENT ADVANCES IN SANDSTONE DIAGENESIS

*10106

Edward D. Pittman
Amoco Production Co., Research Center, Box 591, Tulsa, Oklahoma 74102

INTRODUCTION

Many scientists have studied diagenetic processes and products since 1893 when Walther presented the modern concept of diagenesis. The purpose of this article is to review recent developments, primarily in the last ten years, which have come about because of new instrument technology, experimental studies, a better understanding of chemical processes in rocks, and studies by skilled observers using old techniques.

The seven topics I have picked for review as being of significance undoubtedly reflect my interests: 1. significance of zeolites to burial metamorphism, 2. progress in understanding graywackes, 3. recognition of importance of authigenic clays in sandstone, 4. source of silica for cement of quartzose sandstones, 5. recognition of importance of secondary porosity, 6. diagenesis of red beds, and 7. mathematical modeling of chemical diagenesis. Another petrologist probably would have selected a slightly different list. No significance is attached to the order of presentation of review topics. I will refrain from discussing papers on sandstone diagenesis outside of the above topics; however, many pertinent papers could be cited. The list of references, although long, leaves out important papers. There was no attempt to make a complete historical summary of the topics.

Other recent reviews or discussions of sandstone diagenesis were made by Füchtbauer (1974), Jonas & McBride (1977), and Hayes (1978). Two recent symposia were devoted to sandstone diagenesis: 1. A "state of the art" meeting was held in London in March, 1977 and the proceedings were published in 1978 in the Journal of the Geological Society of London, Volume 135, Part 1 (G. Thomas 1978); 2. A Society of Economic Paleontologists and Mineralogists Symposium on "Diagenesis as it Affects Clastic Reservoirs" was held in April, 1977. Proceedings for this sym-

0084-6597/79/0515-0039$01.00

posium will be published in 1979 as S.E.P.M. Special Publication No. 26 (Scholle & Schluger 1979).

A renewed interest in sandstone diagenesis has been stimulated by the oil industry as the search for oil and gas is extended to sandstone reservoirs with poorer reservoir qualities due to diagenesis. These low quality reservoirs may be shallow or deep. It is now possible through the use of massive hydraulic fracture treatments to complete wells in low permeability, argillaceous sandstones that a few years ago were considered uneconomic.

SIGNIFICANCE OF ZEOLITES TO BURIAL METAMORPHISM

In 1961, Coombs introduced the term "burial metamorphism" into the geological literature. The increased temperature (unrelated to igneous

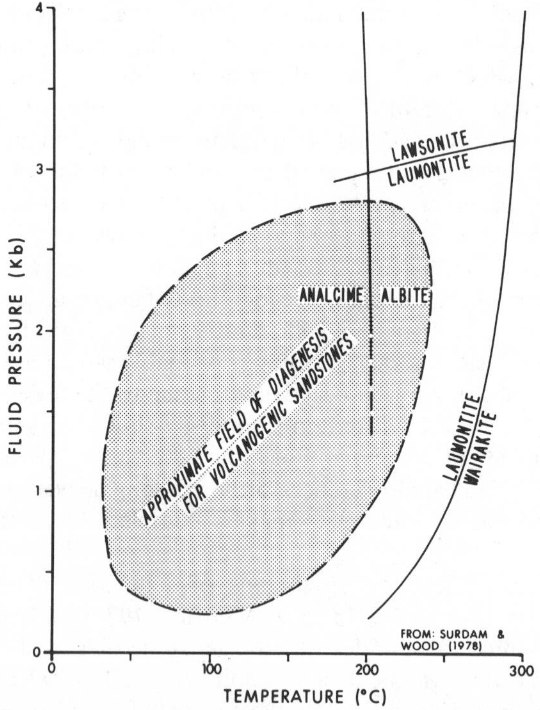

Figure 1 Fluid pressure-temperature diagram for selected equilibrium reactions. Shaded area represents approximate field of diagenesis for volcanogenic sandstones. Original diagram was published in Boles & Coombs (1975) and modified by Surdam and Wood (1978).

intrusions) and pressure associated with burial metamorphism was believed to lead to widespread mineral changes without development of schistosity. In this sense, burial metamorphism differs from "load", "static", or "geothermal" metamorphism which, like "regional" metamorphism, have been applied to processes leading to schistosity.

Fyfe, Turner & Verhoogen (1958) defined the zeolite (metamorphic) facies based on observations made by Coombs (1954) in New Zealand. Later Coombs and co-workers (1959) and Coombs (1971) broadened the original definition of the zeolite facies from a metamorphic facies to a mineral facies, so as to include all co-existing mineral assemblages. The zeolite facies is best regarded as a mineral facies that bridges the transition from diagenesis to metamorphism and includes mineral assemblages produced by diagenetic, metamorphic, and hydrothermal processes (Figure 1).

Many workers (see review by Zen 1974) have reported a variety of interpretations that indicate the complexity of burial metamorphism and the development of the zeolite mineral facies. Studies by Seki (1969), Coombs, Horodyski & Naylor (1970), Surdam (1973), Zen (1974), and Boles & Coombs (1975, 1977) show that the complexity of mineral distribution patterns can be attributed to the interaction of 1. temperature, 2. pressure, 3. rock composition, 4. reaction rates, 5. nucleation kinetics, 6. irreversible reactions, 7. incomplete reactions, 8. permeability, 9. ionic activity ratios in stratal waters, 10. partial pressure of carbon dioxide, and 11. relationship of fluid pressure to total pressure. Because of the interaction of these variables, the zeolite mineral assemblage varies from region to region and even within a single region. For example, in the Southland Syncline of New Zealand, one quarter of the sandstones contain no zeolites and are nondiagnostic of mineral facies despite lithologies similar to zeolite-bearing sandstones elsewhere in the area (Boles & Coombs 1977).

Based on original concepts and definitions of burial metamorphism and zeolite facies, heulandite and analcime were considered to be diagnostic of the diagenetic realm whereas laumontite was supposed to be indicative of metamorphism. Boles & Coombs (1977) show that this approach is unrealistic because of the many variables controlling mineralogy. Laumontite may form at temperatures in the metamorphic realm (up to 300°C), but it also forms as a cement and replacement at temperatures as low as 50°C.

The significance of other zeolite minerals has also been reevaluated by Boles & Coombs (1977). Prehnite, which was once thought to be indicative of metamorphism, is now believed to form in sandstones at inferred temperatures as low as 90°C. Pumpellyite apparently represents higher grade zeolite mineral facies with an estimated lower temperature limit of 190–200°C.

Volcaniclastic sandstones are particularly susceptible to the formation of laumontite, but this zeolite also forms in marine arkosic sandstones that totally lack volcanic detritus. Laumontitization may be either regional or localized both vertically and laterally. According to Stewart & McCulloh (1977), petroleum-bearing arkosic reservoirs are particularly susceptible to laumontitization. The threshold conditions for the formation of laumontite in these rocks appear to be ~120°C at ~300 bars in the presence of water with 17,200–20,000 ppm of sodium chloride and a pH near 7. Laumontite forms most readily in fluvial sandstones with volcanic debris and high permeability, under high geothermal gradients (Stewart & McCulloh 1977).

In summary, because of a variety of chemical and physical factors, zeolite facies development is very complex and does not occur simply in response to increasing pressure and temperature. Many zeolite-bearing sandstones, although they form in the pressure-temperature range traditionally assigned to "burial metamorphism", lack any metamorphic characteristics. These rocks retain the textures and fabrics of sandstones, although they may have up to 25% diagenetic minerals.

PROGRESS IN UNDERSTANDING GRAYWACKES

Graywackes are perhaps the most misunderstood of all sedimentary rocks. Part of this problem stems from inconsistencies in terminology and part from the complexity of these rocks. Probably the only real agreement among workers is that graywackes are "dirty" sandstones. Therefore, the term should be limited to field or informal usage (Dickinson 1970). Problems have arisen because authors have tried to restrict the term "graywacke" to certain textural or compositional characteristics. Confusion regarding graywackes has prevailed for many years because of difficulties in interpreting the origin, texture, mineralogy, and petrographic characteristics.

Cummins (1962) apparently first expressed the view that the matrix in graywackes was diagenetic rather than primary in origin. One of Cummins' arguments against a primary detrital matrix was the failure to find a modern sediment of the graywacke type. Only 2 of 118 cores examined by Hollister & Heezen (1964) were believed to be a graywacke type sand with a primary matrix. Buller & McManus (1973) concluded that diagenetic disintegration of about 30% of the contained sand-size mineral grains would convert a modern turbidite into a graywacke. Based on petrographic evidence, Lovell (1972) and Reimer (1972) interpreted the matrix in graywackes as diagenetic.

Papers by Dickinson (1970) and Galloway (1974) have made significant contributions toward resolving the graywacke problem. A review of this subject by Schluger (1978) provides considerable historical information and succinctly analyzes the situation.

Dickinson (1970) recognized three basic problems in understanding and interpreting the origin of graywackes: 1. identification of detrital grains, especially lithic fragments and other altered grains; 2. interpretation of provenance; and 3. identification of the origin of the matrix. Dickinson identified two types of cement: 1. void filling minerals such as calcite or zeolites; and 2. void lining phyllosilicates, which may blend with the matrix. He also recognized that the matrix could originate in different ways: 1. protomatrix, formed from the original clayey debris; 2. orthomatrix, formed from the recrystallized clay-sized debris; 3. epimatrix, precipitated in open interstices during diagenesis; and 4. pseudomatrix, resulting from plastic deformation of phyllosilicate-rich grains. These matrix types are difficult to distinguish, but often can be identified using Dickinson's criteria. Careful but imaginative petrography is required to make provenance and diagenetic interpretations of graywacke sequences.

Galloway's (1974) major contribution was an interpretation of graywacke diagenesis in Tertiary sandstones of the northeast Pacific region. He recognized that physical diagenesis (compaction) was especially pronounced in the early to intermediate stages of burial where the fabric was not too greatly affected by cementation. Chemical diagenesis was divided into three stages: 1. formation of local very early calcite cement; 2. alteration of labile grains, which makes aluminum and silica available for use in precipitation of clay minerals as pore linings; and 3. precipitation of zeolites and clay minerals as cements. Galloway also pointed out that during chemical diagenesis, replacement of grains and authigenic minerals by newly-formed minerals complicates the diagenetic history and helps obscure the original texture of the rock.

Not all graywackes have abundant authigenic phyllosilicate cement. According to Burns & Ethridge (1979), graywackes in the Eocene Umpqua Formation of Oregon lack abundant phyllosilicate cement although the depth of burial, temperature, and source of reactive materials were adequate. These sandstones contain an abundance of lithic fragments, which deformed plastically to form a pseudomatrix. This pseudomatrix, along with calcite cement and poor sorting, resulted in small pore apertures, reduced permeability, and restricted fluid movement, which probably retarded development of phyllosilicate cement similar to that described by Galloway (1974).

Galloway (1974) showed that temperature is an important variable influencing diagenesis and, hence, porosity gradient. Wells with high geothermal gradients encountered the chemical diagenetic stages discussed above at shallower depths and the sandstones had lower porosity at equivalent depths compared with wells having lower geothermal gradients.

Zeolites occur as cement and replacement of other material in graywackes. Surdam & Boles (1979) showed that zeolites produced by alteration of glass and feldspars are hydration reactions (Figure 2); therefore, zeolite cements and replacements occupy more volume than the reactants, leading to destruction of pore space. They also showed that zeolite to zeolite and zeolite to feldspar reactions are dehydration reactions involving mass transfer associated with fracturing. Secondary porosity could be produced during this stage through dissolution of carbonate minerals.

In summary, the matrix of a graywacke has multiple origins, which may be identified by careful petrography. The term "graywacke" is best used as a field name or informal term. Calcite and zeolites are common as mineral cements, whereas clay forms as authigenic pore linings and

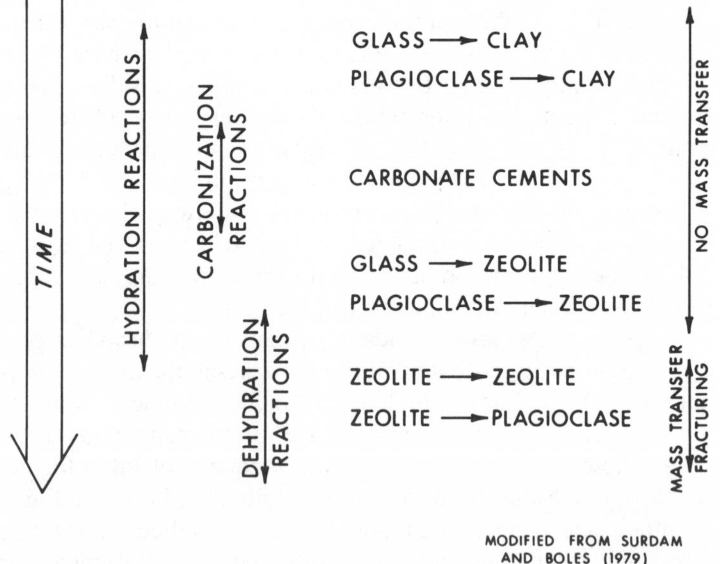

Figure 2 Descriptive framework for the diagenesis of volcanogenic sandstones.

fillings in response to burial diagenesis. Hydration reactions lead to porosity destruction while later dehydration reactions may lead to secondary porosity. Temperature is a particularly important factor influencing diagenesis of graywackes.

RECOGNITION OF IMPORTANCE OF AUTHIGENIC CLAYS IN SANDSTONES

English reviews and translations of Russian literature show that Russian scientists have recognized the importance of authigenic clays of various types in sandstones since at least 1955. Millot (1964; English translation, Millot 1970) in France and Müller (1967) in Germany also were early to appreciate how common authigenic clays are in sandstones. Krynine (1940) perhaps was the first to recognize authigenic clay (other than kaolinite) lining pores. He illustrated this by photomicrograph and sketch. Krynine also recognized that although the clay was not abundant, it occupied a position in the rock whereby the clay would contact and react with fluids. Others who published in English and recognized authigenic clays were Weaver (1955, 1959), Quaide (1956), Lerbekmo (1957), and Milne & Early (1958). Carrigy & Mellon (1964) made an excellent study of authigenic clay in sandstones without the advantages of a scanning electron microscope (SEM). Later, the SEM supplied evidence to convince most skeptics of the importance of authigenic clay. Many workers can be credited with contributing to this facet of sandstone diagenesis including Hayes (1970), Sarkisyan (1971, 1972), Walker & Waugh (1973), Stalder (1973), Gaida, Ruhl & Zimmerle (1973), Flesch & Wilson (1974), Pittman (1974), Wilson (1974), Walker (1974, 1976), Almon, Fullerton & Davies (1976), Almon & Davies (1977), Wilson & Pittman (1977), Neasham (1977), Glennie, Mudd & Nagtegaal (1978), Hancock (1978), Sommer (1978), Hancock & Taylor (1978), Thomas (1978) and Gall, Millot & Gamermann (1978).

Laboratory experiments support the viewpoint that clays in sandstones can form diagenetically. Hawkins & Whetten (1969), Whetten & Hawkins (1970, 1972), Whettan (1971), and Whetten & Hiltabrand (1972) showed that clays are produced by diagenetic changes at temperatures of 250–300°C in loose sand packs of Columbia River sediment (65–70% glassy volcanic rock fragments). Keene, Claque & Nishimori (1973) produced smectite experimentally from alteration of tholeiitic basalt in sea water at 150–250°C and 1 Kb. Divis & McKenzie (1975) experimentally produced diagenetic clay from the detrital components of feldspathic sands at 200°C.

The reader is referred to Wilson & Pittman (1977) for a review of the origin of allogenic and authigenic clay minerals. They also discuss criteria for distinguishing authigenic clays, based on composition, morphology, structure, texture, and distribution. Clay minerals can often be identified by scanning electron microscopy, although mixed layer clays confuse the issue because they may morphologically resemble either of the participating clay minerals.

Authigenic clays occur as 1. pore linings (grain coats), 2. pore fillings, 3. replacement of labile grains, and 4. fracture and vug fillings. The first three are more common. Chlorite, illite, smectite, and mixed layer illite/smectite typically occur as pore linings, which may bridge the contact between detrital sand grains. Kaolinite-dickite and, less commonly, chlorite occur as pore fillings. Smectite, illite, and chlorite are common as replacements. Kaolinite-dickite is best known as an authigenic clay in sandstones because it is often coarse enough to be viewed relatively easily in thin sections. Shelton (1964) compiled a list of occurrences of authigenic kaolinite.

Authigenic clays are common in reservoir rocks of all depositional environments. Chemistry of the formation waters and composition of the rock have a strong influence on growth of authigenic clays. Initial trapped pore water ranges from fresh through brackish to sea water depending on the depositional environment. This water is modified through time by influx of "new" waters, precipitation of minerals (including clays), dissolution of minerals, and cation exchange. Labile components in the rock such as lithic fragments, feldspars, ferromagnesian minerals, and volcanic glass react with formation water to produce clay minerals, which subsequently undergo transformation to more stable forms of clay. Davies & Almon (1977) showed that clay mineral cement formed in volcaniclastic sediments in Guatemala under near surface conditions (25–60°C and 300 m burial) within 2500 years.

Because authigenic clay is so common and can lead to complete filling of pores, it is time to reconsider the origin of argillaceous matrix in sandstone. Most geologists are still thinking in terms of a primary detrital origin for the matrix of sandstones. However, the clay in a sandstone with a bimodal distribution of grain sizes consisting of sand and clay is probably authigenic. Spherical sand-sized grains of quartz (specific gravity 2.65) and platey clay minerals (specific gravity 2.62 or less) are not depositionally compatible, i.e. these particles are not hydraulic equivalents: 0.3 mm quartz sand grains should not be deposited with 1–10 μm clay particles. Pryor (1975) has discussed this problem with regard to the origin of argillaceous sediments. He points out that a current velocity as small as 0.009 cm/sec can maintain a 10 μm diameter particle in suspension during turbulent flow. A 2 μm particle requires six days to fall 1 m through still water, while a current velocity of only 0.0002 cm/sec will keep a 2 μm particle in suspension during turbulent flow. Therefore, to deposit clay particle-by-particle with sand would require totally still non-agitated intervals interspersed with currents of a magnitude suitable to move sand particles.

Pryor (1975) believes that even argillaceous sediment is deposited as floccules rather than particle-by-particle. Clay minerals can be deposited with sand through flocculation. During compaction all traces of the floccule morphology would probably be destroyed.

Undoubtedly, primary depositional matrix occurs in some sandstones. Evidence for such deposits might be poorly sorted sandstones with a complete range of particle sizes from clay through silt to sand. Another line of evidence is sand grains "floating" in a clay matrix. This situation, however, should be analyzed carefully for evidence of replacement.

There are other ways for argillaceous, poorly sorted sandstones to form. One way is through infiltration of clay minerals during pedogenesis (Brewer 1964) or much later during shallow burial (Walker, Waugh & Crone 1978). This process is important in first cycle desert alluvium, which is essentially clay-free when deposited but becomes increasingly clay-rich with age due to infiltration of clay and diagenetic processes. Crone (1974, 1975) provides criteria for the recognition of infiltrated clays based on a scanning electron microscope study of natural and artificial materials. Another important way that argillaceous sands originate is through burrowing by marine organisms, which destroy stratification and mix mud and sand layers. Howard & Reineck (1972) show that extensive homogenization of mud and sand is common.

In summary, authigenic clays are common in sandstones and the distinction between allogenic and authigenic clays is important because it influences the interpretation regarding depositional energy conditions and the depositional environment of the sand.

SOURCE OF SILICA FOR CEMENT OF QUARTZOSE SANDSTONES

For many years, the primary source of silica for quartz cement in sandstones was believed to be pressure solution of quartz grains during burial. This was based on petrographic evidence regarding the nature of the grain contacts. A common observation in quartzose sandstones is the presence of long, concavo-convex, and sutured contacts between detrital quartz grains. These contacts were believed to originate by pressure solution, an application of "Rieckes Principle" whereby a mineral is dissolved at the point of stress with precipitation occurring on surfaces of lower stress. Because quartz cement forms epitaxially and may lack "dust" lines to delineate contacts between nucleus and overgrowth, it is often difficult to determine the presence and amount of cement, which has led to misinterpretations regarding the origin of grain contacts.

Cathodoluminescence techniques pioneered by Sippel (1968) provide a means of resolving this problem because detrital quartz grains luminesce red (most common) or blue in contrast to essentially no luminescence for secondary quartz. Occasionally, some detrital quartz essentially lacks luminescence, and some secondary quartz luminesces red. The color of the luminescence depends on the concentration and interaction of various trace elements as well as on the nature of the host crystal. Sprunt & Nur (1977a) point out that the colors are not infallible indicators: impurities of a few ppm can be dispersed and the defects annealed with time. They believe cathodoluminescence cannot be reliably used to determine the amount of pressure solution in natural sandstones.

Based on cathodoluminescence studies, Sibley & Blatt (1976) believed that pressure solution can account for the source of 30–35% of the quartz cement in the Silurian Tuscarora Formation of the Appalachian Province. One weakness with this study is that Sibley & Blatt (1976) worked with relatively coarse sandstone (0.3 mm mean grain size) because of the difficulties of using high magnification with cathodoluminescence. However, petrographic observations (Heald 1956, Adams 1964) and experimental evidence (Renton, Heald & Cecil 1969, Sprunt & Nur 1976, 1977a, 1977b) indicate that pressure solution increases with a decrease in grain size. Additional cathodoluminescence work needs to be done on very fine-grain sandstones before reaching any generalizations regarding the importance of pressure solution as a source of silica.

Theoretical considerations based on packing of spheres by Manus & Coogan (1974) support Sibley & Blatt's (1976) observations regarding pressure solution. Manus and Coogan, building on research by Rittenhouse (1971), suggest that significant amounts of bulk volume reduction must accompany significant pore reduction for generation of cement by pressure solution. They believe that the common preservation of primary sedimentary features in sandstones precludes significant volume reduction and therefore pressure solution may not be as common as has been generally believed.

Renton, Heald & Cecil (1969), Deelman (1975), DeBoer (1977a, 1977b), and Sprunt & Nur (1976, 1977a, 1977b) have contributed the most recent experimental investigation of pressure solution. Renton, Heald & Cecil (1969) produced pressure solution and quartz overgrowths using a variety of solutions. Angular grains, finer grains, and chert were most susceptible to pressure solution. Deelman (1975) compacted a dry mixture of hard and soft grains and through plastic deformation produced grain contacts similar to those normally interpreted as pressure solution contacts. Deelman believed there is no such thing as pressure solution. DeBoer (1977b) concluded that water is a prerequisite for pressure solution, which may explain Deelman's results on dry samples. DeBoer also concluded that pressure solution is enhanced by increasing temperature, but is scarcely affected by the composition of the aqueous pore solution. Sprunt & Nur (1976, 1977a, 1977b) succeeded in reducing porosity up to 50% by pressure solution without any grain crushing. The rate of the porosity reduction depends on quartz solubility, which is determined by pore pressure and temperature. They determined that the porosity loss increased with a decrease in grain size, which can be accounted for by the increase in specific surface area. Sprunt and Nur believed their most important conclusion to be that porosity reduction requires gross shear stress acting on the rocks. Local heterogeneities at grain contacts apparently are insufficient to reduce porosities rapidly.

Because of doubts concerning the importance of pressure solution, we need to reevaluate the possible sources of silica for cement in sandstones. A survey of the published literature reveals the following possibilities: 1. pressure solution (Waldschmidt 1941); 2. dissolution of siliceous shales; 3. hydration of volcanic glass (Zen 1959, Warner 1965); 4. decomposition of feldspars (Fothergill 1955); 5. replacement of silicates by carbonates (Walker 1960); 6. dissolution of silica secreting organisms such as diatoms, radiolaria, and sponges (Siever 1957); 7. dissolution of eolian quartz abrasion dust in desert sands (Waugh 1970); 8. precipitation from downward percolating ground water (Siever 1959); 9. solubilizing effects of certain naturally occurring complexes on silica (Evans 1964); 10. precipitation directly from sea water (MacKenzie & Gees 1971); and 11. clay mineral diagenesis, e.g. transformation of smectite to illite (Burst 1959).

For sandstones associated with cratonic basins, any of the above could be local sources of silica for quartz cement, but pressure solution, clay mineral diagenesis, replacement of silicates by carbonates, and decomposition of feldspars appear to be most important.

Quartz cement is not important in sandstones deposited in arc trenches; however, the most likely source of silica would be clay mineral diagenesis, alteration of volcanic glass, carbonate replacement of silicate minerals, dissolution of siliceous organisms, and dissolution of siliceous shales.

Precipitation from river waters that percolate downward may be important in some aquifers. Experimental work by Paraguassu (1972) suggests that percolating ground water can dissolve silica, which leads to supersaturated solutions and precipitation of silica as overgrowths. Dissolution of quartz abrasion dust could be important for eolian sandstones. Little is known about the solubilizing effects of organic complexes, especially in regard to quartz precipitation.

In summary, based on petrographic studies, pressure solution is believed to be common in quartzose sandstones. Experimental evidence suggests that pressure solution is a viable process in nature. In contrast, however, cathodoluminescence studies of medium-grained sandstone suggest that volumetrically pressure solution is not as important a source of silica for cement as generally believed; however, more cathodoluminescence studies are needed on finer grained sandstones because petrographic observations and experimental evidence suggest that pressure solution increases with a decrease in grain size. Pressure solution, clay mineral diagenesis, replacement of silicates by carbonates, and decomposition of feldspars appear to be the most important sources of silica for quartz cement in quartzose sandstones.

RECOGNITION OF IMPORTANCE OF SECONDARY POROSITY

Proshlyakov (1960) apparently was first to recognize that secondary porosity is important in sandstones. His observations, however, were largely ignored.

During the early 1970's, several articles were published dealing with the occurrence of dissolution porosity related to feldspars (Morgan & Gordon 1970, Shenhav 1972, Heald & Larese 1973, Rowsell & DeSwardt 1974). Some of these workers favored direct dissolution of feldspars, whereas others believed that the feldspars were replaced by carbonate, which was subsequently dissolved. Porosity originating in this manner forms isolated intragranular or moldic pores unless the feldspar and associated pores are abundant enough to form an interconnecting pore system or intergranular pores interconnect with the dissolution pores associated with the feldspars.

Significant contributions to the study of porosity in sandstones were made by Schmidt, McDonald & Platt (1977), Schmidt & McDonald (1979a, 1979b), and Hayes (1979). In independent parallel studies, they concluded that much of the porosity in hydrocarbon reservoirs worldwide is secondary in origin and results primarily from dissolution of carbonate and/or sulfate minerals occurring as cement, grains, and replacements. Often, this porosity mimics primary intergranular porosity. E. F. McBride and his students (Stanton & McBride 1976, Stanton 1977, McBride 1977, Lindquist 1976, 1977) and Loucks, Bebout & Galloway (1977) also recognized the provincial importance of secondary porosity in Tertiary sandstones of the Gulf Coast.

All of the above studies recognized that primary porosity is destroyed or partially destroyed by mechanical and chemical compaction plus cementation during early burial (Figure 3*A*,*B*). At some depth, secondary porosity may develop as a result of dissolution of soluble material, particularly carbonates. Secondary porosity may form relatively early while considerable primary porosity remains (Figure 3*A*) or later when little or no primary porosity is preserved (Figure 3*B*). The development of secondary porosity may lead to an increase in porosity (short dashed line in Figure 3*A*), or simply be reflected in a lower porosity gradient (long dashed line in Figure 3*A*). Secondary porosity that forms early during active compaction may be destroyed at a faster rate than primary porosity alone because of breakage and rearrangement of grains (long dashed curve, Figure 3*B*).

Secondary porosity that forms late will be destroyed but probably at a slower rate than primary porosity because compaction is of less importance (Figure 3*B*), although some rocks may show evidence of fracturing of grains and collapse of pores developed by dissolution.

Schmidt & McDonald (1979a) and Hayes (1979) call upon an interesting mechanism to create secondary porosity. They postulate that hydrogen ions produced in shales as a product of temperature-controlled kerogen diagenesis dissolve carbonates in sandstones and the flow of water flushes the carbonates updip to be precipitated as early diagenetic cement. As

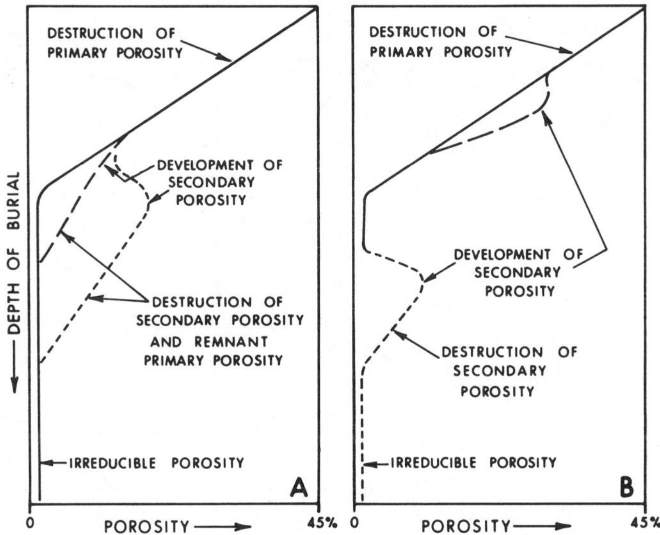

Figure 3 Destruction of primary porosity, development of secondary porosity, and destruction of secondary porosity in sandstones as a function of depth of burial.

these sandstones become more deeply buried, water entering them becomes more acid because of progressive diagenesis, and carbonate is dissolved creating secondary porosity. Thus, carbonate minerals are continually cycled upward as the basin fills and subsides.

Time of migration of hydrocarbons relative to the formation of secon-

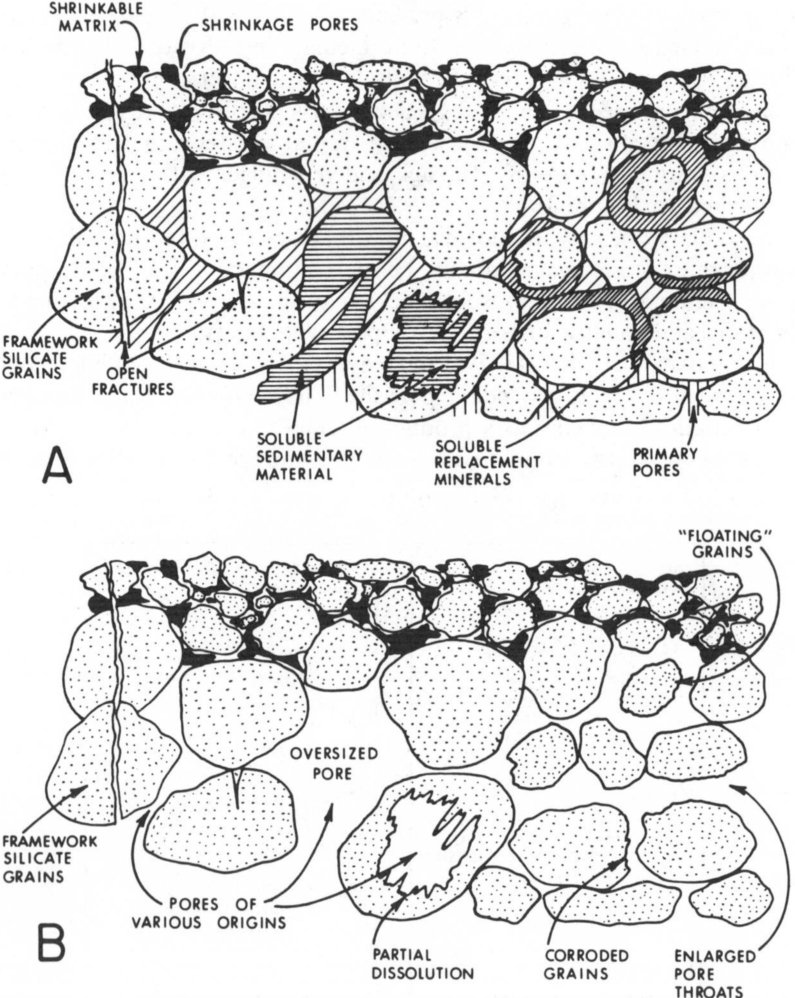

Figure 4 *A*. Sketch showing the origin of various secondary pores in sandstone. Porosity resulting from shrinkable matrix is important only locally. *B*. Result of dissolution of soluble components in *A* is a rock with porosity of variable secondary origin plus primary porosity. Some clues to secondary porosity are oversized pores, partially dissolved grains, corroded grains, enlarged pore throats, and "floating grains".

dary porosity in the reservoir must now be considered by the hydrocarbon explorationist. In the past, this was not considered important because essentially all of the porosity was believed to be primary, and the only concern was for presence or absence of porosity.

Schmidt & McDonald (1979a) provide criteria for identifying various types of secondary porosity in sandstones. They recognize fractures, shrinkage voids, dissolution of sedimentary materials (fossils and other grains), dissolution of authigenic cement, and dissolution of authigenic replacement minerals (Figure 4*A*). Secondary porosity often has multiple origins. Not uncommonly, porosity in a sandstone is a combination of primary and secondary pores (Figure 4*B*).

Pittman (1979) offered a different approach to porosity, which stressed pore geometry, i.e. the size, shape, and distribution of pores in a reservoir. He suggested a classification with four basic types of porosity: intergranular, intragranular-moldic, micro, and fracture. The first three types are related to rock fabric and can be considered end-members of a ternary

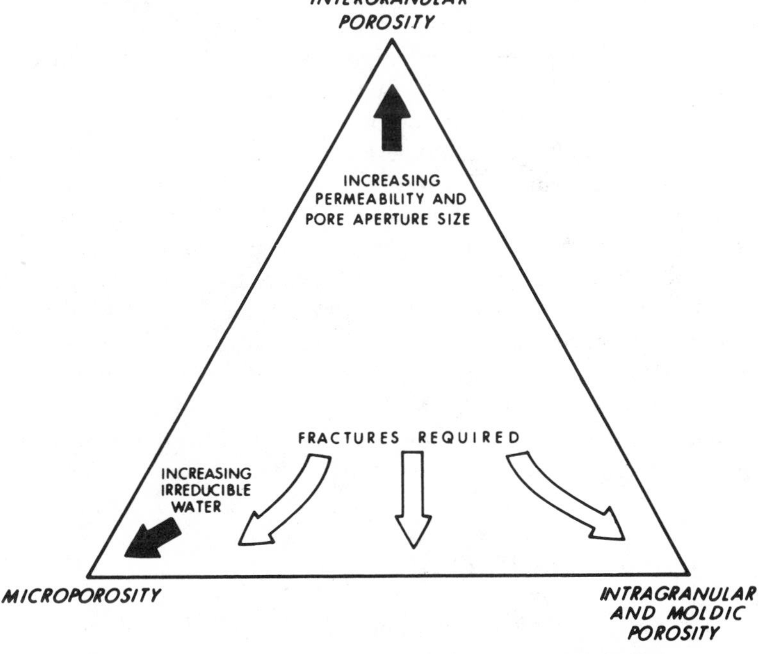

Figure 5 Classification of fabric-related porosity based on pore geometry. Intergranular porosity may be primary or secondary. Intragranular and moldic porosity is always secondary. Microporosity is associated predominantly with clay minerals, which commonly are of authigenic origin. Fracture porosity may be associated with any other porosity type.

classification diagram (Figure 5). Fracture porosity may be associated with any other porosity type.

All sandstones initially have intergranular porosity which if not destroyed often is associated with good permeability, large pore apertures, and prolific hydrocarbon production (Figure 5). Intragranular and moldic porosity results from dissolution of carbonate, feldspar, sulfate, or other soluble material. Sandstone reservoirs with dissolution porosity range from excellent to poor, depending on amount of porosity and interconnection of pores. Excellent reservoirs with secondary intergranular porosity can develop from dissolution of mineral cement. Isolated intragranular and moldic pores result in low permeability. Sandstones with significant amounts of clay minerals have abundant microporosity, high surface area, small pore apertures, low permeability, and high irreducible water saturation (Figure 5). Fracture porosity which contributes no more than a few percent voids to storage space will enhance the deliverability of any reservoir. Open fractures, either natural or induced, are essential for economic deliverability rates from reservoirs with essentially only micropores or isolated intragranular and moldic pores (Figure 5).

Porosity type and/or pore geometry changes with diagenesis: macropores become micropores, minerals dissolve to create voids, and pores are partly to completely occluded by precipitation of minerals. Porosity type seldom is homogeneous in rocks.

Pore geometry influences the type, amount, and rate of fluid produced. For example, a sandstone with abundant authigenic clay may selectively hold irreducible water while yielding hydrocarbons from macropores. A reservoir with fractures or other high permeability zones will behave quite differently than a homogeneous reservoir. These high permeability zones may lead to premature breakthrough of formation water or secondary-tertiary recovery fluids.

In summary, secondary porosity is common and important in sandstone hydrocarbon reservoirs worldwide. Because of the abundance of secondary porosity, hydrocarbon explorationists must consider the time of migration of hydrocarbons in relation to the time of formation of secondary porosity.

DIAGENESIS OF RED BEDS

Largely through the efforts of T. R. Walker and his colleagues and students we now better understand how the red color forms in some sandstones, which has a bearing on the controversy about the origin of red beds. Two major conflicting hypotheses have been used to explain the hematite that provides the red color. One hypothesis, which became popular in the late 1940s, contends that the hematite is detritally derived from lateritic soils of tropical or subtropical regions. Another hypothesis espouses the view that the hematite is authigenic and forms in situ from alteration of iron-bearing minerals.

Walker (1967, 1976) presented strong field, petrographic, and chemical evidence showing that hematite in modern and ancient desert deposits of the southwestern United States and northwestern Mexico formed during diagenesis. Clay minerals migrate into desert fanglomerates and channel sandstones as a result of mechanical infiltration with influent surface water (Crone 1975). This mechanically infiltrated clay, which is iron-bearing (Walker & Honea 1969), characteristically reddens first, within thousands to tens of thousands of years. More intensive red coloration develops as framework ferro-magnesian silicates alter to release iron (Walker, Ribbe & Honea 1967). Thin section and SEM studies showed that these ferro-magnesian minerals alter by dissolution and replacement by clay (Walker & Waugh 1973). This authigenic clay is also iron-bearing and contributes to the red coloration (Walker & Ribbe 1966).

The sediments are not red when deposited, but redden with time. Because the ferro-magnesium minerals are not in equilibrium with the ground waters, they alter diagenetically by reaction with the oxidizing alkaline water.

Iron released by intrastratal hydrolysis is controlled largely by Eh and pH of the water (Garrels & Christ 1965). If the diagenetic environment favors ferrous ions then a ferrous iron-bearing authigenic mineral such as pyrite or siderite will form in a drab colored rock; however, if the diagenetic environment lies in the stability field of hematite, the iron will precipitate as hematite or as a ferric hydrate precursor (Berner 1969, Langmuir 1971) that ultimately converts to hematite. Such sediments are red colored. Because of changes in water chemistry, red beds may bleach, and drab colored ferrous iron-bearing rocks may redden later in their history.

Prolonged intrastratal alteration may remove all traces of the less stable ferro-magnesian minerals. Authigenic hematite, however, may persist indefinitely in a favorable chemical environment.

Recent studies by Turner (1974) in the Silurian of Norway and Hubert & Reed (1978) in the Jurassic of the eastern United States support Walker's views regarding the origin of hematite in sandstone.

To summarize, field, petrographic, and chemical evidence supports the hypothesis that hematite, which provides the red coloration of red beds, is authigenic. The red coloration results from hydrolysis of iron in detrital minerals and clay minerals of infiltration and authigenic origin.

MATHEMATICAL MODELING OF CHEMICAL DIAGENESIS

In 1970, Helfferich and Kachalsky presented a method adaptable for analyzing mineral precipitation in geological systems. They provide a general framework for purely diffusive processes. Building on this work, Wood & Surdam (1979) have tried a mathematical model approach to determine how the products of a reaction between a moving fluid and a porous medium can be characterized spatially and temporally. Using this approach, they modeled processes in terms of differential equations, and sought solutions in terms of analytical functions or computer simulations. The primary advantage of this approach is that the various parameters can be displayed and evaluated. The disadvantage is in solving the differential equations. These equations, describing a geological process, can be set up but often are so numerous and complex that explicit solutions are not possible.

Wood & Surdam conclude that the only feasible mechanism for the precipitation of material from a fluid under isochemical, isobaric conditions is the collision of two or more counter-current mass flows. This is based on the assumption that partial equilibrium exists between the fluid and precipitates (Helgeson 1968). Counter-current flow can be

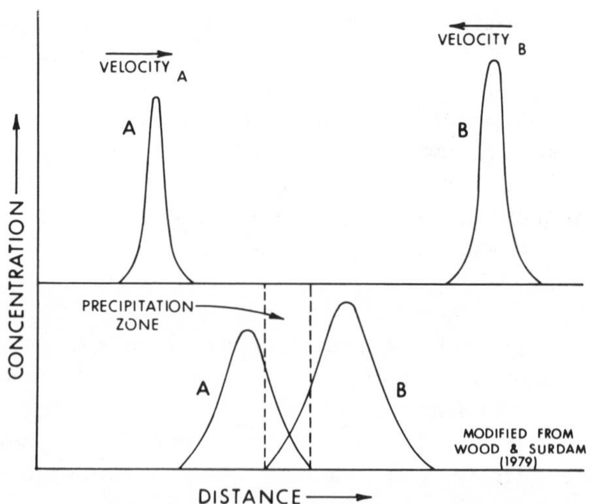

Figure 6 Schematic diagram illustrating counter-current flow as a mechanism for precipitation. Two waves of composition *A* and *B* and differing concentrations move towards each other (upper diagram). When the wave fronts collide they are capable of reacting to form fluid *AB* leading to precipitation where the two waves overlap (lower diagram).

defined as movement of two or more waves of different composition toward one another. This movement can be attributable to diffusion, fluid flow, or a combination of both. For example, a simple two-wave system is diagrammatically depicted in Figure 6. The concentration wave on the left contains dissolved substance *A* at a certain concentration and moves to the right at some velocity. The concentration wave on the right contains dissolved substance *B* at a certain concentration and moves to the left at some velocity. When the wave fronts for *A* and *B* collide, they are capable of reacting to form a fluid *AB*, leading to precipitation where the two waves overlap (lower diagram, Figure 6).

Wood & Surdam treat the counter-current flow system mathematically using Daily & Harleman's (1966) convection-diffusion equation as a starting point. This equation and other derived equations take into account the *X*, *Y*, and *Z* components of fluid velocity, time, molar concentration of the fluid, and diffusion coefficient. An equation is written for each aqueous species that is capable of moving independently in the fluid and a mass action equation is written for any reaction that may occur in the system. The reader is referred to Wood & Surdam (1979) for the mathematical details and solutions for a variety of conditions. Thus far, they have only considered the simplest of geologic systems, but research is continuing. Hayes (1979) also discusses modeling and suggests some logical approaches and geological factors to consider. This line of research is new and intriguing.

In summary, mathematical models have been used to determine how the products of a reaction between a moving fluid and a porous medium can be characterized in time and space. To date, only simple geologic models have been tested, but research is continuing. This research offers interesting possibilities for predicting diagenetic sequences.

CONCLUSIONS

The study of diagenesis has been greatly advanced by new instrumentation. The scanning electron microscope, coupled with the energy dispersive X-ray, has aided many studies but was largely responsible for a better description and understanding of authigenic clays. Cathodoluminescence provided a means of detecting secondary quartz in sandstones, which led to a reevaluation of the source of silica for cement in quartzose sandstones.

A better understanding of chemical processes in sedimentary rocks has advanced our understanding of burial diagenesis, graywackes, and red beds.

Experimental studies have contributed to our understanding of the formation of authigenic clays, pressure solution, and the origin of silica for quartz cement in quartzose sandstones.

Some advances were made possible by skilled observers using proved, reliable techniques. Recognition of the importance of secondary porosity particularly falls in this category. Researchers in this field made a breakthrough using essentially only the petrographic microscope.

Research into mathematical models of chemical diagenesis is in the initial stage but offers hope for advancement of diagenetic phenomena as a predictive tool. Despite our interpretive powers to tell what has happened diagenetically to a sandstone, we still do a poor job of predicting ahead of the drill bit in the subsurface.

Future Research

Research possibilities certainly have not been exhausted in the topics reviewed, for example: 1. additional cathodoluminescence research is needed on very fine- and fine-grain sandstones to better evaluate the origin of silica for cement; 2. we need improved knowledge of graywackes because of their hydrocarbon reservoir potential in unexplored frontiers; 3. as the oil industry drills deeper in search of sandstone reservoirs, a better understanding of the zeolite mineral facies and the transition from diagenesis to metamorphism will become increasingly important; and 4. research on mathematical modeling of chemical diagenesis is just starting.

Future research on diagenesis and porosity of argillaceous sandstone reservoirs appears to be worthwhile because there is but slight coverage in public domain literature. These reservoirs often have severe problems related to physical and chemical properties of the rock that other reservoirs do not have. Another topic where research might be profitable is the relationship between water chemistry and diagenesis of sandstones. Collecting representative water samples for this type of study is a problem. A definite trend exists toward more chemical and quantitative studies of sandstone diagenesis with emphasis on why and how rather than what.

ACKNOWLEDGMENTS

I wish to thank my colleagues J. B. Thomas and R. E. Larese who read various drafts of the manuscript and made valuable contributions. They also have influenced my thinking on this subject through many discussions.

Literature Cited

Adams, W. L. 1964. Diagenetic aspects of Lower Morrowan Pennsylvanian sandstones, northwestern Oklahoma. *Am. Assoc. Petrol. Geol. Bull.* 48:1568–80

Almon, W. R., Davies, D. K. 1977. Understanding diagenetic zones vital. *Oil Gas J.* 75:209–16

Almon, W. R., Fullerton, L. B., Davies, D. K. 1976. Pore space reduction in Cretaceous sandstones through chemical precipitation of clay minerals. *J. Sediment. Petrol.* 46:89–96

Berner, R. A. 1969. Goethite stability and the origin of red beds. *Geochim. Cosmochim. Acta* 33:267–73

Boles, J. R., Coombs, D. S. 1975. Mineral reactions in zeolitic Triassic tuff, Hokonui Hills, New Zealand. *Geol. Soc. Am. Bull.* 86:163–73

Boles, J. R., Coombs, D. S. 1977. Zeolite facies alteration of sandstones in the Southland Syncline, New Zealand. *Am. J. Sci.* 277:982–1012

Brewer, R. 1964. *Fabric and Mineral Analysis of Soils.* New York: John Wiley and Sons. 470 pp.

Buller, A. T., McManus, J. 1973. Modes of turbidite deposition deduced from grain size analysis. *Geol. Mag.* 109:491–500

Burns, L. K., Ethridge, F. G. 1979. Petrology and diagenetic effects of graywacke sandstones: Eocene Umpqua Formation, southwest Oregon. *Soc. Econ. Paleontol. Mineral. Spec. Publ.* No. 26. In press

Burst, J. R. Jr. 1959. Post-diagenetic clay mineral environmental relationships in the Gulf Coast Eocene. *Clays Clay Miner.* 6:327–41

Carrigy, M. A., Mellon, G. B. 1964. Authigenic clay mineral cements in Cretaceous and Tertiary sandstones of Alberta. *J. Sediment. Petrol.* 34:461–72

Coombs, D. S. 1954. The nature and alteration of some Triassic sediments from Southland, New Zealand. *R. Soc. N. Z. Trans.* 82:65–109

Coombs, D. S. 1961. Some recent work on the lower grades of metamorphism. *Aust. J. Sci.* 24:203–15

Coombs, D. S. 1971. Present status of the zeolite facies. In *Molecular Sieve Zeolites*, eds. E. M. Flanigen, L. B. Sand, pp. 317–27. Am. Chem. Soc., *Adv. Chem. Ser. No. 101*

Coombs, D. S., Ellis, A. J., Fyfe, W. S., Taylor, A. M. 1959. The zeolite facies, with comments on the interpretation of hydrothermal synthesis. *Geochim. Cosmochim. Acta* 17:53–107

Coombs, D. S. Horodyski, R. J., Naylor, R. S. 1970. Occurrence of prehnite-pumpellyite facies metamorphism in northern Maine. *Am. J. Sci.* 268:142–56

Crone, A. J. 1974. Experimental studies of mechanically infiltrated clay matrix in sand. *Geol. Soc. Am. Abstr. with Progr.* 6:701

Crone, A. J. 1975. *Laboratory and field studies of mechanically-infiltrated matrix clay in arid fluvial sediments.* PhD thesis. University of Colorado. 162 pp.

Cummins, W. A. 1962. The graywacke problem. *Liverpool Manchester Geol. J.* 3:51–71

Daily, J. W., Harleman, D. R. 1966. *Fluid Dynamics.* Reading, Mass: Addison-Wesley

Davies, D. K., Almon, W. R. 1977. Diagenesis of Tertiary volcaniclastics, Guatemala. *Am. Assoc. Petrol. Geol., Soc. Econ. Paleontol. Mineral., Rocky Mtn. Secs., Prog. with abstr.*, pp. 50

DeBoer, R. B. 1977a. On the thermodynamics of pressure solution—interaction between chemical and mechanical forces. *Geochim. Cosmochim. Acta* 41:249–56

DeBoer, R. B. 1977b. Pressure solution: theory and experiments. *Tectonophysics* 39:287–301

Deelman, J. C. 1975. "Pressure solution" or indentation. *Geology* 3:23–24

Dickinson, W. R. 1970. Interpreting detrital modes of graywacke and Arkose. *J. Sediment. Petrol.* 40:695–707

Divis, A. F., McKenzie, J. 1975. Experimental authigenesis of phyllosilicates from feldspathic sands. *Sedimentology* 22:147–55

Evans, W. D. 1964. The organic solubilization of minerals in sediments. In *Advances in Organic Geochemistry*, ed. U. Colombo, G. D. Hobson. New York: Macmillan. 488 pp.

Flesch, G., Wilson, M. D. 1974. Petrography of Morrison Formation (Jurassic) sandstone of the Ojito Springs Quadrangle, Sandoval County, New Mexico. *Twenty-fifth New Mexico Geol. Soc. Field Conf. Guidebook Central-Northern Mexico*, pp. 197–210

Fothergill, C. A. 1955. The cementation of oil reservoir sands and its origin. Fourth World Petrol. Cong., Sec. 1/B, paper 1, pp. 301–14

Füchtbauer, H. 1974. Some problems of diagenesis in sandstones. *Cent. Rech. Pau. Bull.* 8:391–403

Fyfe, W. S., Turner, F. J., Verhoogen, J. 1958. Metamorphic reactions and metamorphic facies. *Geol. Soc. Am. Mem.* 73, 259 pp.

Gaida, K. H., Ruhl, W., Zimmerle, W. 1973. Rasterelektronenmikroskopische untersuchen des Porenraumes von Sandsteinen. *Erdoel Erdgas Z.* 89:336–43

Gall, J. C., Millot, G., Gamermann, N. 1978. Les argiles authigenes à architecture alvéolaire des grès: morphologie, nature et genèse. *Compt. Rend. H.* 286:587–90

Galloway, W. E. 1974. Deposition and diagenetic alteration of sandstone in northeast Pacific arc-related basins: Implications for graywacke genesis. *Geol. Soc. Am. Bull.* 85:379–90

Garrels, R. M., Christ, C. L. 1965. *Solutions, Minerals and Equilibria.* New York: Harper and Row. 450 pp.

Glennie, K. W., Mudd, G. C., Nagtegaal, P. J. C. 1978. Depositional environment and diagenesis of Permian Rotliegendes sandstones in Leman Bank and Sole Pit areas of the UK southern North Sea. *J. Geol. Soc. London* 135:25–34

Hancock, N. J. 1978. Possible causes of Rotliegend Sandstone diagenesis in northern West Germany. *J. Geol. Soc. London* 135:35–40

Hancock, N. J., Taylor, A. M. 1978. Clay mineral diagenesis and oil migration in the middle Jurassic Brent Sand formation. *J. Geol. Soc. London* 135:69–71

Hawkins, J. W. Jr., Whetten, J. T. 1969. Graywacke matrix: hydrothermal reactions with Columbia River sediments. *Science* 166:868–70

Hayes, J. B. 1970. Polytypism of chlorite in sedimentary rocks. *Clays Clay Miner.* 18:255–306

Hayes, J. B. 1978. Sandstone diagenesis—recent advances and unsolved problems. Notes for Am. Assoc. Petrol. Geol. Clastic Diagenesis Sch., 31 pp.

Hayes, J. B. 1979. Sandstone diagenesis—the hole truth. *Soc. Econ. Paleontol. Mineral. Spec. Publ.* No. 26. In press

Heald, M. T. 1956. Cementation of Simpson and St. Peter Sandstones in parts of Oklahoma, Arkansas, and Missouri. *J. Geol.* 64:16–30

Heald, M. T., Larese, R. E. 1973. The significance of solution of feldspar in porosity development. *J. Sediment. Petrol.* 43: 458–60

Helfferich, F., Katchalsky, A. 1970. A simple model for interdiffusion with precipitation. *J. Phys. Chem.* 74:308–14

Helgeson, H. C. 1968. Evaluation of irreversible reactions in geochemical processes involving minerals and aqueous solutions: 1. thermodynamic relations. *Geochim. Cosmochim. Acta* 32:853–77

Hollister, C. D., Heezen, B. C. 1964. Modern graywacke-type sands. *Science* 146:1523–74

Howard, J. D., Reineck, H. 1972. Georgia coastal region, Sapelo Island, U.S.A., sedimentology and biology, IV. Physical and biogenic sedimentary structures of the nearshore shelf. *Senckenbergiana Maritima* 4:81–123

Hubert, J. F., Reed, A. A. 1978. Red-bed diagenesis in the East Berline Formation, Newark Group, Connecticut Valley. *J. Sediment. Petrol.* 48:175–84

Jonas, E. C., McBride, E. F. 1977. Diagenesis of sandstone and shale: application to exploration for hydrocarbons. *Dept. of Geol. Sci., Univ. of Texas, Austin, Cont. Educ. Prog., Publ. No. 1.* 165 pp.

Keene, J. B., Claque, D. A., Nishimori, R. K. 1973. Experimental hydrothermal alteration of tholeiitic basalt in seawater: resultant mineralogy and textures. Geol. Soc. Am., Cordilleran Sec., Ann. Mtg., Abstr. with prog., pp. 65–66

Krynine, P. D. 1940. Petrology and genesis of the Third Bradford Sand. *Penn. State Univ. Mineral. Ind. Exp. Sta. Bull. 29.* 134 pp.

Langmuir, D. 1971. Particle-size effect on the reaction geothite-hematite plus water. *Am. J. Sci.* 271:147–56

Lerbekmo, J. F. 1957. Authigenic montmorillonoid cement in andesitic sandstone. *J. Sediment. Petrol.* 27:298–305

Lindquist, S. J. 1976. Leached porosity in overpressured sandstones—Frio Formation (Oligocene), south Texas. *Trans. Gulf Coast Assoc. Geol. Soc.* 26:332 (Abstr.)

Lindquist, S. J. 1977. Secondary porosity development and subsequent reduction, overpressured Frio Formation sandstone (Oligocene), south Texas. *Trans. Gulf Coast Assoc. Geol. Soc.* 27:99–107

Loucks, R. G., Bebout, D. G., Galloway, W. E. 1977. Relationship of porosity formation and preservation to sandstone consolidation history—Gulf Coast Lower Tertiary Frio Formation. *Trans. Gulf Coast Assoc. Geol. Soc.* 27:109–120

Lovell, J. P. B. 1972. Dhagenetic origin of graywacke matrix minerals: A discussion. *Sedimentology* 19:141–3

MacKenzie, F. T., Gees, R. 1971. Quartz: synthesis at earth-surface conditions. *Science* 173:533–5

Manus, R. W., Coogan, A. H. 1974. Bulk volume reduction and pressure solution derived cement. *J. Sediment. Petrol.* 44: 466–71

McBride, E. F. 1977. Secondary porosity—importance in sandstone reservoirs in Texas. *Trans. Gulf Coast Assoc. Geol. Soc.* 27:121–2

Millot, G. 1964. *Géologie des argiles: altérations, sédimentologie, geóchimie.* Paris: Masson et cie. 499 pp.

Millot, G. 1970. *Geology of Clays.* Transl. W. R. Farrand, H. Paquet. New York: Springer-Verlag. 429 pp.

Milne, I. H., Early, J. W. 1958. Effect of source and environment on clay minerals. *Am. Assoc. Petrol. Geol. Bull.* 42:328–32

Morgan, J. T., Gordon, D. T. 1970. Influence of pore geometry on water-oil relative permeability. *J. Petrol. Tech.* 22:1199–1208

Müller, G. 1967. Sudoit ("dioktaedrischer Chlorit", "Al-Chlorit") im Cornberger Sandstein von Cornberg, Hessen. *Contr. Mineral. Petrol.* 14:176–89

Neasham, J. W. 1977. The morphology of dispersed clay in sandstone reservoirs and its effect on sandstone shaliness, pore space and fluid flow properties. *Soc. Petrol. Eng. Ann. Mtg., Denver, Colo., Preprint 6858*. 8 pp.

Paraguassu, A. B. 1972. Experimental silicification of sandstone. *Geol. Soc. Am. Bull.* 83:2853–58

Pittman, E. D. 1974. Origin of clay in sandstone—evidence from scanning electron microscopy. Am. Assoc. Petrol. Geol., Soc. Econ. Paleontol. Mineral. Ann. Mtgs. Abstr. 1:71

Pittman, E. D. 1979. Porosity diagenesis and productive capability of sandstone reservoirs. *Soc. Econ. Paleontol. Mineral. Spec. Publ. No. 26*. In press

Proshlyakov, B. K. 1960. Reservoir properties of rocks as a function of their depth and lithology. *Geol. Nefti Gaza* 4:24–29 (In Russian)

Pryor, W. A. 1975. Biogenic sedimentation and alteration of argillaceous sediments in shallow marine environments. *Geol. Soc. Am. Bull.* 86:1244–54

Quaide, W. L. 1956. *Petrography and clay mineralogy of Pliocene sedimentary rocks from the Ventura Basin, California*. PhD thesis. Univ. of Calif., Berkeley, Calif.

Reimer, T. O. 1972. Diagenetic reactions in early Precambrian graywackes of the Barberton Mountain Land (South Africa). *Sediment. Geol.* 7:263–82

Renton, J. J., Heald, M. T., Cecil, C. B. 1969. Experimental investigation of pressure solution of quartz. *J. Sediment. Petrol.* 39:1107–17

Rittenhouse, G. 1971. Pore-space reduction by solution and cementation. *Am. Assoc. Petrol. Geol. Bull.* 55:80–91

Rowsell, D. M., DeSwardt, A. M. J. 1974. Secondary leaching porosity in Middle Ecca sandstones. *Trans. Geol. Soc. S. Afr.* 77:131–40

Sarkisyan, S. G. 1971. Application of the scanning electron microscope in the investigation of oil and gas reservoir rocks. *J. Sediment. Petrol.* 41:289–92

Sarkisyan, S. G. 1972. Origin of authigenic clay minerals and their significance in petroleum geology. *Sediment. Geol.* 7:1–22

Schluger, P. R. 1978. Diagenetic problems in "graywacke" deposits. Notes for Am. Assoc. Petrol. Geol. Clastic Diagenesis Sch. 10 pp.

Schmidt, V., McDonald, D. A., Platt, R. L. 1977. Pore geometry and reservoir aspects of secondary porosity in sandstone. *Bull. Can. Petrol. Geol.* 25:271–90

Schmidt, V., McDonald, D. A. 1979a. The rôle of secondary porosity in the course of sandstone diagenesis. *Soc. Econ. Paleontol. Mineral. Spec. Publ. No. 26*. In press

Schmidt, V., McDonald, D. A. 1979b. Texture and recognition of secondary porosity in sandstone. *Soc. Econ. Paleontol. Mineral. Spec. Publ. No. 26*. In press

Scholle, P. A., Schluger, P. R. eds. 1979. Aspects of diagenesis. *Soc. Econ. Paleontol. Mineral. Spec. Publ. No. 26*. In press

Seki, Y. 1969. Facies series in low-grade metamorphism. *Geol. Soc. Jap.* 75:255–66

Shelton, J. W. 1964. Authigenic kaolinite in sandstone. *J. Sediment. Petrol.* 34:102–11

Shenhav, H. 1972. Lower Cretaceous sandstone reservoirs, Israel; petrography, porosity, permeability. *Am. Assoc. Petrol. Geol. Bull.* 55:2194–2224

Sibley, D. F., Blatt, H. 1976. Intergranular pressure solution and cementation of the Tuscarora Orthoquartzite. *J. Sediment. Petrol.* 46:881–96

Siever, R. 1957. The silica budget in the sedimentary cycle. *Am. Mineral.* 42:821–41

Siever, R. 1959. Petrology and geochemistry of silica cementation in some Pennsylvanian sandstones. In *Silica In Sediments*, ed. H. A. Ireland. *Soc. Econ. Paleontol. Mineral. Spec. Paper. No.* 7, 185 pp.

Sippel, R. F. 1968. Sandstone petrology, evidence from luminescence petrography. *J. Sediment. Petrol.* 38:530–54

Sommer, F. 1978. Diagenesis of Jurassic sandstones in the Viking Graben. *J. Geol. Soc. London* 135:63–68

Sprunt, E. S., Nur, A. 1976. Reduction of porosity by pressure solution: experimental verification. *Geology* 4:463–66

Sprunt, E. S., Nur, A. 1977a. Destruction of porosity through pressure solution. *Geophysics* 42:726–41

Sprunt, E. S., Nur, A. 1977b. Experimental study of the effects of stress on solution rate. *J. Geophys. Res.* 82:3013–22

Stalder, P. J. 1973. Influence of crystallographic habit and aggregate structure of authigenic clay minerals on sandstone permeability. *Geol. Mijnbouw* 52:217–20

Stanton, G. D. 1977. Secondary porosity in sandstones of the lower Wilcox (Eocene), Karnes County, Texas. *Trans. Gulf Coast Assoc. Geol. Soc.* 27:197–207

Stanton, G. D., McBride, E. F. 1976. Factors influencing porosity and permeability of lower Wilcox (Eocene) Sandstone. Karnes County, Texas. *Am. Assoc. Petrol. Geol. Bull.* 60:725–26 (Abstr.)

Stewart, R. J., McCulloh, T. H. 1977. Widespread occurrence of laumontite in the late Mesozoic and Tertiary basins of the Pacific margin. Am. Assoc. Petrol. Geol., Soc. Econ. Paleontol. Mineral., Rocky

Mtn. Secs., Progr. with abstr., pp. 48–49

Surdam, R. C. 1973. Low-grade metamorphism of tuffaceous rocks in the Karmutsen Group, Vancouver Island, British Columbia. *Geol. Soc. Am. Bull.* 84: 1911–22

Surdam, R. C., Boles, J. R. 1979. Diagenesis of volcanic sandstones. *Soc. Econ. Paleontol. Mineral. Spec. Publ. No. 26.* In press

Surdam, R. C., Wood, J. R. 1978. Chemical diagenesis. Notes for Am. Assoc. Petrol. Geol. Clastic Diagenesis Sch., 36 pp.

Thomas, G., ed. 1978. Symposium on sandstone diagenesis: state of the art, 1977. *J. Geol. Soc. London* 135: 1–156

Thomas, J. B. 1978. Diagenetic sequences in low-permeability argillaceous sandstones. *J. Geol. Soc. London* 135: 93–100

Turner, P. 1974. Origin of red beds in the Ringerike Group (Silurian) of Norway. *Sediment. Geol.* 12: 215–35

Waldschmidt, W. A. 1941. Cementing materials in sandstones and their probable influence on migration and accumulation of oil and gas. *Am. Assoc. Petrol. Geol. Bull.* 25: 1839–79

Walker, T. R. 1960. Carbonate replacement of detrital crystalline silicate minerals as a source of authigenic silica in sedimentary rocks. *Geol. Soc. Am. Bull.* 91: 145–52

Walker, T. R. 1967. Formation of red beds in modern and ancient deserts. *Geol. Soc. Am. Bull.* 78: 353–68

Walker, T. R. 1974. Formation of red beds in moist tropical climates: A hypothesis. *Geol. Soc. Am. Bull.* 85: 633–38

Walker, T. R. 1976. Diagenetic origin of red beds. In *The Continental Permian in Central, West, and South Europe*, pp. 240–82, ed. H. Falke. Dordrecht, Holland: Reidel Publishing Company

Walker, T. R., Honea, R. M. 1969. Iron content of modern deposits in the Sonoran Desert: A contribution to the origin of red beds. *Geol. Soc. Am. Bull.* 80: 535–44

Walker, T. R., Ribbe, P. H. 1966. Formation of iron-rich authigenic clay by intrastratal alteration of hornblende in an arid climate: A contribution to the origin of hematite stained matrix in arkosic red beds. *Geol. Soc. Amer. Spec. Pap. No. 101*, pp. 233–34

Walker, T. R., Ribbe, P. H., Honea, R. M. 1967. Geochemistry of hornblende alteration in Pliocene red beds, Baja California, Mexico. *Geol. Soc. Am. Bull.* 78: 1055–60

Walker, T. R., Waugh, B. 1973. Intrastratal alteration of silicate minerals in late Tertiary fluvial arkose, Baja California, Mexico. *Geol. Soc. Am. Abstr. with Prog.* 7: 853

Walker, T. R., Waugh, B., Crone, A. J. 1978. Diagenesis in first-cycle desert alluvium of Cenozoic age southwestern United States and northwestern Mexico. *Geol. Soc. Am. Bull.* 89: 19–32

Walther, J. 1893. *Einleitung in die Geologie als historische Wissenschaft*. Jena: Fischer Verlag. 1055 pp.

Warner, M. M. 1965. Cementation as a clue to structure drainage patterns, permeability, and other factors. *J. Sediment. Petrol.* 35: 797–804

Waugh, B. 1970. Formation of quartz overgrowths in the Penrith Sandstone (Lower Permian) of northwest England as revealed by scanning electron microscopy. *Sedimentology* 14: 309–20

Weaver, C. E. 1955. Mineralogy and petrology of the rocks near the Quadrant-Phosphoria boundary in southwest Montana. *J. Sediment. Petrol.* 25: 163–93

Weaver, C. E. 1959. The clay petrology of sediments. *Clays Clay Miner., 6th Nat. Conf., 1957*, pp. 154–87

Whetten, J. T. 1971. Diagenetic origin of matrix minerals: laboratory and field evidence. *Eighth Int. Sediment. Congr., Heidelberg*, p. 109 (Abstr.)

Whetten, J. T., Hawkins, J. W. Jr. 1972. Diagenetic origin of graywacke matrix. *Sedimentology* 15: 347–61

Whetten, J. T., Hawkins, J. W., Jr. 1971. Diagenetic origin of graywacke matrix minerals: a reply. *Sedimentology* 19: 144–46

Whetten, J. T., Hiltabrand, R. R. 1972. Formation of authigenic clay in detrital sand. *Am. Assoc. Petrol. Geol. Bull.* 56: 662 (Abstr.)

Wilson, M. D. 1974. Common forms of diagenetic clay in sandstones. Am. Assoc. Petrol. Geol., Soc. Econ. Paleontol. Mineral. Ann. Mtgs., Abstr. with progr. 1: 100–1

Wilson, M. D., Pittman, E. D. 1977. Authigenic clays in sandstones: recognition and influence on reservoir properties and paleoenvironmental analysis. *J. Sediment. Petrol.* 47: 3–31

Wood, J. R., Surdam, R. C. 1979. Application of convective-diffusion models to diagenetic processes. *Soc. Econ. Paleontol. Mineral. Spec. Publ. No. 26.* In press

Zen, E-An. 1959. Mineralogy and petrography of marine bottom sediment samples off the coast of Peru and Chile. *J. Sediment. Petrol.* 29: 513–39

Zen, E-An. 1974. Burial Metamorphism. *Can. Mineral.* 12: 445–55

Ann. Rev. Earth Planet. Sci. 1979. 7 : 63–92

INFLUENCES OF MANKIND ON CLIMATE

×10107

William W. Kellogg
National Center for Atmospheric Research,[1] Boulder, Colorado 80307

INTRODUCTION

There is a growing realization, both in and outside of the scientific community, that mankind does indeed have the capability to modify the climate of our planet on a global scale. The purpose of this review is to outline the various human activities that can cause such climatic changes, to indicate the state of our knowledge of the system that determines our climate, and to draw a kind of climatic scenario of the future.

The last step involves a prediction, and it is important to make a distinction between predictions of the natural behavior of the climate on one hand and a specification of its behavior as we inadvertently alter the characteristics of the atmosphere or land surface. We are still very far from a generally accepted technique for making useful predictions of weather and climate for a season or a year ahead, and indeed there are the pessimists who believe that the "noise" of the more-or-less random daily fluctuations of the weather patterns will always stand in the way of long-range weather or climate forecasts.

There is a second kind of prediction, however, that is based on the knowledge that some feature or boundary condition of the climate system will be changing in a specified way, and then estimating how the climate system will respond to that change. The success of such "predictions of the second kind" [Lorenz, Appendix 2.1 in Global Atmospheric Research Programme (GARP) 1975] depends on the degree to which we can specify the course of the future external influence, be it natural or anthropogenic, and the degree to which our theoretical models of the climate system

[1] The National Center for Atmospheric Research (NCAR) is sponsored by the National Science Foundation. Dr. Kellogg is currently on leave of absence from NCAR working with the World Meteorological Organization in Geneva, Switzerland, to plan and implement the World Climate Programme.

0084-6597/79/0515-0063$01.00

simulate the response of the real system to an external change (see Figure 1). These are the two essential ingredients of a climate prediction of the second kind—or "extrapolation," as some prefer to call it—and it should be clear at the outset that both must be fraught with some uncertainties (Smagorinsky 1974, 1977, Kellogg 1977, 1977–78).

In the following sections we review the status of climate modeling and indicate the parts of the system about which we know the least. This is followed by a brief discussion of the various human activities that can potentially influence climate, with emphasis on the effects of adding carbon dioxide to the atmosphere from fossil fuel burning.

We thereby establish the basis for a credible (though imprecise) prediction of the second kind, and proceed to outline a climatic scenario for the next several decades, including some of the more obvious implications of a warmer earth. We conclude by volunteering some nonexpert conjectures about what this all means to the peoples of the world in social and economic terms.

MODELING THE GLOBAL SYSTEM THAT DETERMINES CLIMATE

The climate system, shown schematically in Figure 1, consists of the atmosphere, oceans, ice-and-snow masses (the cryosphere), land surface, and living plus dead organisms on land and in the ocean (the biosphere). All these components are dynamic subsystems and interact with each

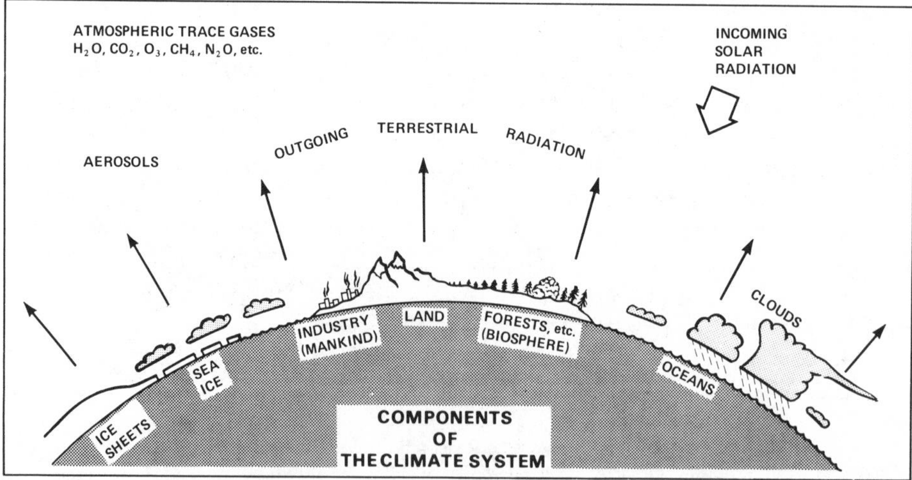

Figure 1 The components of the climate system of the earth, all of which interact with each other in one way or another.

other. The intellectual process of simulating the behavior of this complex system is called "modeling," and for the most part it is done by combining the relevant equations that govern the physical behavior of the components and solving them analytically or integrating them simultaneously on a computer.

Climate models are the most important tools available to study the way in which the real climate system would respond to externally imposed changes, such as those that are now being imposed by mankind. After all, we can at will do experiments with such models to see their sensitivity, but in the case of the earth we must wait for a few decades (at least) to learn the outcome of the inadvertent "experiments" that we are doing—and by then it may be too late to be of value. It is therefore important to understand the various basic approaches that scientists have taken to modeling the climate system. (See also, for example, Schneider & Dickinson 1974, Smagorinsky 1974, 1977, and Gates 1976.)

Considering first the heat available to the planetary system as a whole, the balance between solar radiation received by the earth and outgoing infrared radiation at the top of the atmosphere, F, can be expressed

$$F = \sigma T_e^4 = \bar{Q}_0(1-\alpha),$$

where σ is the Steffan-Boltzmann constant, T_e is the radiative equilibrium temperature of the earth (the "planetary temperature"), $\bar{Q}_0$ is one fourth the solar irradiance at the average distance of the earth from the sun (the factor 1/4 being the ratio of the area of the earth's disc to its spherical surface area), and α is the planetary reflectance, or average albedo. The value of $\bar{Q}_0$ is $\frac{1}{4} \times 1380$ W/m^2 or 345 W/m^2, and α is about 0.28, so T_e turns out to be 257 K.

The average surface temperature, T_s, however, is about 15°C or 288 K, which is some 31 K warmer than the planetary temperature. The difference is due to the fact that there are a number of gases in the atmosphere that absorb infrared radiation, radiation from the surface that would otherwise escape to space (Schneider & Kellogg 1973). This phenomenon is often called "the greenhouse effect," though the analogy to a greenhouse is actually not a very good one. Thus, the earth's surface is maintained at its relatively warm temperature even though the average solar flux absorbed at the surface is only 150–160 W/m^2, taking into account the solar radiation scattered and absorbed by the atmosphere, clouds, and aerosols, and that reflected by the surface itself.

Radiative-Convective Models

A number of calculations have been made to determine the sensitivity of T_s to changes in the infrared absorption by the atmosphere, changes due to the addition of (for example) carbon dioxide or chlorofluoromethanes (CFMs), both of which absorb infrared radiation quite strongly and remain reasonably well mixed throughout the troposphere. In such calculations some assumptions have to be made about other factors in the atmosphere that might vary and, following the lead of Manabe & Wetherald (1967), who pioneered in such "radiative-convective model" studies, it has been customary to keep the temperature lapse rate dT/dz constant at its present average value (6.5 K/km), to let relative humidity remain constant at its present average value as a function of pressure, and to let the fractional amount of cloudiness remain fixed and either cloud-top altitude fixed or cloud-top temperature fixed.

These assumptions are important ones, since they imply that the complex dynamic processes involving the large-scale circulations, meridional transport of heat and water vapor, and vertical mixing by convection continue to operate more or less as they do now in the face of a modest climate change. The assumptions concerning lapse rate and relative humidity, with minor variations (such as those of Cess 1974, Ramanathan 1976, and Augustsson & Ramanathan 1977), seem to be justified by studies of the real atmosphere (Coakley 1977) and by experiments with more complex models (described below), but we are still not sure about the best way to treat cloud-top variations (Schneider 1972, Cess 1976, Augustsson & Ramanathan 1977, Ramanathan 1977, Wang et al 1976). The assumption of fixed cloud-top height yields surface temperature changes that are about 30% less than the same calculation assuming fixed cloud-top temperature (Wang et al 1976).

In any case, the use of one-dimensional (in the vertical) radiative-convective models has been a useful approach to climate modeling, because it permits a comparison of different calculations in which the purely radiative mechanisms can be treated in considerable detail. It must be recognized, however, that such models do not include some other feedback mechanisms that must be taken into account for a more realistic simulation of the complete climate system. Schneider & Dickinson (1974), in their comprehensive review of climate modeling, refer to such simple models as "mechanistic," as distinct from the "simulation" models that try to include as many of the internal interacting processes as possible. This is an important distinction.

Horizontally Varying Energy Balance Models

An important advance in climate modeling was the almost simultaneous publication of the models by Budyko (1969) and Sellers (1969) in which the thermal balance of each latitude band was taken into account, including the meridional convergence of heat transported by the atmosphere-ocean circulations. These two models, and a large number of others that are variations on the same theme (e.g. Saltzman & Vernekar 1972, North 1975, Temkin & Snell 1976, and Drazin & Griffel 1977), have the advantage that they can take into consideration the latitudinal dependence of surface temperature (T_s), surface albedo (α_s), solar flux absorbed at the surface, and so forth. Their shortcomings stem from the fact that all the meridional transport processes must be parameterized in terms of the latitudinal gradient of T_s, and that they tell little about longitudinal variations or land-sea interactions. Furthermore, most models of this kind are annually averaged and without seasonal variations, though this is not an absolute constraint (Suarez & Held 1975).

There is an important feedback mechanism that can be included in such models, however, and that is the interaction between temperature, polar ice/snow, albedo, and solar radiation absorption at the surface. For example, if the mean temperature at the surface in the polar regions were to rise, this would be expected to cause a shrinkage of the area covered by snow, which would cause more solar radiation to be absorbed in summer (because of the decrease in albedo), which would cause a further rise in temperature, and so forth—an amplification of a small temperature change, referred to as a "positive feedback" (Kellogg 1975).

A striking discovery in experiments with the early energy balance models was that a relatively modest decrease in solar energy input (less than 2%) would apparently cause the whole planet to freeze over, and that the snow-covered planet would then be expected to remain in that state, white and snow-covered, even if the sun were to return to its present radiation level. Subsequent studies have explored this surprising behavior and have shown where certain earlier model deficiencies probably lie (Schneider & Gal-Chen 1973, Gal-Chen & Schneider 1976, Cess 1976, Lian & Cess 1977, Su & Hsieh 1976, Lindzen & Farrell 1977).

Extension of this approach by taking longitudinal differences between land and sea have been carried out by Sellers (1973) and Adem (1970), though their approaches to the parameterization of thermal transports by the atmosphere are quite different (Schneider & Dickinson 1974).

Simple Dynamic Models

A number of steps have been taken to develop models that consider the dynamics of the atmosphere in a more explicit way than the energy balance models, some of these actually predating those of Budyko and Sellers (Eliassen 1952, Wiin-Nielsen 1972). These are not the three-dimensional time-dependent general circulation models described in the next section, but are still highly parameterized and in many cases also zonally symmetric. An example of a zonally symmetric model that includes the vertical dimension in some detail is that of MacCracken & Luther (1974).

The variety of approaches to this kind of atmospheric or climate model is so great that it would be difficult to generalize, but it is probably true that most of these were developed in order to explore some aspect of the dynamical behavior of the system rather than to simulate reality (Schneider & Dickinson 1974). Rarely (e.g. MacCracken & Potter 1975) have such models been used in experiments to simulate climatic change, though several of them probably could be adapted for that purpose.

Coupled General Circulation Models

The most elaborate and ambitious approach to atmospheric simulation is the general circulation model, or GCM. In this kind of model the "primitive equations" governing the physical behavior of the atmosphere, most of them time-dependent differential equations, are combined and integrated at each grid point on each height or pressure interval to calculate the change that would occur in a short time interval (the time step). Then the calculations are repeated for the next time step, and so on for as long as the modeler wishes (or until something goes wrong with the model). In a few GCMs isentropic (constant potential temperature) surfaces are taken as the reference levels in the vertical. The more advanced GCMs now involve grid sizes of the order of several hundred kilometers, from two to ten or more vertical intervals, and time steps of five to ten minutes of model atmosphere time. Naturally, it is necessary to carry out such massive "number-crunching" calculations on a large computer with an adequate memory, and to run them several times in order to gain enough data to determine the statistical significance of the result (Chervin & Schneider 1976).

The variables that define the state of the atmosphere are temperature (T), potential temperature (θ), pressure (p), density (ρ), absolute or relative humidity (q or R), and the horizontal components of the atmospheric motion vector, $\mathbf{v}$. All of these are, of course, functions of three-dimensional space and time; and they can be related to each other by a set of seven time-dependent differential equations. There are thus seven equations and seven unknown variables, which constitute a closed set of equations that can be simultaneously integrated numerically, as we have indicated. To these must be added the equations and variables that determine the gain or loss of heat by radiation and other nonadiabatic processes, and the way in which that is accomplished varies from model to model.

The first difficulty with this procedure is that the grid size and height or pressure intervals are necessarily finite, and all the turbulent and convective processes that take place on smaller scales must somehow be taken into account by a set of assumptions about how they act. This is referred to as "sub-grid-scale parameterization," and there are a variety of approaches to it, none of which is considered completely satisfactory.

A second difficulty concerns the parameterization of clouds, which constitute a very important factor in the radiation balance, since they reflect solar radiation and increase the albedo. One has only to look out of one's window to see that clouds are far too complex to ever be described in detail, but it is nevertheless important to assess the overall influence of "cloudiness." It had been feared that the influence of cloudiness might be so great in the course of a climatic change that any model that did not handle it properly would not provide a valid simulation of climatic change (Schneider 1972, Paltridge 1974, Smagorinsky 1974, 1977). While this is still a problem that deserves a great deal of attention (and is indeed receiving it), some recent studies have strongly suggested that the large-scale cloudiness of the real atmosphere does not vary greatly in going from summer to winter (a very large climate change, in a sense) nor with latitude (Cess 1976); and experiments with a GCM suggest that cloudiness response depends on latitude (Schneider, Washington & Chervin 1978). It appears that over a wide range of conditions and considering large portions of the globe the area covered by clouds (where moist air is rising) relative to the clear areas (where air is sinking) does not change by any large factor—a demonstration of the well-known saying that "what goes up must come down."

This is reasonable and reassuring, but there remains the unanswered question about how the average height and temperature of cloud tops may change, and this can have an influence on the outgoing infrared radiation. Also, there are areas where low clouds are formed in the boundary layer (fog and low stratus), and the argument about "air that goes up must come down" does not apply so well to boundary layer cloudiness.

A third and most fundamental problem when general circulation models are considered for climate experiments is the fact that the oceans are just as important as the atmosphere in determining the behavior of the climate system (Bjerkness 1963). For example, ocean currents transport virtually as much heat poleward at middle latitudes as does the atmosphere, and the temperature and orographic contrasts between oceans and continents play a large part in shaping the general circulation of the atmosphere. Beyond that, the oceans are the major source of water vapor for the atmosphere, and their role as a sink for the carbon dioxide we are adding to the atmosphere is now recognized as a most important one (a point discussed further below).

Thus, any effort to use a general circulation model of the atmosphere as a *climate* model must somehow consider the coupling of the atmosphere and oceans. There are ocean circulation models being developed, and some excellent first attempts have already been made to run coupled ocean-atmosphere models (e.g. Manabe & Bryan 1969, Manabe, Bryan & Spelman 1975, Bryan, Manabe & Pocanowski 1975). The degree to which the large-scale ocean circulations have been simulated has so far lagged behind atmospheric simulations, however.

Pending the availability of an adequate ocean circulation model which could be coupled to an atmospheric general circulation model, a simpler and more feasible approach is to use the present sea surface temperature distribution as a lower boundary for the atmosphere, but to let the average ocean temperature rise (or fall) so that the oceans are in quasi-thermal equilibrium in the course of a climate change experiment. In retaining the existing temperature gradients there is the implied assumption that upper ocean circulations would remain unchanged. So far, this kind of experiment has not been performed.

The only experiments with a coupled general circulation model to determine the changes that would ensue with a change of heat input ("solar constant" change) or a change in carbon dioxide amount have been those of Manabe & Wetherald (1975; see also Wetherald & Manabe 1975, Manabe & Holloway 1975). In their model the oceans were assumed to be an infinite source of water vapor but with no heat capacity (like a swamp), cloudiness was fixed, there was just one continent, and it was run without a change of seasons. In spite of such simplifications, these experiments stand as the most sophisticated modeling studies to date of the climatic effects of changing carbon dioxide and heat input.

HUMAN INFLUENCES ON THE CLIMATE SYSTEM

A great variety of human activities modify the factors that determine our climate. Probably the one we have been working at longest is the rearrangement of the surface of the land, which we do by clearing forests for agriculture, by irrigating deserts, by overgrazing marginal lands, by creating large lakes, by building roads and cities, and so forth. All of these result in alterations of the albedo of the surface, hence the surface temperature, and also the flux of water vapor to the atmosphere. Certainly such changes must have had an effect on regional climate (especially at the margins of deserts), and probably a smaller effect on global climate as well, but it is difficult to quantify such effects because the changes have been so slow and so complex (Flohn, Appendix 1.2 in GARP 1975, Otterman 1977, Charney et al 1977, and Kellogg & Schneider 1977).

An effect that is already noticeable in large cities and industrialized areas is the direct warming of the air by the heat released from all the energy being generated and used, sometimes referred to as "thermal pollution." Cities are indeed noticeably warmer than the surrounding countryside, especially in winter when the lower atmosphere is more stable and the energy density of the human sources is a large fraction of the solar radiation absorbed by the ground [Study of Man's Impact on Climate (SMIC) 1971, Landsberg 1974, and Hosler & Landsberg 1977]. However, on a global scale, mankind does not yet compete significantly with the sun, which deposits an average of about 150 W/m^2 at the earth's surface (less over the continents due to the larger albedo of land relative to that of water), and this amounts to roughly 8×10^4 TW for the whole earth (one terawatt is 10^{12} watts). Comparing this with the 8 to 10 TW generated currently in the form of heat, it can be seen that we are only supplying 0.01% of that from the sun, and this can only make a trivial difference to the global energy balance.

It is generally assumed that we will continue to increase our total use of energy for the next century or more, and one should ask whether thermal pollution might become a factor in the global energy balance some day. However, most responsible studies of the possible rate of energy growth have concluded, for a number of reasons, that between 35 and 50 TW by the middle of the next century is probably close to an upper limit on the feasible growth of energy demand and use (Häfele & Sassin 1977, Kahn, Brown & Martel 1976)—and there are many who doubt that such a rapid growth is either feasible or desirable (Lovins 1976, Holdren & Ehrlich 1974). Experiments with climate models (Wetherald & Manabe 1975, Sellers 1973, 1974) indicate that it would require an additional 1% of heat available at the earth's surface to raise the mean surface temperature by about 2 K (Cess 1967), or some 800 TW! This is surely very far in the future and, in any case, beyond the time horizon with which we are dealing here.

It is possible, however, that before such an enormous global outpouring of energy occurs mankind will have been able to shift the mean circulation patterns by its concentrations of heating in a few places. The positions of the semi-permanent high and low pressure areas, such as the Icelandic and Aleutian "lows," are determined by the contrasts between land masses and oceans, and any large new perturbing factor could cause a displacement of those mean flow patterns. Model experiments have suggested that this sort of perturbation could occur as a result of very large quantities of heat being released in the eastern seaboard of the United States (Llewellyn & Washington 1977, Washington & Chervin 1978) or from highly concentrated "power parks" in mid-ocean (Williams, Krömer & Gilchrist 1977, Williams 1978), but here again the effects will almost certainly not be very significant in the next several decades.

A more significant influence, and one which can be more easily quantified, is the change of atmospheric composition as we add carbon dioxide, smoke and smog, and a variety of relatively stable trace gases such as the chlorofluoromethanes, methane, nitrous oxide, carbon disulfide, and so forth. As mentioned in the first part of the previous section, the heat balance of the surface is very greatly influenced by the radiative properties of the atmosphere, since these determine both the scattering and absorption of sunlight and also the absorption of infrared radiation from the surface that would otherwise escape directly to space. The trace gases that absorb such terrestrial infrared radiation also reemit infrared radiation, and this downward flux warms the surface (the "greenhouse effect"). At the same time, to keep a balance between absorbed solar radiation and outward infrared flux at the top of the atmosphere these trace gases tend to cool the stratosphere.

When we compare the magnitudes of the various human effects on climate it seems that the largest single influence is that of adding carbon dioxide. A further and more detailed discussion of this is therefore reserved for the next section; we now deal with some of the other ways by which mankind tampers with his environment on a large scale.

It is obvious that our industrial areas have experienced an increase in aerosols, created directly by burning coal or fuel oil or wood, and also created secondarily from some of the gases that are produced, notably sulfur dioxide and unburned gasoline. These gases are oxidized by a complex set of photochemical reactions in the presence of oxides of nitrogen (also products of combustion) and solar ultraviolet radiation, and after they are oxidized they form the small particles that are often referred to as "smog" (Machta & Telegadas 1974, Kiang & Middleton 1977).

Several studies in the past have concluded that if these aerosols were distributed uniformly over the earth they would increase the earth's overall albedo by scattering sunlight and thereby cause a general cooling (Rasool & Schneider 1971, Yamamoto & Tanaka 1972, Bryson & Wendland 1975, Budyko 1977). The reasons why this is almost surely *not* the case are summarized by Kellogg, Coakley & Grams (1975) (see also Kellogg 1977), and they are briefly restated. First, such industrial aerosols (and the same would apply to agricultural slash-and-burn smoke) do not remain airborne in the lower levels of the atmosphere for more than about five days on the average (Moore, Poet & Martell 1973). That means that they are a regional phenomenon and are limited for the most part to the land areas where they were created. Second, they are now known to be fairly highly absorbing in the visible, apparently because they contain small particles of carbon or black organic substances, and therefore they absorb solar radiation as well as scatter it (Brosset 1976, DeLuisi et al 1976). Detailed calculations taking account of their size distribution, optical properties, and the albedo of the underlying surface (Coakley & Chýlek 1975, Viskanta, Bergstrom & Johnson 1977, Russell & Grams 1975) show that low-lying aerosols will lower the albedo of a typical land surface, thereby causing a warming of the lower atmosphere. (Over a dark ocean surface this is not true, and they may cause an increase in albedo, but most anthropogenic aerosols are over land.) This theoretical conclusion that aerosols over land cause a warming has been verified by actual observations in aerosol-laden air (Idso & Brazel 1977).

The above arguments have applied to aerosols in clear air. When industrial aerosols are mixed into clouds they generally cause the albedo of the clouds to decrease (except for quite thin clouds), and this also causes more absorption of solar radiation and a net warming (Ackerman & Baker 1977).

It has been pointed out by Bryson & Baerreis (1967) that low-level aerosols can increase the stability of the lower layers of the troposphere during the daytime, and that this would suppress convective-type rainfall. This effect is due to the fact that the upper levels of an aerosol layer absorb sunlight and therefore warm that part of the aerosol layer, while less solar radiation reaches the ground. Theoretical studies have subsequently shown that this can probably be a significant effect regionally in industrial areas or wherever there is blowing dust (Wang & Domoto 1974, Atwater 1975, Ackerman 1977).

There are, to be sure, other somewhat more subtle ways by which we can modify the environment on a large scale, though they are probably of less importance than the ones just described. For example, by adding particles that can act as condensation and freezing nuclei and thereby influencing the formation of clouds and precipitation, we are changing "the weather" (Hobbs, Harrison & Robinson 1974), though it cannot be said just how much. This could account in part for the observed increases in summertime rainfall downwind from large cities (Dettwiller & Changnon 1976).

It has been suggested (Boeck 1976) that the addition of radioactive Krypton-85 from the extensive use and reprocessing of nuclear fuels could change the conductivity of the atmosphere, and that this might in turn influence the formation of thunderstorms (Kellogg 1977). This effect might turn out to be a real one, but we know too little about the interactions between the earth's electric field and thunderstorms to be able to say at this time what the final effect on climate might be.

In Table 1 the various kinds of human activities that can have an

Table 1 Human activities influencing climate

Activity	Climatic effect	Scale and importance of the effect
Release of carbon dioxide by burning fossil fuels	Increases the atmospheric absorption and emission of terrestrial infrared radiation (greenhouse effect) resulting in warming of lower atmosphere and cooling of the stratosphere	Global; potentially a major influence on climate
Release of chlorofluoromethanes, nitrous oxide, carbon tetrachloride, carbon disulfide, etc	Same effect as that of carbon dioxide since these, too, are infrared-absorbing and chemically stable trace gases	Global; potentially significant
Release of particles (aerosols) from industry and slash-and-burn practices	These sunlight-absorbing particles probably decrease albedo over land, causing a warming; they also change stability of lower atmosphere	Regional, since aerosols have an average lifetime of only a few days; stability increase may suppress convective rainfall
Release of heat (thermal pollution)	Warms the lower atmosphere directly	Locally important now; will become significant regionally; could modify large-scale circulation
Release of aerosols that act as condensation and freezing nuclei	Influences growth of cloud droplets and ice crystals; may affect precipitation in either direction	Local or (at most) regional influence on precipitation
Upward transport of chlorofluoromethanes and nitrous oxide into the stratosphere	Photochemical reaction of their dissociation products reduces stratospheric ozone	Global; probably small influences on climate; allows more solar ultraviolet radiation to reach the surface
Patterns of land use, e.g. urbanization, agriculture, overgrazing, deforestation, etc	Changes surface albedo and evapotranspiration	Regional; global importance speculative
Release of radioactive Krypton-85 from nuclear reactors and fuel reprocessing plants	Increases conductivity of lower atmosphere, with implications for electric field and precipitation from convective clouds	Global; importance of influence is highly speculative

influence on either regional or global climate are summarized. We turn now to the single influence that appears to be the most significant, the addition of carbon dioxide to the atmosphere.

Effects of Adding Carbon Dioxide

One of the features of the Industrial Revolution that began in the early part of the past century was the extensive use of coal. The present century witnessed a gradual shift from this kind of fossil fuel to petroleum and natural gas, and it seems likely that the world will depend on all of these convenient and still relatively inexpensive sources of energy for many decades to come. The rate of increase of fossil fuel use has remained almost constant at a little over 4% per year since before the turn of the century, with some decreases in this rate during the two world wars and the Great Depression (Baes et al 1977, Rotty & Weinberg 1977).

In Figure 2 is shown the long-term rise in atmospheric carbon dioxide as a result of this combustion of fossil fuels and some extrapolations into the future. In the upper left part of the figure is the more or less continuous record, beginning in 1958, from several parts of the world. This shows that carbon dioxide increase is indeed a global phenomenon, which is reasonable in view of the fact that this trace gas is chemically stable and can remain in the atmosphere long enough to become well mixed from pole to pole. It will be noted that there is slightly more in the northern hemisphere, which reflects the time (about three years) for the carbon dioxide added in the northern hemisphere (where most fossil fuel is burned) to be transported to the southern hemisphere.

The figures on yearly burning of fossil fuels are quite well established (we are now releasing about 20×10^9 tons of carbon dioxide per year), and when these are compared with the total amount that can be accounted for in the atmosphere it turns out that only a bit more than half has remained there (Keeling & Bacastow 1977, Revelle & Munk 1977). Where did the other half go?

It has been thought that some of the "lost" carbon dioxide went into the oceans, which are a potentially large sink, and that some went into the biosphere, on the assumption that many kinds of photosynthetic systems use it faster as the concentration increases (SMIC 1971, Bacastow & Keeling 1973, Machta 1973, Broecker 1975, Oeschger et al 1975, Keeling & Bacastow 1977, Revelle & Munk 1977). Currently our understanding of these sinks is clouded by the realization that we do not know whether the mass of the biosphere is actually growing or shrinking, and there is some evidence that is difficult to refute indicating that deforestation, especially in the tropics, is causing a decrease of this mass (Bolin 1977, Brünig 1977, Woodwell et al 1978). If this were true, then the biosphere would be another source of carbon dioxide as this harvested organic material was burned or allowed to decay, and according to some estimates it could be an even larger source than fossil fuel burning.

The consternation that has been aroused by this possibility is due to the strongly held opinion in the oceanographic community that the oceans could not take up much more carbon dioxide each year than half of that which we have been producing from fossil fuels (Broecker 1975, Stuiver 1978, Siegenthaler & Oeschger 1978). The entire ocean contains some 60 times more carbon dioxide than resides in the atmosphere, so it is potentially an enormous sink for new carbon dioxide; however, only the upper

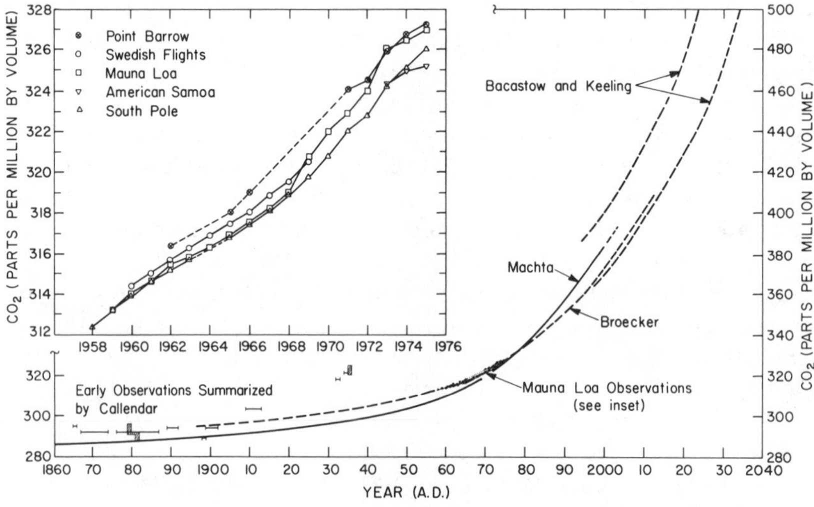

Figure 2 The long-term rise in atmospheric carbon dioxide content, starting at the time of the Industrial Revolution and continuing into the next century. The early data were critically reviewed by Callendar (1958) and subsequently reevaluated by Barrett (1975). The current series of observations for Mauna Loa are those reported by Keeling et al (1976b) and C. D. Keeling (private communication), for South Pole by Keeling et al (1976a) and C. D. Keeling (private communication), for American Samoa and Point Barrow by Miller (1975) and T. Harris (private communication), and for the Swedish aircraft observations by Bolin & Bischof (1970). Note that the carbon dioxide concentrations are given in terms of the "adjusted index values" (for the sake of continuity with the earlier data); it may be necessary to adjust these values upward by about 3 to 4 ppmv, according to Keeling et al (1976b), to obtain the correct mole-fraction but this would not affect the slopes of the curves. The model calculations predicting future carbon dioxide increases by Machta (1973), Broecker (1975), and Bacastow & Keeling (1973) all take account of the takeup of anthropogenic carbon dioxide by oceans and the biosphere (but in somewhat different ways), and assume a nearly exponential increase in the rate of burning of fossil fuels (notably coal) in the next half century or more. It is expected that, in this time period, about half of the new carbon dioxide released will remain in the atmosphere.

well-mixed part of the ocean is in contact with the atmosphere, representing a layer with an average depth of several hundred meters which is only about 10% of the total ocean volume. The rate at which the newly added carbon dioxide can be transported downwards in the oceans by eddy diffusion or larger scale overturning is extremely slow, and the total exchange time to go half way to a new equilibrium is estimated to be 1000 to 1500 years. Knowledge of these mixing and exchange rates is based on studies of the distribution of natural trace constituents in the ocean, and more recently the rate at which radioactive tracers (tritium and carbon-14) from the nuclear tests of the early 1960s have worked their way downwards (Stuiver 1978).

The only thing that is clear at this point is that the school of thought believing in a large shrinking of the biosphere and the school of thought believing in a slow uptake by the oceans cannot both be right—unless someone can think of a new and undiscovered sink somewhere else on the planet. The fact that our data base for assessing the rate of depletion of the tropical forests is so scanty (Elliott & Machta 1977) suggests that one should probably side with the oceanographers until we know more. Furthermore, it is not unreasonable, whatever the truth turns out to be, to assume that the system would continue, for the next several decades at least, to behave as it has in the past, and that a little over one half of our added carbon dioxide would continue to remain in the atmosphere. This, at any rate, is what we will assume here.

In designing scenarios of the future, it has been convenient to estimate when the atmospheric carbon dioxide content will become twice its pre-1900 level, or about 580 parts per million by volume (ppmv) (see Figure 2). This will obviously depend on our future rate of burning of fossil fuels, and that will, in turn, depend on population growth and the course of energy demand throughout the world, the rate of introduction of alternative energy sources such as nuclear and solar power, and technological developments of many kinds. We will attempt to justify what may be considered reasonable upper and lower limits of the rate of production of carbon dioxide.

With the inevitable gradual depletion of natural gas and petroleum reserves the cost of energy will rise, and coal and nuclear power will probably not be able to take their place fast enough, especially if we continue our growth of energy demand at an exponential rate. Furthermore, there will be attempts where possible to substitute other sources for fossil fuel. Thus, the assumption of a continuing 4% per year exponential rate of increase of fossil fuel consumption (it has actually been going up at more nearly 4.3% per year) seems to be a barely credible *upper limit* when carried into the middle of the next century.

A more reasonable assumption would be a gradual decrease in the rate of growth leading to a constant use of fossil fuels some time in the future, very likely followed by a decline as fossil fuels become harder to recover and other sources are exploited. This kind of prediction of fossil fuel use (and carbon dioxide production) can be simply expressed mathematically by letting the rate of growth (r) decrease linearly with time and become zero at some time τ in the future, after which there would be a negative growth rate. Then

$$r(t) = r_0(1 - t/\tau)$$

where t is time in years and r_0 is the present rate of increase of 0.04 per year (4%). Substituting this in the compound interest equation, the future production rate of carbon dioxide can be expressed as

$$P(t) = P_0[1 + r_0(1 - t/\tau)]^t,$$

where P_0 is the present production rate. This reduces to a simple exponen-

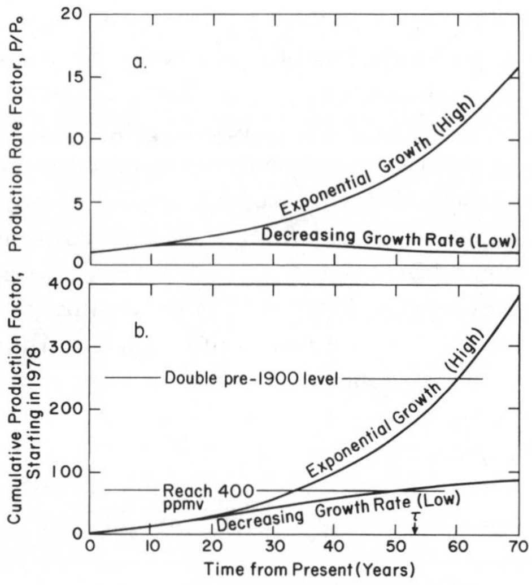

Figure 3 The future change of the carbon dioxide production rate factor (part *a*), assuming an exponential increase of 4% per year (high) or a linearly decreasing rate of increase (see text). The cumulative production factor (part *b*) is shown for the same two assumptions concerning rate of increase; and, since the rate of increase of concentration is currently nearly 1 ppmv per year, the values on the ordinate can also be read directly in terms of parts per million by volume of carbon dioxide added to the present amount (about 330 ppmv). This assumes that the same fraction of the added carbon dioxide, about 55%, remains in the atmosphere. The time τ marks the point when rate of increase per year becomes zero for the "low" case, and thereafter the yearly addition of carbon dioxide to the atmosphere would be less than now and diminishing.

tially increasing function as τ approaches infinity. We will refer to P/P_0 as "the production rate factor."

In Figure 3 possible rates of production and accumulation are plotted as a function of time. The doubling time for rate of production with 4% exponential growth is about 17.5 years, and we have chosen as a kind of "optimistic" *lower limit* of fossil fuel use a reduction of the rate of increase to zero after three such doubling times, or a τ of 52.5 years. The choice of about 50 years is based on the fact that it would take *at least* that long to halt the worldwide construction of new fossil fuel plants and for different energy sources to come into widespread use (Häfele & Sassin 1977).

To justify this as a reasonable lower limit, we argue further that it would probably not be possible to persuade the countries of the world to stop building new coal-burning electric generating plants or to stop buying more cars or reduce the use of fossil fuel in industry any faster than this, in view of the tremendous inertia of our industrialized society and a natural reluctance to give up such a convenient source of energy. This would be especially true in the developing countries, which will undoubtedly claim a larger share of the worldwide energy than they do now (Kahn, Brown & Martel 1976, Häfele & Sassin 1977, Williams 1978). (This is a matter to which we return at the end.)

There are clearly other assumptions that can be made about the future use of fossil fuels, and some of these have been explored by Baes et al (1977), Rotty & Weinberg (1977), Keeling & Bacastow (1977), Häfele & Sassin (1977), Revelle & Munk (1977), and Siegenthaler & Oeschger (1978). It will be seen from Figure 3 that the doubling time for carbon dioxide to a level of 580 ppmv could occur before the middle of the next century, if we consider the upper limit, and in any case it will probably reach 400 ppmv early in the next century. That is the time scale to keep in mind.

The fact that carbon dioxide absorbs infrared radiation and warms the surface by the so-called greenhouse effect (and at the same time cools the stratosphere) has already been discussed. A number of quantitative studies of this phenomenon using radiative-convective models have been made for the global average effect on surface temperature (Manabe & Wetherald 1967, Augustsson & Ramanathan 1977, Wang et al 1976). Sellers (1974) has used his energy balance model to determine the effect as a function of latitude, and one experiment (with the GCM "swamp model") has been carried out by Manabe & Wetherald (1975) which shows the expected change as a function of both latitude and height. These various calculations are now more or less in agreement on the magnitude of the effect of a doubling of carbon dioxide, and the reasons for some different results in earlier calculations have been explained by Schneider (1975). There remains a question about how to treat the possible changes in average cloud-top height and cloud-top temperature (the "swamp model" had fixed cloudiness), but this uncertainty probably will affect the answer by less than 50%. There are almost certainly other feedback mechanisms in the climate system that are not taken into account adequately in these calculations, but none that we can identify now seem to be very important. The polar ice-albedo-temperature feedback that we have referred to is already taken into account.

The best estimate of the "greenhouse effect" due to a doubling of carbon dioxide lies between 2 and 3.5°C increase in average surface temperature (Schneider 1975, Augustsson & Ramanathan 1977, Wang et al 1976), and both the models and the record of the behavior of the real climate show that the change in the polar regions will be greater than this by a factor of from 3 to 5, especially in winter (van Loon & Williams 1976, 1977, Borzenkova et al 1976, Manabe & Wetherald 1975, SMIC 1971). When the level of carbon dioxide has risen to 400 ppmv from its present 330 ppmv, the rise in average surface temperature is estimated to be about 1°C. These figures refer to the effect of carbon dioxide alone and, as was shown in Table 1, many of the other things mankind does also contribute to a warming, so the real effect may be even slightly larger.

This course of possible events is shown in Figure 4, where the shaded portion indicates roughly the range of globally averaged temperatures that have been experienced in the past thousand years or so. It will be seen that by the end of the century, unless some unexpected major climatic event intervenes (for example, a series of large volcanic eruptions

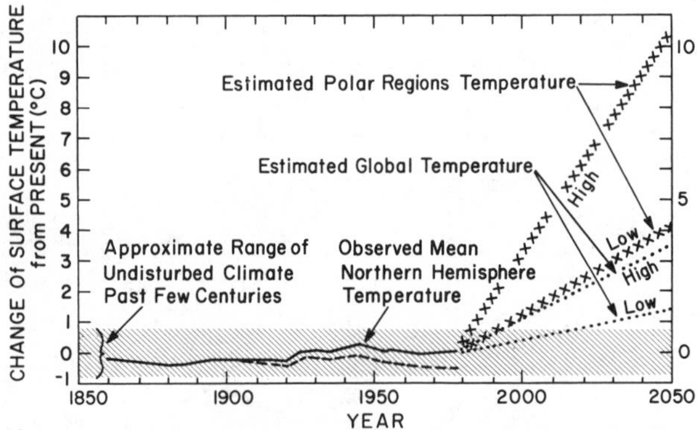

Figure 4 Estimates of past and future changes of surface temperature to be expected for each of the two assumptions about rate of increase of fossil fuel consumption and carbon dioxide production shown in Figure 3. The shaded region is the approximate range of mean surface temperature during the past 1000 years or more. The dashed line indicates the past temperature changes that might have occurred if there had been no carbon dioxide increase.

whose stratospheric particles block the solar radiation from reaching the surface), the average temperature at the surface could be warmer than at any time in the past millenium; and in any case that point should be reached by 2020 AD. Notice that the warming in the polar regions is expected to be even more striking.

One further point should be stressed again, and that is the fact that the oceans can only take up the added carbon dioxide slowly, since the rate of mixing of the upper layers with the deep ocean water is very slow indeed. It is estimated that the time to return halfway to a new equilibrium, should no more carbon dioxide be added to the atmosphere, is 1000 to 1500 years (Keeling & Bacastow 1977, Siegenthaler & Oeschger 1978). Thus, the increased concentration of carbon dioxide and the corresponding temperature increases are for all practical purposes irreversible on the time scale of human affairs, unless we were to take some heroic measures to remove the carbon dioxide from the atmosphere.

IMPLICATIONS OF A WARMER EARTH

We have shown that the most significant effect of mankind's activities in the next century is probably going to be the warming due to the added carbon dioxide, plus some small additional warming from the other things we do. This is useful information, but it is not enough to show what will happen in specific places to the temperature and rainfall or snowfall. After all, it is the *regional* climatic changes that planners must be concerned with. Furthermore, other questions always arise when one imagines what will happen on a warmer earth: What about the polar snow and ice—how much of it will melt, and will sea level change?

Changes of Rainfall

Our theoretical models should, in principle, be able to describe the changes that would occur in some detail, but so far they have not advanced to that point. The latitudinal dependence of temperature and rainfall can be seen in the Manabe & Wetherald (1975) model experiment results, and some further information is available there concerning continent-ocean interactions (Manabe & Holloway 1975), but the fact that it was an annually averaged model makes it hard to deduce anything about the all-important (except in the tropics) changes between winter and summer.

Recalling that both theory and observations of the real atmosphere indicate a larger climatic response in the polar regions than at low latitudes, we can predict that a warmer earth will also have smaller equator-to-pole temperature contrast. Thus, the atmospheric heat engine, with a relatively warmer "condenser" at the polar end, will run more slowly.

In meteorological terms, there will be less available potential energy in the system to convert to kinetic energy.

There is no doubt that this will influence the prevailing large-scale circulation patterns in both summer and winter, and this will cause regional climate changes that are larger than the average for a given latitude belt—it is not unlikely that some regions might even experience a cooling trend, for a while at least. By the same token, precipitation patterns will become very different.

In order to get an idea of the magnitude of these regional changes in precipitation, we can go back to a period in the not-too-distant past when the planet was warmer on the average than now and inquire about the distribution of precipitation then. A warmer atmosphere and ocean will result in more evaporation and more rainfall (a speeded-up hydrological cycle), but the pertinent question is where the rain and snow will fall. The period variously called the Altithermal or Hypsithermal or Atlantic some 4000 to 8000 years ago was generally warmer, and some of the regions where it was wetter or drier than the present are shown in Figure 5.

The information on which this map was based comes from many sources, such as studies of the kinds of pollens at various depths in soils, lake bottoms, and peat bogs, fluctuations in lake size, locations of ancient sand dunes, studies of glacial moraines, rates of deposit of silt in river beds, and archaeological findings showing the habits of ancient peoples. (See, for example, Lamb 1972, 1974, Appendices by Flohn and Kutzbach in GARP 1975, and Mitchell 1976.) The widespread cooling and often dramatic changes of precipitation that took place about 4000 years ago seem to have occurred over a period of only a few centuries (possibly even less time in some places), and historians are now trying to explain many of the migrations and changes of life styles that occurred at about that time as attributable to this climatic change. For example, the subtropical deserts of North Africa (the Sahara), Northwestern India (the Rajastan), and Southwestern North America were more like prairies than deserts, and were inhabited by tribes who must have been forced to abandon their ancestral homes as the rainfall gradually failed. In Central North America the reverse seems to have occurred, and the original prairie that extended from the Rocky Mountains almost to the Appalachian Mountains gave way to advancing forests in the eastern part as the reliability of summertime rains increased (Kellogg 1977, 1977–78).

Figure 5 is presented here not so much as a forecast of things to come as an impressive example of the large and complex changes in circulation and precipitation patterns that must inevitably go with a temperature change. It would be surprising if the Altithermal patterns were exactly reestablished as we warmed the earth again, since the causes of that

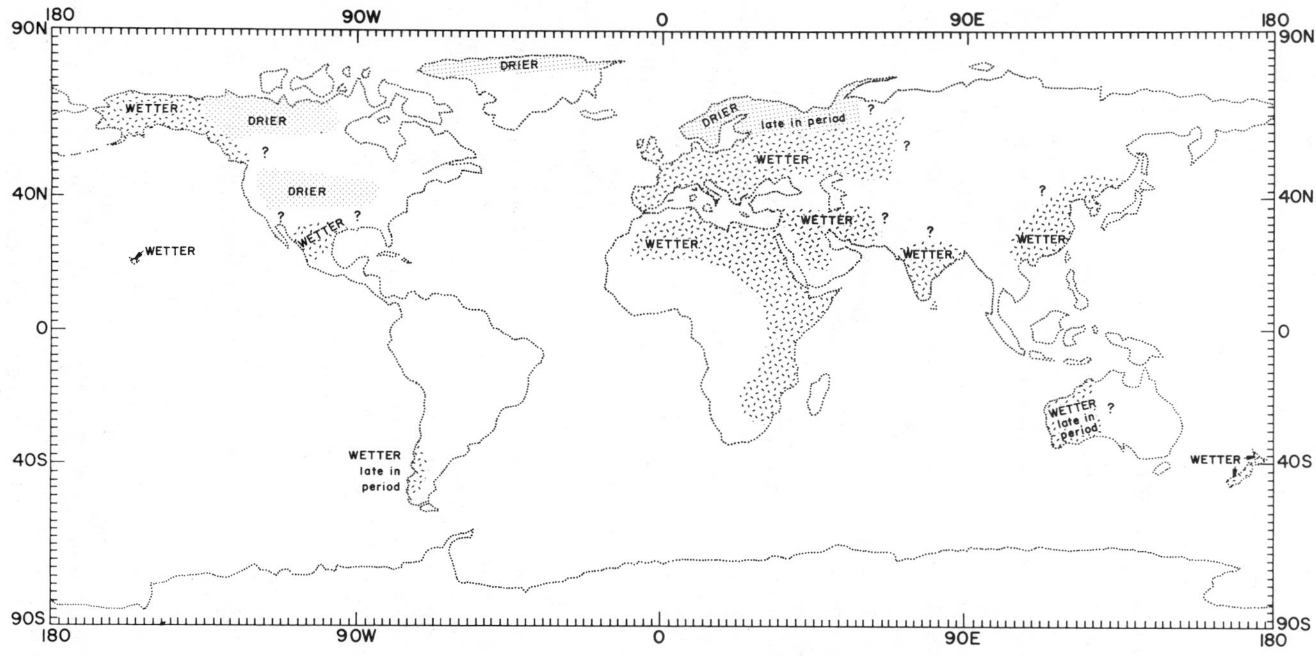

Figure 5 An approximate depiction of the regions in the world which were wetter or drier than the present during the Altithermal Period about 4000 to 8000 years ago. The blank areas on the map are not necessarily regions where there was no change, since our information is far from complete (Kellogg 1977).

warming were different (probably related to changes in the earth's orbit around the sun and the relative distribution of solar radiation at various seasons of the year), and also since the initial patterns of desert and forest and ice fields are different now than they were 8000 years ago at the end of the Wisconsin glaciation.

The Ice of the Polar Regions

If the warming is largest in the polar regions, perhaps approaching as much as 10°C by the middle of the next century (see Figure 4), it is certainly pertinent to ask what this implies. There are five distinct regions of ice and snow: underground permafrost; the winter snowcover on land that melts in summer; floating sea ice, or "pack ice," some of which now survives through the summer in both polar regions; mountain glaciers, that can occur at any latitude; and, by far the largest, the great ice sheets of Greenland and the Antarctic that have remained more or less intact for many millions of years. We concern ourselves here with the pack ice and the ice sheets, for reasons that will be apparent.

In both the Arctic and Antarctic Oceans the areas of pack ice expand in winter and shrink during the summer melting season, but beyond this there are large dissimilarities. The Arctic Ocean is, of course, largely bounded by land, so in winter the pack ice extends solidly to the shores and can only extend in a small way into the Atlantic and Pacific. The Antarctic Ocean has no northern boundary, and the area of pack ice that is frozen each winter and melted the following summer is larger than the entire Arctic Ocean.

A polar warming of the sort that can be anticipated will reduce the formation of sea ice in both polar regions, but a particular possibility that has tantalized climatologists is that the Arctic Ocean ice might disappear and not return (SMIC 1971, Budyko 1971, 1974, Flohn 1977). If this were to occur, then the world would witness as nearly an irreversible change as anything we can think of, since this ocean has not been ice free for over a million years. An ice-free Arctic Ocean would modify the entire climate of the Arctic Basin, providing more water vapor for rain in summer and snow in winter, and one can only begin to speculate about the corresponding ecological changes that would take place on land and in the sea.

The first serious attempt to estimate the degree of warming that would be required to eliminate the Arctic ice pack in summer was Budyko's (1971, 1974), and he reasoned that only a 4°C summertime increase in surface air temperature would suffice (see also Flohn 1977). A much more refined model of the Arctic ice pack that takes into account the heat fluxes in and out of the surface and also the drift of the pack ice has been developed by Parkinson & Washington (1978), and when this was run with conditions roughly simulating a doubling of carbon dioxide, as may occur in less than 70 years, the model showed a complete disappearance of the ice pack for most of the year (C. L. Parkinson, private communication).

There are good reasons to believe that the Arctic Ocean may have just two stable states, a largely frozen-over one (as at present) and an ice-free one. Two of these reasons are, first, that the open ocean would absorb much more solar radiation in summer which would be stored in the upper layers well into the fall and early winter, and, second, that the low-salinity and highly stable layer of water some 30–50 m deep that now exists just under the ice and prevents much flow of heat from the deeper waters would be overturned and eliminated by wave action in an open ocean. With this stable insulating layer gone it would be much harder to form new ice in the middle of the ocean—though ice could presumably still form at the shores in winter and build outward to some extent, as it does in the Antarctic.

The ice sheets of the Antarctic and Greenland contain about one third of all the fresh water on the planet, and have remained more or less intact on Greenland for 2 to 3 million years and on the Antarctic for 10 to 15 million years (Kennett 1977). They are in a sense more like geological formations than active components of the hydrologic system. Speculations about their possible response to a global warming have revolved around two main ideas: first, that they may melt, causing a rise in sea level, and, second, that they may start to slide into the ocean, also causing a rise in sea level.

Unfortunately, at this time the question of how the ice sheets will behave on a warmer earth is still a matter of some controversy, since the theory of their behavior based on ice sheet models is still in an early stage. Some simplistic calculations have been made of the rate of melting and of flow of ice sheets under warmer conditions, but these do not take account of the fact (first pointed out by the explorer Captain Robert F. Scott in 1905) that a warmer atmosphere can contain more moisture, and this will fall as more snow on the tops of these massive ice sheets (typically some 3 km above sea level) resulting in an increase of volume. Which of these two processes would win out in the long run we cannot say, but even the more extreme calculations cannot justify any major change in ice sheet volume due to simple melting for the next 100 years, which is the time horizon of this report (Birchfield 1977, Radok 1978).

The matter of ice sheets sliding and flowing more rapidly than they do now, or "surging," has attracted considerable attention lately, and glaciologists are still debating what it might take to make an ice sheet surge. Mountain glaciers are known to surge, and some mathematical models of a large ice sheet such as the East Antarctic one can be adjusted so that they surge also (Budd & McInnes 1977, Radok 1978). Thus, there is definitely this theoretical possibility.

The West Antarctic ice sheet is unique in that a good bit of it is resting on bedrock below sea level, and sea water flows under its two large extensions, the Ross Ice Shelf and the Filchner Ice Shelf. There is a possibility that with a polar warming sea water could penetrate under more of the ice, the ice shelves would disintegrate, and eventually all the ice not grounded above sea level would break up and melt as it drifted away (Hughes 1973, Mercer 1968, 1978). Note that this is a breaking-up process and not a surge, and Mercer believes it could occur very rapidly once it started. There is no clear sign that anything like this is happening now, but if the West Antarctic ice sheet were to disappear completely into the ocean it would raise sea level by 4–6 m. Thus, even a partial event of this sort would have a serious impact on all the shorelines of the world.

In conclusion, it is clear that the response of the ice masses in those seemingly remote polar regions can have important implications for the rest of the globe (Fletcher 1970). We are learning more about them all the time, but still cannot predict adequately what to expect when the earth gets warmer. Polar research is being armed with new tools to gain the needed information, such as techniques for aerial surveys, for drilling deep into the ice sheets, for satellite determinations of topography (by radar altimeters) and ice motions (by beacons on the surface), all feeding into theoretical efforts to create more realistic models. We can hope that progress here will be rapid and that the answers we badly need will be soon forthcoming.

SOME CONSIDERATIONS OF PRACTICAL ISSUES

While it would probably be out of place here to deal with the socioeconomic aspects of a worldwide climate change, even if we knew how to, we cannot leave the subject without some brief mention of the practical issues that must be faced—and there are a great many indeed. Some further discussions of them will be found elsewhere (e.g. SMIC 1971, Schneider 1976, Newell, Tanaka & Misra 1976, Kellogg & Schneider 1974, 1978, Kellogg & Mead 1977, Kellogg 1977, 1977–78, 1978, National Academy of Sciences 1977).

We have been dealing with the prospect of a gradual change that would be felt on a time scale of several decades, but we must never forget that the natural factors in the climate system are also capable of causing short-term and long-term climatic fluctuations that are very significant. Year-to-year climatic *variability* has indeed been a fact of life since the beginning of time and will continue to be. Thus, we can expect to see droughts in marginal areas such as the Sahel of North Africa, parts of India and China, the western Great Plains of the United States and Canada, and so forth. There will be the "unusually hard" winters and the "unusually hot" summers, which only seem unusual because our tribal memory is short.

Our society must consider the basic issue of the costs and benefits of a global climatic change. It is fairly clear that some peoples will fare better on a warmer earth and others will be worse off, depending on where they live and what they do for a living. For example, mid-latitude farmers can expect about 10 days longer growing season for every 1°C rise in mean summertime temperature (a rough rule of thumb), whereas some places may have too much heat and too little rainfall to continue to grow their accustomed crops. Heating bills in winter would go down, but air conditioning costs in those parts of the world that can afford it would go up. If sea level were to rise, it would cause great distress in the harbors and lowlands of the world, but it would not bother the people living inland, except indirectly.

Such cost-benefit considerations cannot begin in earnest until climatologists can describe the expected changes with more certainty and on a regional basis. As we have shown, the regional changes will be more extreme in general than the average changes for a given latitude belt. Our coupled general circulation models are just getting to the point where they are capable of throwing light on the regional changes that would accompany a general warming, but no experiments have been published to date. Such experiments are, to be sure, expensive in terms of scientific manpower and computer time, since high resolution models are very elaborate and there are only a handful of research organizations in the world capable of developing and running them. Nevertheless, we can probably look forward to some preliminary results in a very few years.

If the scenario presented here turns out to be correct, more or less, and climatic change becomes an accepted reality, what will we decide to do about it? The countries of the world must act together, since the effect is global in character and the cause, primarily the burning of fossil fuels, is an international activity. The situation is unprecedented in the history of mankind: we have a strong premonition of a major alteration of the global environment and we could, at least in principle, do something to change the course of events if we decided to. (One possible precedent is the Biblical story of a prediction by Joseph of a 14-year cycle of drought in Egypt.)

Action on the part of the community of nations could only ensue if two things occurred: first, the climatologists, economists, social scientists, and politicians must understand the future clearly enough to decide that a global warming would indeed be too costly to mankind as a whole to be "acceptable" (a value judgment in the last analysis); and, second, there must be the international machinery to make the final decision to act and to enforce the decision. No such machinery exists, nor do we even see how to set it up (Kellogg & Mead 1977).

This most general conclusion reinforces the point made at the beginning, namely, that a crucial first step is to learn more about the global system that determines our climate so that we can predict the impact of mankind's behavior on the environment.

Literature Cited

Ackerman, T. P. 1977. A model of the effect of aerosols on urban climates with particular applications to the Los Angeles basin. *J. Atmos. Sci.* 34:531–47

Ackerman, T., Baker, M. B. 1977. Shortwave radiative effects of unactivated aerosol particles in clouds. *J. Appl. Meteorol.* 16:63–69

Adem, J. 1970. Incorporation of advection of heat by mean winds and by ocean currents in a thermodynamic model for long-range weather prediction. *Mon. Weather Rev.* 98:776–86

Atwater, M. A. 1975. Thermal changes induced by urbanization and pollutants. *J. Appl. Meteorol.* 14:1061–71

Augustsson, T., Ramanathan, V. 1977. Radiative-convective model study of the CO_2 climate problem. *J. Atmos. Sci.* 34:448–51

Bacastow, R., Keeling, C. D. 1973. Atmospheric carbon dioxide and radiocarbon in the natural carbon cycle: II. Changes from A.D. 1700 to 2070 as deduced from a geochemical model. In *Carbon and the Biosphere*, ed. G. M. Woodwell, E. V. Pecan. US Atomic Energy Comm. CONF-720510

Baes, C. F. Jr., Goeller, H. E., Olson, J. S., Rotty, R. M. 1977. Carbon dioxide and climate: The uncontrolled experiment. *Am. Sci.* 65:316–20

Barrett, E. W. 1975. Inadvertent weather and climate modification. *Crit. Rev. Environ. Control* 6:15–90

Birchfield, G. E. 1977. A study of the stability of a model continental ice sheet subject to periodic variations in heat input. *J. Geophys. Res.* 82:4909–14

Bjerkness, J. 1963. Climatic change as an ocean-atmosphere problem. *Changes of Climate, Proc. Rome WMO-UNESCO Symposium, UNESCO, Arid Zone Research Series XX*, pp. 297–321. Paris

Boeck, W. L. 1976. Meteorological consequences of atmospheric Krypton-85. *Science* 193:195–98

Bolin, B. 1977. Changes of land biota and their importance for the carbon cycle. *Science* 196:613–19

Bolin, B., Bischof, W. 1970. Variations in the carbon dioxide content of the atmosphere of the northern hemisphere. *Tellus* 22:431–42

Borzenkova, I. I., Vinnikov, K. Ya., Spirina, L. P., Stekhnovskiy, D. I. 1976. Change in the air temperature of the northern hemisphere for the period 1881–1975. *Meteorol. Gidrol.* 7:27–35

Broecker, W. S. 1975. Climatic change: Are we on the brink of a pronounced global warming? *Science* 189:460–63

Brosset, C. 1976. Airborne particles: Black and white episodes. *Ambio* 5:157–63

Brünig, E. 1977. The tropical rain forest—a wasted asset or an essential biospheric resource? *Ambio* 6:187–91

Bryan, K. S., Manabe, S., Pocanowski, R. C. 1975. A global ocean-atmosphere climate model: Part II, The ocean circulation. *J. Phys. Oceanogr.* 5:30–47

Bryson, R. A., Baerreis, D. A. 1967. Possibilities of major climatic modification and their implications: Northwest India, a case for study. *Bull. Am. Meteorol. Soc.* 48:136–42

Bryson, R. A., Wendland, W. M. 1975. Climatic effects of atmospheric pollution. In *The Changing Global Environment*, ed. S. F. Singer, pp. 139–48. Dordrecht, Holland: Reidel

Budd, W., McInnes, B. 1977. Modeling of

ice masses: Implications for climatic change. In *Climatic Change and Variability: A Southern Perspective*, ed. A. B. Pittock, L. A. Frakes, D. Jenssen, J. A. Peterson, J. W. Zillman, pp. 228–34. Cambridge University Press

Budyko, M. I. 1969. The effect of solar radiation variations on the climate of the earth. *Tellus* 21:611–19

Budyko, M. I. 1971. *Climate and Life*. Leningrad: Hydrological Publishing House (in Russian)

Budyko, M. I. 1974. *Climate and Life*, ed. D. H. Miller. *Int. Geophys. Ser.* Vol. 18. New York: Academic. 508 pp. (English edition)

Budyko, M. I. 1977. On present-day climatic changes. *Tellus* 29:193–204

Callendar, G. S. 1958. On the amount of carbon dioxide in the atmosphere. *Tellus* 10:243–48

Cess, R. D. 1974. Radiative transfer due to atmospheric water vapor: Global considerations of the earth's energy balance. *J. Quant. Spectros. Radiat. Transfer* 14: 861–71

Cess, R. D. 1976. Climatic change: An appraisal of atmospheric feedback mechanisms employing zonal climatology. *J. Atmos. Sci.* 33:1831–43

Charney, J., Quirk, W. J., Chow, S.-H., Kornfield, J. 1977. A comparative study of the effects of albedo change on drought in semi-arid regions. *J. Atmos. Sci.* 34: 1366–85

Chervin, R. M., Schneider, S. H. 1976. On determining the statistical significance of climate experiments with general circulation models. *J. Atmos. Sci.* 33:405–12

Coakley, J. A. Jr. 1977. Feedbacks in vertical-column energy balance models. *J. Atmos. Sci.* 34:465–70

Coakley, J. A. Jr., Chýlek, P. 1975. The two-stream approximation radiative transfer: Including the angle of the incident radiation. *J. Atmos. Sci.* 32:409–18

DeLuisi, J. J., Furukawa, P. M., Gillette, D. A., Schuster, B. G., Charlson, R. J., Porch, W. M., Fegley, R. W., Herman, B. M., Rabinoff, R. A., Twitty, J. T., Weinman, J. A. 1976. Results of a comprehensive atmospheric aerosol-radiation experiment in the southwestern United States, Part II: Radiation flux measurements and theoretical interpretation. *J. Appl. Meteorol.* 15:455–63

Dettwiller, J., Changnon, S. A. Jr. 1976. Possible urban effects on maximum daily rainfall rates at Paris, St. Louis, and Chicago. *J. Appl. Meteorol.* 15:517–19

Drazin, P. G., Griffel, D. H. 1977. On the branching structure of diffusive climatological models. *J. Atmos. Sci.* 34:1696–1706

Eliassen, A. 1952. Slow thermally or frictionally controlled meridional circulations in a circular vortex. *Astrophys. Norweg.* 5:19–60

Elliott, W. P., Machta, L., eds. 1977. *Proc. of the ERDA Workshop on Environmental Effects of Carbon Dioxide from Fossil Fuel Combustion, Miami.* Washington, DC: Dept. of Energy

Fletcher, J. O. 1970. Polar ice and the global climate machine. *Bull. At. Sci.* 26:39–47

Flohn, H. 1977. Climate and energy: A scenario to the 21st century problem. *Climatic Change* 1:5–20

Gal-Chen, T., Schneider, S. H. 1976. Energy balance climate modeling: Comparison of radiative and dynamic feedback mechanisms. *Tellus* 28:108–21

Global Atmospheric Research Programme. 1975. The physical basis of climate and climate modeling. WMO-ICSU Joint Organizing Committee, *GARP Publ. Series No. 16*. Geneva, Switzerland

Gates, W. L. 1976. An essay on climate dynamics. *Bull. Am. Meteorol. Soc.* 57: 542–47

Häfele, W., Sassin, W. 1977. The global energy system. *Ann. Rev. Energy* 2:1–30

Hobbs, P. V., Harrison, H., Robinson, E. 1974. Atmospheric effects of pollutants. *Science* 183:909–15

Holdren, J. P., Ehrlich, P. R. 1974. Human population and the global environment. *Am. Sci.* 62:282–92

Hosler, C. L., Landsberg, H. E. 1977. The effect of localized man-made heat and moisture sources in mesoscale weather modification. See National Academy of Sciences 1977, pp. 96–105

Hughes, T. 1973. Is the West Antarctic ice sheet disintegrating? *J. Geophys. Res.* 78:7889–910

Idso, S. B., Brazel, A. J. 1977. Planetary radiation balance as a function of atmospheric dust: Climatological consequences. *Science* 198:731–33

Kahn, H., Brown, W., Martel, L. 1976. *The Next 200 Years: A Scenario for America and the World.* New York: Morrow. 241 pp.

Keeling, C. D., Adams, J. A. Jr., Ekdahl, C. A. Jr., Guenther, P. R. 1976a. Atmospheric carbon dioxide variation at the South Pole. *Tellus* 28:552–64

Keeling, C. D., Bacastow, R. B., Bainbridge, A. E., Ekdahl, C. A. Jr., Guenther, P. R., Waterman, L. S. 1976b. Atmospheric carbon dioxide variations at Mauna Loa Observatory, Hawaii. *Tellus* 28:538–51

Keeling, C. D., Bacastow, R. B. 1977. Impact

of industrial gases on climate. See National Academy of Sciences 1977, pp. 72–95

Kellogg, W. W. 1975. Climatic feedback mechanisms involving the polar regions. In *Climate of the Arctic*, ed. G. Weller, S. A. Bowling, pp. 111–16. Fairbanks: Geophys. Inst., University of Alaska

Kellogg, W. W. 1977. Effects of human activities on global climate. *WMO Tech. Note No. 156, WMO No. 486*. Geneva, Switzerland

Kellogg, W. W. 1977–78. Effects of human activities on global climate. *WMO Bull.* Part I. 26:229–40 (1977), Part II. 27: 3–10 (1978)

Kellogg, W. W. 1978. Is mankind warming the earth? *Bull. At. Sci.* 34:10–19

Kellogg, W. W., Coakley, J. A. Jr., Grams, G. W. 1975. Effect of anthropogenic aerosols on the global climate. *Proc. WMO/IAMAP Symp. on Long-Term Climatic Fluctuations, Norwich, U.K., WMO Doc. 421*, pp. 323–30. Geneva, Switzerland

Kellogg, W. W., Mead, M., eds. 1977. *The Atmosphere: Endangered and Endangering. Fogarty Intl. Cent. Proc. No. 39.* Publ. No. (NIH 77-1065). Washington, DC: Natl. Inst. Health

Kellogg, W. W., Schneider, S. H. 1974. Climate stabilization: For better or for worse? *Science* 186:1163–72

Kellogg, W. W., Schneider, S. H. 1977. Climate, desertification, and human activities. In *Desertification*, ed. M. H. Glantz, pp. 141–63. Boulder, Colo: Westview Press

Kellogg, W. W., Schneider, S. H. 1978. Global air pollution and climate change. *IEEE Trans. on Geoscience Electronics*, GE-16 (No. 1), pp. 44–50

Kennett, J. P. 1977. Cenozoic evolution of Antarctic glaciation, the circum-Antarctic Ocean, and their impact on global paleoceanography. *J. Geophys. Res.* 82: 3843–860

Kiang, C. S., Middleton, P. 1977. Formation of secondary sulfuric acid aerosols in urban atmosphere. *Geophys. Res. Lett.* 4:17–20

Lamb, H. H. 1972. *Climate: Present, Past and Future Vol. I: Fundamentals and Climate Now.* London: Methuen. 613 pp.

Lamb, H. H. 1974. Climates and circulation regimes developed over the northern hemisphere during and since the last ice age. *Proc. IAMAP/WMO Symp. on Physical and Dynamic Climatology, Leningrad, August 1971, WMO No. 347*, pp. 233–61

Landsberg, H. 1974. Inadvertent atmospheric modifications through urbanization. In *Weather and Climate Modification*, ed. W. N. Hess, pp. 726–63. New York: Wiley

Lian, M. S., Cess, R. D. 1977. Energy balance climate models: A reappraisal of ice-albedo feedback. *J. Atmos. Sci.* 34: 1058–62

Lindzen, R. S., Farrell, B. 1977. Some realistic modifications of simple climate models. *J. Atmos. Sci.* 34:1487–1501

Llewellyn, R. A., Washington, W. M. 1977. Effluents of energy production: Regional and global aspects. See National Academy of Sciences 1977

Lovins, A. 1976. Energy strategy: The road not taken? *Foreign Affairs* 55:65–96

MacCracken, M. C., Luther, F. M. 1974. Climate studies using a zonal atmospheric model. *Proc. IAMAP/IAPSO Intl. Conf. on Structure, Composition, and General Circulation of the Upper and Lower Atmospheres and Possible Anthropogenic Perturbations, Melbourne*, Vol. II, pp. 1107–28

MacCracken, M. C., Potter, G. L. 1975. Comparative climate impact of increased stratospheric aerosol loading and decreased solar constant in a zonal climate model. *Proc. WMO/IAMAP Symp. on Long-Term Climatic Fluctuations, Norwich, U.K., WMO Doc. 421*, pp. 415–20. Geneva

Machta, L. 1973. Prediction of CO_2 in the atmosphere. In *Carbon and the Biosphere*, ed. G. M. Woodwell, E. V. Pecan, pp. 21–31. US Atomic Energy Comm. CONF-720510

Machta, L., Telegadas, K. 1974. Inadvertent large-scale weather modification. See Landsberg 1974, pp. 681–726. New York: Wiley

Manabe, S., Bryan, K. 1969. Climate calculations with a combined ocean-atmosphere model. *J. Atmos. Sci.* 26: 786–89

Manabe, S., Bryan, K., Spelman, M. J. 1975. A global ocean-atmosphere climate model: Part I, The atmospheric circulation. *J. Phys. Oceanog.* 5:3–29

Manabe, S., Holloway, J. L. Jr. 1975. The seasonal variation of the hydrologic cycle as simulated by a global model of the atmosphere. *J. Geophys. Res.* 80:1617–49

Manabe, S., Wetherald, R. T. 1967. Thermal equilibrium of the atmosphere with a given distribution of relative humidity. *J. Atmos. Sci.* 24:241–59

Manabe, S., Wetherald, R. T. 1975. The effects of doubling the CO_2 concentration on the climate of a general circulation model. *J. Atmos. Sci.* 32:3–15

Mercer, J. H. 1968. Antarctic ice and

Sangamon sea level. *Int. Assoc. Sci. Hydrol. Publ. No. 79*, pp. 217–25
Mercer, J. H. 1978. West Antarctic ice sheet and CO_2 greenhouse effect: A threat of disaster. *Nature* 271:321–25
Miller, J. M., ed. 1975. *Geophysical Monitoring for Climatic Change, No. 3, Summary Report—1974*. Boulder, Colo.: Natl. Oceanic Atmos. Admin.
Mitchell, J. M. Jr. 1976. An overview of climatic variability and its causal mechanisms. *Quat. Res.* 6:481–93
Moore, H. E., Poet, S. E., Martell, E. A. 1973. ^{222}Rn, ^{210}Pb, ^{210}Bi, and ^{210}Po profiles and aerosol residence times versus altitude. *J. Geophys. Res.* 78:7065–75
National Academy of Sciences. 1977. *Energy and Climate*. Washington, DC: Geophys. Res. Board, Natl. Acad. Sci.
Newell, R. E., Tanaka, M. Misra, B. 1976. Climate and food workshop: A report. *Bull. Am. Meteorol. Soc.* 57:192–98
North, G. R. 1975. Theory of energy-balance climate models. *J. Atmos. Sci.* 32:2033–43
Oeschger, H., Siegenthaler, U., Schotterer, U., Gugelman, A. 1975. A box diffusion model to study the carbon dioxide exchange in nature. *Tellus* 27:168–92
Otterman, J. 1977. Anthropogenic impact on the albedo of the earth. *Climatic Change* 1:137–55
Paltridge, G. W. 1974. Global cloud cover and earth surface temperature. *J. Atmos. Sci* 31:1571–76
Parkinson, C. L., Washington, W. M. 1978. A large-scale numerical model of sea ice. *J. Geophys. Res.* In press
Radok, U. 1978. Climatic roles of ice. *Tech. Doc. Hydrol., Int. Hydrol. Progr.* Paris: UNESCO
Ramanathan, V. 1976. Radiative transfer within the earth's troposphere and stratosphere: A simplified radiative convective model. *J. Atmos. Sci.* 33:1330–46
Ramanathan, V. 1977. Interaction between ice-albedo, lapse rate, and cloud-top feedbacks: An analysis of the nonlinear response of a GCM climate model. *J. Atmos. Sci.* 34:1885–97
Rasool, S. I., Schneider, S. H. 1971. Atmospheric carbon dioxide and aerosols: Effects of large increases on global climate. *Science* 173:138–41
Revelle, R., Munk, W. 1977. The carbon dioxide cycle and the biosphere. See National Academy of Sciences 1977, pp. 140–58
Rotty, R. M., Weinberg, A. M. 1977. How long is coal's future? *Climatic Change* 1:45–58
Russell, P. B., Grams, G. W. 1975. Application of soil dust optical properties in analytical models of climate change. *J. Appl. Meteorol.* 14:1037–43
Saltzman, B., Vernekar, A. D. 1972. Global equilibrium solutions for the zonally averaged macroclimate. *J. Geophys. Res.* 77:3936–45
Schneider, S. H. 1972. Cloudiness as a global feedback mechanism: The effect on the radiation balance and surface temperature of variations in cloudiness. *J. Atmos. Sci.* 29:1413–22
Schneider, S. H. 1975. On the carbon dioxide-climate confusion. *J. Atmos. Sci.* 32:2060–66
Schneider, S. H. (with L. E. Mesirow) 1976. *The Genesis Strategy: Climate and Global Survival*. New York: Plenum. 419 pp.
Schneider, S. H., Dickinson, R. E. 1974. Climate modeling. *Rev. Geophys. Space Phys.* 12:447–93
Schneider, S. H., Gal-Chen, T. 1973. Numerical experiments in climate stability. *J. Geophys. Res.* 78:6182–94
Schneider, S. H., Kellogg, W. W. 1973. The chemical basis for climate change. In *Chemistry of the Lower Atmosphere*, ed. S. I. Rasool, pp. 203–50. New York: Plenum
Schneider, S. H., Washington, W. M., Chervin, R. M. 1978. Cloudiness as a climatic feedback mechanism: Effects on cloud amounts of prescribed global and regional surface temperature changes in the NCAR GCM. *J. Atmos. Sci.* In press
Sellers, W. D. 1969. A global climatic model based on the energy balance of the earth-atmosphere system. *J. Appl. Meteorol.* 8:392–400
Sellers, W. D. 1973. A new global climatic model. *J. Appl. Meteorol.* 12:241–54
Sellers, W. D. 1974. A reassessment of the effect of CO_2 variation on a simple global climatic model. *J. Appl. Meteorol.* 13:831–33
Siegenthaler, U., Oeschger, H. 1978. Predicting future atmospheric carbon dioxide levels. *Science* 199:388–95
Smagorinsky, J. 1974. Global atmospheric modeling and numerical simulation of climate. See Landsberg 1974, pp. 633–86
Smagorinsky, J. 1977. Modeling and predictability. See National Academy of Sciences 1977, pp. 133–39
Study of Man's Impact on Climate. 1971. *Inadvertent Climate Modification: Report of the Study of Man's Impact on Climate*. Cambridge, Mass: MIT Press
Stuiver, M. 1978. Atmospheric carbon dioxide and carbon reservoir changes. *Science* 199:253–58
Su, C. H., Hsieh, D. Y. 1976. Stability of

the Budyko climate model. *J. Atmos. Sci.* 33:2273–75

Suarez, M. J., Held, I. M. 1975. The effect of seasonally varying insolation on a simple albedo-feedback model. *Proc. WMO/IAMAP Symp. on Long-Term Climatic Fluctuations, Norwich, U.K., WMO Doc. No. 421*, pp. 407–14. Geneva, Switzerland

Temkin, R. L., Snell, F. M. 1976. An annual zonally averaged hemispherical climatic model with diffuse cloudiness feedback. *J. Atmos. Sci.* 33:1671–85

van Loon, H., Williams, J. 1976. The connection between trends of mean temperature and circulation at the surface: Part I, Winter. *Mon. Weather Rev.* 104:365–80

van Loon, H., Williams, J. 1977. The connection between trends of mean temperature and circulation at the surface: Part IV, Comparison of the surface changes in the northern hemisphere with the upper air and with the Antarctic in winter. *Mon. Weather Rev.* 105:638–47

Viskanta, R., Bergstrom, R. W., Johnson, R. O. 1977. Radiative transfer in a polluted urban planetary layer. *J. Atmos. Sci.* 34:1091–103

Wang, W.-C., Domoto, G. A. 1974. The radiative effect of aerosols in the earth's atmosphere. *J. Appl. Meteorol.* 13:521–34

Wang, W.-C., Yung, Y. L., Lacis, A. A., Mo, T., Hansen, J. E. 1976. Greenhouse effects due to man-made perturbations of trace gases. *Science* 194:685–90

Washington, W. M., Chervin, R. M. 1978. Regional climatic effects of large-scale thermal pollution: Simulation studies with the NCAR general circulation model. *J. Appl. Meteorol.* In press

Wetherald, R. T., Manabe, S. 1975. The effects of changing the solar constant on the climate of a general circulation model. *J. Atmos. Sci.* 32:2044–59

Wiin-Nielsen, A. 1972. Simulations of the annual variation of the zonally averaged state of the atmosphere. *Geofys. Publ.* 28: 1–45

Williams, J. 1978. The effects of climate on energy policy. *Electronics and Power* 24: 261–68

Williams, J., Krömer, G., Gilchrist, A. 1977. Further studies of the impact of waste heat release on simulated global climate, Part 2. *Int. Inst. for Appl. Sys. Anal., RM-77-34.* Laxenburg, Austria

Woodwell, G. M., Whittaker, R. H., Reiners, W. A., Likens, G. E., Delwiche, C. C., Botkin, D. B. 1978. The biota and the world carbon budget. *Science* 199:141–46

Yamamoto, G., Tanaka, M. 1972. Increase of global albedo due to air pollution. *J. Atmos. Sci.* 29:1405–12

Ann. Rev. Earth Planet. Sci. 1979. 7: 93–115

CALORIMETRY: ITS APPLICATION TO PETROLOGY

Alexandra Navrotsky
Department of Chemistry, Arizona State University, Tempe, Arizona 85281

1 INTRODUCTION

Calorimetry has been applied to petrologic problems since the pioneering work of Torgeson & Sahama (1948), King (1952), and Kracek (1953). However, in the last ten years, interest among geologists in the use of thermodynamics to systematize and explain mineral equilibria and the pressure, temperature composition history of rocks has grown tremendously. This interest, coupled with improved calorimetric techniques and a greater awareness of the importance of mineral purity and of structure on an atomic scale (including order-disorder phenomena), has led to increased calorimetric effort which, in many cases, has been quite successful in unraveling some of the subtleties of mineral stability.

In the last five years, several meetings have been held and a number of reviews published on specific aspects of thermodynamics as applied to geology (Navrotsky 1977, Fraser 1977, Kleppa 1972, 1976). The purpose of this review is to present an overview of available calorimetric techniques and their application, with emphasis on new (1975–1978) data and work in progress.

2 EXPERIMENTAL METHODS

Fundamentally, calorimetry provides two types of data. The first is measurement of heat capacities. This leads to values of the standard entropy, S_T°, the enthalpy (or heat content), $H_T - H_{298}$, and enthalpies and entropies of *rapid* phase changes (fusion, vaporization, polymorphism) of a single material. The second type is measurement of heats of chemical reaction (formation from the elements or the oxides, relative stability of competing phase assemblages, mixing in solid and liquid solutions), either by direct reaction or through a thermochemical cycle such as is involved

0084-6597/79/0515-0093$01.00

in solution calorimetry. Available equipment and techniques, and their applicability and limitations, are discussed below.

2.1 *Heat Capacities*

The standard entropy of a phase, exclusive of complications caused by "frozen in" disorder and imperfections, is given by

$$S_T^\circ = \int_0^{T_i} (Cp/T)\,\mathrm{d}T \tag{1}$$

where Cp is the measured heat capacity at atmospheric pressure and the integral is taken from the absolute zero to the temperature of interest. For simple monatomic materials, the contribution to the heat capacity at low temperature from lattice vibrations follows the Debye Law and Cp is proportional to T^3. However, for complex silicates with several kinds of atoms and silicate chains, rings, or frameworks, the heat capacity need not follow a T^3 law and, for an accurate entropy determination, the heat capacity must be measured, rather than estimated, in the 0–50°K range. Recently, progress has been made in estimating low temperature heat capacities of minerals using a more sophisticated semiempirical model for lattice vibrations (S. Kieffer, in preparation), but these estimates do not eliminate the need for measurements. Furthermore, magnetic transitions at low temperature occur, especially in sulfides and oxides. These often lead to λ-type anomalies in the heat capacity and can affect the entropy significantly (Ulbrich & Waldbaum 1976).

Low temperature heat capacities are measured in adiabatic calorimeters; the state of the art has been summarized by Westrum, Furukawa & McCullough (1968), and Robie & Hemingway (1972). Despite advances in electronic temperature control and measurement and in computer-aided data acquisition, the measurement of low temperature heat capacities is still a demanding and relatively time-consuming operation. A serious limitation of low temperature adiabatic calorimetry has been the sample size needed: 70 to 100 grams for normal operations, about 30 grams for a scaled-down version. This has generally limited study of minerals to natural samples, with concomitant problems of impurities, and has not permitted the determination of entropies of high pressure phases. The development by E. F. Westrum (personal communication) of a miniature heat capacity calorimeter which takes about 1 gram of sample is very encouraging in this context.

The commercial availability of differential scanning calorimeters (DSC) provides a new tool for heat capacity measurements at 250–1000°C. Advantages over conventional adiabatic and/or drop calorimetry include ease and rapidity of operation and the use of very small (5–20 mg) samples.

The major disadvantage of the method, at present, is that, unless operations are quite carefully standardized and calibrated, low quality data, which do not look obviously in error, can be produced. At optimum operation this method can yield heat capacities of accuracies (1–2%) comparable to other calorimetric methods. The upper temperature limit of most available instruments, approximately 1000°K, is unfortunately just the temperature range where interesting phenomena (melting, glass transitions, rapid phase transformation) begin to occur in silicates.

Although differential thermal analysis (DTA) generally provides qualitative rather than quantitative data on phase transitions, its application at high pressure and temperature is a significant development for mineral thermodynamics. Studies include the solubility of H_2O and CO_2 in diopside melts (Rosenhauer & Eggler 1975), the incongruent melting of diopside (Rosenhauer 1976), and polymorphism in analcite (Rosenhauer & Mao 1975) by DTA in a piston-cylinder apparatus. Further applications in high pressure research should be very fruitful.

Heat capacities above 1000°K are generally measured by drop calorimetry, in which an encapsulated sample is equilibrated at high temperature in a properly controlled furnace and then is dropped into an appropriate receiving calorimeter at room temperature. This is a classical technique of long standing (Hultgren et al 1959), and some of the original equipment is still in use (Bacon 1977). The measured heat content, $H_T - H_{298}$, is then differentiated to yield values of Cp. The experiments require careful operation, long equilibration times, and large samples ($\sim$100 grams). Heat contents have also been measured by "transposed temperature drop calorimetry" (Holm, Kleppa & Westrum 1967, Navrotsky 1973a, Kasper, Holloway & Navrotsky 1978) in which a sample at room temperature is dropped into a calorimeter near 1000°K. Smaller samples ($\sim$100 mg) are required than for conventional drop calorimetry.

2.2 *Enthalpies of Reaction*

As is well known, most silicates are notoriously unreactive. Therefore, direct measurement of heats of formation and phase transition is not possible and methods based on solution calorimetry, i.e. on measuring the heats of solution of reactants and products in a suitable solvent, have been devised. Two major methods are in use: hydrofluoric acid (HF) solution calorimetry at 25–90°C and oxide melt solution calorimetry at 600–1000°C. Experimental details have been described elsewhere (Torgeson & Sahama 1948, Robie & Hemingway 1972, Kleppa 1972, 1976, Navrotsky 1977). Salient features of the two methods are summarized in Table 1. The methods are complementary in that HF calorimetry handles hydrous phases with relative ease while oxide melt calorimetry handles refractory

Table 1 A comparison of hydrofluoric acid solution calorimetry and high temperature oxide melt calorimetry

Characteristic	HF calorimetry	Oxide melt calorimetry
Principle of calorimeter operation	adiabatic: measure ΔT for a reaction	Calvet-Tian type: measure and integrate heat flow vs time curve
Temperature range	25–90°C	600–1000°C
Size of silicate sample per run	500–2000 mg	20–50 mg
Heat of solution of quartz (typical of magnitude of heat effects)	-33.00 ± 0.02 kcal/mol (Torgeson & Sahama 1948)	-1.23 ± 0.07 kcal/mol (Charlu et al 1975)
Attainable precision in heats of solution	~0.1%	~1%
Attainable precision in heats of formation from oxide components	~0.1 kcal/mol	~0.1 kcal/mol
Effect of dilution and slight changes in solvent composition	important	minor (working in very dilute solution, <1 mol % solute)
Ability to handle refractory oxides (Al_2O_3, MgO, Cr_2O_3) directly	very limited	good
Ability to handle hydrous phases	good	poor (except possibly in new high pressure developments)

oxides easily. Oxide melt calorimetry requires a much smaller amount of sample; 200–300 mg total of a high pressure phase can suffice for several calorimetric runs.

Several recent developments in oxide melt calorimetry should be mentioned. The first is the achievement of high accuracy (0.1 kcal/mole or better) in measurements of heats of solution of 20–40 mg samples of crystalline and glassy materials in the system $CaO\text{-}MgO\text{-}Al_2O_3\text{-}SiO_2$ (Charlu, Newton & Kleppa 1975, 1978, Hon, Weill & Navrotsky, in preparation) by careful control of operating conditions. The second is the ability to handle samples containing Mn^{2+} and Fe^{2+} (Navrotsky & Coons 1976, Navrotsky, Pintchovski & Akimoto, 1979) by careful atmosphere control. The third is extension of the method to include measurements on phosphate- and fluoride-containing minerals (Westrich 1978). These developments extend the applicability of oxide melt calorimetry to additional petrologically interesting systems.

3 APPLICATION TO MINERAL STABILITY

3.1 *Compounds in the System CaO-MgO-Al_2O_3-SiO_2*

Oxide melt calorimetry is ideally suited to the study of phases in this system. Recent work has emphasized the accurate determination of enthalpies of formation and transition. The enthalpies of formation are summarized in Table 2 and some enthalpies of transition are given in Table 3.

3.1.1 ALUMINUM SILICATE STABILITY RELATIONS In addition to numerous phase equilibrium studies, this system has been studied extensively by high temperature solution calorimetry, first by Holm & Kleppa (1966),

Table 2 Enthalpies of formation of silicates in the system CaO-MgO-Al_2O_3-SiO_2 from oxides near 1000 K

Mineral	Formula	Source	ΔH_f° (kcal/mol)
Orthoenstatite	$MgSiO_3$	meteoritic & synthetic	-8.81 ± 0.17[a]
Aluminous enstatite	$Mg_{0.9}Al_{0.2}Si_{0.9}O_3$	synthetic	-7.01 ± 0.15[a]
Sillimanite	Al_2SiO_5	natural	-0.57 ± 0.14[a]
Forsterite	Mg_2SiO_4	synthetic	-14.97 ± 0.27[a]
Spinel	$MgAl_2O_4$	synthetic	-5.38 ± 0.18[a]
Pyrope	$Mg_3Al_2Si_3O_{12}$	synthetic	-20.21 ± 0.38[a]
Cordierite	$Mg_2Al_4Si_5O_{18}$	synthetic	-15.87 ± 0.31[a]
		natural	-16.28 ± 0.48[a]
Sapphirine	$Mg_7Al_{18}Si_3O_{40}$	synthetic	-38.72 ± 1.80[a,b]
		natural	-46.30 ± 2.00[a,b]
Anorthite	$CaAl_2Si_2O_8$	synthetic	-23.14 ± 0.45[c]
		natural	-23.93 ± 0.48[c]
Grossular	$Ca_3Si_2Al_3O_{12}$	synthetic	-77.91 ± 0.67[c]
Diopside	$CaMgSi_2O_6$	natural & synthetic	-34.99 ± 0.41[c]
		natural & synthetic	-34.32 ± 0.39[d]
CATS	$CaAl_2SiO_6$	synthetic	-18.23 ± 0.39[c]
Wollastonite	$CaSiO_3$	natural & synthetic	-21.48 ± 0.36[c]
Pseudowollastonite	$CaSiO_3$	synthetic	-19.92 ± 0.31[c]
High pressure polymorph	$CaSiO_3$	synthetic	-19.84 ± 0.34[c]

[a] Charlu, Newton, & Kleppa (1975).
[b] Average of two samples presented in paper.
[c] Charlu, Newton, & Kleppa (1978).
[d] Navrotsky & Coons (1976).

Table 3 Enthalpies near 980°K of some reactions involving pyroxenes

Compound	Reaction	$\Delta H°$(kcal)
$MgSiO_3$(opx)	formation from oxides	-8.81 ± 0.22[a]
$MnSiO_3$(rhodonite)	formation from oxides	-6.3 ± 0.3[b]
$FeSiO_3$(opx)	formation from oxides	-3.4 ± 0.5[c,f]
$CoSiO_3$(opx)	formation from oxides	-2.3 ± 0.4[c]
$MgCaSi_2O_6$(cpx)	formation from oxides	-34.6 ± 0.5[d]
$FeCaSi_2O_6$(cpx)	formation from oxides	-24.6 ± 0.5[f]
$CoCaSi_2O_6$(cpx)	formation from oxides	-26.7 ± 0.5[b]
$NiCaSi_2O_6$(cpx)	formation from oxides	-27.1 ± 0.5[b]
$CaAl_2SiO_6$(cpx)	formation from oxides	-18.2 ± 0.4[a]
$MgAl_2SiO_6$	formation from oxides	$+0.6 \pm 1.5$[a]
$CaSiO_3$(wo)	formation from oxides	-21.5 ± 0.4[a]
$CaSiO_3$	wo → pwo	$+1.56 \pm 0.21$[a]
$CaSiO_3$	wo → wo II	$+1.64 \pm 0.25$[a]
$MnSiO_3$	rhodonite → pyroxmangite	$+0.06 \pm 0.33$[b]
$FeSiO_3$	ol + q → 2 opx	$+0.24 \pm 0.19$[d]
$CoSiO_3$	ol + q → 2 opx	$+0.89 \pm 0.37$[d]
$NaAlSi_2O_6$	ab → jd + q	$+0.27 \pm 0.50$[e]
$CaAl_2SiO_6$	an → cats + q	$+5.70 \pm 0.36$[a]

[a] Charlu, Newton, & Kleppa (1975).
[b] Navrotsky & Coons (1976).
[c] Navrotsky, Pintchovski, & Akimoto (1979).
[d] Average of values in e and b.
[e] Hlabse & Kleppa (1968).
[f] Pintchovski & Navrotsky (unpublished).

then by Anderson & Kleppa (1969), by Navrotsky, Newton & Kleppa (1973), and by Anderson, Newton & Kleppa (1977). The calorimetric and crystallographic data on sillimanite samples heated at different temperatures (see Figure 1) strongly support the conjectures by Holdaway (1971) and Greenwood (1973) that Al-Si disorder affects the thermodynamic properties of sillimanite. The slopes of both the kyanite-sillimanite and andalusite-sillimanite phase boundaries are sensitive to this effect. The Al_2SiO_5 diagram of Holdaway (1971) and the thermochemical data are entirely consistent when one applies the simple disordering model proposed by Navrotsky, Newton & Kleppa (1973), with an Al-Si interchange enthalpy of 16 kcal as suggested by the calorimetric data (Anderson, Newton & Kleppa 1977).

Mullite, an aluminum silicate of composition varying from 60 to 76 mol percent Al_2O_3 (Aksay & Pask 1975, Cameron 1977), occurs as a high temperature phase. Its positive enthalpy of formation from the oxides (Holm & Kleppa 1966) implies that it is "entropy stabilized" both with respect to Al_2O_3 and SiO_2 and with respect to the Al_2SiO_5 polymorphs. Structural studies on a large number of synthetic and natural mullites .of different compositions and thermal histories have recently been completed (Cameron 1977). These studies suggest that the structures are interrelated by the ordering of oxygen vacancies and of Al and Si. Furthermore, mullite can be related to a metastable form of alumina, t-Al_2O_3, by the coupled substitution 2Si+0 for 2Al+2 vacancies. Heat of solution data obtained a number of years ago on several mullite samples of different preparation (Navrotsky & Kleppa, unpublished) showed a variation even greater than for the heat-treated sillimanites. These variations doubtless reflect differences both in composition

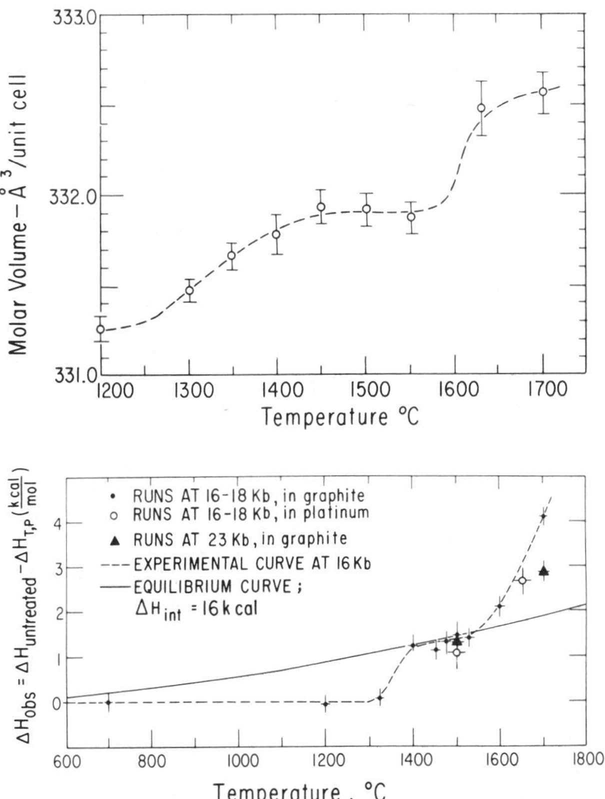

Figure 1 Crystallographic and calorimetric data for sillimanite heat treated at various temperature. (Navrotsky, Newton, & Kleppa 1973)

and in ordering scheme. A careful simultaneous structural and thermochemical study would be needed to clarify these relations.

3.1.2 POSSIBLE DISORDER IN OTHER ALUMINOSILICATE MINERALS The calorimetric data for cordierite, sapphirine, and anorthite (see Table 2) show differences between natural and synthetic samples which can not be ascribed to compositional variations. Rather, these differences are probably caused by differences in structural state, with greater disorder in the synthetic than in the natural samples. When phase boundaries have relatively shallow P-T slopes, then small variations in configurational entropy can have rather large effects on their location. Such cases, to quote Charlu, Newton & Kleppa (1975), "should sharply warn the petrologist against uncritical application of experimental data from synthetic systems to natural situations, even when the natural phases closely approach the compositions of the synthetics." This warning applies equally to the use of calorimetric data, of phase "equilibria" determined in relatively short runs on synthetic systems, and to the generation of self-consistent data banks. However, calorimetry, when combined with structural characterization of the samples used, provides a framework for the understanding of this structural complexity and the prediction of its thermodynamic consequences.

Charlu, Newton & Kleppa (1975) suggested the possibility of a variable heat of solution of pyrope garnet resulting from cation disorder, possibly including octahedral silicon occupancy and leading to an anomalously high entropy. However, additional experiments (Newton, Charlu & Kleppa 1977) show that the high entropy may be due instead to positional disorder and/or an anomalously large vibrational contribution from Mg in the dodecahedral sites. In that context, Neil, Navrotsky & Kleppa (1971) suggested that the high entropy of $CdTiO_3$ perovskite, which leads to a negative dP/dT for the ilmenite-perovskite transition, may be associated with the occupancy of the 12-fold site by a rather small ion. Furthermore, recent calorimetric data (Charlu, Newton & Kleppa 1978) confirm a negative P-T slope for the transition of $CaSiO_3$ (wollastonite) to the denser (wollastonite II) high pressure polymorph, and they suggest that the high pressure polymorph has a higher entropy (by 0.3 cal/K) than wollastonite. The structure of this high pressure polymorph is rather complex but it can be described as having "irregular layers of Ca atoms... interconnected by pairs of Ca atoms and Si_3O_9 rings" (Troger 1969). One might be tempted to assign the high entropy to the larger vibrations of the interlayer Ca pairs. The three cases above can lead to the following hypothesis, of possible use in predicting the slopes of phase transitions in the mantle. When high density in a phase is achieved by a structure in which divalent cations assume a coordination geometry which offers them relatively large metal-oxygen distances and a high coordination number, then that phase will have an anomalously high entropy relative to the low pressure phase.

3.2 *Stability of Pyroxenes and Pyroxenoids*

Thermochemical data, obtained fairly recently by high temperature calorimetry, are summarized in Table 3. In particular, data now are available for a number of transition-metal ortho- and clinopyroxenes. For hedenbergite, $FeCaSi_2O_6$, recent experiments (Pintchovski & Navrotsky, unpublished) are in general agreement with the value of ΔG° estimated in the earlier (Navrotsky & Coons 1976) paper. The thermochemical data offer some insight into crystal chemical systematics. The data for the clinopyroxenes are consistent with the pattern observed for other silicates and germanates (Navrotsky 1971) in which the magnesian end-members are considerably more stable than those containing Fe, Co, or Ni. However, whereas Ni_2SiO_4 is considerably less stable than Co_2SiO_4, and $NiSiO_3$ (which does not appear to exist under any P,T conditions) is much less stable than $FeSiO_3$ and $CoSiO_3$, the opposite trend is observed in the clinopyroxenes, where the enthalpy of formation becomes more negative in the sequence $FeCaSi_2O_6$, $CoCaSi_2O_6$, $NiCaSi_2O_6$. This is probably due to the competition of two factors: (*a*) the general trend toward lower stability of ternary compounds in the series Mg, Mn, Fe, Co, Ni, Cu discussed by Navrotsky (1971), and (*b*) the stabilization of the diopside structure by small cations in the M1 sites. Ghose & Wan (1975) observe that in $MCaSi_2O_6$ clinopyroxenes the average M2-O distance increases approximately linearly with increasing M1-O distance. The thermochemical data suggest that energetically optimum Ca-O distances can be achieved in clinopyroxenes having smaller ions (Mg and Ni) in M1 sites.

The rather complete thermochemical data for transition metal olivines, orthopyroxenes, and clinopyroxenes have been used to calculate M-Mg distribution coefficients between mineral pairs and between mineral and silicate liquid (Navrotsky 1978). That work concludes that the thermochemical data generally predict the observed K_D values quite adequately within the framework of an ideal solution model, except possibly for Mn-Mg partitioning.

The small positive ΔH° for the rhodonite → pyroxmangite transition in $MnSiO_3$ (Navrotsky & Coons 1976) favors the P-T diagram of Akimoto & Syono (1972) in which pyroxmangite is unlikely to be stable below

10 kbar at any temperature rather than other diagrams which suggest a stability field for pyroxmangite at or near atmospheric pressure below 500°C.

High pressure pyroxenes containing Al have been studied by solution calorimetry (see Table 3). Early work (Hlabse & Kleppa 1968) determined the stability of jadeite with respect to albite and quartz. Disorder in albite plays an important role in this equilibrium. Recent calorimetric data (Charlu, Newton & Kleppa 1978) for Ca-Tschermak's pyroxene (CATS, $CaAl_2SiO_6$) are in agreement with values estimated by Wood (in preparation) from equilibria involving CATS, anorthite, and quartz. The rather high entropy calculated for CATS (Charlu, Newton & Kleppa 1978) and the thermochemical behavior of CATS-diopside solid solutions (Newton, Charlu & Kleppa 1977) support the hypothesis of almost complete tetrahedral AlSi disorder in the pyroxene, consistent with the x-ray studies of Okamura, Ghose & Ohashi (1974). From their study of an aluminous orthopyroxene, Charlu, Newton & Kleppa (1975) estimate the enthalpy of formation of $MgAl_2SiO_6$ from the oxides to be +0.6 kcal, and its enthalpy of formation from enstatite and alumina to be +9.4 kcal. This compares with values of −18.2 and +3.3 kcal for the formation of CATS from the oxides and from wollastonite and alumina, respectively. The thermochemical data thus confirm the much greater instability of Mg-Tschermak's pyroxene than of Ca-Tschermak's pyroxene which has been inferred from petrologic evidence.

3.3 *The Olivine-Spinel Transition and Related Equilibria at 30–200 Kbars*

Enthalpies of solution in lead borate have been measured for the olivine (α) and spinel (γ) forms of Ni_2SiO_4 (Navrotsky 1973a), Mg_2GeO_4 (Navrotsky 1973b), Fe_2SiO_4, and Co_2SiO_4, and of the modified spinel (β-phase) form of Co_2SiO_4 (Navrotsky, Pintchovski & Akimoto 1979). The thermochemical data permit the calculation of *P-T* diagrams such as the one for the system "FeO"-SiO_2 shown in Figure 2. This work leads to the following conclusions. 1. The thermochemical data confirm the stable α, β, γ triple point in Co_2SiO_4 and are in reasonable accord with the measured slopes of the α-β, α-γ, and β-γ transitions in Co_2SiO_4 and with the slope of the α-γ transition in Fe_2SiO_4. The calculated boundaries are quite sensitive to the values of thermochemical parameters, to values of compressibility and thermal expansion factors, and to possible small degrees of cation disorder in the spinel and/or modified spinel. 2. β-Co_2SiO_4 and β-Mg_2SiO_4 are more similar to the spinel phase than the olivine phase thermochemically as well as structurally. 3. The entropy of the olivine-spinel transition is not constant for different materials but varies approximately linearly with the volume change. ΔS varies by approximately a factor of 4, becoming more negative in the series Ni_2SiO_4, Mg_2GeO_4, Co_2SiO_4, Fe_2SiO_4, Mg_2SiO_4. Thus the pressure for the olivine-spinel transition (at constant temperature) does not vary in a simple manner with the enthalpy of transition. Crystal chemical arguments explaining the stability of the polymorphs must therefore include entropic as well as energetic considerations. 4. The calorimetric data for $FeSiO_3$ and $CoSiO_3$ provide confirmation of the slopes of the boundary between olivine plus quartz and orthopyroxene. $FeSiO_3$ and $CoSiO_3$ are predicted to decompose to spinel plus stishovite at pressures near the coesite-stishovite transition (80–100 kbar), in accord with experiment. $MgSiO_3$ is predicted to decompose to β-Mg_2SiO_4 + stishovite at pressures of 140–160 kbar. These decomposition reactions have large negative volume and entropy changes, positive dP/dT, and are insensitive

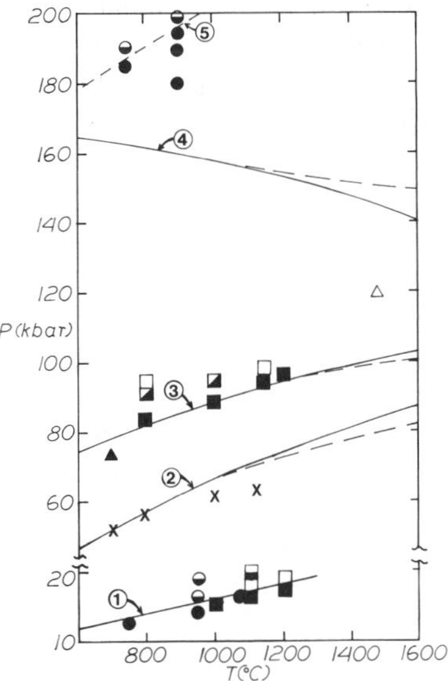

Figure 2 Calculated and experimental high pressure phase relations in the system "FeO"-SiO_2 (Navrotsky, Pintchovski, & Akimoto 1979). Solid curves are calculated from calorimetric and P,V,T data; dashed curves are calculated including possible cation disorder. Curves and data are as follows. (*1*) is fayalite + quartz → ferrosilite. Circles are data of Akimoto et al (1965), squares are from Lindsley (1965). (*2*) is olivine → spinel. Crosses are transition pressures from Akimoto et al (1977). (*3*) is pyroxene → spinel + stishovite. Filled triangle is highest pressure at which decomposition did not occur (Akimoto et al 1965). Open triangle is pressure at which transition occurred (Ringwood & Major 1966). Squares are data of Akimoto & Syono (1970). (*4*) is spinel → rocksalt + stishovite. Circles and dashed boundary (5) are data of Kawada (in preparation).

to possible cation disorder. 5. Disproportionation of Fe_2SiO_4, Co_2SiO_4, and Ni_2SiO_4 spinels to rocksalt plus stishovite is calculated to occur in the 150–200 kbar range and disproportionation of Mg_2SiO_4 near 200 kbar. The thermochemical data predict negative dP/dT values for Fe_2SiO_4, Mg_2SiO_4, and Co_2SiO_4, and a positive dP/dT for Ni_2SiO_4. Uncertainties in the high pressure experiments, and in the thermochemical data and the compressibilities and thermal expansion factors, and disorder in the spinels at high temperatures may partially reconcile the observed positive dP/dT values with the negative predicted ones. The disproportionation of Mg_2SiO_4 is predicted to occur at pressures very similar to those at which post-spinel ilmenite and perövskite phases have been observed.

Thus the calorimetric data support the experimentally determined sequences of spinel and post-spinel reactions except for the reservations noted above concerning the slopes of spinel disproportionation reactions and the possible role of cation disorder. High temperature calorimetric study of phases synthesized at 50–200 kbar should prove very valuable in establishing thermochemical data for mantle silicates. This approach is especially useful because of the uncertainties of experimental pressure and temperature measurements in this range and because of the difficulties in obtaining equilibrium, especially in decomposition reactions involving large-scale diffusion, in high pressure experiments. The amount of sample required for oxide melt calorimetry, 200–300 mg, although it still represents a substantial amount of synthetic labor, is not totally out of scale for current high pressure apparatus in the 50–200 kbar range. Improvements in the near future in relatively large volume, high pressure apparatus should make it easier to pursue such studies. The amount of sample needed for calorimetry might, under favorable conditions, be decreased to 150–200 mg but it seems unlikely at present that a further reduction, say by a factor of 5 to 10, would be attainable.

Table 4 Thermodynamic data for fluoride end-member minerals

Mineral	Standard entropy S°_{298}(cal K^{-1})	Formation from elements[a] $\Delta H^\circ_{T,298}$(kcal)	$\Delta G^\circ_{f,298}$(kcal)
Fluoroapatite	92.70	-1640.7 ± 2.0	-1553.7 ± 3.0
Fluoropargasite	165.80	-3776.3 ± 4.0	-2929.1 ± 6.0
Fluoroapatite	80.38	-1527.4 ± 2.0	-1446.3 ± 3.0

[a] From data of Westrich (1978). Errors are estimated by propagation of error from individual data.

3.4 *Hydrous Minerals and their Fluorine Analogues*

Recently Westrich measured the enthalpies of the following reactions by high temperature oxide melt calorimetry (Westrich 1978)

$$0.5CaF_2 + 1.5Ca_3(PO_4)_2 = Ca_5(PO_4)_3F \text{ (fluoroapatite)} \quad (2)$$
$$\Delta H^\circ_{986} = -17.16 \pm 1.19 \text{ kcal}$$

$$NaF + 2CaO + 3.5MgO + 0.5MgF_2 + 1.5Al_2O_3 + 6SiO_2$$
$$= NaCa_2Mg_4Al_3Si_6O_{22}F_2 \text{ (fluoroparqasite)} \quad (3)$$
$$\Delta H^\circ_{986} = -74.21 \pm 1.43 \text{ kcal}$$

$$KF + 2.5MgO + 0.5MgF_2 + 0.5Al_2O_3 + 3SiO_2$$
$$= KMg_3AlSi_3O_{10}F_2 \text{ (fluorophlogopite)} \quad (4)$$
$$\Delta H^\circ_{986} = -38.32 \pm 1.12 \text{ kcal}$$

From these data, his own measurements of $H_{986} - H_{298}$, and enthalpies and heat capacities in the literature, Westrich recommends the values for standard enthalpies, entropies, and Gibbs free energies shown in Table 4. These thermodynamic data are combined with studies of (F,OH) exchange between the crystalline phases and an aqueous fluid to estimate deviations from ideality for F-OH exchange and to compare the relative stabilities of hydroxy- and fluoro- end-member minerals. These first results are quite encouraging and more work along similar lines would be fruitful.

4 THERMODYNAMICS OF MIXING IN SOLUTION

Mixing properties of solid solutions, of silicate melts, and of high temperature aqueous fluids are important in determining crystallization sequences of minerals and the partitioning of elements among various phases. Although activity-composition relations may often be inferred or back-calculated from observed equilibria, both in synthetic systems and in natural mineral assemblages, such calculations suffer three inherent limitations. First, because pressure, temperature, and composition vary simultaneously for a given equilibrium (e.g. partitioning of a trace element between a solid and liquid phase), it is difficult to separate the effects of these factors, i.e. to uniquely define the chemical potential of a component as a function of pressure, temperature, and composition. Second, because many equilibria (e.g. solid solution-silicate melt) involve variable composition in both coexisting phases, it is easier to obtain relative values of the differences in deviations from ideal behavior in the two phases than it is to obtain unique absolute values for activity coefficients in each phase. Third, because many equilibria can only be measured over a rather small temperature range, it is difficult to separate the effects of enthalpy and entropy. This separation is very useful in understanding, on a microscopic level, the atomic interactions which lead to the observed activity-composition relations.

Direct calorimetric measurements of enthalpies of mixing in a single phase at a given P and T can overcome some of these limitations. These measurements can be made in two ways: by direct mixing of components or by solution calorimetry of a previously prepared solid or glass series. The former method is useful for a study of aqueous solutions, molten salts, and oxide melts of relatively low melting temperature and low viscosity. The latter approach is applicable to solid solutions and to silicate glasses and melts of geologically interesting compositions. Recent applications of calorimetry to solid solutions, silicate glasses and melts, and aqueous fluids at high pressure and temperature are described below.

4.1 *Solid Solutions*

4.1.1 FELDSPARS Hydrofluoric acid solution calorimetry by Waldbaum & Robie (1971) and by Hovis & Waldbaum (1977) provide data on Na-K mixing in $(Na,K)AlSi_3O_8$ feldspars of a high and low degree of Al, Si order respectively. The Si-Al ordered solid solutions show a larger positive heat of Na-K mixing than do the Si-Al disordered series. These data complement earlier calorimetric studies of Al-Si disorder in microcline (Hovis 1974) and in albite (Holm & Kleppa 1968).

4.1.2 PYROXENES AND GARNETS Newton, Charlu & Kleppa (1977) measured the enthalpies of solution in lead borate of solid solutions along the pyrope-grossular and diopside-CATS joins. The garnet solid solutions show positive heats of mixing, which, when fitted to a Margules equation

$$\Delta H_{\text{mixing}} = 3N_{Ca}N_{Mg}(W_{H,Mg}N_{Ca} + W_{H,Ca}N_{Mg}),$$

result in $W_{H,Mg} = 2.0$ kcal, $W_{H,Ca} = 3.8$ kcal. This asymmetry is a result of ionic size, with the substitution of a large ion into a smaller lattice being more unfavorable, energetically, than the opposite process. The calorimetric data are in good agreement at high N_{Mg} with the predictions of Ganguly & Kennedy (1974) and of Hensen, Schmid & Wood (1975). However, the asymmetry in ΔH_{mixing} leads to a correspondingly asymmetric solvus with a critical solution temperature lower than that predicted by Ganguly & Kennedy (1974).

The $CaMgSi_2O_6$-$CaAl_2SiO_6$ join also shows positive heat of mixing, with some asymmetry toward diopside. The calorimetric data in the diopside-rich region are consistent with a model of completely disordered tetrahedral Si and Al which was suggested by Wood (1976) from the equilibria between diopside-rich aluminous pyroxenes, anorthite, and quartz. At more alumina-rich compositions, the calorimetric data appeared to suggest some local order of Al and Si, in order to be consistent with the activities calculated by Wood (1976). However, nonstoichiometry in tschermakitic clinopyroxenes increases with increasing Al-content (Wood & Henderson 1978) and when the nonstoichiometry of pyroxenes coexisting with anorthite and quartz is taken into account, the thermochemical data appear to support the complete disorder model over the entire diopside-CATS join (B. J. Wood 1977, personal communication). Recently, Newton et al (1979) completed a calorimetric study of clinopyroxene $Mg_{1+x}Ca_{1-x}Si_2O_6$ ($0 < x < 0.78$).

4.2 *Silicate Melts and Glasses*

A number of molten oxide mixtures have been studied by high temperature calorimetry. These include the systems PbO-V_2O_5 at 680°C (Yokokawa & Kleppa 1964), PbO-B_2O_3 at 800°C (Holm & Kleppa 1967), PbO-SiO_2 at 900°C (Østvold & Kleppa 1969), PbO-GeO_2 at 900°C (Müller & Kleppa 1973), Na_2O-MoO_3 at 697°C (Navrotsky & Kleppa 1967), Li_2O-B_2O_3 at 940°C, Na_2O-B_2O_3 at 985°C (Østvold & Kleppa 1970), and binary mixtures of alkali metaphosphates at 843°C (Ko & Kleppa 1970). These studies were generally performed by dissolving small amounts of component oxides into a relatively large mount of melt of a given composition; thus the partial molar enthalpies of solution of each component were obtained as a function of composition. The data obtained can be understood in terms of acid-base reactions and polymerization equilibria in the melts.

The structure and thermodynamics of naturally occurring silicate melts is one of the unsolved problems in petrology. Clearly, calorimetric study of silicate and aluminosilicate melts along the lines described above would be very useful. However, the temperatures at which these mixtures are molten and have relatively low viscosities (1100–1500°C) are at the upper limit of calorimetric capability. Both Kleppa & Hong (1974) and Warner, Roye & Jeffes (1973) have reported the construction of calorimeters having all-alumina blocks and capable of precise operation to at least 1300°C. Such instruments might be used for the direct study of heats of mixing in silicate melts.

Another approach is the study of silicate glasses by solution calorimetry in oxide melts near 700°C. Although a glass is not fully representative of the liquid state, it is clearly more so at 700°C (at or near the glass transition temperature) than near room temperature where HF calorimetry is practiced. Such measurements also provide data on the enthalpies of fusion of silicates when combined with measurements of the heat capacities of the solid, glass, and supercooled liquid.

Recently Hon, Weill & Navrotsky (1977, and in preparation) completed a calorimetric study of glasses in the system albite-anorthite-diopside. The heats of mixing, see Figure 3, show fairly complex behavior: positive enthalpies along the albite-diopside join, negative enthalpies along albite-anorthite and anorthite-diopside, and a corresponding heat of mixing surface in the ternary. When the heat of mixing data are corrected for the glass transition and for the heat capacities of solid, liquid, and glass, they can be used to calculate the heats of fusion of albite, anorthite, and diopside as well as the heats of mixing in the silicate melts. The results indicate that the heats of fusion can be 20–40% different from the heats of vitrification because of the large difference in heat capacities of liquid and glass. The heats of mixing of the liquids, on the other hand, do not differ much from those of the glasses. The calorimetric data, when used with a simple two-sublattice model of the entropies of mixing, successfully predict the diopside liquidus in the ternary system, see Figure 4. These results suggest that the simple models for the entropies and heats of mixing in aluminosilicate melts can be made. Work is continuing on other compositions in the systems Na_2O-Al_2O_3-SiO_2, CaO-Al_2O_3-SiO_2, and MgO-Al_2O_3-SiO_2 in order to refine models of heats of mixing.

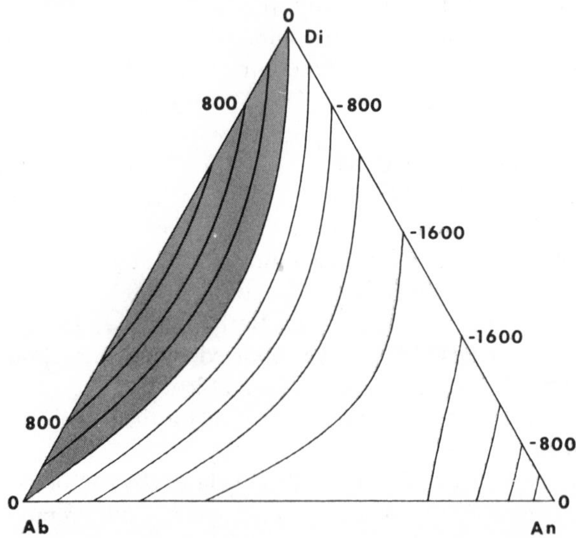

Figure 3 Enthalpies of mixing of liquids in the system diopside-albite-anorthite, based on an extrapolation of binary heat of solution data into the ternary system (Weill, Hon, & Navrotsky, in preparation).

One should note that the polymerization models of Masson, Smith & Whiteway (1970) and others are not readily applied to aluminosilicate melts because of the high degree of polymerization and the ambiguous role of the aluminum ion. Several years ago Burnham (1975) proposed a model for feldspar-related aluminosilicate melts which suggested that, on an equimolar eight-oxygen basis, ideal mixing between aluminosilicate end-member components would be observed in these highly polymerized melts. The enthalpy of mixing data appear to contradict such a model. Rather, the measured data should be used to devise more realistic thermodynamic approximations. A possible major contribution to the energetics appears to be the formation of Na (nonframework) – Al (framework) groupings or complexes in the melt.

4.3 *Aqueous Solutions at High Pressure and Temperature*

Knowledge of the thermodynamic properties of aqueous electrolyte solutions at high pressures and temperatures would permit more accurate calculation of phase equilibria at high temperatures and pressures, and of ore-forming processes.

Most previous thermodynamic studies of aqueous solutions have been performed near room temperature and atmospheric pressure. In particular, calorimetric studies under high temperature conditions have generally been limited by experimental design to experiments along the liquid-vapor curve. Measurements of enthalpy and heat capacity at temperatures above

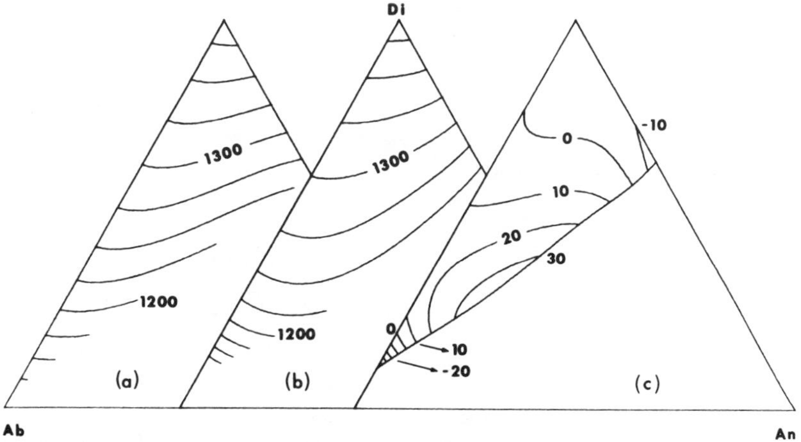

Figure 4 Comparison of calculated and measured diopside saturation surface (°C) (*a*) Experimentally determined, polynomial fit to data. (*b*) Calculated from thermochemical data (Kohler equation approach) and a single two-lattice (Temkin) model of the entropies of mixing. (*c*) $\Delta T = T$calculated − Tmeasured (Weill, Hon, & Navrotsky, in preparation).

473°K have been very difficult when using calorimeters with teflon or other polymeric gaskets.

Kasper, Holloway & Navrotsky (1978) recently developed a new calorimetric technique which can provide direct enthalpy measurements of solutions at temperatures to 1073°K and pressures to 2 kbar. The method utilizes a Tian-Calvet microcalorimeter containing a cold-seal pressure vessel in which the capsule of solution to be studied is confined at constant pressure.

Initial experiments at high T and P consisted of measurements of the isobaric specific enthalpy, h(T)−h(298.15°K), of aqueous NaCl solutions as measured by "transposed-temperature drop calorimetry." Since the pressure is set externally to the solution, and all pressure seals remain near room temperature, the technique is not limited to the liquid-vapor curve and can explore the complete range of pressure, temperature, and composition.

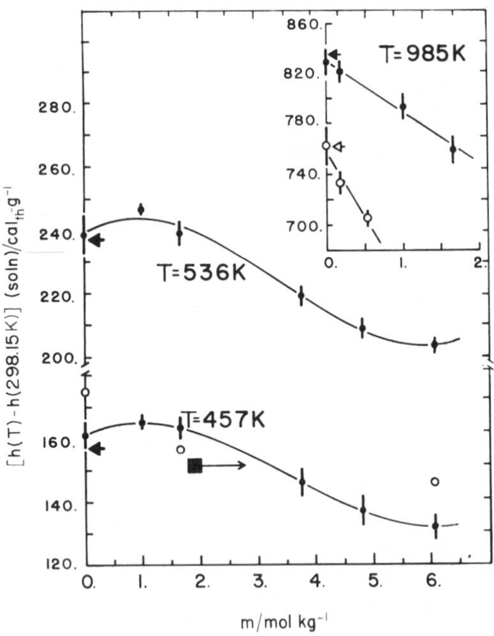

Figure 5 Measured specific enthalpies h(t)−h(298.15°K) of aqueous sodium chloride solutions as functions of molality (mol/kg). Filled circles are measured values for NaCl-H_2O at 510 bar. Error bars are two standard deviations of the mean. Open circles are values for NaCl-H_2O at 1000 bar. The square represents the Salton Sea Brine at 1.88 mol/kg (Na + K + Ca). The long arrow on the square symbol points to the ionic strength of that solution. The short arrows at zero molality represent the specific enthalpy (cal/g) of pure H_2O calculated the P,V,T data (Kasper, Holloway, & Navrotsky 1978).

The results from the system NaCl-H_2O at three temperatures and two pressures are shown in Figure 5, which plots the specific enthalpy, h(T)-h(298°K), in cal/g, versus the molality. For pure H_2O at 510 and 1000 bar, measured enthalpies agree with these calculated from P-V-T data. The measured enthalpy of the solutions changes very slowly with molality in the range 0 to 1.5 mol/kg, showing a shallow maximum near 1.5 mol/kg at 457 and 536°K. The data of Sourirajan & Kennedy (1962) indicate a two-phase field in the NaCl-H_2O system at 985°K at solution concentrations greater than 1.3 mass percent NaCl at 510 bar and greater than 3.8 mass percent NaCl at 1000 bar. In the two-phase field, one expects the enthalpy to vary linearly with the molality of the initial solution since the amounts of the two-phases present are governed by the lever rule. This has indeed been observed calorimetrically. At 457 and 536°K, the data were obtained over a sufficient concentration range to warrant the approximate calculation of partial molal heat contents of the NaCl and H_2O components. The partial enthalpy of NaCl, $(H(T)\text{-}H(298))_{\mathrm{NaCl}}$ shows a minimum near 3.5 mol/kg, suggesting rather complex thermodynamic behavior which requires further study. An isothermal mixing cell, for measuring enthalpies of reaction in the calorimeter at 423–623°K at 500–1000 bar, has been constructed and is being tested. This will be applied to measuring heats of dilution in the system NaCl-H_2O and heats of complexation reaction relevant to mineral transport.

5 CALORIMETRY, INTERNALLY CONSISTENT DATA BASES, AND FUTURE DEVELOPMENTS

In recent years, there has been considerable interest in generating computer-based, self-consistent sets of thermochemical data applicable over a wide P,T range (Kaufman & Nesor 1973, Pelton & Bale 1978, Haas & Fisher 1976, Helgeson, in preparation). Such values for heat capacities, standard entropies and enthalpies, and parameters for the formation of compounds from the elements or oxides, stored as appropriate algebraic function (power series, Maier–Kelley equations for heat capacities, Margules parameters, etc) could then be used to calculate equilibrium phase assemblages and their compositions as functions of pressure, temperature, and bulk composition. For petrologic applications, the computer program could then calculate phase boundaries of dehydration, melting, and phase transformation reactions. Because a given phase will participate in a large number of equilibria, and the P-T slopes of each reaction are quite sensitive to the thermochemical and volume parameters used in the calculation, the large total number of reactions impress a rather stringent set of conditions that the thermochemical data must satisfy to be consistent, simultaneously, with all the equilibria. These constraints can be used, first, to identify and weed out obviously inconsistent equilibrium and thermochemical data. Then regression analysis of the remaining data can be used to generate a set of thermodynamic parameters consistent with all the reversed equilibria and, ideally, within the limits of error of the measured calorimetric data. By these methods, it has been suggested (H. Helgeson, personal communication) that much tighter constraints can be put on ΔH° and ΔS° than are often possible by direct calorimetric measurements.

Nonetheless, this development does not put calorimetrists out of business. Rather, the calculations point to areas needing further work. The dangers of regression analysis are that the derived thermodynamic parameters may depend on the forms of equations used, may not be unique, and may take on physically unreasonable values. Furthermore, a change in one parameter will cause corresponding changes in many other parameters when locked into an internally consistent calculation scheme. This can lead to chronic bouts of recalculation which can in part be averted by accepting key data, for example, some reliably determined heats and entropies of silicates in which complication caused by disorder and nonstoichiometry are very unlikely. Similarly, calorimetry can be used to redetermine thermochemical parameters whose values appear grossly inconsistent with the calculations described above.

Solid solution formation, disorder, and nonstoichiometry in silicate minerals are problems which the calculation schemes above can not readily handle at present. In particular, for many metamorphic and igneous equilibria, although nominally the same phase participates in many equilibria, in fact its composition can vary significantly. The effect of such variations on thermodynamic properties in general is not well known. Calorimetric study of solid solutions is a useful approach to obtaining such information.

Obviously, the prediction of the future contributions of calorimetry to petrology carries some uncertainty. However, the following areas appear likely to bear fruit: 1. low temperature heat capacity measurements on small samples, especially of garnets, pyroxenes, and selected high pressure minerals, 2. high temperature heat capacity measurements, especially on glasses and silicate melts, 3. studies of solid solutions, 4. studies of ultra-high pressure phases and applications to mantle mineralogy, and 5. study of water-containing systems at 200–800°C and 500–2000 bars. The combination of thermodynamic and detailed structural data on complex minerals can be expected to lead to better understanding of phase stability and the P,T history of rocks.

ACKNOWLEDGMENTS

I thank S. Akimoto, R. Hon, K. Kawada, and H. Westrich for the use of data prior to publication. The calorimetric work has been supported by the National Science Foundation.

Literature Cited

Akimoto, S., Katsura, T., Syono, Y., Fujisawa, H., Komada, E. 1965. Polymorphic transition of pyroxenes $FeSiO_3$ and $CoSiO_3$ at high pressures and temperatures. *J. Geophys. Res.* 70:5269–78

Akimoto, S., Syono, Y. 1970. High pressure decomposition of the system $FeSiO_3$-$MgSiO_3$. *Phys. Earth Planet. Inter.* 3: 186–88

Akimoto, S., Syono, Y. 1972. High pressure transformations in $MnSiO_3$. *Am. Mineral.* 57:76–84

Akimoto, S., Yagi, Y., Inoue, K. 1977. High temperature-high pressure phase boundaries in silicate systems using in-situ x-ray diffraction. In *High Pressure Research—Application to Geophysics*, ed. M. Manghnani, S. Akimoto, pp. 586–602. New York: Academic

Aksay, I. A., Pask, J. A. 1975. Stable and metastable equilibria in the system SiO_2-Al_2O_3. *J. Am. Ceram. Soc.* 58:507–12

Anderson, P. A. M., Kleppa, O. J. 1969. The thermochemistry of the kyanite-sillimanite equilibrium. *Am. J. Sci.* 267: 285–90

Anderson, P. A. M., Newton, R. C., Kleppa, O. J. 1977. The enthalpy change of the andalusite-sillimanite reaction and the Al_2SiO_5 diagram. *Am. J. Sci.* 277:585–93

Bacon, C. R. 1977. High temperature heat content and heat capacity of silicate glasses; experimental determination and a model for calculation. *Am. J. Sci.* 277: 109–35

Burnham, C. W. 1975. Water and magmas; a mixing model. *Geochim. Cosmochim. Acta* 39:1077–84

Cameron, W. E. 1977. Mullite: a substituted alumina. *Am. Mineral.* 62:747–53

Charlu, T. V., Newton, R. C., Kleppa, O. J. 1975. Enthalpies of formation at 970 K of compounds in the system MgO-Al_2O_3-SiO_2 from high temperature solution calorimetry. *Geochim. Cosmochim. Acta* 39:1487–98

Charlu, T. V., Newton, R. C., Kleppa, O. J. 1978. Enthalpy of formation of some lime silicates by high temperature solution calorimetry, with discussion of high pressure phase equilibria. *Geochim Cosmochim. Acta* 42:367–75

Fraser, D. G., ed. 1977. *Thermodynamics in Geology*. Dordrecht, Holland: Reidel

Ganguly, J., Kennedy, G. C. 1974. The energetics of natural garnet solid solutions: I. Mixing of the aluminosilicate end-members. *Contrib. Mineral. Petrol.* 48:137–48

Ghose, S., Wan, C. 1975. Crystal structures of $CaCoSi_2O_6$ and $CaNiSi_2O_6$, crystal chemical relations in C2/C pyroxenes. *Eos* 56:1076

Greenwood, H. J. 1973. Al^{IV}-Si^{IV} disorder in sillimanite and its effect on phase relations of the aluminum silicate minerals. *Geol. Soc. Am. Mem.* 132:553–71

Haas, J. L., Fisher, J. B. 1976. Simultaneous evaluation and correlation of thermodynamic data. *Am. J. Sci.* 276:525–45

Hensen, B. J., Schmid, R., Wood, B. J. 1975. Activity-composition relations for pyrope-grassular garnet. *Contrib. Mineral. Petrol.* 51:161–66

Hlabse, T., Kleppa, O. J. 1968. The thermochemistry of jadeite. *Am. Mineral.* 53: 1281–92

Holdaway, M. J. 1971. Stability of andalusite and the aluminum silicate phase diagram. *Am. J. Sci.* 271:91–131

Holm, J. L., Kleppa, O. J. 1966. The thermochemical properties of the aluminum silicates. *Am. Mineral.* 51:1608–22

Holm, J. L., Kleppa, O. J. 1967. Thermochemistry of the liquid system lead(II) oxide-boron oxide at 800°. *Inorg. Chem.* 6:645–48

Holm, J. L., Kleppa, O. J. 1968. Thermodynamics of the disordering process in albite. *Am. Mineral.* 53:123–33

Holm, J. L., Kleppa, O. J., Westrum, E. F. Jr. 1967. Thermodynamics of polymorphic transformations in silica. Thermal properties from 5 to 1070°K and pressure-temperature stability fields for coesite and stishovite. *Geochim. Cosmochim. Acta* 31:2289–307

Hon, R., Weill, D. F., Navrotsky, A. 1977. Enthalpies of mixing of glasses in the system albite-anorthite-diopside. *Eos* 58: 1243 (Abstr.)

Hovis, G. L. 1974. A solution calorimetric and x-ray investigation of Al-Si distribution in monoclinic potassium feldspars.

In *The Feldspars*, ed. W. S. McKenzie, S. Zussman, pp. 114–144. Manchester, England: Univ. of Manchester Press

Hovis, G. L., Waldbaum, D. R. 1977. A solution calorimetric investigation of K-Na mixing in a sanidine-analbite ion-exchange series. *Am. Mineral.* 62:680–86

Hultgren, R., Newcomb, P., Orr, R. L., Warner, L. 1959. A diphenyl ether calorimeter for measuring high temperature heat contents of metals and alloys. *Proc. HMSO Natl. Phys. Lab. Symp., London, No. 9*

Kasper, R. B., Holloway, J. R., Navrotsky, A. 1978. Direct calorimetric measurement of enthalpies of aqueous sodium chloride solutions at high temperatures and pressures. *J. Chem. Thermodyn.* In press

Kaufman, L., Nesor, H. 1973. Theoretical approaches to the determination of phase diagrams. *Ann. Rev. Mater. Sci.* 3:1–30

King, E. G. 1952. Heats of formation of manganous metasilicate (rhodonite) and ferrous orthosilicate (fayalite). *J. Am. Chem. Soc.* 74:4446–48

Kleppa, O. J. 1972. Oxide melt solution calorimetry. *Colloq. Int. CNRS No. 201-Thermochimie*, pp. 119–27

Kleppa, O. J. 1976. Mineralogical applications of high temperature reaction calorimetry. In *The Physics and Chemistry of Rocks and Minerals*, ed. R. G. J. Sterns, pp. 369–88. New York: Wiley

Kleppa, O. J., Hong, K. C. 1974. Enthalpies of mixing in liquid alkaline earth fluoride-alkali fluoride mixtures II. Calcium fluoride with lithium, sodium and potassium fluorides. *Phys. Chem.* 78:1478–81

Ko, H. C., Kleppa, O. J. 1970. Thermochemical studies of liquid alkali metaphosphates. *Inorg. Chem.* 10:771–75

Kracek, F. C. 1953. Thermochemical properties of minerals. *Carnegie Inst. Wash. Yearbook* 1952–53, pp. 69–74

Lindsley, D. H. 1965. Ferrosilite. *Carnegie Inst. Wash. Yearbook* 64:148–50

Masson, C. R., Smith, I. B., Whiteway, S. G. 1970. Molecular size distribution in multi-chain polymers: applications of polymer theory to silicate melts. *Can. J. Chem.* 48:201–10

Müller, F., Kleppa, O. J. 1973. A calorimetric study of the system lead (II) oxide-germanium dioxide at 900°C. *Z. Anorg. Allg. Chem.* 397:171–78

Navrotsky, A. 1971. Thermodynamics of formation of the silicates and germanates of some divalent transition metals and of magnesium. *J. Inorg. Nucl. Chem.* 33: 4035–50

Navrotsky, A. 1973a. Ni_2SiO_4-enthalpy of the olivine-spinel transition by solution calorimetry at 713°. *Earth Planet. Sci. Lett.* 19:474–75

Navrotsky, A. 1973b. Enthalpy of the olivine-spinel transition in magnesium orthogermanate and the thermodynamics of olivine-spinel-phenacite stability relations. *Proceedings of the Conference on Phase Transitions and their Applications in Material Science*, pp. 383–98. Pergamon

Navrotsky, A. 1977. Progress and new directions in high temperature calorimetry. *Phys. Chem. Min.* 2:89–104

Navrotsky, A. 1978. Thermodynamics of element partitioning: (1) systematics of transition metals in crystalline and molten silicates and (2) defect chemistry and the Henry's Law Problem. *Geochim. Cosmochim. Acta* 42:887–902

Navrotsky, A., Coons, W. E. 1976. Thermochemistry of some pyroxenes and related compounds. *Geochim. Cosmochim. Acta* 40:1281–88

Navrotsky, A., Kleppa, O. J. 1967. A calorimetric study of molten Na_2MoO_4-MoO_3 mixtures at 970°K. *Inorg. Chem.* 6:2119–21

Navrotsky, A., Newton, R. C., Kleppa, O. J. 1973. Sillimanite-disordering enthalpy by calorimetry. *Geochim. Cosmochim. Acta* 37:2497–508

Navrotsky, A., Pintchovski, F. S., Akimoto, S. 1979. Calorimetric study of high pressure phases in the systems CoO-SiO_2 and "FeO"-SiO_2 and calculation of MO-SiO_2 phase diagrams. *Phys. Earth Planet. Inter.* In press

Neil, J. M., Navrotsky, A., Kleppa, O. J. 1971. The enthalpy of the ilmenite-perovskite transformation in cadmium titanate. *Inorg. Chem.* 10:2076–77

Newton, R. C., Charlu, T. V., Kleppa, O. J. 1977. Thermochemistry of high pressure garnets and clinopyroxenes in the system CaO-MgO-Al_2O_3-SiO_2. *Geochim. Cosmochim. Acta* 41:369–77

Newton, R. C., Charlu, T. V., Anderson, P. A. M., Kleppa, O. J. 1979. Thermochemistry of diopside-structure clinopyroxenes on the join $CaMgSi_2O_6$-$Mg_2Si_2O_6$. *Geochim. Cosmochim. Acta.* In press

Okamura, F. P., Ghose, S., Ohashi, H. 1974. Structure and crystal chemistry of calcium Tschermak's pyroxene. *Am. Mineral.* 59:549–57

Østvold, T., Kleppa, O. J. 1969. Thermochemistry of the liquid system lead oxide-silica at 900°. *Inorg. Chem.* 8:78–82

Østvold, T., Kleppa, O. J. 1970. Thermochemistry of liquid borates. II. Partial enthalpies of solution of boric oxide in its liquid mixtures with lithium, sodium, and potassium oxides. *Inorg. Chem.* 9: 1395–400

Pelton, A. D., Bale, C. W. 1978. Computational techniques for the treatment of thermodynamic data in multicomponent systems and the calculation of phase equilibria. *Calphad J.* In press

Ringwood, A. E., Major, A. 1966. High pressure transformation of $FeSiO_3$ pyroxene to spinel plus stishovite. *Earth Planet. Sci. Lett.* 1:135–36

Robie, R. A., Hemingway, B. S. 1972. Calorimeters for heat of solution and low-temperature heat capacity measurements. *US Geol. Sur. Prof. Pap. 755*

Rosenhauer, M. 1976. Effect of pressure on the melting enthalpy of diopside under dry and H_2O-saturated conditions. *Carnegie Inst. Wash. Yearbook* 75:648–50

Rosenhauer, M., Eggler, D. H. 1975. Solution of H_2O and CO_2 in diopside melt. *Carnegie Inst. Wash. Yearbook* 74:474–79

Rosenhauer, M., Mao, H. K. 1975. Studies on the high-pressure polymorphism of analcite by powder x-ray diffraction and differential thermal analysis methods. *Carnegie Inst. Wash. Yearbook* 74:413–17

Sourirajan, S., Kennedy, G. C. 1962. The system H_2O-NaCl at elevated temperatures and pressures. *Am. J. Sci.* 260:115–41

Torgeson, O., Sahama, T. 1948. A hydrofluoric acid solution calorimeter and the determination of the heats of formation of Mg_2SiO_4, $MgSiO_3$, and $CaSiO_3$. *J. Am. Chem. Soc.* 70:2156–60

Troger, F. J. 1969. The crystal structure of a high pressure form of $CaSiO_3$. *Zeit. Kristallogr.* 130:185–206

Ulbrich, H. H., Waldbaum, D. R. 1976. Structure and other contributions to the third law entropies of silicates. *Geochim. Cosmochim. Acta* 40:1–24

Waldbaum, D. R., Robie, R. A. 1971. Calorimetric investigation of Na-K mixing and polymorphism in the alkali feldspars. *Zeit. Kristallogr.* 134:381–420

Warner, A. E. M., Roye, M. P., Jeffes, J. H. E. 1973. Alumina-block high-temperature differential calorimeter, enthalpy and heat of fusion of PbO. *Trans. Inst. Mining Met.* 82:C246–C248

Westrich, H. 1978. *Fluoride-hydroxyl exchange equilibria in several hydrous minerals.* PhD thesis. Arizona State University, Tempe.

Westrum, E. F. Jr., Furukawa, G. T., McCullough, J. P. 1968. Adiabatic low temperature calorimetry. In *Experimental Thermodynamics, Vol. I, Calorimetry of Non-Reacting Systems*, ed. J. P. McCullough, D. W. Scott, pp. 133–214. New York: Plenum

Wood, B. J. 1976. Mixing properties of tschermakite clinopyroxenes. *Am. Mineral.* 61:599–602

Wood, B. J., Henderson, C. M. B. 1978. Composition and unit cell parameters of synthetic non-stoichiometric tschermakitic clinopyroxenes. *Am. Mineral.* 63:66–72

Yokokawa, T., Kleppa, O. J. 1964. A calorimetric study of the lead (II) oxide-vanadium (V) oxide system at 680°. *Inorg. Chem.* 3:954–57

Ann. Rev. Earth Planet. Sci. 1979. 7: 117–61

THE DYNAMICAL STRUCTURE AND EVOLUTION OF THUNDERSTORMS AND SQUALL LINES

Douglas K. Lilly
National Center for Atmospheric Research, P.O. Box 3000, Boulder, Colorado 80307

1 INTRODUCTION

General Nature of Thunderstorms and Squall Lines

The *Glossary of Meteorology* (Huschke 1970) defines the thunderstorm as "a local storm produced by a cumulonimbus cloud, always accompanied by lightning and thunder." This definition emphasizes the electrical activity characteristic of these storms. The emphasis in the present article is, however, not on the electrical phenomena, which are energetically a minor detail, but on the dynamic and thermodynamic structure, evolution, and processes involved in deep convective clouds and cloud systems. I also largely neglect the detailed particle physics processes which allow water vapor to condense into cloud droplets and the latter to aggregate into precipitable hydrometeors, rain, snow, and hail. The importance of these microphysical processes relative to the continuum fluid dynamics has been and remains a subject of continuing research. This article by no means resolves those issues, but in it I attempt to present the dynamical viewpoint.

The Glossary also defines a squall line as "a non-frontal line or narrow band of thunderstorms." Here I tend to neglect the distinction between frontal and non-frontal lines of thunderstorms. There is often almost as large a contrast in atmospheric properties across a strong "non-frontal" squall line as there is across a front, and frontal contrasts are probably enhanced if a strong thunderstorm line forms along the front. Thus the distinction is at least partly semantic.

0084-6597/79/0515-0117$01.00

A number of fairly recent reviews contain material relevant to these subject areas. Cotton (1975) reviews the theoretical dynamics of cumuliform clouds, which include thunderstorms at the large amplitude end. Cotton, Jiusto, & Srivastava (1975) present a brief review and bibliography of progress in cloud physics and radar meteorology. Lilly (1975) reviews the state of knowledge of severe storms and storm systems, with special emphasis on the needs for future observational and modeling research. Simpson (1976) reviews the understanding and application of rain augmentation from cumulus clouds and cloud systems. The present review may be considered as an update of all of these, but especially of Cotton's and Lilly's reviews, and is specifically oriented toward consideration of thunderstorms and squall lines as natural thermodynamic engines which ingest and process large quantities of air and water vapor, discharging them in a different state and creating impacts on nearby and remote environments.

The basic function of deep convective clouds, including thunderstorms, is to transfer heat from the lower to the higher levels of the troposphere. Such transfer is necessary because the atmosphere is largely transparent to incoming solar radiation but largely opaque to the infrared radiation emitted by the earth's surface. If radiative processes acted alone, much of the atmosphere would be statically unstable even without moisture, and the presence of moisture evaporated from the surface and mixed through the turbulent boundary layer greatly enhances instability and the energy available for convective overturning. In winter a large horizontal temperature gradient exists between the tropics and polar regions, allowing for modes of quasi-horizontal convection which are slower to develop but much less subject to turbulent dissipation than are thunderstorms. These modes, the extratropical cyclones and frontal systems, dominate the middle latitude meteorological circulations and tend to suppress direct convective overturning. In various other regions and for various reasons deep convection and thunderstorms are rare, for example over the subtropical west coasts, over much of the mid-latitude oceans in summer, in the trade wind regions, and in most deserts. They are most common over continents in summer, over mountainous regions, and near the equator.

Even in regions where the deep convective mode is dominant, such as in Florida in the summer, the fraction of time any given location is under an active thunderstorm circulation is a few percent at most. Thunderstorms are so efficient at their job that there is only room for a few. Over the earth's surface there are about 2000 thunderstorms active at any given time (Brooks 1925). If each of these covers an area of 100–200 km^2, the fraction of the earth's atmosphere participating in thunderstorms at any given time is less than 0.1%.

Observational Tools and Techniques

Through the last three decades the most important observational tool for the study of thunderstorms has been microwave radar. Over the usual meteorological radar wave length range, about 2–20 cm, radar is highly sensitive to rain drops, with the reflected signal amplitude proportional to the 6th power of drop size. Thunderstorms typically extend to 10 km above the surface or higher, so that it is possible to observe them from a ground based radar 300 km distant. Incoherent radar principally observes reflectivity patterns and intensities and their changes with time. This information is of limited value to dynamicists, since the liquid water distribution tends to lag and imperfectly reflect the dynamic evolution. With the advent in the last few years of practical techniques for monitoring the velocity field from the doppler shift of echoes from coherent transmission, radar has gained a new dimension of effectiveness for dynamic studies. Two doppler radars scanning a common volume can directly measure two velocity components in precipitating clouds, and the mass continuity equation can be used to estimate the third velocity component (see e.g. Kropfli & Miller 1976). With three or more radars it is possible in principle to determine the three air velocity components as well as the fall speed of the hydrometeors. In practice there remain many technical problems and uncertainties in accuracy, but active technological development is continuing.

Numerous other remote sensing techniques are utilized for observing characteristics of thunderstorms, including photography, passive microwaves, acoustic and light beam radar, and various ways of monitoring electrical activity and thunder. Visible and infrared light imagery from satellites is especially useful in surveying convective storm evolution and motion over large areas simultaneously.

Direct sensing of thunderstorm parameters is difficult, except at the ground, where measurements have been upgraded in recent years by development of digitally recording or telemetering station networks (Brock & Govind 1977). Balloon-borne sondes, which measure temperature, pressure, humidity, and wind, are the observational foundation for middle latitude large scale meteorology, but are less practicable for convective cloud systems because the systems are small, relatively infrequent, and often move rapidly. In addition the strong updrafts and downdrafts, high liquid water contents, and icing make sonde measurements in a thunderstorm of uncertain accuracy. Networks of sondes at spacings of 20–100 km have, however, been used successfully to observe the immediate storm environment on time scales of order 1–2 hours (Barnes, Henderson & Ketchum 1971, Fankhauser 1974). Aircraft penetrations of thunderstorms are to be avoided on safety grounds for all but the most ruggedly constructed aircraft, and only then under close control of a radar-equipped ground station, but aircraft are used frequently to observe the thunderstorm environment and lower portions of the updraft.

Some of the important parameters of cloud dynamics, including temperature, pressure, and liquid water content, seem to defy reliable measurement in thunderstorm conditions. Direct measurements of these quantities are subject to the problems of sonde and aircraft positioning and data interpretation. Fortunately, techniques, described in the next subsection, are under development which may allow accurate determination of both temperature and pressure from doppler radar velocity fields. Attempts to estimate liquid water content from radar reflectivity seem to be subject to factor of two uncertainties, however.

Theoretical Foundations and Techniques

The basic differential equations for the dynamics and thermodynamics of deep cloud convection are fairly well understood and accepted, and a variety of techniques are available for manipulating and solving them, at least in principle, though some uncertainty exists as to the adequacy of the available turbulence approximations. Neglecting effects of molecular viscosity and diffusion, earth's rotation, and precipitation, equations for rates of change of the principal dependent variables may be written as

$$d\mathbf{V}/dt = -\nabla p/\rho - g\mathbf{k}, \qquad \text{Motion} \qquad (1.1)$$

$$d\rho/dt = -\rho\nabla\cdot\mathbf{V}, \qquad \text{Mass continuity} \qquad (1.2)$$

$$d\theta_e/dt = 0, \qquad \text{Thermodynamic energy} \qquad (1.3)$$

$$d(q_v + q_l + q_i)/dt = 0, \qquad \text{Water substance} \qquad (1.4)$$

where $d/dt = \partial/\partial t + \mathbf{V}\cdot\nabla$, $\mathbf{V}$ is the three-dimensional velocity vector, p is pressure, ρ is density, g the acceleration of gravity, θ_e the equivalent potential temperature, and q_v, q_l, and q_i the mass fractions of water vapor, liquid water, and ice, respectively. The pressure is the sum of the partial pressures of dry air, p_a, and of water vapor, p_v, where each is considered as a perfect gas, i.e.

$$p = p_a + p_v = \rho_a R_a T + \rho_v R_v T, \qquad (1.5)$$

where R_a and R_v are the corresponding gas constants and T is the temperature. The density is the sum of the partial densities of dry air and the various forms of water substance, i.e.

$$\rho = \rho_a + \rho_v + \rho_l + \rho_i = \rho_a + \rho(q_v + q_l + q_i). \qquad (1.6)$$

The equivalent potential temperature is the temperature which the air would have if its pressure were reduced to zero adiabatically, the condensed and frozen water removed, and the dry air then recompressed adiabatically to a nominal sea level pressure, usually 10^4Pa. It is defined differentially by

$$\frac{d\theta_e}{\theta_e} = \frac{dT}{T} - \frac{R}{c_p}\frac{dp}{p} + \frac{L_v + L_f}{T}\,dq_v + \frac{L_f}{T}\,dq_l \tag{1.7}$$

where R and c_p are the effective gas constant and specific heat at constant pressure of the air mixture and L_v and L_f are the latent heats of vaporization and freezing of water substance, respectively. Since the total water mass ratio and θ_e are conserved, from (1.3) and (1.4), Equation (1.7) shows that a positive increment of liquid water at constant pressure may either occur from condensation, in which case $dq_v = -dq_l$ and temperature increases, from melting of ice, when $dq_l = -dq_i$ and temperature decreases, or a combination of these processes. Equations (1.3) and (1.7) imply, from the first law of thermodynamics, that no external heat sources are being considered. In particular, radiative heat exchange is usually neglected in thunderstorm dynamics, though it is important in setting up the initial instability.

The buoyancy terms on the right of Equation (1.1) are often expressed in ways more convenient for further analysis or solution. First by applying Equation (1.7) they may be written as

$$\frac{\nabla p}{\rho} + g\mathbf{k} = \nabla[c_p T + gz + (L_v + L_f)q_v + L_f q_i] - \frac{c_p T}{\theta_e}\nabla\theta_e \tag{1.8}$$

which is convenient for forming energy equations. Another approach, which illustrates more clearly the effects of water substance on buoyancy, is obtained by first rewriting the pressure gradient as

$$\frac{\nabla p}{\rho} = \left[\left(1 + \frac{R_v}{R_a} - 1\right)q_v - q_l - q_i\right]\frac{R_a T \nabla p}{p} \tag{1.9}$$

where Equations (1.5) and (1.6) have been used. A reference state dry atmosphere of constant θ_e is then assumed with $\bar{T} = T_0 - (g/c_p)z$ and $\bar{p} = p_0(1 - gz/c_p T_0)^{c_p/R_a}$. This state satisfies the dry hydrostatic equation, $R_a \bar{T}\partial(\ln \bar{p})/\partial z + g = 0$, which comes from requiring each side of Equation (1.1) to vanish. Taking now $T = \bar{T} + T'$ and $p = \bar{p} + p'$ and expanding Equation (1.11) to first order terms, one obtains the approximation

$$\frac{\nabla p}{\rho} + g\mathbf{k} \approx \nabla(p'/\bar{\rho}) - g\mathbf{k}[\theta'/\bar{\theta} + (R_v/R_a - 1)q_v - q_l - q_i] \tag{1.10}$$

where θ, the dry potential temperature, is the temperature the air would have if compressed adiabatically to nominal sea level pressure, i.e.

$$\theta = T(p_0/p)^{R/c_p}. \tag{1.11}$$

The ratio of gas constants is the inverse ratio of molecular weights, so that $R_v/R_a - 1 \approx 0.61$. Upon entering Equation (1.10) into the equation of motion, (1.1), it is evident that positive increments of potential temperature and water vapor produce upward acceleration while positive liquid and solid water increments produce downward acceleration. The latter effect may be quite substantial in a thunderstorm. Liquid water contents greater than 10 grams per kilogram of air occur, which are equivalent to a cooling of more than three degrees in their effect on buoyancy.

Since the buoyancy terms depend on the relative amounts of water in vapor and condensed forms as well as on the latent heat released by condensation and/or freezing it is necessary to evaluate the exchanges of water between its phases. The simplest reasonably accurate assumption is that all water is in vapor form so long as the vapor pressure is less than the saturation vapor pressure, p_{vs}, given by solution of the Clausius-Clapeyron equation,

$$\mathrm{d}p_{vs}/\mathrm{d}T = L_v p_{vs}/R_v T^2. \tag{1.12}$$

Any excess of total water substance over that corresponding to $p_v = p_{vs}$ is assumed to be immediately transformed to liquid form, and if liquid water exists its amount is regulated to maintain saturation conditions. If the temperature is below freezing, transformation to ice may be assumed to occur, with additional release of latent heat. Since freezing only occurs on appropriate nuclei, however, most cloud water remains liquid until the temperature falls below -10 to -20°C. This process thus requires more detailed microphysical evaluation.

Detailed microphysical treatment is also required for evaluation of precipitation onset and rate, since the terminal velocity of a raindrop is a strong function of its radius. In principle it is necessary to assume a spectrum of drop sizes and to consider growth of the spectrum through condensation and the exchange of water mass between one drop size and another through coalescence and drop break-up. A major simplification of this procedure, due to Kessler (1969), is to define two classes of liquid water, cloud water and rain water. The former is assumed to remain suspended like a gas, while the latter falls at an assumed terminal velocity determined in some empirical or theoretical way. Formulae for estimating the rate of conversion between cloud and liquid water have been proposed on empirical or theoretical grounds, and that rate is usually considered to be zero until the total q_l attains some threshold value.

In the following sections I discuss some linear and nonlinear quasi-analytic models for deep convection. The analysis is largely selfcontained but it is appropriate to make a few preliminary comments. Linear perturbation analysis, used effectively in many areas of fluid dynamics, is difficult to apply to deep convection, even for determination of initial stability, because of the special nature of the moist buoyancy process. Figure 1 is a standard form of atmospheric thermodynamic diagram (tephigram), in which the abscissa is temperature and the ordinate potential temperature plotted on a logarithmic scale. Lines of constant pressure are curved diagonals. A dry parcel ascending or descending adiabatically moves horizontally on this diagram. A saturated parcel rising adiabatically follows a curved line (moist adiabat) along which θ_e is constant. The figure shows the ambient atmospheric stratification for a typical convective storm environment, with a constant temperature stratosphere above the 20 KPa level. A parcel of air rising from the surface and maintaining the same pressure as the environment will cool relative to it until condensation occurs (point *A*), then start to warm, become positively buoyant at point *B*, continue to warm until the slope of the moist adiabat is steeper than that of the environment, and finally become negatively buoyant at a much higher level (point *C*). A descending parcel, starting at ambient conditions in the mid-troposphere (point *D*) will warm relative to the environment if it is dry and will cool if it is saturated, say by evaporating rain falling through it. The term "latent instability" is associated with

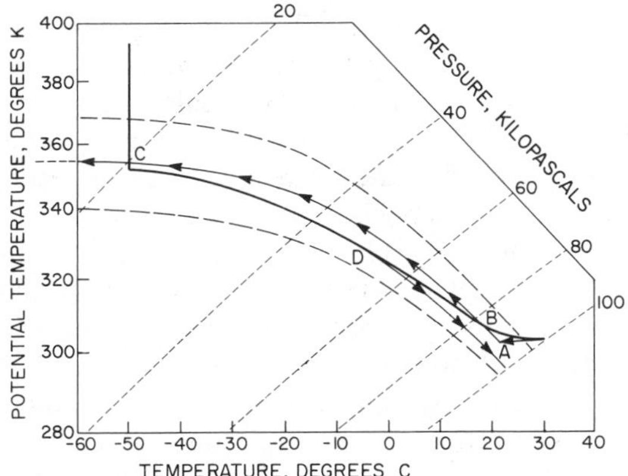

Figure 1 Tephigram showing typical environmental profile (heavy curve) of temperature (abscissa) vs potential temperature (ordinate) and pressure (dotted diagonals). The curve with upward pointing arrows would be followed by an undiluted parcel rising from the surface and reaching its condensation level at point *A*. The curve with downward pointing arrows would be followed by a parcel descending from point *D* and remaining saturated in its descent. Dashed curves are additional moist adiabats.

the small area of cooling between the surface and point B, which is sufficient to stop the upward motion of weak disturbances. A strong initial updraft or temperature perturbation will overcome this small potential barrier and release buoyant energy from the large region between points B and C where the parcel temperature is greater than that of the environment. In applying a typical linear perturbation approach, one is almost forced to ignore the latently unstable aspect, and also to make bold and usually unwarranted assumptions as to the availability of liquid water for evaporation in a downdraft.

A sometimes more profitable approach than linear analysis has been the development of nonlinear time-dependent or steady-state models of the flow in and around an isolated convective element. The simplest of these are merely extensions of the classical one-dimensional similarity solutions for plumes (steady-state) or bubble-like thermals (time-dependent) of buoyant fluid rising through a neutral or stable environment, while the more complex two- and three-dimensional models are "brute force" attempts to numerically simulate complete cloud systems. The foundations of the plume and thermal models are adequately described by Turner (1973) and some of the more important cloud convection applications are summarized by Cotton (1975) and Simpson (1976). Early models of this kind concentrated on predicting the maximum height and vertical velocity attained by a cloud element within a particular environment. For research purposes it is generally recognized that the 1-d models should defer to 2-d or 3-d models for realism. The former are used as frameworks, however, for more detailed computations of microphysical interactions than can presently be carried out in the more complex dynamical models, for example hail growth calculations. In addition 1-d models provide the basis for estimating the transports of heat, moisture, and momentum by sub-grid scale convective elements within larger scale prediction and simulation models.

Two-dimensional numerical cloud simulation models, using either slab (x-z) or axial (r-z) symmetry, have been in existence for about 15 years, while the first 3-d results appeared about five years ago. The equations integrated in 2-d and 3-d models are similar to Equations (1.1)–(1.7) above, with the buoyancy terms typically evaluated similarly to (1.10). This set allows, however, generation and propagation of sound waves, and the maintenance of computational stability would then require use of very short time steps. To avoid this most modelers have used the "anelastic" approximation (Ogura & Phillips 1962, Wilhelmson & Ogura 1972) or some variant thereof. This involves removal of the partial derivative with respect to time in Equation (1.2) and the replacement of density by a reference mean profile $\bar{\rho}(z)$ everywhere except in the buoyancy terms.

Thus Equation (1.2) becomes

$$\nabla \cdot (\bar{\rho}\mathbf{V}) = 0 \tag{1.13}$$

This constrains the solutions of Equation (1.1), in that a Poisson equation for either stream function (in the 2-d case) or for pressure must be solved, subject to appropriate boundary conditions. The solution of this equation and some of the extra boundary conditions required can be burdensome. As another approach, Klemp & Wilhelmson (1978a) use the fully elastic equations but incorporate a time step splitting technique. Short time steps which allow the sound waves to propagate involve linearization of the governing equations, while only the longer time steps simulate the important buoyancy-related physical processes. In terms of computational efficiency and accuracy these two approaches are apparently comparable.

Eddy diffusion and dissipation are important processes for any cloud simulation, though in a sense these processes are only artificial mathematical consequences of our inability to measure and calculate fluid motions with sufficient resolution. It is normally appropriate to assume that each variable corresponds to an average over the mesh box volume surrounding its nominal position. Assuming, then that $\bar{\rho}$ is the mean density over that volume, the continuity equation (1.2) may be written

$$\frac{\partial \bar{\rho}}{\partial t} + \nabla \cdot (\bar{\rho}\tilde{\mathbf{V}}) = 0, \tag{1.14}$$

where $\tilde{\mathbf{V}} = \overline{\rho \mathbf{V}}/\bar{\rho}$. The mass averaged total derivative for any other variable, say F, may be written as

$$\frac{\mathrm{d}\tilde{F}}{\mathrm{d}t} = \frac{\partial \tilde{F}}{\partial t} + \tilde{\mathbf{V}} \cdot \nabla \tilde{F} - \frac{1}{\bar{\rho}} \nabla \cdot [\bar{\rho}(\tilde{\mathbf{V}}\tilde{F} - \widetilde{\mathbf{V}F})], \tag{1.15}$$

with the term inside brackets identified as the eddy or Reynolds flux. It is common to write that term as $-\overline{\rho \mathbf{V}'F'}$, where $F' = F - \tilde{F}$ and $\mathbf{V}' = \mathbf{V} - \tilde{\mathbf{V}}$, but this is not accurate for running spatial averages, since the Reynolds postulates cannot be justified for this case (see Leonard 1973). Nevertheless it is possible to develop an ordered hierarchy of approximations for the term (Lilly 1967) in which the first order approximation involves an eddy coefficient, K, i.e.

$$\tilde{\mathbf{V}}\tilde{F} - \widetilde{\mathbf{V}F} = K\nabla \tilde{F}, \tag{1.16}$$

and higher order approximations require time-dependent equations for the eddy fluxes. Each order of approximation involves rapidly increasing complexity and increasing numbers of assumptions and approximations, but it is believed that the higher order approximations deliver more accurate and consistent results.

Most modelers use Equation (1.16), with K either held constant or set proportional to the magnitude of the deformation tensor of **V**. The latter formulation, introduced by Smagorinsky (1963) and modified by Lilly (1962) to incorporate buoyancy effects, is a consistent first order approximation for three-dimensional flows. Klemp & Wilhelmson (1978a) have used a "$1\frac{1}{2}$th order" approximation, with turbulent kinetic energy, E, computed from a time dependent equation and the eddy diffusion coefficients set proportional to $E^{1/2}$. Deardorff (1973, 1974) used a second order approximation for simulating boundary layer flows and compared the results to those using a consistent first order approximation. In my opinion the constant K assumption is inadequate and the Smagorinsky-Lilly formulation is marginal. The second order approximation may be reliable for most conditions, but it requires tripling the number of variables to be integrated and stored in high speed accessible memory. It seems likely that the "$1\frac{1}{2}$th order" approximations will also be satisfactory at considerably less cost, but the results have not been fully calibrated against a better approximation. For two-dimensional flows it is not clear which is the most relevant approximation to any order, but this is somewhat academic since real cloud convection always involves three-dimensional turbulence. Lilly (1970) developed an approximate method for simulating the effects of three-dimensional turbulence in two dimensions, which was used with apparent success by Drake, Coyle & Anderson (1974).

Other important elements of any cloud convection simulation are the numerical accuracy and the boundary conditions. Most modelers continue to use finite difference approximations to spatial and temporal derivatives, rather than the theoretically more powerful spectral or Galerkin techniques, principally because of their simplicity and apparently greater flexibility in fitting boundary conditions. In most cases the spatial finite difference approximations are of either second or fourth order accuracy, with the latter probably giving the best accuracy for the programming and computational effort involved.

The lateral boundary conditions are apparently an important factor in the success of deep convection and thunderstorm simulation models. There is normally a net upward motion in active convective clouds, which must be compensated by subsidence somewhere in the atmosphere. If the outer boundaries of a numerical model are closed or periodic, however, then all of the subsidence is forced to occur in the immediate vicinity of the cloud. In reality a thunderstorm draws moist air from the boundary layer a considerable distance away from the storm, and sends outflow cirrus streams hundreds of kilometers away. The development of boundary conditions which are as open as possible to outflow and inflow motions seems important if the thermodynamics are not to be severely constrained by excess environmental subsidence.

The proper initial conditions for starting a cloud model are uncertain, since there are few detailed observations of pre-thunderstorm conditions. Most modelers use a perturbation of roughly hemispheric shape in the temperature, humidity, or vertical velocity field, or some combination. In some circumstances the ultimate evolution is found not to depend strongly on the initial disturbance, provided it is strong enough to release the latent instability of the atmosphere as discussed above.

In the future the complete dynamic equations may be utilized in the numerical analysis of doppler radar data. If a thunderstorm model is to be compared in detail with the atmosphere, or if an atmospheric state observed by doppler data is to be used to initialize such a model, it is necessary to determine the thermodynamic properties of the storm in the same detail as the radar observes its kinematic state. In principle this can be done with the aid of the dynamic equations, provided that estimates of velocity time derivatives can be made, and recognizing that important parts of the storm circulation do not contain large enough drops to reflect radar signals. Hane & Scott (1978) and Gal-Chen (1978) have carried out analyses of the problem and made sample computations with idealized data sources. An analysis similar to Gal-Chen's follows.

The raw equations of motion cannot be used directly to obtain pressure and density, because there is no assurance that observed wind fields will satisfy the continuity equation (1.2), without assuming unrealistically large values of $\partial\rho/\partial t$. Instead a Poisson equation for pressure is obtained by taking the horizontal divergence ($\nabla_H \cdot$) of the horizontal equations of motion, from (1.1), with the pressure and buoyancy terms replaced by (1.10), i.e.

$$\nabla_H^2(p/\bar{\rho}) = -\nabla_H \cdot (\mathrm{d}\mathbf{V}_H/\mathrm{d}t). \tag{1.17}$$

The two-dimensional Poisson equation is then solved for $p'/\bar{\rho}$, using Neumann boundary conditions along a lateral boundary, i.e.

$$\frac{\partial}{\partial n}\left(\frac{p'}{\bar{\rho}}\right) = -\frac{\mathrm{d}V_n}{\mathrm{d}t} \tag{1.18}$$

where V_n and $\partial/\partial n$ are horizontal components perpendicular to the lateral boundary. The acceleration terms in (1.17) and (1.18) are obtained from the doppler radar data. The buoyancy variable may then be obtained from the vertical equation of motion, i.e.

$$g\left[\frac{\theta'}{\bar{\theta}} + \left(\frac{R_v}{R_a} - 1\right) q_v - q_l\right] = \frac{\mathrm{d}w}{\mathrm{d}t} + \frac{\partial}{\partial z}\left(\frac{p'}{\bar{\rho}}\right). \tag{1.19}$$

Since solutions to the Neumann problem are known only to within an arbitrary constant, the left side of Equation (1.19) is obtainable only to within an unknown function of height. To eliminate this uncertainty it is necessary and sufficient to have a vertical sounding profile of buoyancy somewhere within the doppler velocity field. If the sounding is made from a sonde or aircraft profile of temperature and moisture against pressure, reduction of the sounding may also require the use of doppler radar data to evaluate the vertical acceleration terms, and some estimate of liquid and/or solid water content may be required.

Gal-Chen's analysis incorporates terms for the Reynolds stresses, which can be partially evaluated from doppler measurements of small-scale velocity variance. He suggests that the reduction techniques for determining velocity fields from doppler radar data may significantly affect the results of the thermodynamic recovery methods, through compounding of velocity field errors in the nonlinear terms. He recommends that the appropriate velocity products be formed before smoothing is carried out. Further conclusions must await experiments using real data.

2 THE ISOLATED THUNDERSTORM

No Shear Case

Thunderstorms tend to occur in crowds. The most severe and damaging storms, especially, usually occur in some form of organized array. Nonetheless isolated thunderstorms do occur, and are perhaps rightly regarded as a simpler situation to study. In this section I present and discuss the observational and theoretical evidence regarding the mechanism of, first, storms developing in an atmosphere without much ambient wind shear. Second I consider storms which develop in the presence of strong mean shear, and last I discuss a theory of thunderstorm development based on gravity wave dynamics.

The Thunderstorm Project, conducted by scientists at the University of Chicago in 1946 and 1947, was the first organized field project aimed specifically at elucidating thunderstorm structure and mechanisms. For the simple "air mass" thunderstorms the results (Byers & Braham 1949) are still regarded as, in some respects, definitive. It was found that a typical continental thunderstorm (in Ohio and Florida, the sites of the experiment) consists of several identifiable circulation cells, each roughly cylindrical in shape, which individually go through a fairly predictable life cycle of about 30 minutes in duration.

At the beginning of the cycle the motion in the cell is principally upward, as moist air from the sub-cloud boundary layer is positively buoyant once it is lifted a few hundred meters above its condensation level, as indicated in Figure 1. After the updraft has developed a cloud several kilometers deep, which typically requires about 15 minutes, a variety of microphysical processes operate to produce water drops large enough to have fall velocities equal to or larger than that of the updraft, of order 5–10 ms^{-1}. Thus rather large amounts of liquid water begin to accumulate in and fall through the middle levels of the cloud. The weight of this water increases the effective air density by up to a percent or so, enough to substantially reduce or eliminate the cloud buoyancy in the lower half of the cloud. At about the same time the upper part of the updraft, having shed most of its water, has reached the level at which its buoyancy vanishes. Figure 1 shows that equilibrium level as point *C*, which here lies approximately at the tropopause. Any substantial entrainment of environmental air into the updraft will, however, lower that level. Because of its vertical momentum, the updraft air tends to overshoot its equilibrium level, but soon is forced back and spreads out into an anvil-shaped top.

From the mid-tropospheric loss of buoyancy due to water accumulation, a downdraft begins to develop in the same region previously occupied by rising air and cloud water. The accumulated water falls rapidly to the ground as a heavy shower, accompanied by a cold evaporating downdraft which hits the ground and fans out in all directions. The downdraft air is much colder than that which originally rose in the updraft because the rain continues to cool it by evaporation below cloud base and also because in both the updraft and downdraft parts of the cycle substantial amounts of dry air from the tropospheric environment are entrained into the cloud, mixed with it, and cooled by evaporation. Thus Figure 1 shows the downdraft air starting at ambient temperature (point *D*) and falling down a moist adiabat to the surface. Because of the combined effects of water accumulation, precipitation, and turbulent entrainment, the thunderstorm cell cannot be sustained as a steady-state process. The downdraft air spreads out as a density current or micro-cold front, which cuts off the supply of warm air feeding the updraft in the original cell. At the same time this surge of cold air is capable of lifting moist air in the adjacent environment and thereby triggering new cell formation. The process may repeat itself several times, so that the thunderstorm lifetime is typically two hours or so.

The individual cell evolution has now been simulated with reasonable fidelity and success in models developed by a number of investigators (e.g. Soong & Ogura 1973, Takeda 1971, Soong 1974, Hill 1977). Figure 2 shows a sequence of profiles of streamlines and rainwater for Takeda's two-dimensional simulation without shear. In this simulation the microphysical calculations were moderately sophisticated, with equations for rates of growth of a number of drop spectral classes. The lateral boundaries were closed at a distance of 25 km from the center of the convective element, shown as the right-hand boundary on the figures. The profiles show a rapid growth in the cloud top rainwater content, followed by rapid descent and collapse of the lower part of the cloud. Secondary development followed outside of the original disturbance but, according to Takeda, did not attain the amplitude of the primary cloud.

To my knowledge the evolution of a multi-cell thunderstorm without shear has yet to be simulated adequately. Takeda attributed the failure of his secondary cell to reach full amplitude to two-dimensionality and the closed lateral boundaries. Orville & Sloan (1970) were able to simulate sustained or repeating storms, but only by forcing their development by uplift over a simulated mountain. Miller (1978) showed development of a second cell in a three-dimensional model with weak shear, but the environment was unusually unstable. Based on these results it seems likely that some environmental convergence independent of the individual storm dynamics may be necessary in order to maintain thunderstorms for more than one or two cell lifetimes. Convergence predicted by a mesoscale boundary layer model has been shown by Pielke (1974) to correlate well with thunderstorm development over the Florida peninsula, and has also been observed from surface data to precede thunderstorm development by an hour or more (Ulanski & Garstang 1978).

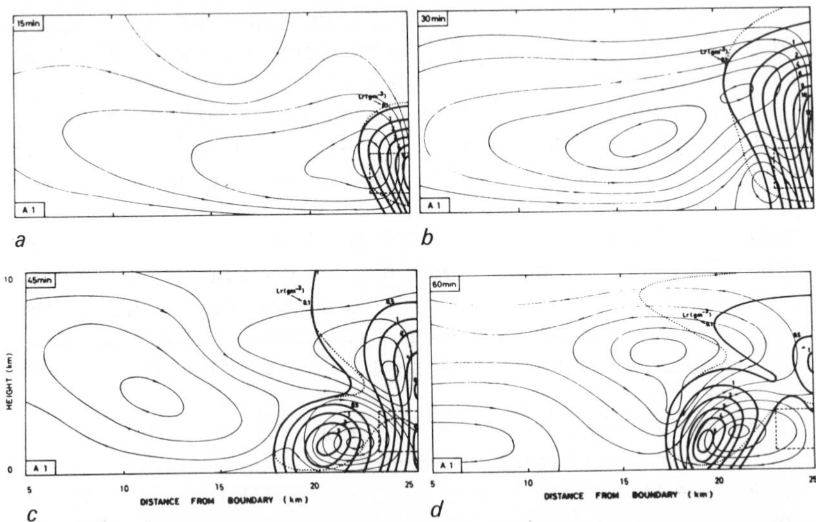

Figure 2a–d Rainwater contents (gm m^{-3}) and streamlines for Takeda's case *A1* at 15, 30, 45 and 60 min. Dotted lines and the square regions outlined by dashed lines indicate cloud boundaries ($L_c = 0.01$ gm m^{-3}) and the size of the initial disturbance, respectively. From Takeda (1971).

Effects of Shear

I turn now to the dynamically interesting case where the evolution and maintenance of thunderstorms are substantially affected by vertical shear of the mean flow. For many years it has been recognized that the most severe and longest lived storms typically form in strongly sheared environments, and that most damaging tornadoes arise from such storms, Some attempts to explain this phenomenon from fundamental principles have utilized linear stability theory, but with little success. Linearized convective instability lines growing in a sheared environment are apparently always stabilized by the shear except when it is parallel to the lines, where there is no effect. As a somewhat paradoxical result, however, shear-induced instability occurs in the form of amplifying Kelvin-Helmholtz waves which line up perpendicular to the shear. Because of the strong preference of these two common kinds of instability for orthogonal orientations, they are apparently almost unable to combine and reinforce each other.

Numerical simulations of two-dimensional atmospheric convection tend to support and extend some of the conclusions of linear theory, although the reasoning is couched in different terms. Both non-precipitating cumulus and non-severe thunderstorm cells simulated in the presence of mean shear are found to develop more slowly and less efficiently, and to rise to a lower maximum height than those without shear (Asai 1964, Takeda 1971). This can be explained as a consequence of two-dimensional turbulence theory. In three-dimensional turbulence a source of kinetic energy introduced on some large scale (such as a mean shear) tends to cascade down to smaller scales and eventually dissipate through molecular viscosity. In the case of turbulent flow confined to two spatial dimensions there is a similar downscale cascade of enstrophy (squared vorticity), but kinetic energy propagates mostly to larger scales. As a consequence a two-dimensional energy-containing element, such as a convection line, tends to give up energy to any existing mean shear, rather than abstracting energy from it. Hane (1973) found that a two-dimensionally simulated thunderstorm with strong instability was somewhat more persistent and intense in the presence of the low level shear, apparently due to a partial separation of the updraft and downdraft elements. Shear at higher levels tended to carry away the top of the updraft and reduce the storm intensity, however.

Results of linear theory suggest that three-dimensional convective elements would probably grow more slowly in shear than in uniform flow. Observations indicate, however, that very strong and long lasting thunderstorms develop in sheared environments and that their extended life cycle is associated with a form of storm organization which allows the updraft region to be spatially separated from the evaporating downdraft and rain. The high level flow is apparently able to pass around these storms without shearing off their tops, and there is some indication that the storms are able to make use of the shear to augment their kinetic energy.

Figure 3 shows the velocity field in three planes, as deduced from dual doppler measurements of a thunderstorm in shearing conditions. The highly three-dimensional nature of the circulation and evidence of centers of rotation are noteworthy.

Three-dimensional numerical simulation of thunderstorms has begun to be feasible in the last few years with the growth of computer power and numerical technology. The results to be described indicate that this development, together with the detailed observations of storm kinematics

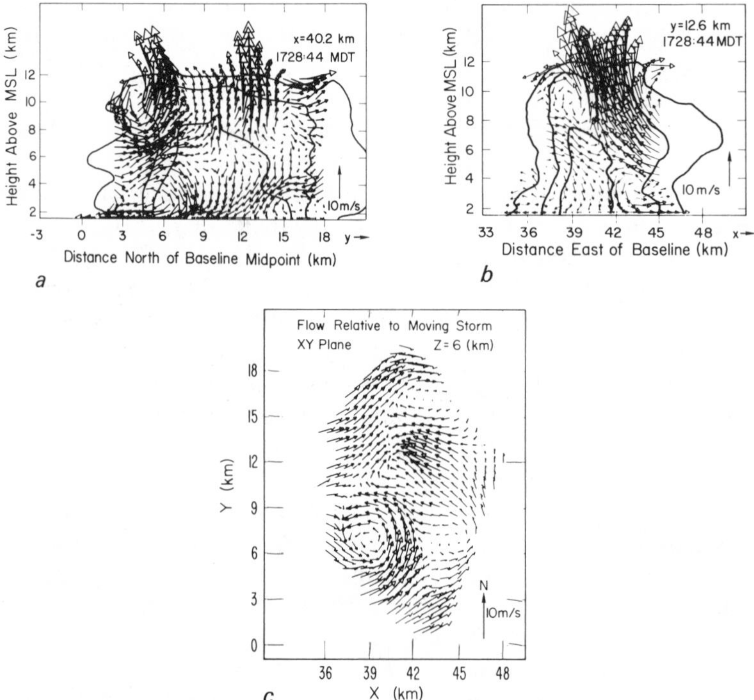

Figure 3 Cross-section of flow deduced from doppler signal analysis of radar echoes from a thunderstorm in Colorado: (*a*) y–z section at $x = 40.2$ km; (*b*) x–z section at $y = 12.6$ km; (*c*) x–y section at $z = 6.0$ km above the surface (7.5 km above MSL). All of the flow vectors are measured relative to the storm movement, which was 3 m s^{-1} toward positive x (approximately eastward). The arrow tail indicates location of the measurement. Note indications of counter-rotating vortices on (*c*). From Kropfli & Miller (1976).

by doppler radar, will produce accelerated progress in understanding storm structure and mechanism. These new observational and theoretical tools are both being dedicated, to a large extent, to the problem of understanding and predicting the severe and persistent storms which develop in strongly sheared environments.

Pastushkov (1975) showed results of an early attempt to define the effects of vertical wind shear on the evolution of three-dimensionally simulated convective clouds. Computer limitations and some rather severe lateral boundary constraints apparently limited the development of the simulated storms. Nevertheless Pastushkov found that for sufficiently strong buoyancy conditions, the existence of moderate shear did not necessarily suppress the storm development, and that a tendency for persistence was accompanied by a separation of the updrafts and downdrafts.

A much clearer picture of this effect has arisen from the work of Schlesinger (1975, 1978), Wilhelmson (1974), Klemp & Wilhelmson (1978a,b), and Wilhelmson & Klemp (1978). Schlesinger used a horizontal mesh spacing of 1.8 km with 27 × 27 points in the horizontal and a vertical resolution of 0.7 km with 20 levels, and his lateral boundary conditions were less constraining than Pastushkov's. Figure 4 shows the initial environmental sounding used for one of Schlesinger's cases. Note the strong directional shear of the wind in the lowest 3 km, which was adapted from a sounding made shortly before passage of a tornadic thunderstorm in Oklahoma. Schlesinger also used the same temperature and humidity sounding with no mean wind. Both cases were started with an initially cylindrical buoyant disturbance, and their evolution for the first 15 minutes

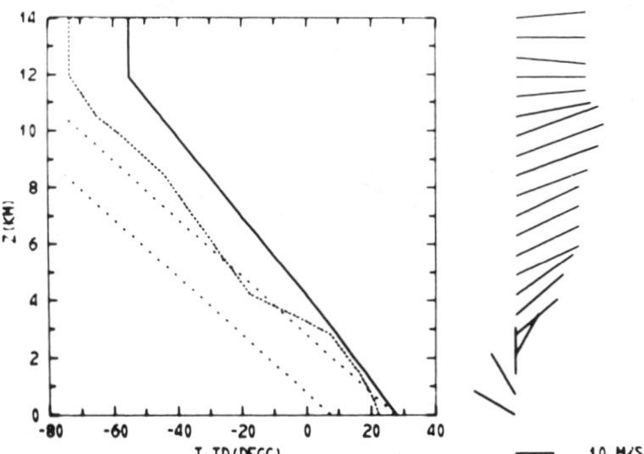

Figure 4 Initial environmental sounding for Schlesinger's Run 3. Solid curve is temperature (°C) and dashed curve is dew point (°C); dotted curves are dry adiabats for $\theta = 280$ K (*left*) and 300 K (*right*). Solid line segments at right of diagram represent horizontal wind vectors at 700 m height intervals from 0 to 14 km; rightward corresponds to eastward and upward to northward. From Schlesinger (1978).

was similar. After that the no wind case grew somewhat faster but started a rapid precipitating collapse, as expected, while the strong shear case developed a completely new structure.

Figure 5 shows maps of the vertical velocity at two height levels, 0.35 km and 3.15 km, for both the sheared and no wind cases at 32 minutes after start of the integrations, when the no wind case was collapsing rapidly and the sheared case was near its peak amplitude. For the no wind case the central downdraft has started at the lowest level, while the updraft remains strong above that, with weak quadrilaterally symmetric downdrafts in the clear air outside the cloud. For the sheared case, the updraft and downdraft have separated longitudinally at the low level, and both have split laterally aloft. To interpret these figures correctly,

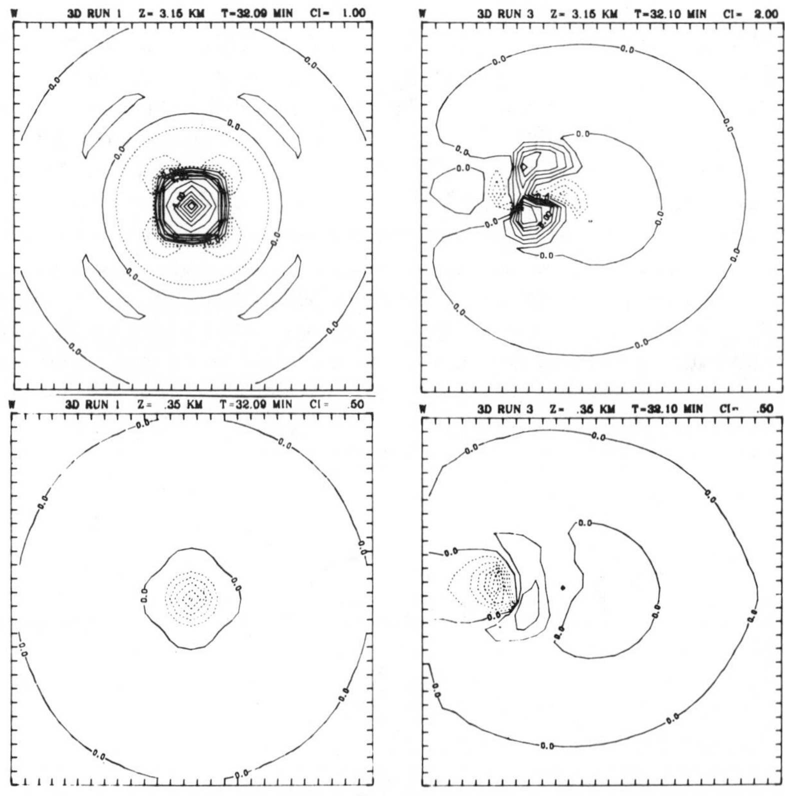

Figure 5 *x–y* sections of vertical velocity for Schlesinger's no-wind case (*left*) and strong shear (*right*) at the 3.15 km (*upper panels*) and 0.35 km (*lower panels*) levels. Positive contours are solid, negative dashed. Horizontal and vertical ticks are 1.8 km apart and the total dimension is 48.6 km square. From Schlesinger (1978).

note that the entire box is moving at the average tropospheric wind speed. The principal downdraft at the 0.35 km level is at the rear of the storm, apparently having been formed from fairly low level air. At the higher level a new downdraft is forming at the front, which from other parts of Schlesinger's presentation can be seen to consist of air which has previously moved around the storm. This newly formed downdraft is splitting the primary updraft into left and right parts which are somewhat unequal, the right side being stronger. Schlesinger carried out his calculations to 48 minutes, at which time the updraft splitting was essentially complete, with the two halves diverging fairly rapidly. The amplitude of most features was at that time slowly diminishing.

The simulations presented by Klemp and Wilhelmson confirm and substantially extend Schlesinger's results. They used computational resolution and domain size similar to Schlesinger's, while the depth and magnitude of instability in the assumed ambient atmosphere were somewhat smaller. Experiments were conducted with several environmental wind profiles, but the most revealing were those using either a rectilinear shear over some depth of the troposphere or two layers of rectilinear shear with a shear direction change between them. Thus the wind speed hodographs (lines connecting the heads of wind vectors at different levels) consisted of either a straight line or two straight line segments, as shown in Figure 6.

In all cases the early results followed the same sequence as seen from Pastushkov's and Schlesinger's calculations. The initial buoyant element rose somewhat less rapidly and not as high as a similar element in an unsheared environment. After about 30 minutes the updraft weakened somewhat and split in two under the influence of a downdraft developing, in contrast to Schlesinger's result, directly under the initial updraft position. In the next 30 minutes the central downdraft also split and the two halves began to follow the updraft pair, as they diverged further from center line. As the paired circulations became separated from each other, they reintensified and subsequently developed steady state three-dimensional flow patterns. The right moving updraft rotated counter-clockwise, the right moving downdraft clockwise, and the left moving pair the reverse. After separation was complete, there was no tendency of the circulations to decay with time, in apparent contrast to Schlesinger's result. While a number of moderately significant differences existed in the numerical formulations which could lead to this contrast, it is possible that Schlesinger's storm would approach a steady state amplitude at a later time.

The development of quasi-steady rotating circulations moving obliquely to the mean flow is a feature commonly observed in severe storms, especially those which produce tornadoes. I believe that it is a correct predic-

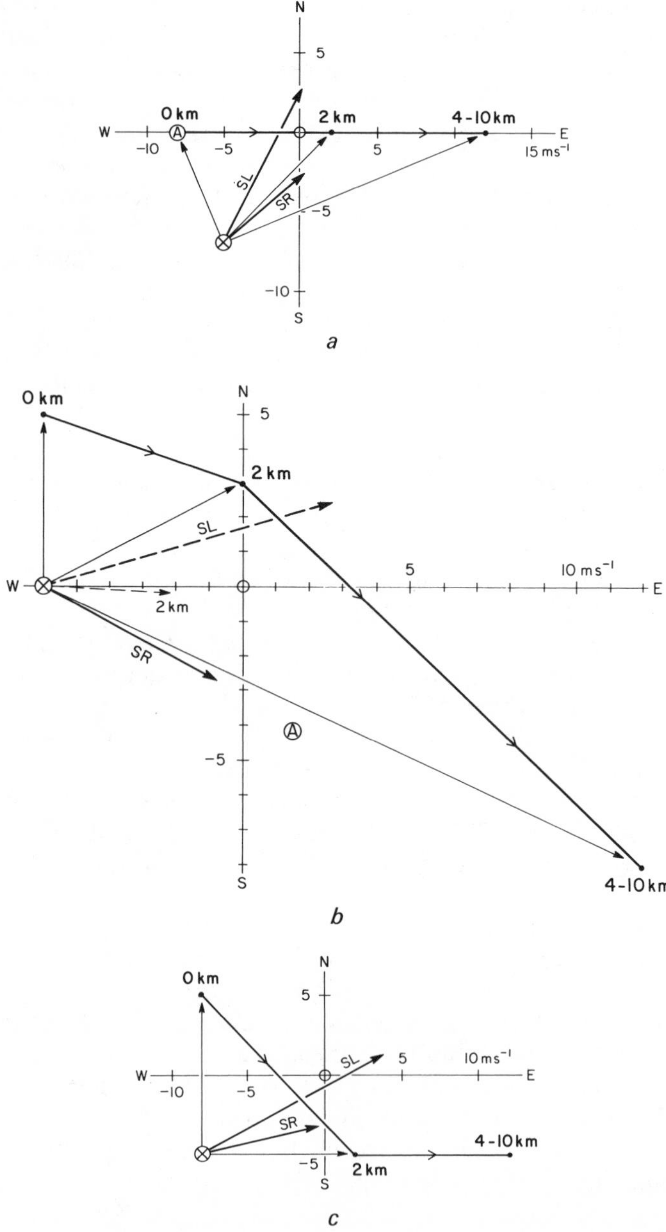

Figure 6 Wind hodographs depicting the environmental shear and motion of simulated left- and right-moving storms (SL and SR, respectively) for (*a*) a rectilinear shear; (*b*) a clockwise shear; and (*c*) a counter-clockwise turning shear. From Klemp & Wilhelmson (1978a).

tion of the K-W model and thus important to diagnose and understand. The critical dynamical feature is evidently the development of rotating eddies on the left and right sides of the initial updraft and the consequent introduction of midtropospheric dry air into the center from the front of the storm. As postulated from observational data by Barnes (1968, 1970), rotation is initially a product of the twisting terms in the vertical vorticity equation, and can be evaluated from a linearized version of that equation as follows:

$$\frac{\partial \zeta}{\partial t} \approx -\bar{\mathbf{V}} \cdot \nabla \zeta - \mathbf{k} \cdot \nabla w \cdot \partial \bar{\mathbf{V}} / \partial z \tag{2.1}$$

Figure 7 shows a sequence of horizontal cross-sections of wind vectors and vertical velocities at the 1.75 km level for a case where the mean shear was unidirectional and had a value of about $4 \times 10^{-3} \mathrm{s}^{-1}$ at that level. The left (positive y) side of the storm is a mirror image of the depicted right side. On the right side of the updraft the gradient of vertical velocity is of order $10^{-3} \mathrm{s}^{-1}$, so that the vorticity generation from the second term on the right of Equation (2.1) is about $5 \times 10^{-6} \mathrm{s}^{-2}$. By 30 minutes this generation term has begun to form centers of rotation on the outside edges of the primary updraft. Barnes observed very similar magnitudes in a severe right-moving thunderstorm in Oklahoma.

Coincident with the development of rotation, the simulated droplet physics of the model produces a high liquid water accumulation in the primary updraft, which reverses the buoyancy, induces a strong downdraft and surface outflow, and starts raining out at 45 minutes. This outflow splits the primary updraft into two halves which follow the upper vorticity maxima. At the 1.75 km level the circulation then begins feeding dry environmental air around the front (positive x side) of the storm and into the falling rain region, forming a continuing evaporating downdraft, while the vorticity maximum continues to propagate outward. At 60 minutes a vorticity center of opposite sign has developed to the southwest of the downdraft, and the downdraft maximum then separates from the center and follows the vorticity maximum. As the two halves of the storm separate into migrating vortex pairs, Equation (2.1) remains approximately valid, rewritten in the form

$$(\bar{u} - c_x)\partial \zeta / \partial x - c_y \partial \zeta / \partial y \approx \frac{\partial w}{\partial y} \frac{\mathrm{d}\bar{u}}{\mathrm{d}z}. \tag{2.2}$$

At the 1.75 km level the first term on the left is approximately zero, so that the fields of vorticity and vertical velocity become proportional, with

$$\zeta \approx -\frac{\mathrm{d}\bar{u}/\mathrm{d}z}{c_y} w. \tag{2.3}$$

Figure 8 shows the approximate coincidence of these fields at 75 minutes. The ratio between their maxima corresponds to a propagation rate about twice that observed, however. The difference is probably associated with some of the nonlinear and diffusive terms neglected here.

The development of a quasi-steady pair of obliquely moving storms with

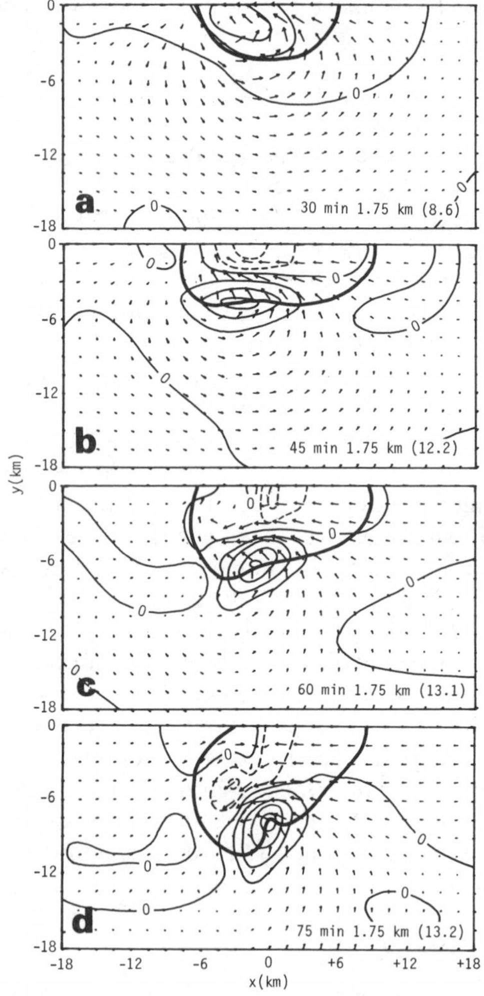

Figure 7 Horizontal cross sections at 30, 45, 60 and 75 min showing the horizontal vector winds and the vertical velocity at 1.75 km. Solid contours are used in updrafts and dashed contours in downdrafts. The contour interval is 2 m s^{-1}. The maximum horizontal wind speed in m s^{-1} is indicated for each level in parentheses. The thick line is the outline of the rainwater region. From Wilhelmson & Klemp (1978).

rotation evidently involves some rather delicate balances between various processes. In the K-W model either halving or doubling the shear prevents splitting, as does lifting the level of the principal shear zone. On the other hand the evolution seems somewhat insensitive to moderate changes in the admittedly crude parameterization of the raindrop growth and fallout process, though liquid water accumulation is necessary to produce the steady-state updraft-downdraft couplet.

Klemp and Wilhelmson carried out an experiment in which the rainwater accumulation mechanism was suppressed, although condensation and cloud water formation were allowed. In this case rotation developed as before, but because the liquid water content was insufficient to produce negative buoyancy, no central downdraft developed and splitting did not

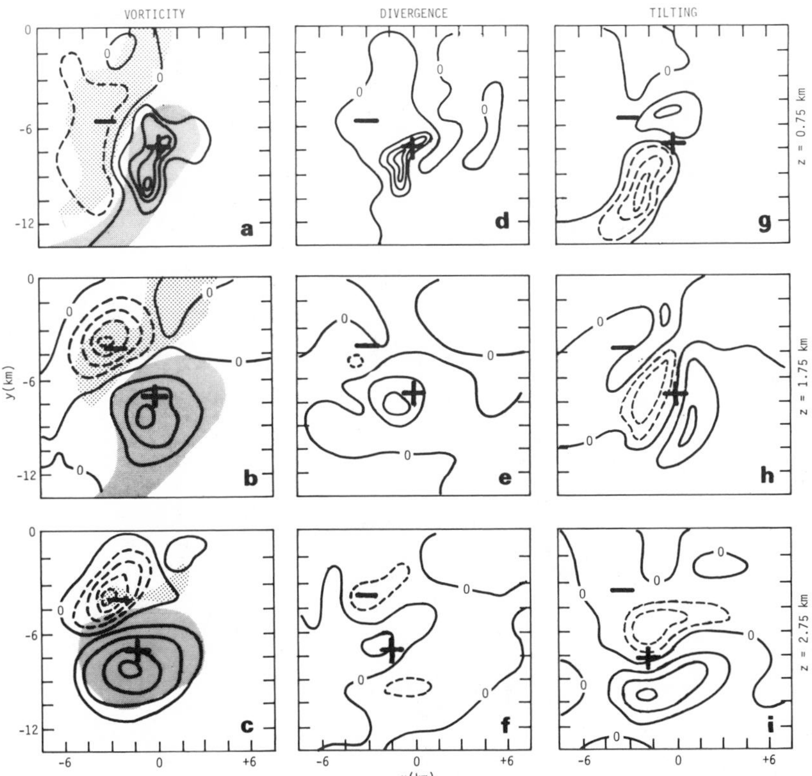

Figure 8 Horizontal cross-section of vertical vorticity at 75 min and at a height of 1.75 km, corresponding to Figure 7*d*. The contour interval is 2×10^{-3} s^{-1}. In the coarsely shaded area vertical velocity is < -1 m s^{-1} and in the finely shaded area it is > 1 m s^{-1}, with the extrema shown by the large $\pm$ signs. From Wilhelmson & Klemp (1978).

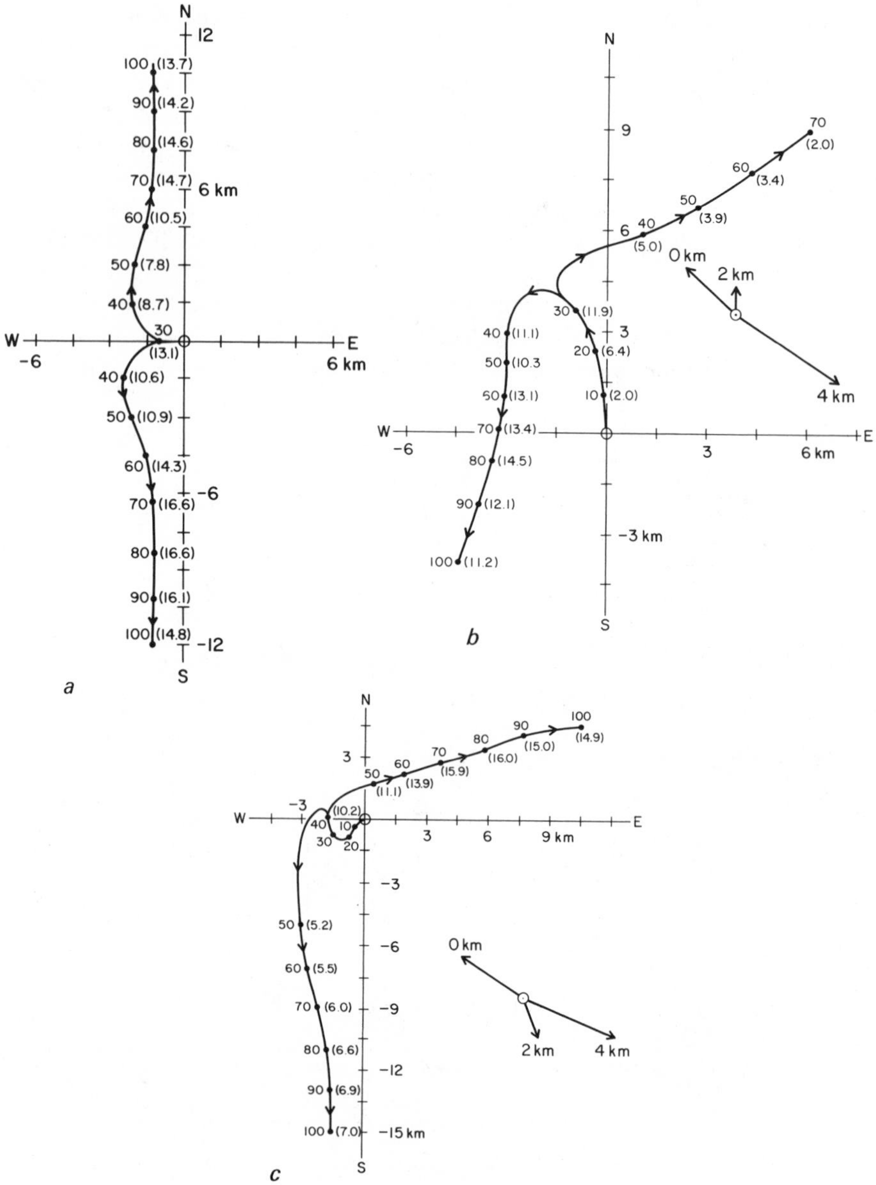

Figure 9 Propagation of the location of maximum updraft for the left- and right-moving storms produced by the hodographs in Figure 6*a–c*. Storm positions are labeled at 10 min intervals with the maximum updraft velocities in m s^{-1} indicated in parentheses. Environmental wind vectors relative to the coordinate origin in the numerical simulation are also included for reference. From Klemp & Wilhelmson (1978b).

occur. In another test rain water was allowed to form but not to evaporate on descent. In this case splitting occurred, but the downdraft intensity was considerably reduced, as was the total storm strength.

Perhaps the most important sensitivity discovered by Klemp and Wilhelmson, also partially evident from Schlesinger's results, is that arising from vertical variation in the direction of the ambient wind shear. The effect of such a wind turning was to suppress one of the diverging vortex pairs, specifically that on the convex side of the curved hodograph. Thus the clockwise rotating hodograph (Figure 6*b*) led to almost complete suppression of the left moving updraft-downdraft pair while the counter-clockwise rotating hodograph (Figure 6*c*) led to a substantial, though less extensive, suppression of the right moving pair. Figure 9 shows plots of the propagation and intensity of the points of maximum vertical velocity for the straight, clockwise, and counterclockwise turning hodographs. For the straight hodograph case the left and right sides are not identical. This is because the Coriolis (earth's rotation) terms were included in the equations of horizontal motion and generated positive vorticity in both updraft regions, thus amplifying the right moving and diminishing the left moving systems. The effect of the turning of the hodograph is, however, much more profound. For the clockwise turning case, the left moving storm is greatly weakened immediately, and soon virtually ceases to exist as a well defined entity. In the counterclockwise case, the right moving storm is similarly weakened, but not as much, and persists as a system with a defined updraft, downdraft, and rain. This asymmetry of results is due to the detailed hodograph shapes, which are not mirror images, and partly to the existence of Coriolis terms. In both cases the favored side maintains about the same amplitude as both sides do in the symmetric case. In interpreting the motion of the split storms note that the total wind vector was arbitrarily specified so as to keep the storms within the computational domain as long as possible. Thus the separation velocity of the storms, rather than the total velocity, is the key result of the simulation.

The above results seem to bear close relationships to the observed behavior of severe thunderstorms which develop in strong shear. Splitting of storms has been reported frequently, as has the apparent transformation of a nonrotating storm to one with rotation and motion oblique to the mean flow (see Raymond 1975, Klemp & Wilhelmson 1978b, and Wilhelmson & Klemp 1978 for citations). The latter situation can now be tentatively interpreted, on the basis of the numerical results, as one where splitting occurs, but where one of the split pair becomes almost immediately indistinguishable by radar observations. Other elements of the structure of the rotating storms were shown by Wilhelmson and Klemp to be closely analogous to those reported in the literature.

Once rotation develops in a storm, most of the water which enters the storm and coalesces into raindrops is transported inward from the updraft region, where its weight plus evaporative cooling will develop a separated downdraft. Since the vorticity generation mechanism also acts to propagate the downdraft-induced vortex outward, the consequent water trajectories evidently force the downdraft and updraft to follow along. The apparent effect of the turning of the hodograph with height is to increase or decrease the rate of transport of water mass away from the updraft. From Figure 6 the unfavored storm propagates much more slowly than the favored one relative to the mean flow at about 2 km, which is near the altitude of maximum downdraft and (probably) maximum rain rate. Thus the cause of the asymmetry in storm strengths may be the inability of the unfavored storm to separate its updraft and downdraft, though that hypothesis has not yet been clearly verified.

Very recently Thorpe & Miller (1978) have presented results of three-dimensional simulations showing both new cell generation and storm splitting in sheared environments. These results appear to be somewhat different from and more complex than those of Klemp and Wilhelmson, but are based on a more complex environmental wind structure.

Gravity Wave Aspects of Thunderstorms

Gravity waves include the class of oscillatory motions which develop in a stably stratified atmosphere in response to any form of unbalanced vertical displacement. Since the upper and outer environments of a thunderstorm are normally stably stratified and the storm produces large vertical displacements, it is typically surrounded by waves of measurable amplitude. The top of a growing convective cloud, in particular, becomes the source of waves propagating outward in apparently similar fashion to those produced by throwing a stone into a pond. Such waves may occasionally trigger new cloud developments at a distance by means of the vertical displacement they produce.

In the remainder of this section I discuss a theoretical approach to the problem of thunderstorm generation and propagation which considers the entire storm as a forced gravity wave and uses linear theory to predict the modes of maximum growth rate. The approach is called "Wave CISK" and is essentially due to Lindzen (1974), Stevens & Lindzen (1978), and Raymond (1975, 1976). CISK stands for "Conditional Instability of the Second Kind," so titled by Charney & Eliassen (1964), and developed with special reference to hurricane formation. In its original concept, especially as formulated by Ooyama (1964), the CISK hypothesis was designed to account for the fact that the hurricane is a large direct heat engine, like a thunderstorm, but with a horizontal scale much greater than that which convection theory indicates will grow most rapidly and use up the available instability first. It was hypothesized that the net result of an ensemble of raining convective cells is a restratification of the atmosphere leading to lower surface pressure, rotation, and boundary layer convergence. The increased and convergent surface winds then advect in and evaporate from the sea surface new sources of moisture, which enhance the convection, and so on. The first reasonably correct models of hurricane development were formulated using the CISK hypothesis.

The application of CISK involves simplification and some distortion of the cloud convection process and its ultimate result. It is assumed that moisture which is brought into and lifted out of the boundary layer by a low level convergence field is then rained out and the resulting heat of condensation distributed vertically in some specified way. As applied to Lindzen's and Raymond's Wave CISK, the source of convergence is a traveling internal gravity wave, propagating in accordance with the mean atmospheric temperature and wind profiles and dry adiabatic processes. The condensation heating and evaporative cooling are released and distributed, however, by the CISK mechanism, and this heat exchange modulates the gravity wave propagation velocity and may cause the wave to amplify. Thus the statically stable parts of a storm circulation, perhaps especially the subsidence in the near environment, are regarded as capable of partially controlling its evolution and movement. In the following I have reproduced a simplified form of Raymond's analysis for the case of no environmental shear.

I will use two-dimensional (x–z) equations of motion, linearized about a state of zero mean velocity, with the buoyancy terms restated as in Equation (1.10), but without consideration of the moisture terms, i.e.

$$\frac{\partial u'}{\partial t} + \frac{\partial}{\partial x}(p'/\bar{\rho}) \tag{2.4}$$

$$\frac{\partial w'}{\partial t} + \frac{\partial}{\partial z}(p'/\bar{\rho}) = g\theta'/\overline{\theta}. \tag{2.5}$$

The continuity equation is replaced by an assumption of incompressibility,

$$\frac{\partial u'}{\partial x} + \frac{\partial w'}{\partial z} = 0, \tag{2.6}$$

and the thermodynamic equation is linearized around a state of constant static stability, but with a forcing term on the right, i.e.

$$\frac{\partial \theta'}{\partial t} + \frac{w' N^2 \overline{\theta}}{g} = F(x,z) \tag{2.7}$$

where $N^2 = (g/\overline{\theta})\,d\overline{\theta}/dz$. If the perturbation (primed) variables are assumed to be proportional to exp $[ik(x-ct)]$, then Equations (2.4)–(2.7) may be reduced to a single ordinary differential equation, i.e.

$$\frac{d^2W}{dz^2} + \left(\frac{N^2}{c^2} - k^2\right)W = \frac{g}{\overline{\theta}}\frac{F}{c^2}\exp\,[-ik(x-ct)] \tag{2.8}$$

where $w = W(z)\exp\,[ik(x-ct)]$. Raymond now assumes that the rate of heat release by moist convection is equal to the vertical velocity at cloud base multiplied by the potential temperature difference between cloud base and cloud top, and is distributed uniformly. Since it is assumed that the top of the cloud is the level where the environmental θ_e is the same as that at the surface, the potential temperature difference is essentially the available heat of condensation of moisture lifted from the surface level, i.e. θ at point C minus θ at point A of Figure 1. The vertical velocity at cloud base is taken to equal the surface convergence multiplied by the height of cloud base. Thus

$$F = b\left(\frac{\partial\overline{\theta}}{\partial z}\frac{\partial w}{\partial z}\right)_{z=0} \quad \text{for} \quad b \leqq z \leqq h \tag{2.9}$$

and therefore

$$\frac{d^2W}{dz^2} + \left(\frac{N^2}{c^2} - k^2\right)W = \begin{cases} \dfrac{N^2 b}{c^2}\dfrac{dW(0)}{dz} & \text{for} \quad b \leqq z \leqq h \\ 0 & \text{elsewhere} \end{cases} \tag{2.10}$$

where b and h are the cloud base and top heights, respectively.

In his derivation, Raymond multiplies the right side of (2.10) by a factor between 1 and 2, which he justifies by qualitative consideration of closed processes. Raymond then shows that this equation is of an eigenvalue type, in that solutions only exist for certain values of c, the complex phase speed. Since he is seeking the most unstable mode he appropriately prescribes a vanishing upper boundary condition, i.e.

$$W = 0 \text{ at } z = \infty. \tag{2.11}$$

The eigenvectors are rather complicated, but as might be expected the solution for vertical velocity has a maximum in the region $b < z < h$ and above that level looks like a damped gravity wave, with phase lines tilted in the direction of propagation. Infinite families of eigenmodes exist, but for those of greatest physical interest the cloud top to base height ratios lie within the range $1 < h/b < 10$ or so. From considering the real and imaginary parts of the wave dispersal relation it can be shown that neutral eigenvalues must satisfy the relation

$$2\left(\frac{Nb}{c}\right)^2 \frac{\sin\,[(Nb/c)^2 - k^2b^2]^{1/2}}{[(Nb/c)^2 - k^2b^2]^{1/2}} = 1. \qquad (2.12)$$

For the long wave length ($k = 0$) case, this has solutions at $Nb/c = 0.741$, 2.973, and at intervals of about π thereafter. The most unstable mode has amplifying solutions for $h/b > 7.48$. For a typical case I calculated a complex phase speed $c = Nb\,(\pm 1.659 + 0.112i)$ with $h/b = 9.08$. Assuming typical atmospheric values of $N = 10^{-2}\ s^{-1}$, $b = 1$ km, the predicted cloud top height ~9 km, the phase speed ~17 m s^{-1}, and an e-folding amplification occurs in about the time required to move two wave lengths downstream. The effect of finite wave length ($k > 0$) is almost negligible if that wave length is greater than about 10 km.

Thus Raymond's theory predicts that for an incompressible ambient atmosphere with constant static stability and no shear, growing modes can propagate in any direction, with the faster growing modes moving at 10–20 ms^{-1} and e-folding in an hour or so. Raymond concentrated most attention on cases taken from real data, where the atmosphere had a complicated velocity profile. He found that typically there were just one or two propagation vectors with maximal growth rate, instead of the circle found for the no-shear case. In his second paper (Raymond 1976) he synthesized flow patterns from summation of the eigenfunctions, and found evidence of storm splitting in certain circumstances. For several of his cases the predicted propagation velocity and splitting characteristics agreed well with observational data.

In my view the Wave-CISK theory has not yet been tested adequately. Initial comparisons with observations and 3-d model simulations suggest both agreements and disagreements. There is not much evidence to suggest reality of the predicted propagation velocities for the no-shear case. Raymond states (personal communication) that his model will not predict splitting if the wind hodograph is straight, in contrast to the Klemp-Wilhelmson model. Both models predict amplification of storms on the concave side of a curved hodograph. More fundamental aspects of the models have yet to be compared. Substantial control of storm propagation by the stable forced motions around it would be a rather remarkable result, if true. To me it is somewhat implausible, analogous to requiring the tail to wag the dog.

3 SQUALL LINES

General Description

The tendency of thunderstorms to be oriented in lines has long been noted but never completely explained. Probably there is no single explanation, since environmental conditions, and internal and intra-storm dynamics are likely to contribute in varying degrees to the tendency. Also it seems likely that the often assumed linear symmetry is overemphasized, especially in the radar analyses produced as fascimile charts for forecasters, pilots, etc. Examination of the original radar records from which these analyses are prepared does not always show as well organized a line as might be supposed to exist.

Many squall lines form along discontinuities in the larger scale atmospheric structure, including particularly cold fronts and dry lines. The classical picture of a cold front as an undercutting wedge of cold air, producing convergence and uplift along its leading edge, remains valid in many cases. Details of the circulation, including generation of deep convection, have been recently simulated by a two-dimensional model (Orlanski & Ross 1977, Ross & Orlanski 1978). The dry line situation is a little more subtle. In the southern plains of the U.S. in spring and summer a sharp demarcation line can usually be observed between warm moist tropical air moving northward from the Gulf of Mexico and hot dry desert air to the west (Schaefer 1974). In contrast to the cold frontal case the density difference across the dry line is small but there are large differences in the depth of the boundary layer and its diurnal cycle. Severe thunderstorm outbreaks often occur in a squall line configuration just to the east of the dry line. Both the thermodynamic structure of the environment and the uplift which triggers these outbreaks may be associated with turbulent mixing in the deep boundary layer west of the dry line (Danielsen 1975).

Squall lines also form well ahead of or apparently unrelated to major line-oriented discontinuities in the large scale environment. For the frequent case when a line forms ahead of but parallel to a cold front the possibility that traveling gravity waves initiated by the frontal circulation could later trigger squall line development was first proposed by Tepper (1950). Traveling gravity waves have since been implicated in many observed cases of thunderstorm and squall line generation and propagation (e.g. Uccelini 1975), but an adequate explanation of the dynamic structure and evolution has yet to be shown. One of the problems is that the larger scales of gravity waves, with wave length 50 km or greater, are not readily trapped by normal atmospheric stratification, but instead propagate vertically and should be imperceptible within an hour or so after generation. Of course the hypothesized Wave-CISK model provides a continuing generation mechanism, but some adjustments must be made to the theory. One typically observes an apparent wave moving rapidly through and beyond the storms which it seems to sporadically trigger, so that the phase speeds of the storms and the wave differ markedly.

Radar depictions of thunderstorm evolution frequently suggest that linear arrangements of thunderstorms form out of originally isolated storms. Some recent numerical calculations by Klemp and Wilhelmson (personal communication) also show evidence of squall line formation from inter-storm dynamics.

The Moncrieff Models

Once a squall line is generated, by whatever means, it often persists for several hours. Thus an assumption of steady state dynamics is not unreasonable a priori. What is not clear is the extent to which line symmetry can be imposed. In many cases thunderstorms maintain their three-dimensional identity within the line, with significant gaps between them. If this were not so, long distance air travel in the U.S. would be severely handicapped in the spring and summer, since few aircraft can climb over the top of a large thunderstorm. Comparisons of two- and three-dimensional model simulations in shearing conditions suggest that the existence of these gaps may be an inherent and critical part of the maintenance of the systems. To examine this question further, I will now consider two closely related quasi-analytic theoretical models by Moncrieff and collaborators, addressed to the questions of the structure and motion of squall lines. Both of these models consider the convective system as steady state and nonlinear, and they purposely ignore the effects of turbulent mixing and dissipation. This neglect must surely be questioned. Observations have shown that extremely high values of dissipation, of order 1 $m^2 s^{-3}$, occur in thunderstorms (Frisch & Strauch 1976, Waldteufel 1976), so that the turbulent kinetic energy would have a half-life of only a few minutes if not continually regenerated. Except for this discrepancy, however, and some problems with downdraft dynamics found in most analytic models, the theoretical treatments appear to consider most of the ingredients that seem to be important to the behavior of the real atmosphere and to the more detailed numerical models.

Figure 10 shows schematic flow patterns proposed by Moncrieff & Green (1972) and Moncrieff & Miller (1976), respectively, which I refer to subsequently as MG and MM. The first was an attempt to create a strictly two-dimensional flow model in strongly sheared environments similar to that proposed on observational grounds by Browning & Ludlam (1962). The model requires that the air flowing into each side of the line reverse its direction before flowing out, when measured in coordinates moving with the line. The upshear tilt of the interface between updraft and downdraft branches would permit rain falling out of the updraft to saturate the downdraft and produce negatively buoyant air and downward acceleration within it. In a recent paper, however, Moncrieff (1978, designated here as M2) showed that for realistic environments and processes, solutions could only be obtained for downshear tilting interfaces, which removes the source of thermodynamic energy for the downdraft and also removes much (perhaps not all) of the model's physical interest. The MM model was intended to apply to relatively weak shear environments, including those often found in the moist tropics. The mean flow is assumed to move through the convective circulation at all levels. The upstream and downstream states are taken to be two-dimensional, but in order to avoid streamlines crossing each other the circulation is presumed to be three-dimensional. The principal products of both models are the line propagation velocity and the outflow temperature and velocity profiles, which are allowed to differ from the environment. I will later show evidence that the MM model is consistent with the 3-d storm simulation models in the strong shear case, as well as with observational data.

In spite of their severe physical limitations the MG-M2 solutions are interesting because they represent, to my knowledge, the first application

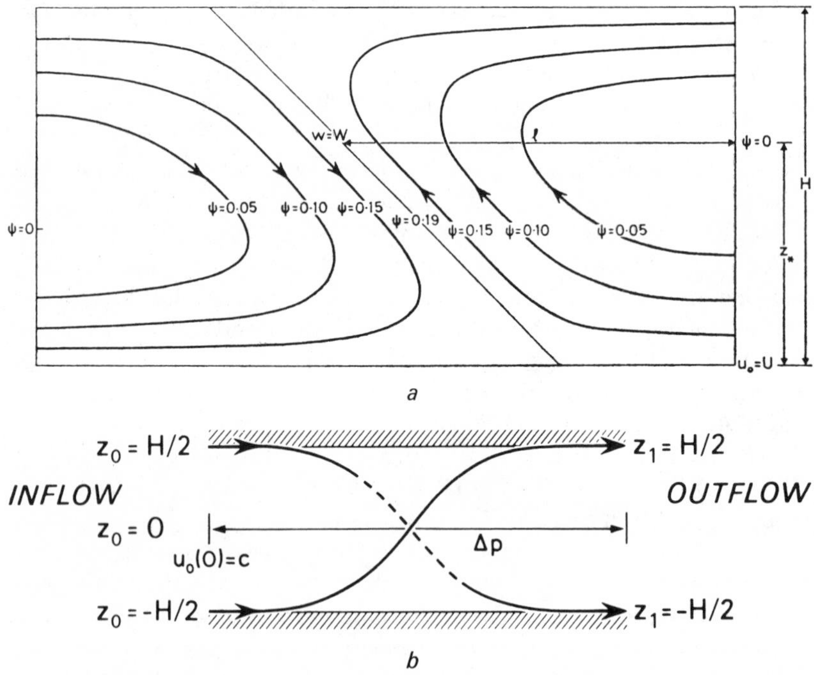

Figure 10 Schematic flows proposed (*a*) by Moncrieff & Green (1972) for a line-symmetric storm in strong environmental shear; and (*b*) by Moncrieff & Miller (1976) for a squall line requiring 3-dimensional internal circulations with or without environmental shear.

to unstably stratified flows of the techniques earlier developed for solution of finite amplitude mountain wave problems. In the following discussion, I rederive a portion of their results from a somewhat different viewpoint, which also introduces the basis for the MM model.

Equation (1.1) in its steady state form may be transformed by vector identities to the form

$$\nabla\left(\frac{V^2}{2}+gz\right)-\mathbf{V}\cdot(\nabla\cdot\mathbf{V})+\nabla p/\rho=0. \tag{3.1}$$

In the case of two-dimensional (x–z) flow, the continuity equation, (1.2), in steady state form allows definition of a scalar stream function, ψ, as follows

$$\rho u=\partial\psi/\partial z,\ \rho w=-\partial\psi/\partial x \tag{3.2}$$

so that Equation (2.1) may be written in the form

$$\nabla\left(\frac{V^2}{2}+gz\right)-\frac{\eta}{\rho}\nabla\psi+\nabla p/\rho=0 \tag{3.3}$$

where $\eta=\nabla\cdot(\nabla\psi/\rho)$ is the y-component of vorticity. Upon forming the inner product of Equation (3.3) with $\mathbf{V}$, the second term disappears, so that

$$\mathbf{V}\cdot\nabla\left(\frac{V^2}{2}+gz\right)+\mathbf{V}\cdot\nabla p/\rho=0. \tag{3.4}$$

Pressure gradients are now replaced by use of Equation (1.8), so that Equation (3.4) becomes

$$\mathbf{V}\cdot\nabla\left[\frac{V^2}{2}+gz+c_pT+(L_v+L_f)q_v+L_fq_l\right]=\frac{c_pT}{\theta_e}\mathbf{V}\cdot\nabla\theta_e. \tag{3.5}$$

Since θ_e is assumed to be constant along a streamline, the quantity in brackets is also, so that Equation (3.3) may be integrated and rearranged to the form

$$\frac{\eta}{\rho}=\frac{\mathrm{d}F}{\mathrm{d}\psi}-\frac{c_pT}{\theta_e}\frac{\mathrm{d}\theta_e}{\mathrm{d}\psi} \tag{3.6}$$

where

$$F=[V^2/2+gz+c_pT+(L_v+L_f)q_v+L_fq_l]=F(\psi). \tag{3.7}$$

The functions of ψ on the right of Equation (3.6) may be evaluated by considering conditions at an inflow region, say as $x\to\infty$. With the inflow variables designed by subscript 0, where the flow is assumed to be horizontal, Equation (3.6) becomes

$$\frac{\eta}{\rho}=\frac{\mathrm{d}u_0/\mathrm{d}z_0}{\rho_0}+\frac{c_p}{\theta_e}\frac{\mathrm{d}\theta_e}{\mathrm{d}z_0}\frac{T_0-T}{\rho_0 u_0}. \tag{3.8}$$

All the quantities on the right side are now functions of ψ except the temperature T. The objective is then to express T as a function of ψ and z only. Use of Equation (3.7) at the inflow boundary yields

$$c_p(T-T_0)+(L_v+L_f)(q_v-q_{v0})+L_f(q_l-q_{l0}) = -g(z-z_0)+\frac{u_0^2-u^2-w^2}{2}. \tag{3.9}$$

Further consideration of the moist compressible case requires defining q_v and q_l under saturation conditions from the Clausius-Clapeyron equation as functions of T and ρ. Probably ρ would initially (or even finally) be approximated as a known function of height. After that Equation (3.8) could be integrated numerically to obtain solutions for $\psi(x,z)$ which satisfy the inflow boundary condition. MG carried out an equivalent procedure, using a somewhat different formulation of the buoyancy terms.

A simplified but revealing analysis can be made for the incompressible case without moisture, where θ replaces θ_e as a conservative variable and ρ is considered constant in Equation (3.8) and removed from the definition of stream function. Here Equations (3.8) and (3.9) may be combined into the relation

$$\frac{\partial^2\psi}{\partial x^2}+\frac{\partial^2\psi}{\partial z^2}=\frac{\mathrm{d}u_0}{\mathrm{d}z_0}+\frac{N_0^2}{u_0}\left\{z-z_0+\left[\left(\frac{\partial\psi}{\partial x}\right)^2+\left(\frac{\partial\psi}{\partial z}\right)^2-u_0^2\right]\Big/2g\right\} \tag{3.10}$$

where $N_0^2=(g/\theta_0)\,\mathrm{d}\theta_0/\mathrm{d}z_0$, the upstream static stability. Equation (3.10) is nonlinear but almost negligibly so. The terms in square brackets on the right could be removed by a minor recasting, such that $\mathbf{k}\cdot\nabla\psi/\theta^{1/2}$ becomes the velocity variable. Instead, following MG, one may simply neglect the bracketed terms, which is appropriate for a Boussinesq type formulation, so that Equation (3.10) becomes

$$\frac{\partial^2\psi}{\partial x^2}+\frac{\partial^2\psi}{\partial z^2}=\frac{\mathrm{d}u_0}{\mathrm{d}z_0}+\frac{N_0^2}{u_0}(z-z_0). \tag{3.11}$$

The next task is to write the mean wind, static stability, and z_0 as explicit functions of ψ. MG considered the case of constant (negative) static stability and constant shear S in the inflow region, so that

$$u_0=u_{00}+Sz_0,\ \psi=u_{00}z_0+Sz_0^2/2,\ \theta=\theta_{00}[1+(N_0^2/g)z_0] \tag{3.12}$$

where the subscript 00 refers to inflow conditions at the lower boundary. Solving these for z_0 and u_0 yields

$$z_0 = -\frac{u_{00}}{S} - \sqrt{\left(\frac{u_{00}}{S}\right)^2 + \frac{2\psi}{S}}, \quad u_0 = \pm S\sqrt{\left(\frac{u_{00}}{S}\right)^2 + \frac{2\psi}{S}}. \tag{3.13}$$

The surface velocity u_{00} is unknown and must actually be a product of the total solution, since all velocities are defined relative to the propagation rate. In choosing the sign of the root it is assumed, in accordance with Figure 9, that u_{00} and ψ are negative and S positive everywhere, and that u_0 is negative in the lower levels and positive aloft. Thus Equation (3.11) may be written explicitly in ψ and z, as

$$\frac{\partial^2\psi}{\partial x^2} + \frac{\partial^2\psi}{\partial z^2} = S - N_0^2\left[\frac{1}{S} \pm \frac{z + u_{00}/S}{\sqrt{u_{00}^2 + 2S\psi}}\right]. \tag{3.14}$$

Equation (3.14) is a nonlinear Helmholtz equation, which must be solved subject to the known inflow conditions, assumed lower and upper boundary conditions (horizontal flow at $z = 0$ and $z =$ H), and an interface condition between the updraft and downdraft branches (continuity of pressure and streamlines parallel to the interface). Most of the MG and M2 solutions were obtained for a constant negative static stability and an arbitrarily chosen capping height H, which is difficult to interpret meteorologically. In one case in M2, however, a sounding is constructed with a thin lower stable layer, instability throughout most of the region, and a thin upper stable layer. This kind of profile approaches the stability structure of the sounding shown on Figure 1. The streamfunction solution for this case is shown on Figure 11 for the thickness of each stable layer

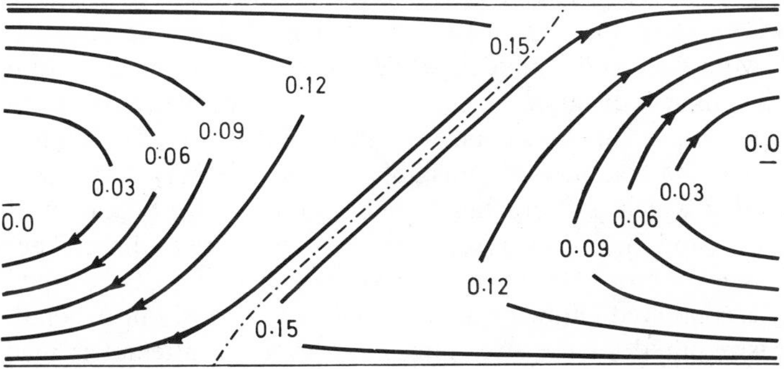

Figure 11 Moncrieff's solution to the incompressible free boundary problem with stable regions near the upper and lower boundary. Streamlines are labeled in units of $2AH^2$, where A is the mean shear and H the total height of the model. From Moncrieff (1978).

equal to 0.1H and for the characteristic negative Richardson number R given by $R = -N_0^2/S^2 = 0.5$. The most important characteristics of Moncrieff's solutions are determined by the magnitude of R. In M2 it is shown that complete solutions are only possible for $-\frac{1}{4} < R < 1$. For cases when buoyancy drives the flow ($R > 0$) the interfacial streamline dividing updraft from downdraft slopes downshear as shown, eliminating the possibility of maintaining an evaporating downdraft by rain falling out of the updraft. In addition the velocity is discontinuous across the interface, so that the flow in this region would be subject to shearing instability.

Because of the above limitations and the general lack of resemblance of the environmental conditions and flow patterns to most severe thunderstorms, it is unlikely that the MG-M2 model is a general enough framework for consideration of the effects of shear on strong convection. Occasionally, however, there is evidence of a storm with updraft features similar to those of the model. One that comes to mind is the storm that produced a severe flash flood on the Big Thompson River in Colorado in 1976. As shown by Maddox et al (1978) this storm appeared to have a nearly steady state structure for several hours, with a strong inflow from the east, apparent outflow aloft toward the north or northeast, an elongated north-south axis, and no evidence of a strong downdraft. The storm was located over the eastern slopes of the Front Range of the Rockies, which may have allowed the development of the upflow branch of a circulation like that of Figure 11, with the higher mountains tending to prevent mixing or other interference from the west. If the buoyant instability is sufficiently strong, the presence of rain falling through and evaporating into the inflow would not necessarily suppress convection, since θ_e above cloud base remains unchanged. Maddox et al show evidence that the storm which produced a similarly severe flash flood in Rapid City, South Dakota, in 1972 also had a similar structure.

The Moncrieff-Miller model, while less ambitious in attempting to create a complete dynamical framework, is equally interesting in its creative approach, and appears to be more successful in application. The basic idea is that the conditionally unstable atmosphere may find a mode of overturning which is energetically much more efficient than turbulent convection, and the process results of such a mode are outlined. The MM model is again based on the assumption of Equation (3.7), i.e. that total energy is conserved following a streamline. Since three-dimensional geometry is required for the streamlines, however, no attempt is made to provide a complete flow solution. Instead a unique and rather powerful energy optimization argument is used to close the problem. The result is a set of predictions of outflow profiles and a propagation rate as functions of the inflow conditions. Again there is a problem with the assumption of moist adiabatic downdrafts, though not as severe as in MG. The results are supported to some extent by three-dimensional numerical simulations and also by observational data.

The basic argument may be stated with reference to Figure 10*b*. It is again assumed that the entire layer is conditionally unstable, with a constant negative moist static stability

$$N^2 = (g/\theta_e)\, d\theta_e/dz = \text{constant} < 0. \tag{3.15}$$

MM then pose the following question. Given an inflow velocity profile and the constraints of mass continuity and energy conservation along streamlines, is there an outflow mass and velocity distribution which optimizes the upward buoyancy flux and the consequent release of kinetic energy? The answer seems to be yes, and the result is indicated by Figure 12, with the first two examples determined from analytic solutions. The outflow velocities are increased substantially from those of the inflow at the top and bottom, and decreased to zero at the center. This roughly parabolic velocity distribution allows for an approximately cubic buoyancy distribution in the outflow, so that most of the flux of both high and low θ_e air away from the convective region occurs in shallow layers close to the boundaries. The middle part of the convective "wake" is filled with almost stationary air (relative to storm movement) originating from the middle level upstream. The propagation velocity, relative to the upstream velocity at the mid-level, turns out to be a fraction of the total potential energy available for release from a parcel rising from the surface, i.e.

$$c = u_0 = 0.32 \int_0^H \int_0^z -N^2\, dz'\, dz \tag{3.16}$$

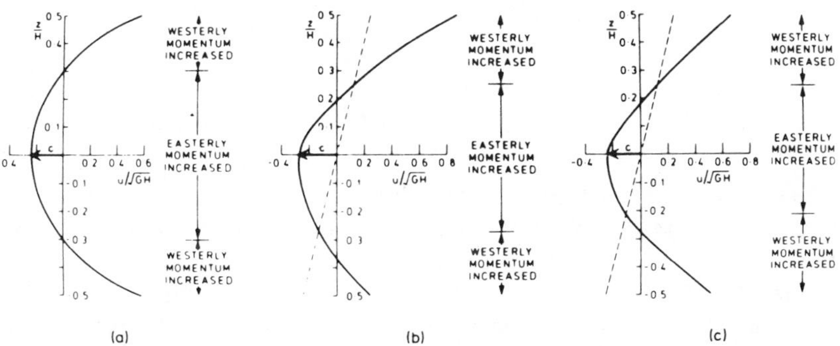

Figure 12 The change in the outflow velocity profiles produced by convection (from Moncrieff & Miller 1976). The downstream and upstream profiles are represented by full and broken lines, respectively: (*a*) unsheared incompressible flow; (*b*) sheared incompressible flow; (*c*) sheared compressible flow.

The integral is proportional to the area between the surface moist adiabat and the sounding on a thermodynamic chart like Figure 1. Within the limitations of the theory, propagation can occur in any direction, and the solutions seem to be somewhat insensitive to the mean shear, at least when it is constant. Betts, Grover & Moncrieff (1976) found that Equation (3.16) predicted the motion of squall lines in Venezuela with reasonable accuracy.

Some of the physical implications of the solutions are interesting. A central streamline starting from $x = -\infty$, $z = 0$ on Figure 10*b* would be level, indicating no exchange of kinetic and potential energy. As shown by Figure 12*a*, however, the velocity of the central streamline vanishes, relative to the moving convective system. This is associated with a pressure jump across the center, given by $\Delta p/\rho = u_0^2/2 = c^2/2$. This pressure jump must be even larger at the top and bottom, since the hydrostatic pressure should be reduced in the center due to warmer air overhead. Thus not all of the energy released from buoyancy goes into kinetic energy. Some is returned as work done on the atmosphere, which goes to change its thermodynamic state both within and above the nominal boundaries of the model. For a non-dissipative system in steady state, the energy equation may be stated as a conservation of horizontal flux, i.e.

$$\int [u(K+P+p)]\, \mathrm{d}z = \text{constant} \tag{3.17}$$

where K and P are the kinetic and potential energies and the integral extends over the vertical extent of the domain. For the MM system $K = \bar{\rho}u^2/2$, and $P = -\bar{\rho}gz\theta_e'/\theta_e = -\bar{\rho}N^2z^2$, where z is measured from the middle of the box. Evaluation of the fluxes upstream and downstream from the MM solution for an incompressible zero-mean shear atmosphere yields the following results

$$\begin{aligned} &\frac{1}{\bar{\rho}}\int u_0K_0\, \mathrm{d}z_0 = 0.5\, H\, u_0^3, \qquad \frac{1}{\bar{\rho}}\int u_1K_1\, \mathrm{d}z_1 = 2.37\, H\, u_0^3 \\ &\frac{1}{\bar{\rho}}\int u_0P_0\, \mathrm{d}z_0 = 1.58\, H\, u_0^3, \qquad \frac{1}{\bar{\rho}}\int u_1P_1\, \mathrm{d}z_1 = -8.10\, H\, u_0^3. \end{aligned} \tag{3.18}$$

Thus the increase in kinetic energy flux is only 15% of the decrease in potential energy flux, with the remainder going into the work required to raise the mean pressure. It is also interesting that the potential energy released by this process is more than six times what would be nominally available from a simple mixing process, since it puts all the potentially warm air very close to the upper boundary and all the potentially cold air very close to the lower boundary.

Seemingly the most arbitrary and questionable assumption, aside from the problems with moist adiabatic descent, is that of constant static instability with an implied lid on convective motion. A more physically plausible assumption leads to some conceptual changes, as shown by Figure 13. An hypothesized parabolic upstream profile of θ_e is shown on the left and the (presumably) optimized downstream profile on the right, with schematic streamlines indicated for transporting air from one side to the other. Those streamlines starting from the lowest levels all rise and release kinetic energy, while those in the middle levels (low θ_e) descend and also release energy. The streamlines in the top third or so descend but must absorb kinetic energy, since they will always be warmer than their upstream environment. The peculiarity of this configuration is that each point on the outflow profile consists of contributions from two separate inflow levels having the same θ_e, so that the velocity field on the right must be the result of a mixture of contributions from two separate energy conserving processes. Nevertheless, solutions for the displacements seem to be accessible for this case, although not in analytic form. The expected change in the horizontal velocity profile is indicated on the right. The maximum increase is at the top, where the maximum kinetic energy is available from parcels rising from the surface. A smaller increase is expected at the surface, while the maximum decrease is to be found in the lower troposphere, where one streamline arrives horizontally and the other comes from the upper troposphere.

The critical question with the MM theory is whether the atmosphere is really capable of operating in reasonable accord with the assumed energy optimization assumption. MM show some results of a three-dimensional numerical simulation which suggest at least that long-lived convective systems are possible. The simulation was for an isolated storm, however, and was carried out with environmental conditions including a fairly strong and complex mean shear. Thus the results are probably more analogous to those of Schlesinger and of Klemp and Wilhelmson than

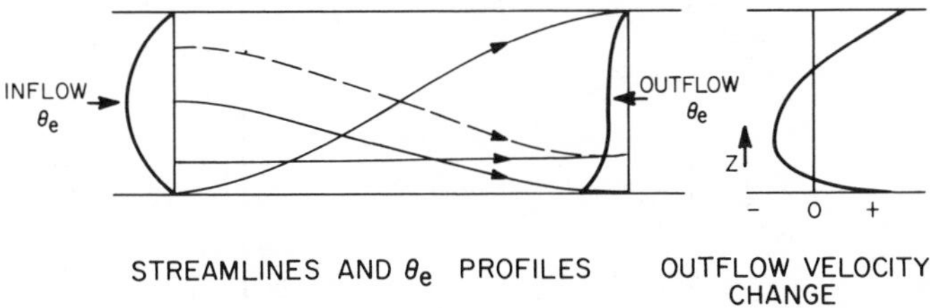

Figure 13 A schematic flow pattern and mean velocity changes for a Moncrieff-Miller type solution with the environment unstable only in the lower half of the region. The assumed profiles of θ_e in the inflow and outflow region are shown at the ends of the trajectories.

to the MM model directly. The evidence from all these model results suggests, however, that shear is the key element which allows development of fairly efficient and durable clouds and makes the MM assumption reasonably valid.

To illustrate the above point, Figure 14 shows a hodograph similar to that of the Klemp-Wilhelmson right moving storm case, with velocity vectors plotted relative to the storm propagation rate. If the storm is regarded as a member of a squall line ensemble with the squall line oriented roughly parallel to the vector shear from 0–4 km, both the upper and lower velocity vectors have inflow components relative to the line, as required by MM. A proposed squall line axis is indicated, with schematic storms superimposed and streamlines showing the principal convergence lines between inflow and low level outflow at the surface. It is evident that in this orientation the storms do not significantly interfere with each other, though it is not obvious that they reinforce each other in any significant way.

Further encouragement for an attempt to coalesce the MM squall line model with the rotating thunderstorm model comes from observational analyses by Newton (1963), Zipser (1969, 1977), and Betts, Grover & Moncrieff (1976). The first used data from a squall line passage through the Thunderstorm Project, the second and third dealt with squall lines

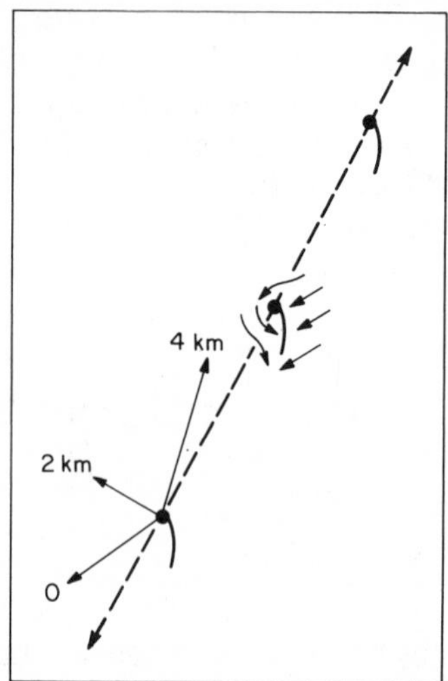

Figure 14 Conjectural squall line produced by aligning a row of Klemp-Wilhelmson right-moving storms approximately parallel to the mean shear vector from 0–4 km. The gust front (convergence line) for each storm is indicated by an arc, and the low level streamlines (relative to storm motion) are shown for one storm.

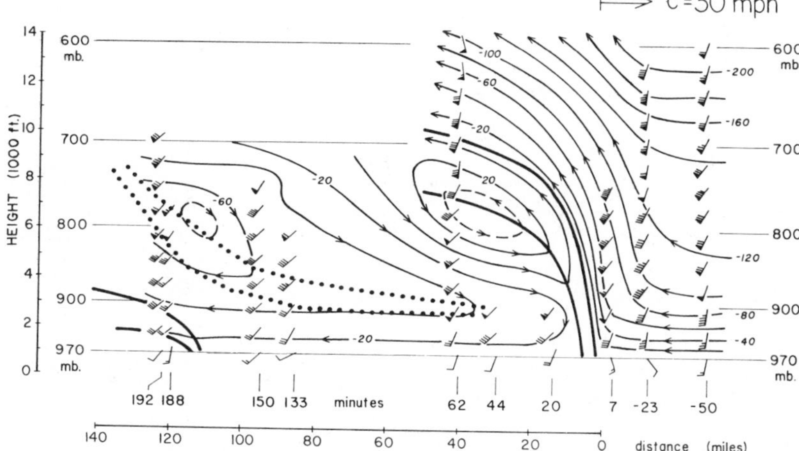

Figure 15 Streamline relative to moving squall front (heavy lines on right) on 29 May 1947. The streamlines are labeled in units of kg s^{-1}. The wind barbs are in ordinary meteorological notation, flying "with the wind," with each straight barb signifying 5 m s^{-1} and a triangular barb 25 m s^{-1}. From Newton (1963).

observed in the tropical Atlantic and Venezuela, respectively. Newton did not consider the possibility of a substantial three-dimensional circulation but, as shown on Figure 15, he found that on the back side of the line the air in the mid-troposphere was either stationary or moving faster than the line, while on the front side it was everywhere moving slower than the line. The principal shear orientation was nearly parallel to the line orientation and the hodograph in the lower levels rotated clockwise. In the tropical case, the stagnation region in the wake of the squall line is more obvious, as shown by Figure 16. The apparent necessity of a three-dimensional circulation is asserted by the identification of a "cross-over zone" where up and downdrafts must alternate along the line. Betts, Grover and Moncrieff obtained similar results. Shears in the tropical environment are typically weaker than in mid-latitudes, but so is the buoyant energy though not the cloud top height or rainfall. Zipser states a belief (personal communication) that substantial shear must exist for tropical squall lines to develop, though they may maintain an identity as they move into low shear regions. No evidence of rotation or splitting of tropical squall line storm cells has been presented in the literature, but Thorpe & Miller (1978) report that such phenomena have been observed.

For many mid-latitude squall lines the high level cirrus outflow runs out ahead of the line, apparently contrary to the MM model. The significant upper level inflow is, however, at the level where air is fed into the

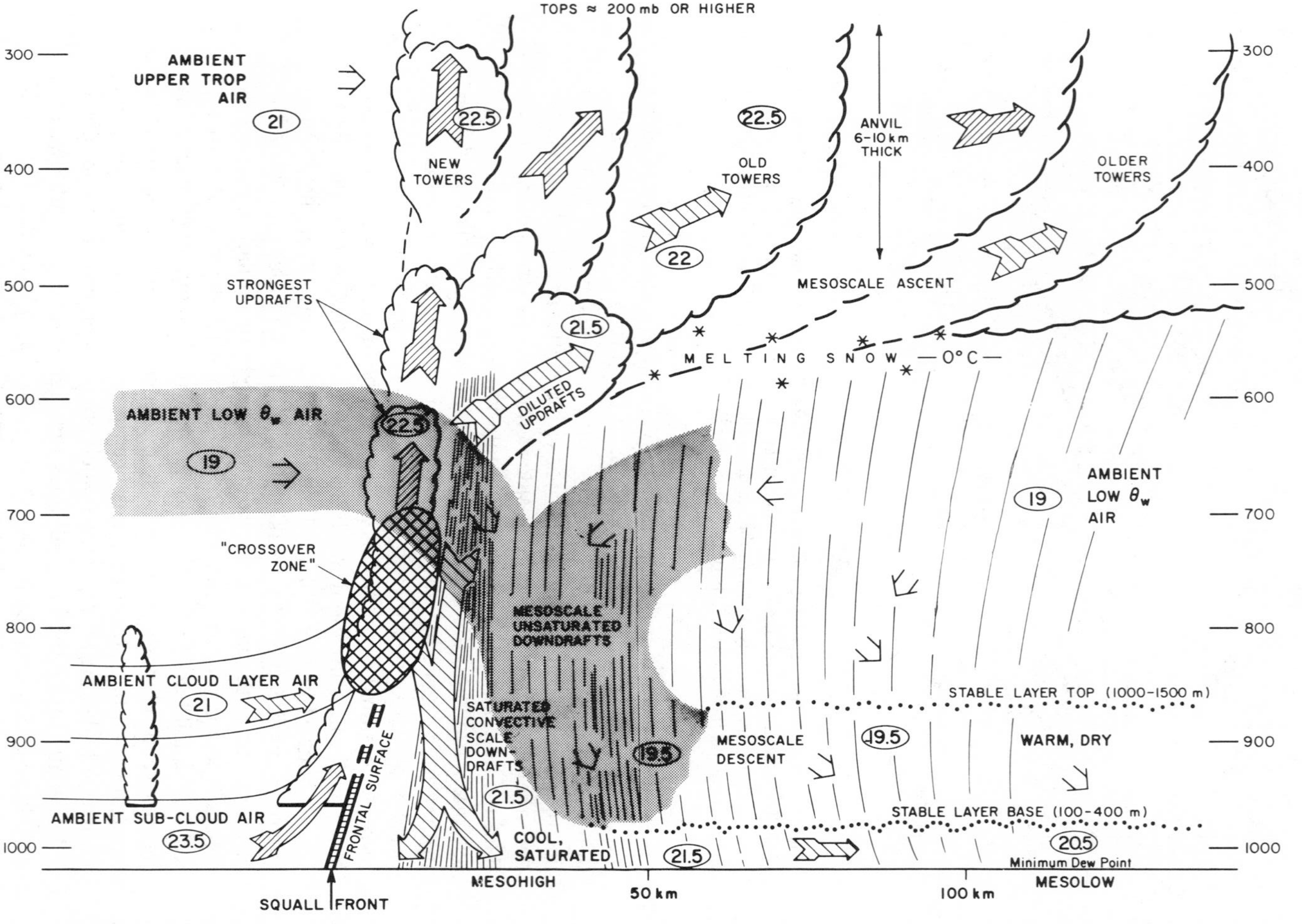

Figure 16 Schematic cross section through a class of squall system. All flow is relative to the squall line which is moving from right to left. Circled numbers are typical values of θ_w in °C. From Zipser (1977).

downdrafts in the mid-troposphere. Klemp and Wilhelmson have found that the environmental velocities at the cirrus outflow level do not significantly affect the propagation rates or energetic structure of their stimulated storms.

4 CONCLUSIONS

In this review I have concentrated on the mechanisms by which thunderstorms and squall lines may propagate relative to their environment, and the consequences of this propagation to their persistence and ability to develop large energy densities and severe storm characteristics. The results of three-dimensional simulations of individual storm evolution in strong shear are regarded as definitive, though still requiring further interpretation and determination of their sensitivity to various modeling simplifications. The Wave-CISK predictions are somewhat of an enigma, remaining to be more fully tested. The Moncrieff-Green model for two-dimensional squall lines seems a doubtful candidate for general application, though it may be partially valid under special circumstances. The Moncrieff-Miller model, which envisions three-dimensional convective elements embedded within a squall line, seems to agree with observational data and to be consistent with predictions of three-dimensional thunderstorm simulations, except in the non-shearing case (which was its original emphasis). The apparent success of the model is regarded as due to the energy optimization assumption, which leads directly to the prediction of relative flow stagnation in the wake. Since most of the buoyant energy release goes into increasing pressure in the wake, the predictions may not be so sensitive to the neglect of dissipation as are those of MG, where all buoyant energy release must go into permanently altering the kinetic energy.

ACKNOWLEDGMENTS

I have benefited from discussions or correspondence with J. Klemp, T. Gal-Chen, M. J. Miller, M. W. Moncrieff, D. J. Raymond, and E. Zipser, although most of these authors have reservations about some of my interpretations of their work. The National Center for Atmospheric Research is sponsored by the National Science Foundation.

Literature Cited

Asai, T. 1964. Cumulus convection in the atmosphere with vertical wind shear—numerical experiment. *J. Meteorol. Soc. Jpn.* 42:245–59

Barnes, S. L. 1968. On the source of thunderstorm rotation. *ESSA Tech. Memo. NSSL 38*. Norman, Okla.

Barnes, S. L. 1970. Some aspects of a severe,

right moving thunderstorm deduced from mesonetwork observations. *J. Atmos. Sci.* 27:634–48

Barnes, S. L., Henderson, J. H., Ketchum, R. S. 1971. Rawinsonde observation and processing techniques of the National Severe Storms Laboratory. *NOAA Tech. Memo. ERL NSSL-53*

Betts, A. K., Grover, R. W., Moncrieff, M. W. 1976. Structure and motion of tropical squall lines over Venezuela. *Q. J. R. Meteorol. Soc.* 102:395–404

Brock, R. V., Govind, P. K. 1977. Portable automated mesonet in operation. *J. Appl. Meteorol.* 16:299–310

Brooks, C. E. P. 1925. The distribution of thunderstorms over the globe. *Geophys. Mem. Lond.* 24:264–74

Browning, K. A., Ludlam, F. H. 1962. Airflow in convective storms. *Q. J. R. Meteorol. Soc.* 88:117–35

Byers, H. R., Braham, R. R. Jr. 1949. *The Thunderstorm.* Washington: Govt. Print. Off. 287 pp.

Charney, J. G., Eliassen, A. 1964. On the growth of the hurricane depression. *J. Atmos. Sci.* 21:68–75

Cotton, W. R. 1975. Theoretical cumulus dynamics. *Rev. Geophys. Space Phys.* 13: 419–48

Cotton, W. R., Jiusto, J. E., Srivastava, R. C. 1975. Cloud physics and radar meteorology. *Rev. Geophys. Space Phys.* 13:753–60

Danielsen, E. F. 1975. The generation and triggering of severe convective storms by large-scale motions. *Proc. SESAME Open Meeting, Sept 4–6, 1974*, ed. D. K. Lilly, pp. 165–85. Natl. Oceanogr. Atmos. Adm., Environ. Res. Lab., Boulder, Colo.

Deardorff, J. W. 1973. The use of subgrid transport equations in a three-dimensional model of atmospheric turbulence. *J. Fluids Eng.* 95:429–38

Deardorff, J. W. 1974. Three-dimensional numerical study of the height and mean structure of a heated planetary boundary layer. *Boundary Layer Meteorol.* 7:81–106

Drake, R. L., Coyle, P. D., Anderson, D. P. 1974. The effects of nonlinear eddy coefficients on rising line thermals. *J. Atmos. Sci.* 31:2046–57

Fankhauser, J. C. 1974. The derivation of consistent fields of wind and geopotential height from mesoscale rawinsonde data. *J. Appl. Meteorol.* 13:637–46

Frisch, A. S., Strauch, R. B. 1976. Doppler radar measurements of turbulent kinetic energy dissipation rates in a northeastern Colorado convective storm. *J. Appl. Meteorol.* 15:1012–17

Gal-Chen, T. 1978. A method for the initialization of the anelastic equations: implications for matching models with observations. *Mon. Weather Rev.* 106: 587–606

Hane, C. E. 1973. The squall line thunderstorm: numerical experimentation. *J. Atmos. Sci.* 30:1672–90

Hane, C. E., Scott, B. C. 1978. Temperature and pressure perturbations derived from detailed air motion information: preliminary testing. *Mon. Weather Rev.* 106: 654–61

Hill, G. E. 1977. Initiation mechanisms and development of cumulus convection. *J. Atmos. Sci.* 34:1934–41

Huschke, R. E., ed. 1970. *Glossary of Meteorology.* Boston, Mass.: Am. Meteorol. Soc. 638 pp.

Kessler, E. 1969. On the distribution and continuity of water substance in atmospheric circulation. *Meteorol. Monogr. 10*, pp. III–84. Am. Meteorol. Soc., Boston, Mass.

Klemp, J. B., Wilhelmson, R. B. 1978a. The simulation of three-dimensional convective storm dynamics. *J. Atmos. Sci.* 35:1070–96

Klemp, J. B., Wilhelmson, R. B. 1978b. Simulations of right and left moving storms produced through storm splitting. *J. Atmos. Sci.* 35:1097–110

Kropfli, R. A., Miller, L. J. 1976. Kinematic structure and flux quantities in a convective storm from dual-Doppler radar observations. *J. Atmos. Sci.* 33:520–29

Leonard, A. 1973. On the energy cascade in large-eddy simulations of turbulent fluid flow. *Adv. Geophys.* 18a:237–48

Lilly, D. K. 1962. On the numerical simulation of buoyant convection. *Tellus XIV* 2:148–72

Lilly, D. K. 1967. The representation of small scale turbulence in numerical simulation experiments. *Proc. IBM Scientific Computing Symp. on Environ. Sci., White Plains, N.Y.*, pp. 195–210

Lilly, D. K. 1970. The numerical simulation of three-dimensional turbulence in two dimensions. Numerical Solution of Field Problems in Continuum Physics. *Volume II SIAM-AMS Proc.*, ed. G. Birkhoff, R. S. Varga, pp. 41–53. Am. Math. Soc., Providence, R. I.

Lilly, D. K. 1975. Severe storms and storm systems: Scientific background, methods, and critical questions. *Pure Appl. Geophys.* 113:713–34

Lindzen, R. S. 1974. Wave-CISK in the tropics. *J. Atmos. Sci.* 31:156–79

Maddox, R. A., Hoxit, L. R., Chappell, C. F., Caracena, F. 1978. Comparison of meteorological aspects of the Big Thompson and Rapid City flash floods. *Mon. Weather Rev.* 106:375–89

Miller, M. J. 1978. The Hampstead storm:

a numerical simulation of a quasi-stationary cumulonimbus system. *Q. J. R. Meteorol. Soc.* 104:413–27

Moncrieff, M. W. 1978. The dynamical structure of two-dimensional steady convection in constant vertical shear. *Q. J. R. Meteorol. Soc.* 104:543–67

Moncrieff, M. W., Green, J. S. A. 1972. The propagation and transfer properties of steady convective overturning in shear. *Q. J. R. Meteorol. Soc.* 98:336–53

Moncrieff, M. W., Miller, M. J. 1976. The dynamics and simulation of tropical squall lines. *Q. J. R. Meteorol. Soc.* 102:373–94

Newton, C. W. 1963. Dynamics of severe convective storms. *Meteorol. Monogr.* 5 (27):33–55

Ogura, Y., Phillips, N. A. 1962. Scale analysis of deep and shallow convection in the atmosphere. *J. Atmos. Sci.* 19:173–79

Ooyama, K. 1964. A dynamical model for the study of tropical cyclone development. *Geofis. Int.* 4:187–98

Orlanski, I., Ross, B. B. 1977. The circulation associated with a cold front. Part I: dry case. *J. Atmos. Sci.* 34:1619–33

Orville, H. C., Sloan, L. J. 1970. A numerical simulation of the life history of a rainstorm. *J. Atmos. Sci.* 27:1148–59

Pastushkov, R. S. 1975. The effects of vertical wind shear on the evolution of convective clouds. *Q. J. R. Meteorol. Soc.* 101:281–91

Pielke, R. A. 1974. A three-dimensional numerical model of the sea breezes over South Florida. *Mon. Weather Rev.* 102: 115–39

Raymond, D. J. 1975. A model for predicting the movement of continuously propagating convective storms. *J. Atmos. Sci.* 32: 1308–17

Raymond, D. J. 1976. Wave-CISK and convective mesosystems. *J. Atmos. Sci.* 33: 2392–98

Ross, B. B., Orlanski, I. 1978. The circulation associated with a cold front. Part II: moist case. *J. Atmos. Sci.* 35:445–65

Schaefer, J. T. 1974. A simulative model of dry line motion. *J. Atmos. Sci.* 31:956–64

Schlesinger, R. E. 1975. A three-dimensional numerical model of an isolated deep convective cloud: Preliminary results. *J. Atmos. Sci.* 32:934–57

Schlesinger, R. E. 1978. A three-dimensional numerical model of an isolated thunderstorm: Part I. Comparative experiments for variable ambient wind shear. *J. Atmos. Sci.* 35:690–713

Simpson, J. 1976. Precipitation augmentation from cumulus clouds and systems: scientific and technological foundations, 1975. *Adv. Geophys.* 19:1–72

Smagorinsky, J. 1963. General circulation experiments with the primitive equations: 1. The basic experiment. *Mon. Weather Rev.* 91:99–164

Soong, S.-T. 1974. Numerical simulation of warm rain development in an axisymmetric cloud model. *J. Atmos. Sci.* 31: 1262–85

Soong, S.-T., Ogura, Y. 1973. A comparison between axi-symmetric and slab-symmetric cumulus cloud models. *J. Atmos. Sci.* 30:879–83

Stevens, D. E., Lindzen, R. S. 1978. Tropical Wave-CISK with a moisture budget and cumulus friction. *J. Atmos. Sci.* 35:940–61

Takeda, T. 1971. Numerical simulation of a precipitating convective cloud: the formation of a "long-lasting" cloud. *J. Atmos. Sci.* 28:350–76

Tepper, M. 1950. A proposed mechanism of squall lines: the pressure jump line. *J. Meteorol.* 8:24–29

Thorpe, A. J., Miller, M. J. 1978. Numerical simulations showing the role of the downdraught in cumulonimbus motion and splitting. To be published in *Q. J. R. Meteorol. Soc.*

Turner, J. S. 1973. *Buoyancy Effects in Fluids*. Cambridge Univ. Press. 367 pp.

Uccelini, L. W. 1975. A case study of apparent gravity wave initiation of severe convective storms. *Mon. Weather Rev.* 103:497–513

Ulanski, S. L., Garstang, M. 1978. The role of surface divergence and vorticity in the life cycle of convective rainfall. Part I. Observations and analysis. *J. Atmos. Sci.* 35:1047–62

Waldteufel, P. 1976. An analysis of weather spectra variance in a tornadic storm. *NOAA Tech. Memo. ERL-NSSL-76*. Natl. Severe Storms Lab., Norman, Oklahoma. 80 pp.

Wilhelmson, R. 1974. The life cycle of a thunderstorm in three dimensions. *J. Atmos. Sci.* 31:1629–51

Wilhelmson, R., Ogura, Y. 1972. The pressure perturbation and the numerical modeling of a cloud. *J. Atmos. Sci.* 29: 1295–1307

Wilhelmson, R. B., Klemp, J. B. 1978. A numerical study of storm splitting that leads to long-lived storms. *J. Atmos. Sci.* 35:1974–86

Zipser, E. J. 1969. The role of organized unsaturated convective downdrafts in the structure and rapid decay of an equatorial disturbance. *J. Appl. Meteorol.* 8:799–814

Zipser, E. J. 1977. Mesoscale and convective scale downdrafts as distinct components of squall-line structure. *Mon. Weather Rev.* 105:1568–89

Ann. Rev. Earth Planet. Sci. 1979. 7 : 163–82

THE ENIGMA OF THE EXTINCTION OF THE DINOSAURS

D. A. Russell

Paleobiology Division, National Museum of Natural Sciences, National Museums of Canada, Ottawa, Canada K1A 0M8

INTRODUCTION

One of the more interesting contributions of paleontology to general knowledge is evidence that giant reptiles were once the dominant life forms on our planet. During the past two centuries some 5000 skeletal fragments of these creatures have been excavated from Mesozoic sediments. Many are objects of public wonder in museum displays around the world.

It is difficult to contemplate the skeletons of dinosaurs or large marine reptiles without also being reminded of their mysterious disappearance. In spite of a voluminous literature on the subject, there is no consensus regarding the stresses which drove the giant reptiles to extinction. It has become apparent, however, that they vanished in a general biotic crisis, and in recent years more attention has been focused on theoretical models which would produce stresses on a global scale.

The problem of the extinction of the dinosaurs is thus clearly an interdisciplinary one. Unfortunately, the competence of a single author is necessarily limited. The review that follows represents the perspective of one student of the great reptiles who is greatly vexed by their apparently sudden disappearance at the end of the Mesozoic Era. The peculiarities of this extinction are summarized and various possible agents of biotic stress are briefly discussed.

The Cretaceous-Paleocene (Mesozoic-Cenozoic) boundary, corresponding to the extinction event during which the dinosaurs disappeared, occurred about 65 million years ago. The Maastrichtian Stage is the final major biostratigraphic division of the Cretaceous and the Danian Stage the first of the Paleocene. For examples of current geochronologic scales, see Van Hinte (1976) and Berggren (1971), noting that the absolute ages given are only approximations of actual values (Jeletzky 1978).

0084-6597/79/0515-0163$01.00

THE TERMINAL CRETACEOUS EXTINCTIONS

Upper Cretaceous sediments, representing the final 30 million years of the Mesozoic Era, have yielded abundant remains of large reptiles, both terrestrial and marine, around the globe. The following generalizations are drawn from this record:

1. There is no evidence of a decline in diversity or of a meridional restriction in the geographic distribution of large reptiles toward the end of Cretaceous time. Their skeletal remains have been recovered from terminal Cretaceous (Maastrichtian) sediments in southern Argentina and in the Queen Elizabeth Islands of arctic Canada (Russell 1975, 1979).
2. The occurrence of large terrestrial reptiles was probably more precisely correlated with high rainfall and abundant plant growth than with swampy environments near the sea (Béland & Russell 1978). Dinosaurs occur in diversity and abundance in sediments deposited over 1500 km from the nearest strand line in central Asia (Russell 1970).
3. Remains of small, unspecialized, and often relatively intelligent dinosaurs (e.g. possessing enlarged endocranial cavities; Hopson 1977) are abundant (although usually disarticulated) in late Upper Cretaceous sediments and show no evidence of a terminal Cretaceous decline. (Remains include members of the Caenagnathidae, Dromaeosauridae, Hypsilophodontidae, Ornithomimidae, Pachycephalosauridae, Protoceratopsidae, Saurornithoididae, and Thescelosauridae; see Béland & Russell 1978, Galton 1974, Gradzinski et al 1977, and references cited therein.)

Vertebrate microfossil evidence, which is interpreted as indicating a gradual replacement of the terminal Cretaceous dinosaurian community by a basal Paleocene mammalian community over a period of approximately 100,000 years, is presented by Van Valen & Sloan (1977) and Archibald & Clemens (1977). The model is not strengthened by the occurrence of articulated dinosaur remains and vertebrate microfossil associations typical of terminal Cretaceous sediments in strata deposited contemporaneously with "transitional" microfossil assemblages (cf Béland & Russell 1978, Russell 1979). It could be argued with equal validity that several microvertebrate communities co-existed in terminal Cretaceous time, and suffered unequally in the subsequent extinctions.

The highest level of articulated dinosaurian remains in terrestrial sequences can be traced laterally into marine sections in several localities. There it coincides with a profound marine extinction event marking the Maastrichtian-Danian boundary. Large marine reptiles are also found throughout the world in Upper Maastrichtian strata, but unreworked remains have never been recovered from sediments of Danian age (Russell

1975, 1979, and references cited therein). The marine extinctions were evidently more severe than those which took place in terrestrial environments (Table 1), probably because they were exacerbated by profound secondary trophic disturbances.

In both continental and marine sections the extinction datum is often accompanied by lithological changes (Russell 1977, Figures 3–8). An abrupt coarsening of clastic particles in continental strata could be interpreted as evidence of more tenuous plant cover, for the basal Paleocene sea was transgressive in the northern interior of the United States (Frye 1969) and evidence of relatively lower-energy fluviatile sedimentation would otherwise be expected. Interestingly, Davies et al (1977) link low abyssal sedimentation rates during Paleocene time with continental aridity and low rates of river discharge into the oceans. They also document sustained low abyssal sedimentation rates later in Cenozoic time, unaccompanied by a comparably severe depression in biotic diversity.

Changes in deep oceanic circulation are postulated to have caused fluctuations in basinal carbonate compensation surfaces (and carbonate deposition-solution patterns) both prior to and following the terminal Cretaceous extinction event. However, a genetic link between the two phenomena is not obvious (Thierstein & Okada 1979). Other lithologic changes may be related to a widespread late Maastrichtian–early Danian withdrawal of epicontinental seas (Naidin 1976; see also section on regressions).

It is clear that the biosphere exerts an influence on rates of terrestrial erosion and biogenic deposition in the oceans. Sedimentological processes probably would respond to the partial collapse in planetary ecosystems, and physical evidence of this response should be carefully sought at the Cretaceous-Paleocene boundary.

The extinction event can often be identified with a bedding plane in sedimentary successions, and it is generally recognized to have taken place within a relatively short period of geologic time. Current estimates of the local duration of the extinctions vary between essentially instantaneous to 10^4–10^5 yrs (Thierstein & Okada 1979, Van Valen & Sloan 1977, Kent 1977). On the basis of the occurrence of the Cretaceous-Paleocene boundary in a zone of negative geomagnetic polarity in northern Italy, and a zone of positive polarity in New Mexico, Kent (1977) suggested that the extinction occurred 500,000 years later in the latter region. However, the boundary also occurs in a zone of positive polarity in Alabama and at several other localities within the western interior of North America (G. P. L. Danis, J. A. Foster, D. A. Russell, manuscript in preparation). It is probable that the time scale of biotic and geomagnetic events during the Cretaceous-Paleocene transition is at the limit of stratigraphic resolution.

Table 1 Number of genera of fossil organisms currently recognized as having lived prior to and following the terminal Cretaceous extinctions (for sources of information see Russell 1977, p. 15)

	A: Before extinctions[a]	B: After extinctions	B/A × 100
Fresh-water organisms			
cartilaginous fishes	4	2	
bony fishes	11	7	
amphibians	9	10	
reptiles	12	16	
	36	35	97
Terrestrial organisms (including fresh-water organisms)			
higher plants	100	90	
snails	16	18	
bivalves	10	7	
cartilaginous fishes	4	2	
bony fishes	11	7	
amphibians	9	10	
reptiles	54	24	
mammals	22	25	
	226	183	81
Floating marine microorganisms			
acritarchs	28	10	
coccoliths	43	4	
dinoflagellates	57	43	
diatoms	10	10	
radiolarians	63	63	
foraminifers	18	3	
ostracods	79	40	
	298	173	58
Bottom-dwelling marine organisms			
calcareous algae	41	35	
sponges	261	81	
foraminifers	95	93	
corals	87	31	
bryozoans	337	204	
brachiopods	28	22	
snails	300	150	
bivalves	399	193	
barnacles	32	24	
malacostracans	69	52	
sea lilies	100	30	
echinoids	190	69	
asteroids	37	28	
	1976	1012	51

Table 1 (*continued*)

	A: Before extinctions[a]	B: After extinctions	B/A × 100
Swimming marine organisms			
ammonites	34	0	
nautiloids	10	7	
belemnites	4	0	
cartilaginous fishes	70	50	
bony fishes	185	39	
reptiles	29	3	
	332	99	30
Totals overall	2868	1502	52

[a] *A* generally signifies the last 20 million years of the Cretaceous; *B* the first 10 million years of the Cenozoic. In the case of marine microfossils and terrestrial organisms the interval is usually much closer to the extinction event, but in several groups of marine macrofossils it is larger. The record of terrestrial organisms is here limited to North America; for marine organisms coverage is global, although existing information is more complete from North America and Europe.

There are very tenuous indications of two nearly synchronous pulses of extinction with different biologic effects. An earlier pulse may have more strongly affected terrestrial plants and some planktonic organisms; a later pulse, other planktonic organisms and terrestrial vertebrates (Boersma & Shackleton 1979, Doerenkamp et al 1976, Russell & Singh 1978). Foraminiferal assemblages of a peculiar but transitional aspect apparently occur within a narrow (30 cm or less) stratigraphic interval at the boundary in some marine sections (Boersma & Shackleton 1979). That the stresses which caused the extinction were of relatively short duration is also suggested by the rapid diversification of surviving organisms in early Cenozoic time (see Béland 1977).

Newell (1962, Figure 6; 1966, Figures 2–5) graphically demonstrated that the terminal Cretaceous extinctions constituted one of the most dramatic extinction events known to have occurred during the last 500 million years. A survey of paleontologic literature suggests a general drop in generic diversity across the Cretaceous-Paleocene boundary of approximately one-half (Table 1, Russell 1977). Because the number of species within genera also declined, Russell (1977) postulated that a total reduction of 75% took place in species diversity during the extinctions. These figures imply a crisis far more severe than that which took place on a scale of 10^4 yrs from the end of the Pleistocene through to the present (Berggren & Van Couvering 1974; Webb 1969, Figure 2; Russell 1976), or that which would follow an essentially instantaneous discharge of one-half of the world's stockpile of nuclear weapons (Nier et al 1975).

SOURCES OF BIOTIC STRESS

There are at least two widely shared points of view among paleontologists concerning the extinction of the dinosaurs: 1. The reason for their demise is unknown, and none of the explanations which have been advanced is supported by a convincing body of evidence (Jepsen 1964). 2. The dinosaurs died as a result of a complex of interrelated and mutually reinforcing stresses which have not been satisfactorily separated and identified in the geologic record (Newell 1975). In the present state of knowledge both points of view can be defended. In this review sources of biologic stress on a planet-wide basis are evaluated with respect to what is known of the Mesozoic-Cenozoic boundary crisis. It is very doubtful that only one of them was operative during the extinction interval and therefore uniquely responsible for the totality of events which took place.

Trophic Effects

Phytoplankton lie at the base of the food chain in the open oceans. Extinction among this group of organisms (cf Tappan & Loeblich 1972, Gartner 1977, Russell 1977) was probably a factor in the simultaneous disappearance of some organisms in higher trophic levels, including marine reptiles. This proposition is consistent with the survival of most of the fresh-water vertebrates, which derived much of their nourishment from terrestrial plant communities (see Table 1). Enormous reserves of oxygen in the atmosphere (Reid 1977, pp. 142–43), coupled with the relatively greater productivity of terrestrial plants (Woodwell et al 1978), render it unlikely that a decline in oxygen production due to marine phytoplankton extinctions would threaten the survival of terrestrial vertebrates. In turn, it would also seem improbable that during the Cretaceous-Paleocene transition continental relief was so low and tectonic activity so reduced (see sections on regressions, volcanicity) that land-derived nutrients dwindled everywhere to the point of causing an abrupt, world-wide extinction of a great number of planktonic organisms in the sea.

Although many taxa of terrestrial plants became extinct (Tschudy 1971), the total change was much less dramatic than in marine phytoplankton (see Jarzen 1977, Hickey 1977, Gartner 1977). In central Alberta Canada, terminal Cretaceous dinosaurs may have briefly survived the palynofloral change elsewhere associated with the Cretaceous-Paleocene boundary within the western interior of North America. The floral change in itself, we can infer, was not severe enough to eliminate large terrestrial vertebrates (Russell & Singh 1978). One could argue whether minor changes in terrestrial floras were as important to dinosaurian consumers as the cessation of dinosaurian browsing, with the extinction of the large herbivores, was to prevailing patterns of ecological succession in plant communities. The demise of several lineages of marsupials, however, could have resulted from a removal of carrion from the food web after the extinction of the giant reptiles (cf. Archibald & Clemens 1977).

The diversification of flowering plants through Upper Cretaceous time (Norris et al 1975) was accompanied by a simultaneous diversification of herbivorous dinosaurs (Ceratopsidae, Hadrosauridae). With no evidence for a late Cretaceous decline in dinosaurian diversity, and in view of the possibility that they survived the palynofloral change associated with the Cretaceous-Paleocene boundary, there is little basis for supposing that dinosaurs became extinct for want of nontoxic plant fodder (Swain 1976).

Marine Regressions

It is widely held that the epicontinental seas of the globe declined greatly in extent from late Cretaceous through Cenozoic time (Hallam 1977), and that this regressive trend constituted a major environmental change which has profoundly affected the biosphere (Newell 1971). Superimposed on the general pattern of withdrawal of seas from the continents, a relatively brief but profound regressive event corresponding to a sea level drop of about 100 m (inferred from Jeletzky 1978, p. 8, Figures 3–7), and coinciding with the terminal Cretaceous–basal Paleocene interval, is often postulated to have produced a global environmental deterioration culminating in the extinctions which brought the age of reptiles to a close (Cooper 1977, Kauffman 1977).

The precision with which global changes in sea level are reflected in regional strand line movements has been questioned by some. Yanshin's (1973) graphs, based on the distribution of epeiric seas over much of the Earth's continental lithosphere, show no pronounced, long-term regressive trend through Upper Cretaceous and Cenozoic time. There would seem to be little parallelism in degree of continental emergence. The graphs relate the position of ancient strand lines to the limits of sialic crust, rather than to existing strand lines (cf Hallam 1977). Bond (1978) has also presented evidence suggesting that widespread post-Miocene epeirogenic uplift ($\sim$100 m) has occurred in Africa, while Europe, Australia, and the Americas remained relatively stable.

Short-term eustatic changes in sea level, such as the terminal Cretaceous withdrawal of the seas, are frequently, if not usually, masked by regional tectonic activity (Bond 1976, Jeletzky 1978). Even geologically brief, glacially induced changes in sea level (85 m, CLIMAP Group 1976) may be locally modified by changing undulations in the equipotential surface of the Earth's gravity field (−70 to +50 m, Mörner 1977). Jeletzky (1978, p. 36), affirming the widespread evidence for a terminal Cretaceous (terminal Maastrichtian) regression, noted that it has been preserved not so much due to its unusual amplitude, but because it occurred during a tectonically quiescent period. Furthermore, according to Reyment & Mörner (1977), the terminal Maastrichtian–basal Danian interval was characterized by advancing seas in the lands bordering the South Atlantic.

The terminal Maastrichtian regression was preceded by a comparably well-defined regressive event during early Campanian time (Jeletzky 1978, p. 16) which was not accompanied by extinctions of exceptional severity. Adams et al (1977) postulate that a global eustatic regression of 40–70 m took place during the late Miocene Messinian crisis, coinciding with hiatuses in carbonate sequences and a general cooling. However, no major discontinuity occurred in Miocene-Pliocene faunal development (Berggren & Van Couvering 1974, Webb 1977). There is little compelling evidence that changes in sea level comparable to those observed during the Maastrichtian-Danian regression would produce large-scale extinctions. Indeed, the availability of large areas of continental shelf to colonization by highly productive terrestrial plant formations should not in itself have constituted an event detrimental to the survival of dinosaurs.

Temperature

Oxygen isotope temperature studies of marine microfossils near the Cretaceous equator in the Pacific (Douglas & Savin 1974, Savin et al 1975) and of belemnites in Cretaceous mid-latitudes in Europe (Spaeth et al 1971) and New Zealand (Stevens & Clayton 1971) indicate a thermal minimum during mid–early Cretaceous time, followed by an early–Upper Cretaceous (Cenomanian-Turonian) maximum and a return to levels comparable to those prevailing during the early Cretaceous minimum near the close of the Cretaceous (Lower Maastrichtian). Their data suggest that tropical surface ocean temperatures during the Lower Maastrichtian were similar to those existing in low latitudes today (28°C), with minimum high-latitude temperatures of approximately 10°C.

Isotopic evidence is in harmony with a late Maastrichtian warming of epicontinental seas in many regions (Stevens & Clayton 1971, and references cited therein). This trend is supported by paleontologic and chemical evidence in northern Europe (Voigt 1964, Jorgensen 1975) and microfaunal data in the western interior of North America (North & Caldwell

1970), although isotopic paleotemperature determinations remain constant in the latter region (Tourtelot & Rye 1969). Saito & van Donk (1974) report a 5°C cooling of oceanic waters during Upper Maastrichtian time in the South Atlantic, and isotopic data suggestive of temperature fluctuations at the Cretaceous-Paleocene boundary in the southern oceans was cited by Margolis et al (1975). Boersma & Shackleton (1979), on the basis of isotopic ratios preserved in foraminiferal tests, found evidence of a thermal rise across the Cretaceous-Paleocene boundary in Atlantic waters, where basal Paleocene temperatures were significantly warmer than during the latest Cretaceous.

It has been postulated (Hughes 1976, McLean 1978) that massive extinctions were precipitated when an Upper Cretaceous warming trend ("radmax") passed a biologically critical threshold at the end of the period. Large reptiles are known to be particularly vulnerable to heat stress (Cloudsley-Thompson & Butt 1977). However, isotopic temperature data suggest that the "radmax" occurred during early Upper Cretaceous time, although a very brief, terminal Cretaceous heating event would not appear to be ruled out.

Changes in terrestrial floras across the Cretaceous-Paleocene boundary have not been viewed as indicative of warmer climates. Instead, a slight but general cooling is usually postulated, accompanied by increased seasonal fluctuation in temperature and rainfall (e.g. Tschudy 1971, Hickey 1977). As evidence of this Krassilov (1975) noted that smooth-margined leaves, typical today of perennially warm, moist environments, are more abundant in terminal Cretaceous strata of mid-latitude sections than in supradjacent sediments of basal Paleocene age, where they tend to be replaced by serrated-margined leaves. Jarzen (1977) cited a shift from animal- to wind-pollination across the boundary as an effect of increased seasonality. The possibility that the decline in the proportion of plant taxa with entire-margined leaves could be the result of local increases in seasonal aridity (cf Gentry 1969) should be examined.

Information sufficient to document consistent global paleotemperature patterns through Maastrichtian-Danian time is, accordingly, unavailable at present. Accepting the uniform validity of existing interpretations, it could be concluded that thermal minima occurred at different times in different regions, and did not necessarily coincide with the terminal Cretaceous extinction datum. Thermal declines of this general magnitude had occurred earlier in Mesozoic time, but were not associated with comparably severe extinctions.

With this in mind, it may be useful to examine the possible consequences of a brief ($\sim$100,000 yrs) but marked thermal decline (10–5°C) in oceanic bottom waters near the Eocene-Oligocene boundary about 38 million years ago (Kennett & Shackleton 1976). In contrast to the rapid and massive extinctions of planktonic organisms at the Cretaceous-Paleocene boundary, it was the benthonic forms that suffered an abrupt decline in diversity. Differences between Eocene and Oligocene marine macrofaunal assemblages are gradational in character and strongly influenced by lithofacies changes (Berggren 1971). In northern mid-latitudes, meridional displacements in the distributions of arborescent plant taxa have been interpreted as resulting from rapidly declining mean annual temperatures (Wolfe 1971, 1972). The floral changes in turn are generally seen as responsible for concomitantly high turnover rates, which were unaccompanied by significant diversity changes, in terrestrial mammalian communities (Webb 1977, and references cited therein). The pattern and severity of the terminal Cretaceous extinctions were evidently not duplicated.

Volcanism

The thermal effect of volcanic dust veils has often been invoked as a reason for a proposed temperature decline at the end of the Cretaceous, although at present such veils have not been known to depress mean annual temperatures by as much as 1°C (Miles & Gildersleeves 1978). Late Cretaceous levels of volcanicity were presumably inferior to those existing during the Pleistocene (Vogt 1972) and mean temperatures departed more greatly from the biologically critical 0°C threshold. It has been suggested that the injection of biologically deleterious substances into the biosphere through volcanism may have brought about the terminal Cretaceous extinctions (e.g. Vogt 1972, Feldman 1977, pp. 137–38). Sediments could be examined for the appropriate chemical residues, but the gradual, episodic nature of terrestrial volcanicity is not in harmony with the sudden and sharp biostratigraphic datum marking the Maastrichtian-Danian boundary.

The occurrence of pathologic and relatively thin shells in dinosaur eggs from the higher levels of the Cretaceous in southern France has been provisionally linked either to the passage of unusually severe atmospheric cold fronts (Thaler 1965) or to the appearance of toxic substances in the environment (Erben 1972). R. Dughi and F. Sirugue (personal communication, 1975), however, note that pathological eggs occur throughout late Cretaceous exposures in the region, and that eggs with thin shells belong to different dinosaurian taxa than do those with thick shells. Thus, according to them, the apparent thinning of shells toward the upper limit of the Cretaceous implies faunal change, not necessarily increasing environmental degradation.

Collisions of Comets and Large Meteorites

Urey (1973) cited a series of biological stresses which might follow the collision of a comet with the Earth's oceans. Among these are a sudden and possibly lethal heating of the atmosphere and oceanic surface waters. However, Urey also noted that a tektite shower, which occurred in the Australasian region about seven hundred thousand years ago and suggested an impact energy consonant with his calculations, produced only a minor stratigraphic discontinuity. Evidence of a cometary collision could be sought in tektite fields and concentrations of certain rare-earth elements (cf Vostrukhov 1977) at the Cretaceous-Paleocene boundary.

The collision of a large meteorite has also been considered as a possible agent of mass extinction. The Popigai crater is the largest Phanerozoic meteor crater known. It measures 95–100 km in diameter and impacted sometime between 29 and 47 million years ago (Masaitis 1975). Although the blanket of ejecta and the air blast may have been lethal over a region two to three times the diameter of the crater (M.R. Dence, personal communication, 1978), no permanent floral or faunal change has yet been linked to the event. The distribution of ancient craters on cratonic areas suggests that a large asteroid (about 4 km in diameter) should fall into the oceans on an average of one per ten million to 100 million years, creating an impact wave which could easily be 5 km high (M. R. Dence, personal communication, 1978). The effects of such an event, as suggested by McLaren (1970, p. 812),

> would certainly spread to all shelf and epicontinental areas connected with the open ocean. The turbulence of the tidal wave and accompanying wind, followed by the gigantic runoff from the land would induce a turbid environment far longer than could be survived by bottom dwelling filter-feeders, and their larvae. The hypothesis of meteoric impact in the ocean explains equally the nonextinction of many other forms of both marine and terrestrial life. There must have been many regions where plankton and nekton could have survived, (as well as) epiplanktonic and rocky-bottom attached organisms . . . (and) land plants and animals . . .

The pattern does not appear to conform to the survivorship pattern following the Cretaceous-Paleocene crisis. However, the biological consequences of collisions with large extraterrestrial objects deserve careful study, as several such events have probably occurred since the beginning of the Phanerozoic.

Ultraviolet Radiation

It is well established that the Earth's ozone layer shields the biosphere from potentially lethal solar ultraviolet radiation (Margulis et al 1976). The ozone layer is vulnerable to partial or complete destruction as a result of the coincidence of a colossal solar flare (10,000 times more intense than any so far observed) with a reversal of the terrestrial magnetosphere, or of the arrival of a blast of high energy radiation from a nearby supernova. The biosphere could thereby be subjected to essentially unattenuated solar ultraviolet radiation, short term atmospheric instability, a reduction in the transmission of visible light through nitrogen dioxide absorption, and a global cooling as much as 0.5°C. These effects would endure for only about 10 years after the arrival of an initial pulse of radiation (Reid 1977, Roy 1977, Tucker 1977, K-TEC Group 1977, pp. 144–48, Hunt 1978). It is unlikely that the cooling effect would have stimulated the formation of continental glaciers during late Cretaceous–early Paleocene time, for high latitude temperatures exceeded 0°C much more than at present (Savin et al 1975).

The reaction of planetary ecosystems to increased levels of UV exposure is very poorly understood. It is possible that plastids of green plants are more vulnerable to UV radiation than the host cell or animal cells (Margulis 1968). Transient reduction in atmospheric ozone in the wake of nuclear detonations on an unprecedented scale would apparently produce a greater incidence of UV damage among terrestrial and aquatic plants than among terrestrial vertebrates (Nier et al 1975, pp. 90–93, 141–51). Electromagnetic (but not particulate) radiation from an extraterrestrial source capable of damaging the ozone layer would not reach ground level (Tucker 1977, Clark et al 1977). This mechanism could account for some of the disturbances in terrestrial and aquatic plant communities which approximately coincided with the extinction of the dinosaurs.

Ionizing Radiation

Tucker (1977, p. 121) estimated that a nearby supernova could generate a rise in background radiation at the base of the atmosphere to levels of the order of 300 roentgens per year. These levels would be sustained for about ten years, coinciding with the passage of the relativistic blast wave.

Accepting the general accuracy of this estimate, the ensuing disturbances in marine ecosystems would probably be less than on land (Nier et al 1975, pp. 126–41). The most vulnerable terrestrial plant formation, the forest, would for the most part respond by little more than a slight inhibition of growth (Woodwell 1963, 1967). However, some conifers (Taxaceae, Pinaceae, Cupressaceae; cf Dugle & El-Lakany 1971) would be more severely affected, for their greater chromosome volume reduces resistance to ionizing radiation. Seeds, including those of conifers, are significantly less vulnerable than parent plants.

Large vertebrates (*Homo sapiens*) are about as radiosensitive as are the above conifers (Woodwell 1963). Here, however, the relationship between chromosomal volume and radiosensitivity tends to be obscured by the greater physiological complexity of the organisms (cf Vorontsov 1958, United Nations 1958, Bond et al 1965). An inverse correlation between reproductive rates and increasing size nevertheless militates against resistance to chronic ionizing radiation (G. M. Woodwell, personal communication, 1978). This would imply that dinosaurs would have been more vulnerable than, for example, lizards. No terrestrial vertebrate exceeding about 25 kg in body weight is known to have survived the terminal Cretaceous extinctions.

On the balance, then, it is probable that exposure rates quoted by Tucker would be seriously detrimental to some terrestrial organisms (Woodwell, personal communication, 1978), whether resulting from a supernova or some other cause. Further, it is not unreasonable to expect that large terrestrial vertebrates would be more adversely affected than terrestrial plants. The dinosaurian extinctions would seem to be in conformity with an increase in background radiation to levels suggested above.

It has been proposed that the presence of high concentrations of radioactive minerals in fossil bones of late Cretaceous age was the result of a contemporaneous increase in background radiation (Salop 1977). There is in fact no correlation between bone radioactivity and stratigraphic proximity to the Cretaceous-Paloecene boundary (Jaworowski & Pensko 1967, Bell et al 1976).

Supernovae

For about two weeks these colossal stellar explosions radiate as much energy as about ten billion suns, or as much as all the stars in a galaxy combined. Terrestrial environments could be affected at distances of up to one hundred light years. Several lines of astrophysical evidence suggest that these explosions occur at a rate of about one every fifty years per galaxy, showing some association with spiral arms. This translates into an average of one supernova within 50 light years of the Earth every 70 million years (Tucker 1977, Clark et al 1977).

At a distance of 50 light years the stresses exerted on the biosphere by a nearby supernova would not be instantaneous. An initial burst of electromagnetic radiation would be injected into the atmosphere over an interval of a few hours. It would be followed in three to thirty years by an intense flux of cosmic rays at the leading edge of a relativistic blast wave. High levels of background radiation ($\sim$300 r) would persist for about 10 years. Some 3,000 to 30,000 years later, the Earth would be immersed in the expanding shell of the supernova remnant, implying radiation doses of about ten roentgen per year for thousands of years (Tucker 1977).

The effects of the initial electromagnetic wave and the following pulse of cosmic radiation might be correlated respectively with disturbances in plant communities and the extermination of the dinosaurs. Presumably the damage sustained by the ozone layer in the wake of the cosmic ray blast (Tucker 1977) would be somewhat less effective, biologically, due to the previous elimination of UV sensitive forms when the layer was disrupted by the initial wave of electromagnetic radiation. There is as yet no compelling astrophysical evidence for a 65 million year old supernova remnant near the solar system, but it is rather unlikely that one would have remained over such a great length of time (Feldman 1977).

Periodic Galactic Events

The solar system revolves around the center of the galaxy once in approximately 200 million years. During the galactic orbit the Earth passes through the equatorial plane of the galaxy several times, where it is exposed to increased levels of cosmic radiation for hundreds of thousands of years (Hatfield & Camp 1970). Terrestrial temperatures are postulated to increase during the perigalactic passage and decline near the apogalactic point, producing long-term climatic cycles (Meyerhoff 1973, Steiner & Grillmair 1973). Encounters with major spiral arms of the galaxy, occurring at 100 million year intervals, may expose the Earth to short-term environmental stresses. The brief immersion of the solar system within dense clouds of interstellar matter could affect terrestrial climates (Talbot et al 1976). Clark et al (1977) note that in the course of the ten million year transit through a spiral arm the Earth is exposed to a high incidence of nearby supernovae, which result from the explosion of massive stars created in the compression lane of the arm.

Thus the galactic environment may produce both episodic and acute stresses within the biosphere. The effects of the galactic milieu in Earth history will continue to be an interesting area of research

CONCLUSIONS

The stratigraphic record suggests that well over half of the species of organisms inhabiting the planet became extinct in a general biotic crisis at the end of the Mesozoic Era. Although secondary trophic effects are difficult to isolate, it would appear that large terrestrial vertebrates, such as dinosaurs, pterosaurs, and giant crocodilians, and planktonic marine organisms were exterminated as a direct result of the primary causes of the extinction. The length of the crisis is currently estimated to be within the range of one-half million to a few tens of years. It may have occurred in two, nearly simultaneous pulses, one affecting terrestrial and marine plants more severely and another affecting terrestrial vertebrates more severely. Extinctions on the scale of those which took place during the Cretaceous-Paleocene transition are rare events in Earth history.

A coincidence of terrestrial stresses, such as marine regressions, global temperature changes, and increased volcanicity, has been invoked to account for the terminal Cretaceous extinctions (Ager 1976). This is unlikely, for the time scale of such changes is substantially greater than that of the extinctions themselves. During the last two million years the biosphere has been subjected to relatively rapid alternations of glacial and interglacial climatic regimes. The period was characterized by changes of strand line and temperature, as well as levels of volcanism, which were at least as great as those which occurred during the late Cretaceous and early Paleocene. Rates of turnover among large terrestrial mammals were high in North America, but diversity remained approximately constant until postglacial time (Webb 1969). There were no massive extinctions of planktonic marine organisms (cf Berggren & Van Couvering 1974).

The biosphere is probably well adapted to coping with the type of biological stresses that occur relatively frequently through geologic time. Episodic changes of the kind described in the preceding paragraph may be considered to belong to this category, as are fluctuations in solar radiation and stresses generated by the infall of comets and asteroids. The solar system, on the whole, has provided an environment favorable for the long-term survival and development of life on our planet.

The solar system is in turn protected by its isolation within the vastness of the galaxy. In human experience there is little visible evidence of a cosmos beyond the sun other than the strangely disquieting beauty of a starlit night. Through the centuries supernovae have occasionally appeared in the sky, even during the day (Stephenson & Clark 1976). They represent the only objects known in our region of the galaxy capable of damaging the terrestrial biosphere even at great distances.

The major stresses resulting from a nearby supernova are postulated to occur in two pulses, each of about a decade's duration, separated by a somewhat longer, relatively stress free interval. A chronic increase in background radiation may endure for several thousands of years. Because the occurrence of a nearby supernova is such a rare event and its effects would be so sudden and so massive, there is little likelihood of the biosphere developing an adaptive response in the way that it is able to adjust to episodic environmental changes.

Unfortunately, when organisms are confronted with any environmental pressure which they can neither tolerate nor evade they can only respond by dying. Even communities of organisms respond to a variety of pressures (cooling, aridity, chemical pollution, radiation) by returning to a formation resembling an earlier stage in the development of a successional climax (Whittaker & Woodwell 1971). Accordingly, the fact of the disappearance of an organism or a community of organisms may not reveal the stresses which brought about its demise. The spectrum of metabolic systems represented by organisms that became extinct at the end of the Cretaceous greatly exceeds the extremes postulated in the current debate on dinosaurian metabolism (Marx 1978). Whether large dinosaurs were warm- or cold-blooded is of uncertain relevance to the problem of their extinction.

There is, consequently, ample justification for geophysical studies of the Cretaceous-Paleocene boundary to more precisely identify agents of biological stresses. For example, increases in cosmic ray intensities from nearby supernovae may produce concentrations of cosmogenic isotopes in the sedimentary record (Higdon & Lingenfelter 1973). Evidence of meteorite impact can be sought through trace element analyses (see Christensen et al 1973), and cometary collisions may be revealed through the identification of tektite fields. Several series of sedimentary samples spanning the Cretaceous-Paleocene boundary in North America, Europe, and New Zealand are housed in the Paleobiology Division, National Museum of Natural Sciences (Ottawa) and are available for such studies. The time period over which the extinctions took place should also be more precisely delimited through detailed magnetostratigraphic studies (see Foster 1977).

Of all the hypotheses which have been advanced to account for the extinction of the dinosaurs, only one involving the explosion of a nearby supernova currently seems plausible to this writer. However, firm astrophysical evidence of a supernova in the vicinity of the solar system about 65 million years ago is as yet unavailable. The probable effects of the collision of supernova ejecta with the atmosphere, and the physical reflection of these effects within sedimentary processes, must be carefully evaluated. Only then can the stratigraphic record be examined meaningfully for physical data which will uphold or negate the hypothesis that a nearby supernova coincided with the biologic catastrophe which occurred at the end of Cretaceous time. In the present state of uncertainty it would not be too surprising if a fundamental deficiency were found in the supernova model. If such should prove to be the case, the disappearance of the dinosaurs would remain an outstanding mystery of the geologic record.

Acknowledgments

It is a pleasure to acknowledge the expert counsel and constructive comments received in the course of preparation of this manuscript from

Drs. P. Béland, D. M. Jarzen, and K. A. Pirozynski, all of the Paleobiology Division, National Museum of Natural Sciences, Ottawa; and Drs. A. Boersma, Lamont Geological Observatory; M. R. Dence, Earth Physics Branch, Department of Energy, Mines and Resources (Canada); P. A. Feldman, Hertzberg Institute of Astrophysics; J. A. Jeletzky, Geological Survey of Canada; L. S. Russell, Royal Ontario Museum; H. R. Thierstein, Scripps Institution of Oceanography; W. H. Tucker, Bonsall, California; and G. M. Woodwell, Ecosystems Center, Woods Hole Marine Biological Laboratory. Unanimous endorsement of all of the opinions expressed in the text is not to be presumed, although the author suspects few of his colleagues would disagree that the extinction of the dinosaurs represents an interval in Earth history of more than usual significance.

Literature Cited

Adams, C. G., Benson, R. H., Kidd, R. B., Ryan, W. B. F., Wright, R. C. 1977. The Messinian salinity crisis and evidence of late Miocene eustatic changes in the world ocean. *Nature* 269:383–86

Ager, D. V. 1976. The nature of the fossil record. *Proc. Geol. Assoc.* 87:131–60

Archibald, J. D., Clemens, W. A. 1977. The beginnings of the age of mammals. *J. Paleontol.* 51 (*Suppl.* 2): 1 (Abstr.)

Béland, P. 1977. Models for the collapse of terrestrial communities of large vertebrates. *Syllogeus* 12:25–37

Béland, P., Russell, D. 1978. Paleoecology of Dinosaur Provincial Park (Cretaceous), Alberta, interpreted from the distribution of articulated vertebrate remains. *Can. J. Earth Sci.* 15:1012–24

Bell, R. T., Steacy, H. R., Zimmerman, J. B. 1976. Uranium-bearing bone occurrences. *Geol. Surv. Can. Pap.* 76-1A:339–40

Berggren, W. A. 1971. Tertiary boundaries and correlation. In *Micropalaeontology of the Oceans*, ed. B. M. Funnell, W. R. Riedell, pp. 693–809. Cambridge Univ. Press. 828 pp.

Berggren, W. A., Van Couvering, J. A. 1974. The late Neogene. *Palaeogeogr. Palaeoclimatol. Palaeoecol.* 16:1–216

Boersma, A., Shackleton, N. 1979. Campanian through Paleocene paleotemperatures and carbon isotope gradients related to faunal evolution and the Cretaceous/Tertiary boundary in the Atlantic Ocean. In *The New Uniformitarianism*, ed. W. A. Berggren, J. A. Van Couvering. Princeton Univ. Press. In press

Bond, G. 1976. Evidence for continental subsidence in North America during the Late Cretaceous global submergence. *Geology* 4:557–60

Bond, G. 1978. Evidence for late Tertiary uplift of Africa relative to North America, South America, Australia and Europe. *J. Geol.* 86:47–65

Bond, V. P., Fliedner, T. M., Archambeau, J. O. 1965. *Mammalian Radiation Lethality*. New York: Academic Press. 340 pp.

Christensen, L., Fregerslev, A., Simonsen, A., Thiede, J. 1973. Sedimentology and depositional environment of lower Danian Fish Clay from Stevns Klint, Denmark. *Geol. Soc. Den. Bull.* 22:193–212

Clark, D. H., McCrea, W. H., Stephenson, F. R. 1977. Frequency of nearby supernovae and climate and biological catastrophes. *Nature* 265:318–19

CLIMAP Group. 1976. The surface of the Ice-Age Earth. *Science* 191:1131–37

Cloudsley-Thompson, J. L., Butt, D. K. 1977. Thermal balance in the tortoise and its relevance to dinosaur extinction. *Br. J. Herpetol.* 5:641–47

Cooper, M. R. 1977. Eustacy during the Cretaceous: its implications and importance. *Palaeogeogr. Palaeoclimatol. Palaeoecol.* 22:1–60

Davies, T. A., Hay, W. W., Southam, J. R., Worsley, T. R. 1977. Estimates of Cenozoic oceanic sedimentation rates. *Science* 197: 53–55

Doerenkamp, A., Jardine, S., Moreau, P. 1976. Cretaceous and Tertiary palynomorph assemblages from Banks Island and adjacent areas (N.W.T.). *Bull. Can. Pet. Geol.* 24:372–417

Douglas, R. G., Savin, S. M. 1974. Marine temperatures during the Cretaceous. *Geol. Soc. Amer. Abstr. Programs* 6:714 (Abstr.)

Dugle, J. R., El-Lakany, M. H. 1971. Check list revisions of the plants of the Whiteshell area, Manitoba including a summary of their published radiosensitivities. *Whiteshell Nuclear Res. Establ. Publ.* AECL-3678, Pinawa, Manitoba. 31 pp.

Erben, H. K. 1972. Ultrastrukturen und Dicke der Wand pathologischer Eischalen. *Akad. Wiss. Lit. Mainz Abh. Math-Naturwiss. Kl.* 6: 193–216

Feldman, P. A. 1977. Astronomical evidence bearing on the supernova hypothesis for the mass extinctions at the end of the Cretaceous. *Syllogeus* 12: 125–35

Foster, J. H. 1977. The magnetic field and the Cretaceous-Tertiary extinctions. *Syllogeus* 12: 63–74

Frye, C. I. 1969. Stratigraphy of the Hell Creek Formation in North Dakota. *N.D. Geol. Surv. Bull.* 54. 65 pp.

Galton, P. M. 1974. Notes on *Thescelosaurus*, a conservative ornithopod dinosaur from the Upper Cretaceous of North America, with comments on ornithopod classification. *J. Paleontol.* 48: 1048–67

Gartner, S. 1977. Nannofossils and biostratigraphy: an overview. *Earth Sci. Rev.* 13: 227–50

Gentry, A. H. 1969. A comparison of some leaf characteristics of tropical dry forest and tropical wet forest in Costa Rica. *Turrialba* 19: 419–28

Gradzinski, R., Kielan-Jaworowska, Z., Maryanska, T. 1977. Upper Cretaceous Djadochta, Barun Goyot and Nemegt formations of Mongolia, including remarks on recent subdivisions. *Acta Geol. Pol.* 27: 281–318

Hallam, A. 1977. Secular changes in marine inundation of USSR and North America through the Phanerozoic. *Nature* 269: 769–72

Hatfield, C. B., Camp, M. J. 1970. Mass extinctions correlated with periodic galactic events. *Geol. Soc. Am. Bull.* 81: 911–14

Hickey, L. J. 1977. Changes in angiosperm flora across the Cretaceous-Paleocene boundary. *J. Paleontol.* 51 (*Suppl.* 2): 14–15 (Abstr.)

Higdon, J. C., Lingenfelter, R. E. 1973. Sea sediments, cosmic rays, and pulsars. *Nature* 246: 403–5

Hopson, J. A. 1977. Relative brain size and behavior in archosaurian reptiles. *Ann. Rev. Ecol. Syst.* 8: 429–48

Hughes, N. F. 1976. *Palaeobiology of angiosperm origins.* Cambridge Univ. Press. 242 pp.

Hunt, G. E. 1978. Possible climatic and biological impact of nearby supernovae. *Nature* 271: 430–31

Jarzen, D. M. 1977. Angiosperm pollen as indicators of Cretaceous-Tertiary environments. *Syllogeus* 12: 39–49

Jaworowski, Z., Pensko, J. 1967. Unusually radioactive bones from Mongolia. *Nature* 214: 161–63

Jeletzky, J. A. 1978. Causes of Cretaceous oscillations of sea level in western and arctic Canada and some general geotectonic implications. *Geol. Surv. Can. Pap. 77-18.* 44 pp.

Jepsen, G. L. 1964. Riddles of the terrible lizards. *Am. Sci.* 52: 227–46

Jorgensen, N. O. 1975. Mg/Sr distribution and diagenesis of Maastrichtian white chalk and Danian bryozoan limestone from Jylland, Denmark. *Geol. Soc. Den. Bull.* 24: 299–325

Kauffman, E. G. 1977. Cretaceous extinction and collapse of marine trophic structure. *J. Paleontol.* 51 (*Suppl.* 2): 16 (Abstr.)

Kennett, J. P., Shackleton, N. J. 1976. Oxygen isotope evidence for the development of the psychrosphere 38 myr ago. *Nature* 260: 513–15

Kent, D. V. 1977. An estimate of the duration of the faunal change at the Cretaceous-Tertiary boundary. *Geology* 5: 769–71

Krassilov, V. A. 1975. Climatic changes in eastern Asia as indicated by fossil floras II. *Palaeogeogr. Palaeoclimatol. Palaeoecol.* 17: 157–72

K-TEC Group. 1977. Cretaceous-Tertiary extinctions and possible terrestrial and extraterrestrial causes. *Syllogeus* 12. 162 pp.

Margolis, S. V., Kroopnick, P. M., Goodney, D. E. 1975. Cenozoic and late Cretaceous high latitude paleotemperatures as determined by oxygen and carbon isotopes of calcareous fossils. *Geol. Soc. Am. Abstr. Programs* 7: 1189 (Abstr.)

Margulis, L. 1968. Visible light: mutagen or killer? *Science* 160: 1255–56

Margulis, L., Walker, J. C. G., Rambler, M. 1976. Reassessment of roles of oxygen and ultraviolet light in Precambrian evolution. *Nature* 264: 620–24

Marx, J. L. 1978. Warm-blooded dinosaurs: evidence pro and con. *Science* 199: 1424–26

Masaitis, V. L. 1975. Astroblemes in the Soviet Union. *Sov. Geol.* 11: 52–64 (In Russian)

McLaren, D. J. 1970. Time, life and boundaries. *J. Paleontol.* 44: 801–15

McLean, D. M. 1978. A terminal Mesozoic "Greenhouse": lessons from the past. *Science* 101: 401–6

Meyerhoff, A. A. 1973. Mass extinctions, world climate changes and galactic motions: possible interractions. *Mem. Can. Pet. Geol.* 2: 745–58

Miles, M. K., Gildersleeves, P. B. 1978. Volcanic dust and changes in northern hemisphere temperature. *Nature* 271: 735–36

Mörner, N. A. 1977. Eustacy and instability of the geoid configuration. *Geol. Foeren. Stockholm. Foerh.* 99:369–76

Naidin, D. P. 1976. Epeirogeny and eustacy. *Vestn. Mosk. Univ. Geol.* 1976(2):3–16 (In Russian)

Newell, N. D. 1962. Paleontological gaps and geochronology. *J. Paleontol.* 36: 592–610

Newell, N. D. 1966. Problems of geochronology. *Proc. Acad. Natl. Sci. Philadelphia* 118:63–89

Newell, N. D. 1971. An outline history of tropical organic reefs. *Am. Mus. Novit.* 2465. 37 pp.

Newell, N. D. 1975. Complexities of extinction. *Nat. Hist.* 84(2):8

Nier, A. O. C., Friend, J. P., Hempelmann, L. H., McCormick, J. F., Parker, D. R., Reiter, E. R., Seymour, A. H., Waggoner, P. E. 1975. *Long-term worldwide effects of multiple nuclear-weapons detonations.* Washington, D.C.: Natl. Acad. Sci. 212 pp.

Norris, G., Jarzen, D. M., Awai-Thorne, B. V. 1975. Evolution of the Cretaceous palynoflora in western Canada. *Geol. Assoc. Can. Spec. Pap.* 13:333–64

North, B. R., Caldwell, W. G. E. 1970. Foraminifera from the late Cretaceous Bearpaw Formation in the South Saskatchewan River Valley. *Sask. Res. Coun. Geol. Div. Rep.* 9. 117 pp.

Reid, G. C. 1977. Stratospheric aeronomy and the Cretaceous-Tertiary extinctions. *Syllogeus* 12:75–88

Reyment, R. A., Mörner, N. A. 1977. Cretaceous transgressions and regressions exemplified by the South Atlantic. *Paleontol. Soc. Jpn. Spec. Pap.* 21:248–61

Roy, J. R. 1977. Variations of the luminosity of the sun and "super" solar flares: possible causes of extinctions. *Syllogeus* 12:89–110

Russell, D. A. 1970. The dinosaurs of central Asia. *Can. Geogr. J.* 81:208–15

Russell, D. A. 1975. Reptilian diversity and the Cretaceous-Tertiary boundary in North America. *Geol. Assoc. Can. Spec. Pap.* 13:119–36

Russell, D. A. 1976. Mass extinctions of dinosaurs and mammals. *Nat. Can.* 5: 18–24

Russell, D. A. 1977. The biotic crisis at the end of the Cretaceous Period. *Syllogeus* 12:11–23

Russell, D. A. 1979. Terminal Cretaceous extinctions of large reptiles. In *The New Uniformitarianism*, ed. W. A. Berggren, J. A. Van Couvering. Princeton Univ. Press. In press

Russell, D. A., Singh, C. 1978. The Cretaceous-Tertiary boundary in south-central Alberta—a reappraisal based on dinosaurian and microfloral extinctions. *Can. J. Earth Sci.* 15:284–92

Saito, T., Van Donk, J. 1974. Oxygen and carbon isotope measurements of late Cretaceous and early Tertiary foraminifera. *Micropaleontology* 20:152–77

Salop, L. I. 1977. Relationship of glaciations and rapid changes in organic life to events in outer space. *Int. Geol. Rev.* 19: 1271–91

Savin, S. M., Douglas, R. G., Stehli, F. G. 1975. Tertiary marine paleotemperatures. *Bull. Geol. Soc. Am.* 86:1499–1510

Spaeth, C., Hoefs, J., Vetter, V. 1971. Some aspects of isotopic composition of belemnites and related paleotemperatures. *Bull. Geol. Soc. Am.* 82:3139–50

Steiner, J., Grillmair, E. 1973. Possible galactic causes for periodic and episodic glaciations. *Bull. Geol. Soc. Am.* 84:1003–18

Stephenson, F. R., Clark, D. H. 1976. Historical supernovas. *Sci. Am.* 234(6):100–7

Stevens, G. R., Clayton, R. N. 1971. Oxygen isotopic studies on Jurassic and Cretaceous belemnites from New Zealand and their biogeographic significance. *N.Z.J. Geol. Geophys.* 14:829–97

Swain, T. 1976. Angiosperm-reptile coevolution. *Linn. Soc. Symp. Ser.* 3:107–22

Talbot, R. J., Butler, D. M., Newman, M. J. 1976. Climatic effects during passage of the solar system through interstellar clouds. *Nature* 262:561–63

Tappan, H., Loeblich, A. R. 1972. Fluctuating rates of protistan evolution, diversification and extinction. *Int. Geol. Congr. 24. Sess.* 7:205–13

Thaler, L. 1965. Les oeufs des dinosaures du Midi de la France livrent le secret de leur extinction. *Sci. Prog. Nat.* Feb. 1965:41–48

Thierstein, H. R., Okada, H. 1979. The Cretaceous/Tertiary boundary event in the North Atlantic. *Initial Rep. Deep Sea Drilling Proj.* Washington, D.C.: Govt. Printing Office. In press

Tourtelot, H. A., Rye, R. O. 1969. Distribution of oxygen and carbon isotopes in fossils of late Cretaceous age, western interior region of North America. *Bull. Geol. Soc. Am.* 80:1903–22

Tschudy, R. H. 1971. Palynology of the Cretaceous-Tertiary boundary in the northern Rocky Mountain and Mississippi Embayment regions. *Geol. Soc. Am. Spec. Pap.* 127:65–111

Tucker, W. H. 1977. The effect of a nearby supernova explosion on the Cretaceous-Tertiary environment. *Syllogeus* 12:111–24

United Nations. 1958. Report of the Scientific Committee on the Effects of Atomic Radiation. *U.N. Official Rec.* 13 *Sess. Suppl.* 17. 228 pp.

Urey, H. C. 1973. Cometary collisions and geological periods. *Nature* 242: 32–33

Van Hinte, J. E. 1976. A Cretaceous time scale. *Bull. Am. Assoc. Petrol. Geol.* 60: 498–516

Van Valen, L., Sloan, R. E. 1977. Ecology and the extinction of the dinosaurs. *Evol. Theory* 2: 37–64

Vogt, P. R. 1972. Evidence for global synchronism in mantle plume convection, and possible significance for geology. *Nature* 240: 338–42

Voigt, E. 1964. Zur Temperatur-Kurve der oberen Kreide in Europa. *Geol. Rundschau* 54: 270–317

Vorontsov, N. N. 1958. The significance of the study of chromosomal constitution for mammalian taxonomy. *Byull. Mosk. Ova. Ispyt. Prir. Otd. Biol.* 63: 5–36 (In Russian)

Vostrukhov, Y. 1977. Recent results from the U.S.S.R. (on the Tunguska explosion in 1908). *New Sci.* 75(1064): 347

Webb, S. D. 1969. Extinction-origination equilibria in late Cenozoic land mammals of North America. *Evolution* 23: 688–702

Webb, S. D. 1977. A history of savanna vertebrates in the new world. Part I. North America. *Ann. Rev. Ecol. Syst.* 8: 355–80

Whittaker, R. H., Woodwell, G. M. 1971. Evolution of natural communities. *Oreg. State Univ. Biol. Colloq.* 31: 137–59

Wolfe, J. A. 1971. Tertiary climatic fluctuations and methods of analysis of Tertiary floras. *Palaeogeogr. Palaeoclimatol. Palaeoecol.* 9: 27–57

Wolfe, J. A. 1972. An interpretation of Alaskan Tertiary floras. In *Floristics and Paleofloristics of Asia and eastern North America*, ed. A. Graham, pp. 201–33. Amsterdam: Elsevier. 278 pp.

Woodwell, G. M. 1963. The ecological effects of radiation. *Sci. Am.* 208(6): 40–49

Woodwell, G. M. 1967. Radiation and the patterns of nature. *Science* 156: 461–70

Woodwell, G. M., Whittaker, R. H., Reiners, W. A., Likens, G. E., Delwiche, C. C., Botkin, D. B. 1978. The biota and the world carbon budget. *Science* 199: 141–46

Yanshin, A. L. 1973. On the recognition of global transgressions and regressions. *Byull. Mosk. Ova. Ispyt. Prir. Otd. Geol.* 48: 9–45 (In Russian)

Ann. Rev. Earth Planet. Sci. 7: 183–98

CLAY MINERAL CATALYSIS AND PETROLEUM GENERATION

William D. Johns
Department of Geology, University of Missouri, Columbia, Missouri 65201

INTRODUCTION

In recent years clay mineralogists and sedimentologists have studied in increasing detail the mineralogical and chemical changes experienced by pelitic sediments during burial diagenesis and early metamorphism (Dunoyer de Segonzac 1970, Heling 1974, Hower et al 1976, Bronson & Hower 1976). Similarly, organic geochemists and petroleum geologists have made considerable progress in understanding the chemical changes involved in the diagenetic transformation of dispersed organic matter in pelitic sediments into liquid and gaseous hydrocarbons (Welte 1965, Henderson et al 1968, Tissot, Oudin & Pelet 1972, Albrecht, Vandenbroucke & Mandengue 1976).

For the most part, only recently have clay mineralogists and organic geochemists coordinated their interests in diagenesis and organic maturation. A number of investigators (Powers 1967, Burst 1969, Perry & Hower 1970) have focused attention on the late-stage dehydration which accompanies smectite to illite transformation during burial diagenesis, suggesting that this water release creates a flushing action responsible for the migration of petroleum hydrocarbons from the source rock. Other investigators, following this suggestion, have attempted to use clay mineral indicators to evaluate degree of oil generating potential of shales, by relating degree of diagenesis with vitrinite reflectance values (Foscolos, Powell & Gunther 1976, Foscolos & Powell 1979).

The direct involvement of clays as natural catalysts in promoting, at low temperatures, the transformation of dispersed organic matter into liquid and gaseous hydrocarbons is the subject of increasing interest. This review focuses on the direct participation of clay mineral catalysis in petroleum formation and maturation, and begins with a brief summary of some of the basic organic geochemistry of petroleum formation.

0084-6597/79/0515-0183$01.00

PRECURSORS OF PETROLEUM

A portion of petroleum seems to have been derived directly from biologically produced lipid precursors such as fats and waxes or their fatty acid derivatives (Cooper 1962, Cooper & Bray 1963, Kvenvolden 1970) and from higher molecular weight hydrocarbon precursors (Bray & Evans 1961, Bendaraitis, Brown & Hepner 1963, Hedberg 1968).

It is now generally believed, however, that the immediate precursor for the bulk of liquid and gaseous hydrocarbons is kerogen, which is the predominant organic constituent of sediments and sedimentary rocks. Kerogen is a complex high molecular weight, organic and acid insoluble organic polymer. Its precursors are also largely plant lipids which polymerize very early in the diagenetic history of the sediment. Most of the fatty acids and their derivatives which survive in organic debris are quickly incorporated into the kerogen polymer.

Most kerogen occurs finely dispersed within the clay matrix of claystones. Most claystones are in fact siltstones containing 15–30% clay. Such a sediment, containing, for example, only 1% organic matter, will have a kerogen content of up to 10% within the clay matrix. Therefore, it is possible and often convenient to isolate and concentrate the kerogen component of a sediment by first separating the clay fraction by mechanical means.

That the organic constituents of sediments and the petroleum produced therefrom were ultimately derived from the lipid fractions of marine and, to a lesser extent, terrestrial plants is documented by studies of the stable carbon isotopes, C^{12} and C^{13}, in the organic matter of sediments and petroleum (Silverman & Epstein 1958, Silverman 1963, 1967). Sedimentary carbonaceous matter consistently has low C^{13}/C^{12} ratios in comparison to higher ratios for sedimentary carbonate. The C^{12} enrichment of petroleum and kerogen is even greater than that of the plant materials from which they originate, but is similar to that of the lipid fractions of these plants.

The C^{12} enrichment occurs initially during photosynthesis (Park & Epstein 1960), so that marine plants are up to 10 per mil lower than atmospheric CO_2, and 20 per mil lower than marine carbonate in C^{13}/C^{12} ratio. Further fractionation and C^{12} enrichment takes place in sediments as selective preservation of the lipid fraction of detrital organic matter results in a further 4–8 per mil lowering of the C^{13}/C^{12} ratio (Silverman 1963).

The organic kerogen and uncombined lipid constituents, fairly highly concentrated within the clay mineral matrix of a sediment, undergo diagenesis along the geothermal gradient during subsequent burial. Among the products of organic diagenesis are liquid petroleum hydrocarbons and natural gas.

ORGANIC DIAGENESIS AND PETROLEUM MATURATION

It has been established that as a sediment and its dispersed organic matter are exposed to increasing temperature during burial in a subsiding basin the organic matter is altered continuously. Organic soluble constituents (bitumen) are produced from the included kerogen and both of these change in character with depth and increasing temperature. Earlier studies of comparative organic geochemistry, where alkane fractions of the bitumen extracts of biological materials, soils, young and ancient sediments, and petroleum were compared and contrasted, revealed systematic changes in paraffin hydrocarbons attributable to maturation with time and temperature (Stevens et al 1956, Bray & Evans 1961, Cooper 1962, Cooper & Bray 1963). These studies suggested that petroleum hydrocarbons were generated from fatty acids, which would involve acid decarboxylation and subsequent hydrocarbon cracking.

Studies of bitumen formation and distribution with depth in individual profiles in a number of Tertiary basins have revealed significant increases in bitumen content below certain depths (Philippi 1965, Louis & Tissot 1967, Connan 1974, Albrecht, Vandenbroucke & Mandengue 1976). This increase corresponds to beginning of abundant oil generation.

In addition studies of kerogen alteration, either in natural profiles (LaPlante 1974, LeTran et al 1974) or by laboratory pyrolysis experiments (Peters, Ishiwatari & Kaplan 1977, Ishiwatari, Rohrback & Kaplan 1977), show that the major onset of bitumen and liquid hydrocarbon formation coincides with the loss of carboxyl and hydroxyl groups and alkane side chains from the bitumen. There seems to be little doubt that liquid hydrocarbon and bitumen production resulted from the degradation of kerogen.

The initial bitumen and liquid hydrocarbon production is soon followed at depth by a rapid decline in amount as these constituents are fragmented or cracked to give volatile hydrocarbons, chiefly methane. Thus, as Pusey (1973) has observed, the concept of a liquid petroleum window applies to nearly all the major fields of the world, the lower and upper temperature limits being about 65 and 150°C, respectively.

The remarkable fact that emerges from those studies is the low temperature (65°) at which significant hydrocarbon generation proceeds, since the major chemical reactions involved in kerogen degradation are decarboxylation and carbon bond cleavage. Carbon bond cleavage energies have been calculated to vary between about 60–90 kcal/mole (Abelson 1963, Benson 1965). Simple kinetic calculations show that at a temperature of 100°C or less thermal cracking cannot be significant in the time framework with which we are concerned, Tertiary or younger (McNab, Smith & Betts 1952). These kinds of considerations have led numerous investigators to conclude that petroleum-forming reactions must be assisted by catalysis. Connan (1974), applying kinetic considerations to the threshold temperatures of intense oil generation, as observed in twelve sedimentary basins worldwide, estimated an activation energy for hydrocarbon production in nature of only about 12–14 kcal/mole. Tissot (1969) earlier had calculated similar activation energies for Paris Basin hydrocarbon formation. It is now with a high degree of certainty that one can evoke catalysis as a contributory factor in petroleum genesis in Tertiary sediments, at least at temperatures below about 100°C.

CLAY MINERAL CATALYSIS

The idea of clay mineral catalysis is not new. Grim (1947) suggested that the clay minerals in shales concentrate organic constituents by absorption, and later act as catalysts in petroleum conversions. Similarly Brooks (1948, 1952) and Louis (1966) showed that various clay minerals could catalyze changes in petroleum. The widespread use of smectite, or expanding clays, as versatile catalysts focused attention on this clay mineral as a likely candidate for catalyzing organic reactions in natural sediments. It is common practice to acid-activate smectitic clay to produce commercial acid catalysts. Cracking processes could then be explained by carbonium ion mechanisms, which require protons to initiate carbonium ion formation (Thomas 1949, Greensfelder, Voge & Good 1949). Since catalytic activity of clays was thought of in terms of Brønsted acidity, it was difficult to see how natural smectitic clays, deposited in an alkaline marine environment, could be naturally acid-activated.

In the intervening years a great deal of research has been carried out on the formation and stability of clay-organic complexes. This work has been summarized in an excellent review by Theng (1974). Also laboratory experiments, using clay catalysts to convert organic compounds to petroleum-like hydrocarbon mixtures, have continued. Galwey (1970, 1972) studied reactions of alcohols and hydrocarbons on montmorillonite surfaces; Jurg & Eisma (1964) and Shimoyama & Johns (1971) have studied the conversion of fatty acids to petroleum hydrocarbons; Henderson et al (1968), using smectite (bentonite) catalysts, studied the cracking and disproportionation of n-octacosane ($C_{28}H_{58}$).

Consideration of the many types of organic reactions catalyzed by clays has revealed those special properties of smectite clays which give them their unique catalytic properties. By virtue of their small particle size, high surface area, and ability to complex with polar organic substances they achieve, even in the natural state, intimate mixing with organic substances and produce highly dispersed clay-organic systems.

Two general types of catalysis associated with two types of catalytic sites have been revealed in smectitic clays. In particular, the work of Solomon and his colleagues (Solomon & Rosser 1965, Solomon 1968, Solomon, Swift & Murphy 1971, Theng 1971) has shown that Lewis acid sites, characterized by octahedrally coordinated Al^{+3} and/or Fe^{+3} ions, exposed at the edges of smectite crystallites, may act as electron acceptors promoting certain kinds of organic reactions. One result is the formation of free radicals which can react further, undergo rearrangement, and lead to C–C bond cleavage or cracking.

A second source of catalytic activity unique to smectite clays is related to their special structural attributes. Negatively charged silicate layers have their charges neutralized by exchangeable counter cations located between contiguous silicate layers. In addition these cations are hydrated, as are the silicate surface layers with which they are associated. Cation exchange smectites can act as strong Brønsted acids, the acidity stemming from dissociation of the interlayer water molecules because of polarization by the exchangeable cations (Mortland 1968, Fripiat 1970, Fripiat & Cruz-Cumplido 1974, Fripiat et al 1965). We can express the situation by the following equilibrium, keeping in mind that this is the condition existing locally between opposing parallel silicate sheet surfaces:

$$n[M(H_2O)_x]^{z+} \overset{K_1}{\rightleftharpoons} n[M(H_2O)_{x-1}(OH)]^{(z-1)+} + nH^+.$$

M is the exchangeable cation of valency z; x represents the water directly coordinated to the cation; n represents the number of exchangeable cations, and thus nz is the interlayer charge on the adjacent silicate surfaces. K_1 is the ionization (dissociation) constant for the interlayer system. The proton-donating ability of the clay has been found to depend on the following factors: 1. the polarizing effect of exchangeable cation (increase with increasing charge and decreasing size); 2. the number of exchangeable cations, n; and 3. the source of isomorphous replacement which gives rise to interlayer charge (greater acidity associated with tetrahedral than octahedral substitution).

In addition the experiments of Mortland & Raman (1968) show that the Brønsted acidity is influenced by the degree of hydration of the clay. When the clays are expanded to more than a single monomolecular layer of water, the polarization effect of the cation is dispersed among a large number of water molecules and the pH approaches that of water in aqueous solution (Theng 1974). At low water contents the polarization acts on the few, residual water molecules causing a significant increase in their dissociation and proton-donating ability. The dissociation constant may be of the order of 10^6 times higher than for normal water (Fripiat 1970). The proton-donating character of these clays can thus promote carbonium ion induced organic reactions.

Looking ahead to effects of diagenesis on clay mineral catalysts, we note that factors which affect dehydration, tetrahedral substitution, and increased interlayer charge of smectite clays will promote their ability to enter into carbonium ion organic reaction.

KINETIC STUDIES AND REACTION MECHANISMS

Decarboxylation

Inasmuch as a first stage in petroleum hydrocarbon formation appears to involve decarboxylation of lipid materials, either free or combined in kerogen, Jurg & Eisma (1964) studied catalytic degradation of docosanoic acid using an uncharacterized "bentonite" clay under conditions which produced no hydrocarbons in the absence of a catalyst. Shimoyama & Johns (1971) also studied the catalytic conversion of n-docosanoic acid

Table 1 Amounts of n-paraffins (μg) produced after heating n-docosanoic acid (50 mg) with Ca-montmorillonite at 200 and 250°C for various times

	50 h		150 h		300 h		500 h	
Carbon no.	200°	250°	200°	250°	200°	250°	200°	250°
16	—[a]	0.3	5.0	0.6	10.5	10.8	11.3	11.1
17	1.9	0.6	4.5	1.2	7.2	8.0	7.4	7.9
18	1.4	2.8	6.8	1.9	6.9	9.5	6.1	9.4
19	4.4	7.0	9.0	6.7	9.5	10.0	9.0	8.5
20	8.5	15.7	14.0	14.4	13.3	15.7	13.4	15.4
21	41.2	107.8	60.9	111.1	59.9	37.0	51.3	43.7
22	—	0.8	—	—	—	0.7	—	0.6
23	—	—	0.6	0.2	—	0.6	—	0.5
24	—	0.5	2.3	0.1	—	2.2	0.4	1.8
25	—	0.1	0.7	—	—	0.2	—	0.1
26	—	0.5	1.5	0.2	—	1.3	—	0.9
27	—	0.3	1.0	—	—	0.8	0.1	0.2
28	—	0.2	1.1	0.6	—	1.7	0.2	1.0
Total	49.1	136.6	107.4	137.1	107.3	98.5	99.2	101.1

[a] —, not detected.

($C_{21}H_{43}COOH$) to petroleum-like paraffins, using Ca-montmorillonite containing 12.4% absorbed water. Analyses of the paraffin reaction products are tabulated in Table 1 for a series of experiments. It is evident from these data that the major reaction involves decarboxylation of the C_{22} acid to C_{21} n-paraffin, although much lesser amounts of paraffin in both the C_{16}–C_{20} and C_{22}–C_{28} ranges are formed. The most interesting aspect of this study is the two-stage reaction, that is, initial decarboxylation of fatty acid and subsequent but delayed decomposition or partial cracking of C_{21} paraffin to form C_{19} and shorter chain paraffins. These data were originally interpreted as showing that the hydrocarbon cracking reaction had a substantially greater activation energy than decarboxylation. Subsequent studies show this is not true when smectite catalysts are used. This is now interpreted to indicate that the cracking reaction awaits activation of the catalyst. That is, two different types of catalytic sites are responsible for the two kinds of reactions. From these experiments activation energies were estimated for fatty acid decarboxylation and alkane cracking (Johns & Shimoyama 1973). When utilized in a kinetic diagenetic model, which takes into account geothermal gradient and geosynclinal subsidence rate (Gulf Coast, USA), the depth range over which decarboxylation and later hydrocarbon cracking take place could be predicted.

Shimoyama & Johns (1975) and Almon & Johns (1976, 1977) studied this reaction further, focusing on the nature of the catalytic sites and the reaction mechanism. By determining rate constants at fixed temperature, using a variety of smectites and related layer silicates of variable composition, it was established that the catalytic activity was associated with Al^{+3} and/or Fe^{+3} ions in octahedral coordination, exposed at the edges of crystallites. This suggested the role of electron acceptor sites in promoting a free radical mechanism as discussed in the previous section. A radical mechanism is likewise suggested by the low ratio of branched to normal isomers (0.114, Table 2) and the formation of alkanes of chain length greater than the starting fatty acid. Radical formation was likewise confirmed by noting the increase in rate constant by adding a radical initiator to the reactants and a decrease by adding a free radical inhibitor. The free radical intermediate could be detected, but not identified in electron paramagnetic resonance spectra. The activation energy of decarboxylation was determined at 31.1 kcal/mole, using Ca-montmorillonite as catalyst. The decarboxylation reaction mechanism is represented schematically in Figure 1. The electron acceptor at the crystallite edge extracts an electron from the acylate ion, forming an acylate radical, which through rearrangement and loss of CO_2 gives rise to an alkyl radical. This is one of the likely decarboxylation reactions suggested by Cooper & Bray (1963), who did not invoke the intervention of a catalyst.

Table 2 Amounts (ug) of alkanes produced by the decarboxylation of behenic acid in the presence of Ca-montmorillonite and excess water at 250°C at various times

Carbon number	25 h normal	25 h branched	150 h normal	150 h branched
14	—	—	6.1	0.5
15	—	—	7.5	1.3
16	0.1	—	9.9	0.7
17	0.4	—	8.8	0.5
18	1.2	0.2	9.1	1.4
19	3.4	0.5	17.2	2.0
20	9.8	0.9	36.1	5.7
21	68.2	7.9	208.3	20.9
22	0.4	—	2.5	0.7
23	0.1	0.1	0.6	0.3
24	0.3	—	1.7	0.4
25	—	—	0.3	0.1
26	—	—	0.9	0.3
27	—	—	0.9	0.2
28	—	—	0.5	0.2
Total	83.9	9.6	310.4	35.2
Branching Ratio	0.114		0.113	

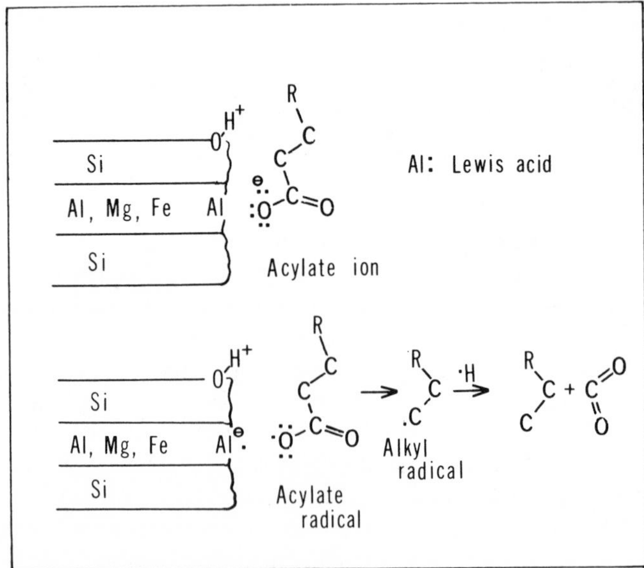

Figure 1 Schematic model of edge-site catalysis of fatty acid functions by radical intermediate formation (hydrogen atoms excluded).

Hydrocarbon Cracking

Since C–C bond cleavage and hydrocarbon cracking are important contributory reactions in petroleum hydrocarbon formation, it is not surprising that they have been the subject of considerable study. Greensfelder, Voge & Good (1949) and Thomas (1949) proposed the carbonium ion mechanism for hydrocarbon cracking, relating this to the proton-donating ability of aluminosilicate catalysts, including clays. Henderson et al (1968) studied and compared the products produced by the pyrolysis of n-octacosane ($C_{28}H_{58}$), thermally and in the presence of bentonite catalyst. They noted that the bentonite brought about a ten-fold increase in alteration compared to the thermally induced cracking. The most interesting result was the very high concentration of branched and cyclic alkanes within the alkane fraction from the treatment involving bentonite. This is certainly suggestive of at least some involvement of carbonium ion intermediates.

The author and his students have been studying the kinetics of clay-catalyzed hydrocarbon cracking. Banet (1976), using Ca-montmorillonite, investigated the kinetics of heneicosane (n-$C_{21}H_{44}$) and stearic acid (n-$C_{17}H_{36}$COOH) mixtures over the temperature range 200–350°C. The activation energy for cracking pure hydrocarbon was only 6.1 kcal/mole, and for the hydrocarbon-stearic acid mixture, 6.4 kcal/mole.

Similarly Jones (1977) carried out a kinetic study of heneicosane using as catalysts the clay fraction (<2.0 μ) extracted from Gulf Coast well cuttings from 4,200 and 15,000 feet. The shallow sample contained primarily a smectite-illite mixed layer clay in the proportions 54:46. The deeper clay was a three component mixed layer clay consisting of smectite, illite, and "vermiculite" (high charge expanding component), in the proportions 13:58:29 (Baumann 1975). Baumann's (1975) chemical characterization of these mixed layer clays showed the following contrasts:

	Interlayer charge	Tetrahedral charge
Shallow clay	0.94	0.49
Deep clay	1.27	0.75

The activation energies determined for hydrocarbon cracking, for shallow and deep clay, were 7.1 and 5.5 kcal/mole respectively. These three experimentally determined activation energies are less but of the same order of magnitude as those determined in the natural setting by Tissot (1969) and Connan (1974). There is no doubt that smectite clays catalyze carbon bond cleavage reactions.

The study of carbon isotope effects during hydrocarbon cracking reactions can also throw light on the cracking mechanism. It has been known for some time and confirmed by Sackett, Nakaparksin & Dalrymple

(1970) that methane produced by hydrocarbon cracking is markedly enriched in C^{12} relative to the parent hydrocarbon. Substitution of C^{13} for C^{12} lowers the vibrational frequency of a C–C bond. Thus less energy is required to break a C^{12}–C^{12} than a C^{13}–C^{12} bond. C^{12}–C^{12} bonds are selectively broken and terminal CH_3 groups give rise to C^{12} enriched methane. In a recent study Sackett (1978) compared isotopically the methane produced from n-octadecane thermally, with and without a Na-montmorillonite catalyst. Thermal cracking alone produced C^{12} enriched methane as expected. Remarkably, the experiments carried out in the presence of smectite clay produced methane showing little, if any, carbon isotope fractionation. This is attributed to a carbonium ion mechanism induced by the clay (and its absorbed water).

That smectite clays are effective hydrocarbon cracking catalysts is well established. It appears likely that they act as acid catalysts through dissociation of interlayer water, thus promoting carbonium ion reactions. It should be possible to confirm this by further kinetic studies using a series of either natural or synthetic smectite catalysts with different exchangeable cations, different interlayer charges, and varying amounts of octahedral and tetrahedral substitution.

NATURAL DIAGENESIS

In an effort to evaluate the role of clay mineral catalysis in natural processes it is necessary to look at the rocks themselves. To this end the author has been attempting to correlate diagenetic changes in clay mineralogy with organic geochemistry of coexisting organic matter in Tertiary profiles from the Gulf Coast, USA. If the clay mineral catalyst changes during its diagenesis, then so must its catalytic properties. If these diagenetic changes can be appropriately characterized it should be possible to predict how the catalytic properties also change.

Although work is still in progress, certain data are available which enable one to make correlations between likely catalytic properties and distribution of certain organic constituents. Working with well cuttings provided by Mobil Research and Development Corporation from a Gulf Coast well sampled over a depth of 2,000–16,000 ft (600–5000 m) and bottoming in Miocene sediments, concurrent studies of clay mineralogy and organic and carbon isotope geochemistry are being carried out. Some of these data are shown diagrammatically in Figure 2.

Throughout the entire profile the sediments are clayey siltstones. The mineralogy is essentially monotonous over the entire depth range, consisting predominantly of quartz, detrital mica, minor K-feldspar, and a clay fraction composed of kaolinite and an illite-montmorillonite mixed layer mineral.

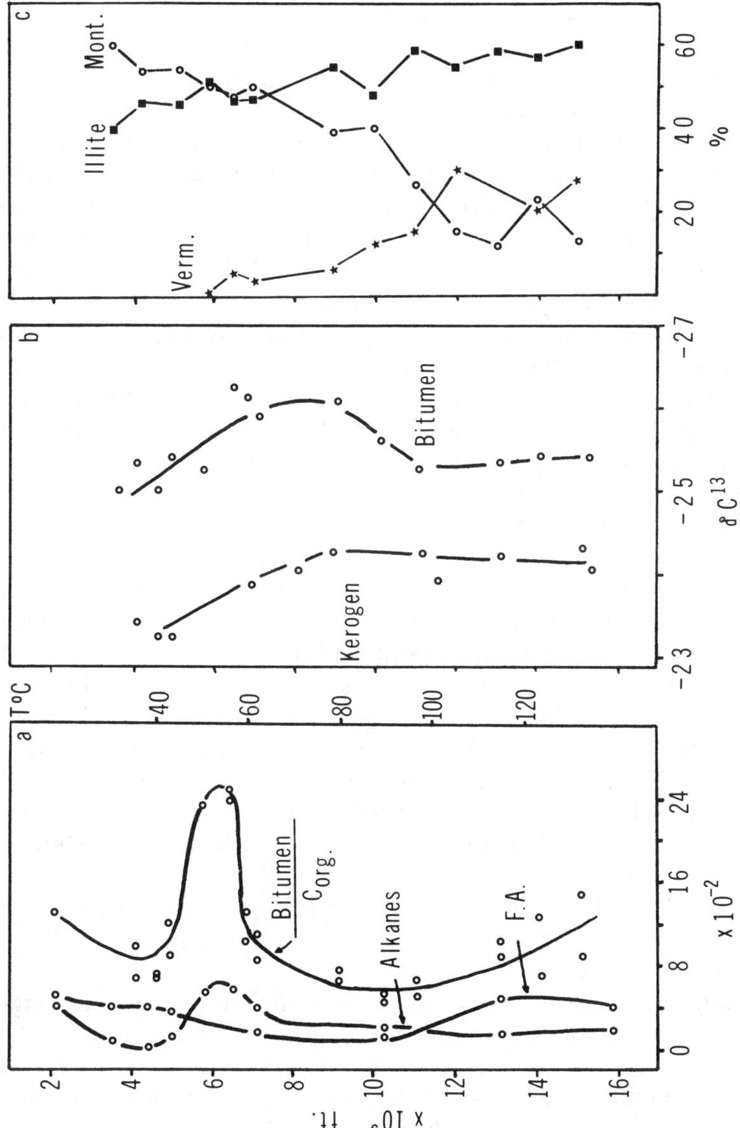

Figure 2 Variation with depth and temperature of (*a*) bitumen, alkane, and fatty acids; (*b*) carbon isotope ratios of kerogen and bitumen; and (*c*) clay mineralogy in a Gulf Coast, USA, profile.

The most significant mineralogical change with depth involves the progressive transformation of the montmorillonite component of the mixed layer sequence as has been noted earlier (Powers 1967, Hower et al 1976). The interpretation of the mixed layer diffraction patterns was made utilizing the MacEwan transform method (MacEwan, Ruiz Amil & Brown 1961). The clay mineralogical transformations are summarized in Figure 2*c*, where the gradual illitization of the montmorillonite component is seen. Below 6000 feet a third component appears in the mixed layer sequence, which is characterized as a high layer charge but partially expandable or vermiculite phase. This is interpreted as an intermediate phase in the formation of illite. Interpretation of the chemical analyses of these mixed layer clays indicates that the initial illitization to 6000 feet results from increasing interlayer charge due to reduction of ferric ion in the octahedral layer. Significantly, the appearance of vermiculite coincides with appreciable tetrahedral substitution of Al^{+3} for Sl^{+4} as well as octahedral Al^{+3} replacement in the mixed layer phase. Although not confirmed in this study, it appears that the aluminum and potassium necessary for illitization are supplied by decomposition of detrital K-feldspar (Hower et al 1976).

In Figure 2*a* are summarized some of the organic geochemical data obtained from the same series of samples. The bitumen (methanol-benzene extract) distribution shows the maximum typical of many Tertiary basins (Connan 1974). The n-alkane fraction, which makes up only a small fraction of the total bitumen, shows a similar distribution. It is evident that the n-fatty acid component is too small to generate the total hydrocarbon content, so the bitumen must be generated from the more abundant kerogen. Studies of the kerogens extracted from these same sediments (Kemmer 1978) reveal initial decarboxylation of acid groups accompanied by loss of alkyl functions. The elemental composition of these kerogens suggests derivation to a large extent from terrestrial plants (Tissot, Durand & Espitale 1974).

It is significant that bitumen production and subsequent degradation can be correlated directly with the presumed catalytic properties of the admixed clay mineral phase.

The significant generation of kerogen over the low temperature range of 50–60°C certainly signifies the contribution of catalysis. It can be rationalized that this is initiated by the decarboxylation of fatty acid groups within the kerogen polymer leading to free radical formation and subsequent C bond cleavage, degrading the kerogen and producing bitumen. The montmorillonitic component of the mixed layer clay serves as the electron acceptor to promote this reaction. Since the fatty acid groups are in this case a part of a solid organic polymer, fabric changes are necessary to continually bring acid groups into contact with catalytic sites at the edges of clay particles. Continuous compaction and readjustment of clay-kerogen interfaces would promote the catalytic effect. Recently Reed & Oertel (1978) have presented evidence that compaction of shales does in fact promote organic diagenesis by creating fresh catalytic sites.

It is further significant that loss of bitumen due to cracking and production of volatile hydrocarbons coincides with the formation of the vermiculite layers in the mixed layer clay. Since a montmorillonite to vermiculite transformation results from partial dehydration of interlayers, increased tetrahedral substitution, and interlayer charge, all factors which promote increased interlayer water dissociation, it appears likely that the clay at this stage acts as an acid catalyst, promoting carbonium ion cracking of the bitumen material.

This interpretation receives support from the C isotope measurements made on the coexisting bitumens and kerogens extracted from this series of sediments. The data are plotted in Figure 2*b* where δC^{13} values are expressed relative to PDB standard. The kerogens are consistently slightly isotopically heavier than the bitumens, apparently due to preferential C^{12}–C^{12} bond breakage which prevails during free radical reaction. The slight enrichment of the kerogen in C^{12} during bitumen production is in accordance with the proposed role of decarboxylation at this stage. In the fatty acid precursors carboxyl carbon is known to be isotopically heavier than associated reduced carbon. As decarboxylation proceeds the kerogen residue should become isotopically lighter as heavier CO_2 is evolved. It is also noteworthy that at greater depth no further fractionation occurs, in spite of continued kerogen degradation. This is analogous to Sackett's (1978) experiments on cracking with clay catalysts, which he attributed to carbonium ion reaction, induced by the clay catalyst.

The bitumen shows similar trends in changing δC^{13} values, indicating that the initially produced bitumen also contains carboxyl groups which give off CO_2. The decrease of δC^{13} values during cracking indicates that thermal cracking becomes more significant at greater depths and higher temperatures in the case of bitumen.

One can conclude from these considerations that the sediments themselves provide compelling evidence that indicates the role of clay mineral catalysts in promoting petroleum hydrocarbon formation.

HYDROCARBON MIGRATION

There is still very little known about the manner in which hydrocarbons formed in argillaceous source rocks migrate and accumulate in porous reservoirs. Some evidence exists, however, that the clay mineral-kerogen complex plays a role in modifying hydrocarbon compositions during migration.

Some time ago Legate & Johns (1964) used gas chromatography to measure the affinity of montmorillonite clays for hydrocarbons of differing polarity, and suggested that during the migration of petroleum chromatographic effects might modify their composition. Recently Young & McIver (1977) developed the chromatographic technique further and convincingly showed that they could in numerous instances predict oil compositions following migration, where the clay-kerogen complex was the chromatographic agent. Yariv (1976) has proposed a chemical model for the structure of the electrical double layer in the presence of organic molecules, and has proposed "organo-pores" shown to be hydrocarbon permeable. These may become migration paths for hydrocarbons in argillaceous source rocks. Further studies of the properties of clay-kerogen complexes during the compaction of sediments are greatly needed.

ACKNOWLEDGMENTS

Support for portions of this work by NSF grant 00206-A01 and ACS-PRF grant 10052-AC2 are gratefully acknowledged. Carbon isotope measurements of the bitumens were carried out by the author when on leave as a von Humboldt senior US Scientist Awardee at the Geochemical Institute, University of Göttingen. The assistance and hospitality of Profs. K. H. Wedepohl and J. Hoefs at this institute are deeply appreciated. Isotopic measurements of the kerogens were made by A. Kemmer at the same institute.

Literature Cited

Abelson, P. H. 1963. Organic geochemistry and the formation of petroleum. *Proc. World Petrol. Cong., 6th Frankfurt*, Sect. I: 397–407

Albrecht, P., Vandenbroucke, M., Mandengue, M. 1976. Geochemical studies on the organic matter from the Douala Basin (Cameroon)—I. Evolution of the extractable organic matter and the formation of petroleum. *Geochim. Cosmochim. Acta* 40: 791–99

Almon, W. R., Johns, W. D. 1976. Petroleum forming reactions: Clay catalyzed fatty acid decarboxylation and paraffin cracking. *Proc. Int. Clay Conf., Mexico City, 1975*, pp. 399–409

Almon, W. R., Johns, W. D. 1977. Petroleum forming reactions: The mechanism and rate of clay catalyzed fatty acid decarboxylation. *Proc. Int. Meet. Org. Geochem., 7th*, ed. R. Campos, J. Goni, pp. 157–71. Revista Espanola de Micropaleontologia, Madrid

Banet, A. C. 1976. *Kinetics of a clay catalyzed cracking reaction.* MA thesis. Univ. of Missouri-Columbia. 77 pp.

Baumann, D. K. 1975. *Chemical diagenesis during burial of Tertiary Gulf Coast pelitic sediments.* MA thesis. Univ. of Missouri-Columbia. 72 pp.

Bendaraitis, J. G., Brown, B. L., Hepner, L. S. 1963. Isolation and identification of isoprenoids in petroleum. *Proc. World Petrol. Cong., 6th Frankfurt*, Sect. V: 13–29

Benson, S. W. 1965. Bond energies. *J. Chem. Educ.* 42: 502–18

Bray, E. E., Evans, E. D. 1961. Distribution of n-paraffins as a clue to recognition of source beds. *Geochim. Cosmochim. Acta* 22: 2–15

Bronson, J. L., Hower, J. 1976. Mechanism of burial metamorphism of argillaceous sediment. 2. Radiogenic evidence. *Geol. Soc. Am. Bull.* 87:735–44

Brooks, B. T. 1948. Active-surface catalysts in formation of petroleum, Pt. 1. *Am. Assoc. Petrol. Geol. Bull.* 32:2269–86

Brooks, B. T. 1952. Evidence of catalytic action in petroleum formation. *Ind. Eng. Chem.* 44:2570–77

Burst, J. F. 1969. Diagenesis of Gulf Coast clayey sediments and its possible relation to petroleum migration. *Am. Assoc. Petrol. Geol. Bull.* 53:73–93

Connan, J. 1974. Time-temperature relation in oil genesis. *Am. Assoc. Petrol. Geol. Bull.* 58:2516–21

Cooper, J. E. 1962. Fatty acids in Recent and ancient sediments and petroleum reservoir waters. *Nature* 193:744–46

Cooper, J. E., Bray, E. E. 1963. A postulated role of fatty acids in petroleum formation. *Geochim. Cosmochim. Acta* 27:1113–27

Dunoyer de Segonzac, G. 1970. Transformation of clay minerals during diagenesis and low-grade metamorphism: A review. *Sedimentology* 15:281–346

Foscolos, A. E., Powell, T. G. 1979. Mineralogical and geochemical transformation of clays during burial diagenesis (catagenesis): Relation to oil generation. *Proc. Int. Clay Conference, Oxford 1978.* In press

Foscolos, A. E., Powell, T. C., Gunther, P. R. 1976. The use of clay minerals and inorganic and organic indicators for evaluating the degree of diagenesis and oil generating potential of shales. *Geochim. Cosmochim. Acta* 40:953–66

Fripiat, J. J. 1970. Interacción aqua-arcilla. *Madrid Anales Reunión Hispana-Belga Minerales Arcilla, CSIC*: 1–9

Fripiat, J. J., Cruz-Cumplido, M. I. 1974. Clays as catalysts for natural processes. *Ann. Rev. Earth Planet. Sci.* 2:239–56

Fripiat, J. J., Telli, A. M., Poncelet, G., Andre, T. 1965. Thermodynamic properties of absorbed water molecules and electrical conduction in montmorillonites and silica. *J. Phys. Chem.* 69:2185–97

Galwey, A. K. 1970. Reactions of alcohols and of hydrocarbons on montmorillonite surfaces. *J. Catal.* 19:330–42

Galwey, A. K. 1972. The rate of hydrocarbon desorption from mineral surfaces and the contribution of heterogeneous catalytic-type processes to petroleum genesis. *Geochim. Cosmochim. Acta* 36: 1115–30

Greensfelder, B. S., Voge, H. H., Good, G. M. 1949. Catalytic and thermal cracking of pure hydrocarbons. *Ind. Eng. Chem.* 41:2573

Grim, R. E. 1947. Relation of clay mineralogy to origin and recovery of petroleum. *Am. Assoc. Petrol. Geol. Bull.* 31:1491–99

Hedberg, H. D. 1968. Significance of high wax oils in the genesis of petroleum. *Am. Assoc. Petrol. Geol. Bull.* 52:736–50

Heling, D. 1974. Diagenetic alteration of smectite in argillaceous sediments of the Rhinegraben (SW Germany). *Sedimentology* 21:463–72

Henderson, W., Eglington, G., Simmonds, P., Lovelock, J. E. 1968. Thermal alteration as a contributory process to the genesis of petroleum. *Nature* 219:1012–16

Hower, J., Eslinger, E. V., Hower, M. E., Perry, E. A. 1976. Mechanism of burial metamorphism of argillaceous sediment. 1. Mineralogical and chemical evidence. *Geol. Soc. Am. Bull.* 87:725–37

Ishiwatari, R., Rohrback, B. G., Kaplan, I. R. 1977. Hydrocarbon generation by thermal alteration of kerogen from different sediments. *Am. Assoc. Petrol. Geol. Bull.* 62:687–92

Johns, W. D., Shimoyama, A. 1973. Clay minerals and petroleum-forming reactions during burial and diagenesis. *Am. Assoc. Petrol. Geol. Bull.* 56:2160–67

Jones, R. T. 1977. *Effect of clay mineral diagenesis on hydrocarbon cracking ability of clay catalysts.* MA thesis. Univ. of Missouri-Columbia. 69 pp.

Jurg, J. W., Eisma, E. 1964. Petroleum hydrocarbons: generation from fatty acid. *Science* 144:1451–52

Kemmer, D. A. 1978. *Characterization of kerogen from Gulf Coast Tertiary sediments.* MA thesis. Univ. of Missouri-Columbia. 88 pp.

Kvenvolden, K. A. 1970. Evidence for transformations of normal fatty acids in sediments. *Adv. Org. Geochem.*, pp. 335–66

La Plante, R. E. 1974. Hydrocarbon generation in Gulf Coast Tertiary sediments. *Am. Assoc. Petrol. Geol. Bull.* 58:1281–89

Legate, C. E., Johns, W. D. 1964. Gaschromatographische Untersuchung einiger Systeme aus Tonmineralien und organischen Stoffen: Bestimmung der Aktivitaetskoeffizienten and Absorptions waermen. *Beitr. Mineral. Petrogr.* 10: 60–69

LeTran, K., Connan, J., Van der Weide, B. M. 1974. Diagenesis of organic matter and occurrence of hydrocarbons and hydrogen sulfide in the SW Aquitaine Basin (France). *Cent. Rech. Pau. Bull.* 8:111–37

Louis, M. 1966. Essais sur l' évolution de pétrole à faible temperature en présence de minéraux. *Adv. Org. Chem.* pp. 261–78

Louis, M. C., Tissot, B. P. 1967. Influence de la temperature et de la pression sur la formation des hydrocarbures dans les argiles à kérogene. *Proc. World Petrol. Cong., 7th, Mexico* 2:47–60

MacEwan, D. M. C., Ruiz Amil, A., Brown, G. 1961. Interstratified clay minerals. In *The X-Ray Identification and Crystal Structures of Clay Minerals*, ed. G. Brown, pp. 393–445. Min. Soc. London

McNab, J. G., Smith, P. V., Betts, R. L. 1952. The evolution of petroleum. *Ind. Eng. Chem.* 44:2556–63

Mortland, M. M. 1968. Protonation of compounds at clay mineral surfaces. *Trans. Int. Cong. Soil Sci., 9th* 1:691–99

Mortland, M. M., Raman, K. V. 1968. Surface acidity of smectites in relation to hydration, exchangeable cation and structure. *Clays Clay Miner.* 16:393–98

Park, R., Epstein, S. 1960. Carbon isotope fractionation during photosynthesis. *Geochim. Cosmochim. Acta* 21:110–26

Perry, E. A., Hower, J. 1970. Burial diagenesis in Gulf Coast pelitic sediments. *Clays Clay Miner.* 18:165–77

Peters, K. E., Ishiwatari, R., Kaplan, I. R. 1977. Color of kerogen as index of organic maturity. *Am. Assoc. Petrol. Geol. Bull.* 61:504–10

Philippi, G. T. 1965. On the depth, time and mechanism of petroleum generation. *Geochim. Cosmochim. Acta* 29:1021–49

Powers, M. C. 1967. Fluid-release mechanisms in compacting marine mudrocks and their importance in oil exploration. *Am. Assoc. Petrol. Geol. Bull.* 51:1240–54

Pusey, W. C. 1973. Paleotemperatures in the Gulf Coast using the ESR-kerogen method. *Gulf Coast Assoc. Geol. Trans.* 23:195–202

Reed, W. E., Oertel, G. 1978. Is compaction a factor in organic diagenesis? *Geol. Soc. Am. Bull.* 89:658–62

Sackett, W. M. 1978. Carbon and hydrogen isotope effects during the thermocatalytic production of hydrocarbons in laboratory simulation experiments. *Geochim. Cosmochim. Acta* 42:571–80

Sackett, W. M., Nakaparksin, S., Dalrymple, D. 1970. Carbon isotope effects of methane production by thermal cracking. *Adv. Org. Geochem.*, 1966, pp. 37–53

Shimoyama, A., Johns, W. D. 1971. Catalytic conversion of fatty acids to petroleum like paraffins and their maturation. *Nature Phys. Sci.* 232:140–44

Shimoyama, A., Johns, W. D. 1975. Clay minerals and petroleum hydrocarbon formation. *Contrib. Clay Mineral.* (Japan) pp. 257–61

Silverman, S. R. 1963. Investigations of petroleum origin and evolution mechanisms by carbon isotope studies. In *Isotope and Cosmic Chemistry*, ed H. Craig, S. L. Miller, G. J. Wasserburg, pp. 92–101. Amsterdam: North-Holland

Silverman, S. R. 1967. Carbon isotope evidence for the role of lipids in petroleum formation. *J. Am. Oil Chem. Soc.* 14:691–95

Silverman, S. R., Epstein, S. 1958. Carbon isotope compositions of petroleum and other sedimentary organic materials. *Am. Assoc. Petrol. Geol. Bull.* 42:998–1012

Solomon, D. H. 1968. Clay minerals as electron acceptors and electron donors in organic reactions. *Clays Clay Miner.* 16:31–39

Solomon, D. H., Rosser, M. J. 1965. Reactions catalyzed by minerals, I. Polymerization of styrene. *J. Appl. Polym. Sci.* 9:1261–71

Solomon, D. H., Swift, J. D., Murphy, A. J. 1971. Acidity of clay minerals in polymerization and related reactions. *J. Macromol. Sci. Chem.* 5:587–601

Stevens, N. P., Bray, E. E., Evans, E. D. 1956. Hydrocarbons in the sediments of the Gulf of Mexico. *Am. Assoc. Petrol. Geol. Bull.* 40:975–83

Theng, B. K. G. 1971. Mechanisms of formation of colored clay-organic complexes. A review. *Clays Clay Miner.* 19:383–90

Theng, B. K. G. 1974. *The Chemistry of Clay-Organic Reactions.* New York: Wiley. 343 pp.

Thomas, C. C. 1949. Chemistry of cracking catalysts. *Ind. Eng. Chem.* 41:2564–73

Tissot, B. 1969. Premières données sur les mécanismes et la cinétique de la formation du pétrole dan les sédiments: simulation d'un schéma réactionnel sur ordinateur. *Inst. Français Pétrole Rev.* 24:470–501

Tissot, B., Durand, B., Espitalie, J. 1974. Influence of nature and diagenesis of organic matter in formation of petroleum. *Am. Assoc. Petrol. Geol. Bull.* 58:499–506

Tissot, B., Oudin, J. L., Pelet, R. 1972. Critères d'origine et d'évolution des pétroles. Application à l'étude géochimique des bassins sédimentaires. *Adv. Org. Geochem.* pp. 113–34

Welte, D. H. 1965. Relation between petroleum and source rock. *Am. Assoc. Petrol. Geol. Bull.* 49:2246–68

Yariv, S. 1976. Organophilic pores as proposed primary migration media for hydrocarbons in argillaceous rocks. *Clay Sci.* 5:19–29

Young, A., McIver, R. D. 1977. Distribution of hydrocarbons between oils and associated fine-grained sedimentary rocks-Physical chemistry applied to petroleum geology. *Am. Assoc. Petrol. Geol. Bull.* 61:1407–36

Ann. Rev. Earth Planet. Sci. 1979. 7: 199–225

ATMOSPHERIC TIDES

*10112

R. S. Lindzen
Naval Research Laboratory, Washington, D.C. 20390 and Center for Earth and Planetary Physics, Harvard University, Cambridge, Massachusetts 02138

1 INTRODUCTION

Atmospheric tides refer to those oscillations in the atmosphere whose periods are integral fractions of a lunar or solar day. The 24-hour Fourier component is referred to as a diurnal tide, the 12-hour component as a semidiurnal tide. The total tidal variation is referred to as the daily variation. Although atmospheric tides are, in small measure, gravitationally forced, they are primarily forced by daily variations in solar insolation. Tides, as defined above, exist on a variety of scales; generally only those tidal oscillations on a global scale are considered in tidal theory. Moreover, one distinguishes between migrating tides (which are functions of local time) and non-migrating tides (which depend on both local time and longitude). Our primary interest has traditionally been in migrating tides. To the extent that migrating tides dominate, tides may be observationally determined, for a given latitude, from a time series at a single longitude. Unfortunately, below 30-km altitude non-migratory diurnal tides are important and, hence, migrating diurnal tides can only be determined from data around a latitude circle.

For the reader with no background in this subject, an extensive review up to 1969 can be found in the monograph by Chapman & Lindzen (1970; subsequently referred to as CL). Chapter 1 of CL is non-mathematical and historical in emphasis. Prior to the time CL was prepared, the primary problem in atmospheric tides was to account for the daily variation in surface pressure which is dominated by a solar semidiurnal oscillation of amplitude 1.15 mb at the equator (with maxima at 0940 and 2140 LT); the solar diurnal tide is about half as large and far more irregular. The solution (to the extent it then existed) consisted in identifying two main sources of thermal excitation [absorption of sunlight by

0084-6597/79/0515-0199$01.00

water (Siebert 1961) and ozone (Butler & Small 1963)], and in noting that diurnal tides, in large measure, were incapable of propagating vertically (Lindzen 1966, 1967). However, by 1969 it was already realized that solar tides were an important part of the overall meteorology of the atmosphere, playing a dominant role above 100 km altitude.

This review primarily concerns observational and theoretical developments in solar tides of the neutral atmosphere since the publication of CL (see also Lindzen & Chapman 1969, Lindzen 1971a, and Lindzen 1972a for related reviews). Such topics as tidal variations in thermospheric composition (Mayr & Harris 1977, Forbes 1978) and tidally induced geomagnetic variations (Forbes & Lindzen 1976a, 1976b, 1977, Richmond et al 1976, Forbes & Garrett 1978b) are not treated here though current reviews may be found in the cited references. Finally, equations and analyses given in CL and other references are freely cited without reproduction in the present review. Equations are generally avoided also. There is, however, one concept in traditional tidal theory that is referred to so frequently in this review that it bears brief description before proceeding: the concept of tidal modes.

In classical tidal theory (as described in detail in CL) the equations of motion (consisting in the linearized equations for tidal perturbations on a basic state with neither mean flow nor horizontal temperature gradients in an inviscid, non-conducting atmosphere) can be reduced to a single equation for, let us say, geopotential height; the choice is obviously not unique. The resulting equation (for a given choice of zonal wavenumber and frequency) is separable in its latitude and altitude dependence. The equation for the latitude dependence is Laplace's Tidal Equation whose solution consists in an infinite set of eigenfunctions (called Hough functions) and associated eigenvalues or separation constants (called equivalent depths). One expands the forcing in terms of these Hough functions and uses the vertical equation to calculate the response to the forcing for each Hough function. The equivalent depth determines the nature of the vertical structure. Positive equivalent depths smaller than a certain value ($\sim$8 km) imply vertical wave propagation while large positive equivalent depths and negative equivalent depths imply vertical trapping. The totality of all meteorological fields (velocity, temperature, etc) associated with each Hough function constitute a Hough mode. For the migrating semidiurnal tide (zonal wavenumber equals 2) the Hough functions are similar to associated Legendre Polynomials of order two and are numbered similarly: $\theta_{2,2}(\theta)$, $\theta_{2,3}(\theta)$, $\theta_{2,4}(\theta), \ldots$ where $\theta =$ colatitude. $\Theta_{2,2}$ is the gravest symmetric semidiurnal Hough function; $\Theta_{2,3}$ is the gravest antisymmetric function; increasing indices involve increasing number of zeros.

All semidiurnal Hough functions have positive equivalent depths. The 2,2 mode has an almost infinite vertical wavelength; the 2,3 has a vertical wavelength of about 80 km; the 2,4 about 50 km; higher order modes have progressively shorter vertical wavelengths. For migrating diurnal tides (zonal wavenumber 1) the situation is more complicated; there are two distinct sets of Hough functions: one set with relatively small positive equivalent depths is restricted primarily to latitudes equatorward of 30° and the other with negative equivalent depths is restricted mostly (except for the lowest order modes) to latitudes poleward of 30°. The positive equivalent depth functions are $\Theta_{1,1}$, $\Theta_{1,2}, \ldots$ beginning with the main symmetric function; the negative equivalent depth functions are $\Theta_{1,-2}$, $\Theta_{1,-3}, \ldots$ where $\Theta_{1,-2}$ is the main trapped symmetric function. There is an antisymmetric Hough function $\Theta_{1,-1}$ which has an almost infinite equivalent depth. The function $\Theta_{1,1}$ is associated with a vertical wavelength of about 25 km, while higher order positive equivalent depth modes have progressively shorter wavelengths. The negative equivalent depth modes are trapped by the earth's rotation; poleward of 30° the Coriolis parameter exceeds the diurnal frequency. It should finally be mentioned that vertically propagating atmospheric waves tend to have amplitudes which increase with height inversely with the square root of the basic unperturbed pressure. This accounts, in part, for the importance of tides in the upper atmosphere.

Of necessity, the work reviewed will not constitute a fully integrated treatment. For example, some of the new data described in Section 2 has led to theoretical questions which have not yet been fully dealt with. Similarly, some of the theoretical work described in Section 3 (which deals with the effects of mean winds and horizontal temperature gradients) and in Section 4 (which deals with the realistic inclusion of damping due to viscosity, thermal conductivity, ion drag, and infrared radiation) addresses questions arising from older observations and observational analyses. Section 5 deals with some current combined theoretical and empirical approaches to the latest data. Section 6 reconsiders the assumed sources of tidal forcing, suggesting an additional thermal source (daily variations in cumulonimbus convection) which has not yet been incorporated into theoretical models described in Sections 3 and 4. Section 7 briefly mentions a variety of topics such as nonlinear tidal calculations and the generation of mean flows by tides which are still in a very preliminary state of development or which are considered peripheral to the main concerns of this review. There is, in this review, no attempt at a comprehensive citation of all recent work in this field. Indeed, where recent reviews of specific areas (like rocket observations, for example) exist, these are cited in preference to individual papers. Moreover, in reviewing theoretical work, emphasis is placed on the work of the present author and his colleagues.

Ten years ago classical tidal theory was in reasonable agreement with then existing data below 100 km. There were, however, some notable discrepancies for the semidiurnal tide: 1. A predicted 180° phase shift for horizontal wind oscillations predicted to occur near 28 km was not observed there. 2. The surface pressure oscillation maximum occurs 30 minutes to an hour later than predicted. In addition, calculations of the lunar gravitational atmospheric tide disclosed a profound sensitivity to details of the mean temperature structure—for which there was no observational evidence. This situation is described in detail in CL. The currently held approach to these discrepancies is described in Sections 3 and 6. In the last ten years greatly increased and improved data from above 80 km has revealed new, complex modes of tidal behavior wherein diurnal and semidiurnal oscillations alternate in importance in different regions of the atmosphere; where hitherto neglected higher order semidiurnal modes assume considerable importance; and where thermospheric EUV (extreme ultraviolet) forcing and lower atmospheric forcing compete in importance. Satisfactory explanations for much of this behavior already exist, though uncertainties are substantial. Sections 3–5 deal with these matters.

2 OBSERVATIONS AND OBSERVATIONAL ANALYSES

In the last decade advances have been made both in the acquisition and analyses of new data as well as in the analysis of existing data. As concerns surface pressure (our most extensively analyzed field), little has occurred to alter the relatively satisfactory state of observations and analyses. A recent review of this material is given by Haurwitz & Cowley (1973). Interesting relations between tides and rainfall have been noted, but discussion of these results is deferred to Section 6. This section dwells on recent advances in observations above the surface.

Rawinsonde Data

The collection of rawinsonde data (wind observations from radar tracking of meteorological balloons) illustrates how the intelligent utilization of existing data has led to useful information. Most notably, Wallace & Hartranft (1969) utilized 12-hourly data to estimate the global structure of the diurnal tide in horizontal wind between the 900-mb and 15-mb levels. This and similar utilizations of rawinsonde data were described in CL. More recently, Wallace & Tadd (1974) updated and improved the analyses in Wallace & Hartranft (1969) (long term averages of 12-hour wind differences were considered as opposed to the differences of long term averages of data taken at 0000 and 1200 GMT) for the diurnal tide and utilized stations with six-hourly data to evaluate semidiurnal oscillations. In general, important non-migrating diurnal components were

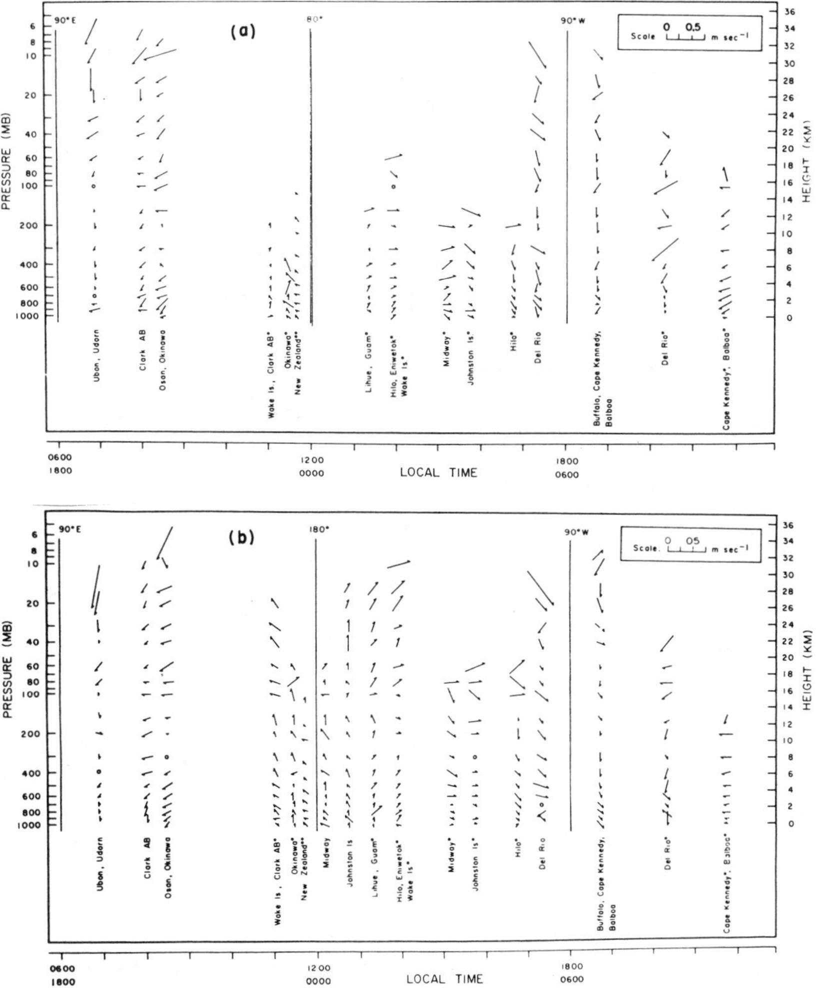

Figure 1 Longitude-height or time height section for the semidiurnal wind. Arrows are centered on the longitude of the stations or the mean longitudes of the groups of stations. Direction of flow is indicated by direction of arrows, with upwards corresponding to northwards (except for New Zealand where upwards indicates to the south). (*a*) December-February; (*b*) June-August. (From Wallace & Tadd 1974.)

found. Consistent with classical tidal theory, non-migratory components at upper levels were found primarily equatorwards of 30° lat. (i.e. where diurnal modes propagate vertically). By 30 mb non-migratory components are of large horizontal scale and are estimated to have vertical wavelength of approximately 10 km, thus making effective propagation into the mesosphere unlikely. Virtually no non-migrating semidiurnal tides were found, thus rendering longitude variation equivalent to local time variation. Figure 1 shows time (longitude) height cross sections for horizontal wind. Although data above 30 mb are sparse, there is no evidence of a null at 28 km, though a modest minimum in amplitude at this altitude can be discerned. This behavior is consistent with recent theoretical developments described in Section 6.

Rocket Data

Rockets via a variety of techniques (radar tracking of falling spheres and chaff, acoustic monitoring of sequential grenade explosions, and visual tracking of luminescent trails are examples) can be used to observe atmospheric winds (also density and temperature) between 30 km and 110 km. For economic reasons, however, it has been rare for sufficient rockets (in frequency and length of record) to be fired to properly delineate tides at the higher levels (above 60 km). Notable exceptions exist such as a series of 13 rocket grenade launchings at Kourou on 19–22 September 1971 (Smith et al 1974) and a series of rocket grenade launchings at Natal in 1966–68 (Smith et al 1968, 1969, 1970). Results from these tropical stations showed diurnal oscillations in horizontal winds with vertical wavelengths of about 25 km and amplitudes increasing from 10 m/s near 60 km to about 40 m/s near 80 km; semidiurnal oscillations were found with extremely large vertical wavelengths and amplitudes increasing from 5 m/s at 60 km to 10–20 m/s at 80 km. These results are completely consistent with the theory described in CL. The data is of considerable value since data between 60 and 80 km is particularly rare. Below 60 km small meteorological rockets provide substantial data on tides. The analyses of such data by Reed et al (1969) and Reed (1967) were reviewed in CL. Reed (1972) subsequently considered more recent data and found it produced little change in his earlier analysis. Similarly, a major series of 70 rocket launchings at eight western hemisphere stations during 19–20 March 1974 yielded results consistent with earlier studies (Schmidlin et al 1975). Groves (1976, 1978) has extensively reviewed recent rocket data and the reader is referred to these reviews for further information. Groves (1978) attempts to analyze rocket data for high order tidal modes, but since periods of observation are generally small compared to set up times for such modes, the analyses are open to considerable question.

Mention should also be made of a series of chemiluminscent probes of the lower thermosphere over Eglin, Florida (Zimmerman et al 1974). These results only cover 1800–0500 LT and are, therefore, difficult to decompose into diurnal and semidiurnal components. The results, however, are consistent with a dominant diurnal component below 110 km with amplitudes of about 80 m/s and dominant semidiurnal oscillations between 110 and 160 km with amplitudes of about 40–100 m/s.

Radiometeor Data

One of the most important developments in upper atmosphere wind observations has been the improvement of the doppler (radio frequency) radar observations of micrometeors to delineate horizontal winds with vertical resolutions of 1–2 km in the range 80–105 km. Results obtained with this method are reviewed by Glass & Spizzichino (1974). Figures 2 and 3 show results for semidiurnal and diurnal tides over Garchy,

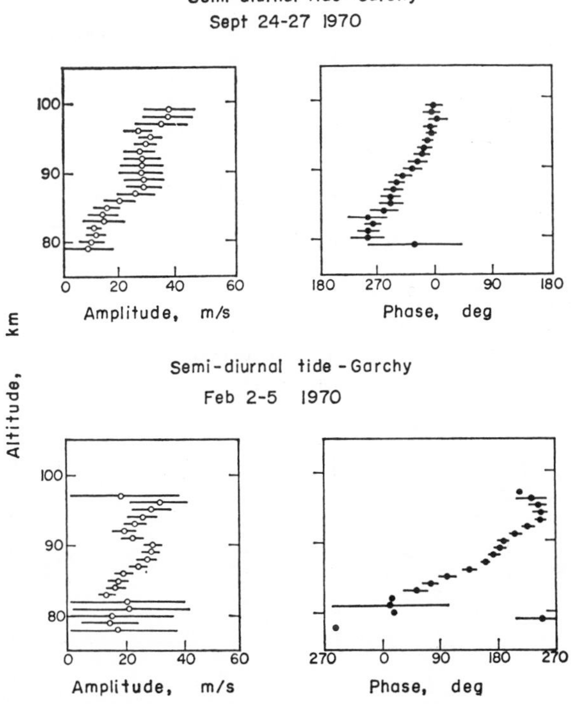

Figure 2 Amplitude and phase of the semidiurnal component of the eastward velocity over Garchy as observed by meteor radar during 24–27 September 1970 and 2–5 February 1970. (From Glass & Spizzichino 1974.)

France. The results for 24–27 September 1970 are relatively typical and are consistent with the joint presence of various lower order modes (2, 2; 2, 3; 2, 4; and 2, 5). However, occasionally, as on 2–5 February 1970, higher order modes with vertical wavelengths of about 30 km dominate. The diurnal analysis for 29 April 1970 is fairly typical and displays a propagating 1, 1 mode. The behavior for 25 July 1971 is unusual and appears to be dominated by the vertically trapped 1, −2 mode. Diurnal amplitudes at Garchy are generally weaker than those found at Adelaide, consistent with the discussion of earlier data in CL. The discovery that semidiurnal modes other than 2, 2 are of importance is, perhaps, the most important recent observation.

Incoherent Backscatter Data

Incoherent backscatter of powerful radar signals by ions and electrons in the thermosphere above 100 km has, over the last decade, become an

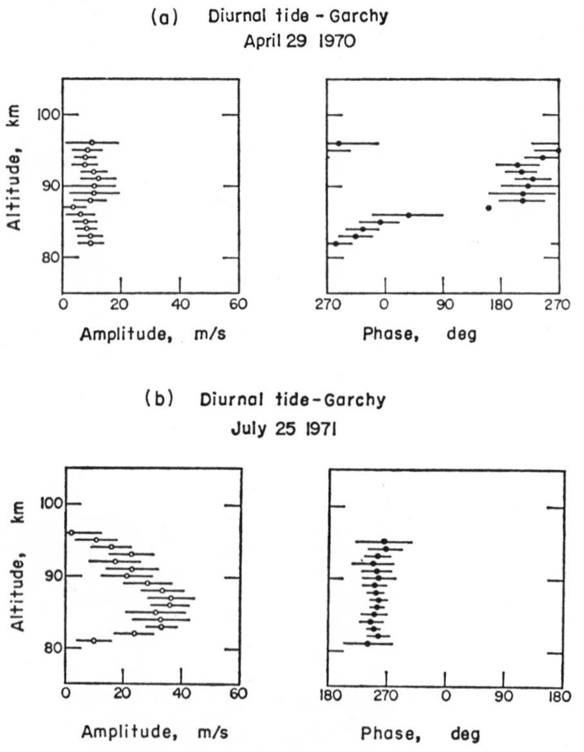

Figure 3 Amplitude and phase of the diurnal component of the eastward velocity over Garchy as observed by meteor radar. Data for 29 April 1970 are typical, while those for 25 July 1971 are unusual. (From Glass & Spizzichino 1974.)

extremely useful method of observing temperatures and winds. The subject has been reviewed at length by Evans (1972, 1975, 1978) and Amayenc et al (1973). Neutral temperatures can be measured with a 3–5-km height resolution and 20–30-minute time resolution in the height interval 100–130 km during the day. Data of this type are currently available from Millstone Hill near Boston, St. Santin in France, Arecibo, Puerto Rico, and Jicamaraca in Peru. Incoherent backscatter may also be used to measure E-region ion drifts; ion motions along magnetic field lines generally follow neutral winds. Such measurements have been made between 100 km and 165 km.

Although only daytime data are available from this technique, the predominance of the semidiurnal component between 110 and 130 km allows one to estimate semidiurnal amplitudes and phases by fitting a 12-hour wave to radar observations. Examples of results so obtained are shown in Figure 4. Results from St. Santin are shown in Figure 5. The two peaks in semidiurnal amplitude (near 115 km and 160 km) are characteristic, as is the renewed dominance of the diurnal component above 150 km.

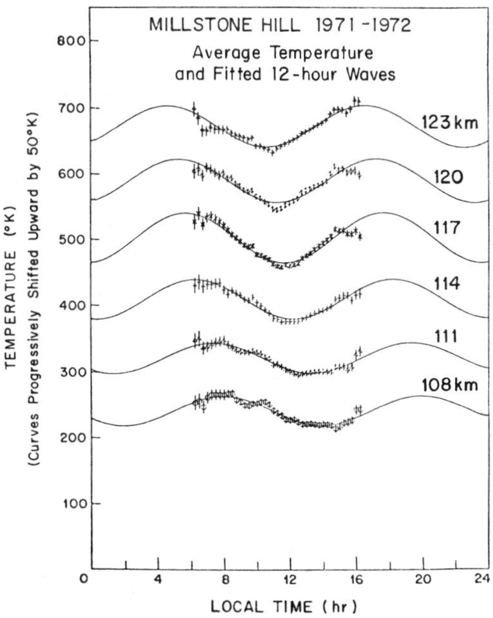

Figure 4 Average variation of the E-region temperature observed over Millstone Hill on 45 days in 1971–1972. Also shown are 12-hour waves fitted to the results. (From Salah & Wand 1974.)

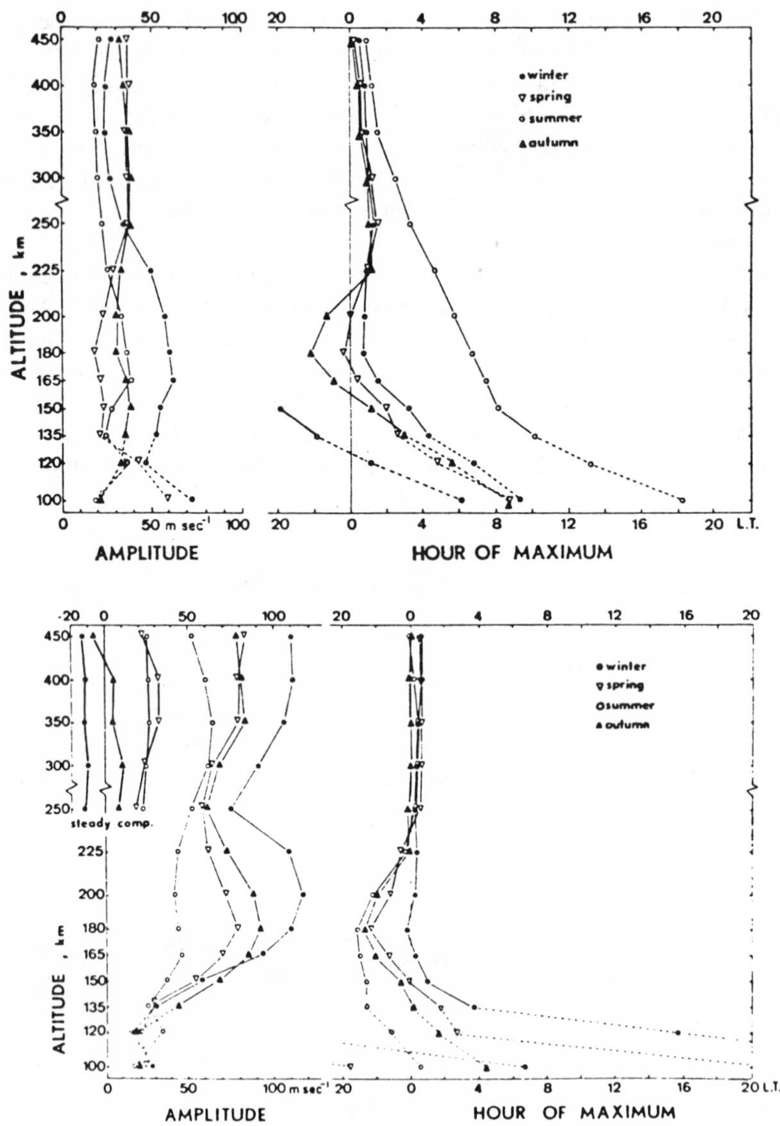

Figure 5 Mean seasonal vertical structures of amplitude and phase of the southward neutral wind from 1971–1972 observations at St. Santin (45° N). (*top*) Semidiurnal component; (*bottom*) steady and diurnal components. (From Amayenc 1974.)

Other Methods

The above by no means exhausts the techniques used to observe atmospheric tides. Partial reflections from the D-region have been used to measure winds between 70 and 100 km (Manson et al 1973). Inferences concerning tides have also been made from observations of quiet day (Sq) geomagnetic variations by means of dynamo theory (Richmond et al 1976, Forbes & Lindzen 1976a, 1976b, Forbes & Garrett 1978b). Also, satellite drag data on diurnal variations in thermospheric density remain useful (Jacchia & Slowey 1968). By and large, the above data are consistent with the more extensive observations described in Section 2.

3 THEORETICAL DEVELOPMENTS: EFFECTS OF MEAN WINDS AND HORIZONTAL TEMPERATURE GRADIENTS

Since mean meridional temperature variations and mean zonal winds are related via geostrophy, their effects are not strictly separable. We shall usually, however, speak of the effects of mean winds.

In CL it was noted that a measure of the effect of mean winds on tides is given by the ratio of the mean zonal wind to the linear rotation speed of the earth. Since the latter is approximately 466 m/s at the equator, little effect was expected for tropical latitudes. Thus, for the 1, 1 mode, which is confined to the tropics, Lindzen (1972b) was able to evaluate the small effects of mean winds by perturbation techniques. The consequences of mean wind effects on trapped diurnal modes should also be small. The situation for semidiurnal tides is somewhat different; higher order modes are significant at high latitudes where the linear rotation speed of the earth is much smaller. The inclusion of mean zonal winds varying with height and latitude leads to tidal equations whose height and latitude dependence are no longer mathematically separable.[1] A detailed numerical treatment of the non-separable problem has been presented by Lindzen & Hong (1974). A prominent feature of these results is mode coupling wherein a given mode interacts with the mean flow to contribute to the generation of other modes. Despite mode coupling, individual modes retain their integrity as concerns vertical propagation, etc. The calculations of Lindzen & Hong (1974) used the same thermal forcing given in CL. The mean winds used corresponded to observationally based models with westerlies in the winter mesosphere and easterlies in the summer hemisphere. Ex-

[1] In the unrealistic case of mean winds independent of height, separability does remain. This case has been studied by Ivanovsky & Semenovsky (1971).

amples of some of the results in Lindzen & Hong are shown in Figures 6 and 7. We see that the phase shift, which in the absence of mean winds occurs near 28 km, is moved upwards at higher latitudes in summer—consistent with the observational analysis of Reed (1972). However, no such shift is predicted in winter—in apparent disagreement with the observational analysis of Wallace & Tadd (1974). We return to this matter in Section 6. The theoretical shift in summer is due to the enhanced production of the 2, 3 mode by mode coupling with the 2, 2 mode induced by the antisymmetric mean wind distribution. We also see the prominent role assumed by higher order modes at high altitudes—consistent with recent data. The higher order modes (characterized by shorter vertical wavelengths) are generated to a significant extent by mode coupling. However, their prominence is due as well to the attenuation of the 2, 2 mode both by trapping in the mesosphere (see CL) and by losses incurred through mode coupling.

Lindzen & Hong (1974) also considered the lunar gravitational semidiurnal tide which they found was no longer sensitive to details of the mean temperature. The reason for this is that the reflecting surface in the mesosphere, which in classical theory is perfectly horizontal, is now bent and hence does not produce significant interference effects.

It should be noted in passing that the phase of the surface pressure oscillation is unaffected by mean winds and remains 30 minutes to an hour earlier than observed.

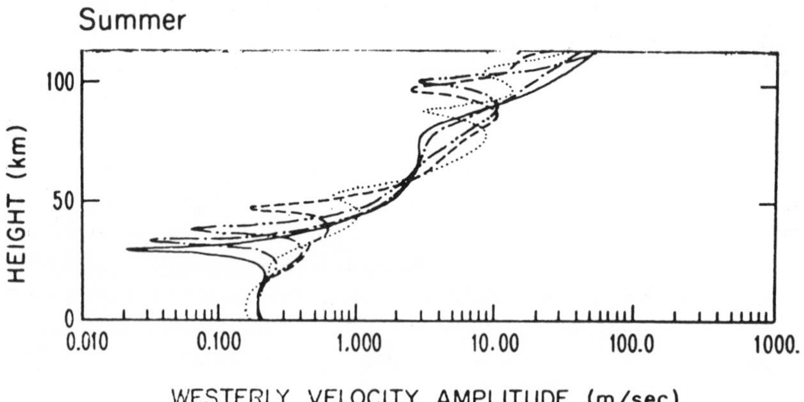

Figure 6 Amplitude of the solar semidiurnal component of westerly wind velocity as a function of height at various latitudes during summer ($y = \cos\theta$, θ = colatitude): — $y = 0.0$; —·— $y = 0.2$; —··— $y = 0.4$; ----- $y = 0.6$; ······ $y = 0.8$. (From Lindzen & Hong 1974.)

4 THEORETICAL DEVELOPMENTS: EFFECTS OF MOLECULAR VISCOSITY AND THERMAL CONDUCTIVITY AND ION DRAG IN THE THERMOSPHERE

In CL it was noted that the treatment of tides in the thermosphere would require the inclusion of ion drag, and molecular viscosity and thermal conductivity, the relative importance of the latter processes increasing inversely with air density. The theoretical understanding of thermospheric tides is an important task, since tides form the dominant motion system in the thermosphere (Dickinson 1975). On the other hand, the inclusion of viscosity leads to non-separable equations which are of eighth order in the vertical. The difficulties in the numerical treatment of this problem have been largely handled in recent efforts by Hong & Lindzen (1976) and Forbes & Garrett (1976, 1978a). However, much insight into the problem was obtained in earlier, simplified approaches.

The simplified approaches took various forms. One set of approaches essentially followed Pitteway & Hines (1963) in approximating viscous diffusion by linear friction (wherein damping is linearly proportional to velocity). The equations so obtained are still non-separable, but are now of only second order in the vertical. Various non-systematic approaches to the non-separability were attempted. These are described in Volland & Mayr (1972a, 1972b, 1974) and reviewed in Volland & Mayr (1977).

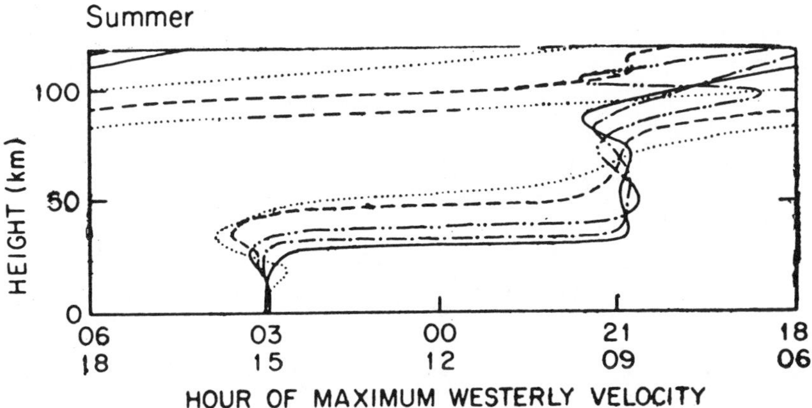

Figure 7 Phase (hour of maximum) of the solar semidiurnal component of westerly velocity as a function of height at various latitudes during summer. Profile designations given in Figure 6. (From Lindzen & Hong 1974.)

Another approach involved noting that the viscous equations remained separable in planar geometries (rotating and non-rotating) and that such geometries could approximate spherical results in restricted latitude belts. This has led to the equivalent gravity mode (EGM) formalism described at length for the non-rotating case (applicable to the equatorial region with 30% accuracy) in Lindzen (1970a, 1971b) and Lindzen & Blake (1971), though the first use of this approach appears in Nunn (1967). Recently, this formalism has been extended to rotating systems (Murata 1972, Richmond 1975); Forbes & Garrett have shown such extensions significantly improve accuracy. The EGM formalism assumes that modes retain their identity at all heights and thus fail to provide a global description of how these modes, in fact, change shape with height due to the actual non-separability. Nevertheless, for many purposes this approach provides an economical means of predicting thermospheric behavior with specifiable accuracy. Some basic effects of viscosity and conductivity obtained by Lindzen (1970a) are shown in Figure 8, for which a basic state with an 800°K thermosphere (corresponding to sunspot minimum conditions) was used. The following features in the figures are characteristic:

1. At some sufficiently great height all modes approach constant amplitude and phase (even modes which are trapped in the inviscid limit—such as the main diurnal trapped mode 1, −2). The height of this transition increases with the wavelength or trapping scale of the inviscid mode. It also increases with increasing mean thermospheric temperature.
2. Modes with long inviscid vertical wavelengths (like the main semidiurnal mode 2, 2) make the transition from growing approximately as (density)$^{-1/2}$ to constancy with little or no reduction in amplitude.
3. Modes with short vertical wavelengths (compared to $2\pi \times$ scale height) such as the main propagating diurnal mode (1, 1) undergo significant amplitude attenuation before asymptoting to a constant.
4. As viscosity becomes dominant the attenuation of trapped modes (like the main diurnal trapped mode, 1, −2) ceases. The transition is accompanied by a region of marked phase change. A more detailed discussion of these properties may be found in Lindzen (1971a).

The above results suggest that diurnal tides excited below the thermosphere will not contribute significantly to thermospheric tides. Lindzen (1971b) shows that forcing due to EUV and UV radiative heating within the thermosphere is sufficient to account for observed tides. These results are substantially confirmed in the careful spherical calculations of Forbes & Garrett (1976, 1978a). The results of Forbes & Garrett give comprehensive global results. They moreover show, somewhat surprisingly, that under sunspot maximum conditions diurnal temperature amplitudes are almost three times larger than at solar minimum ($|\delta T| \sim 150°K$ vs $|\delta T| \sim 50°K$) despite the fact that EUV heating only doubles; moreover, horizontal wind amplitudes barely change. These effects are due to ion drag which also increases with increasing solar activity thus inhibiting the growth of horizontal winds. This in turn diminishes the horizontal convergence; horizontal convergence produces vertical velocities which give rise to adiabatic cooling which generally counteracts direct diabatic

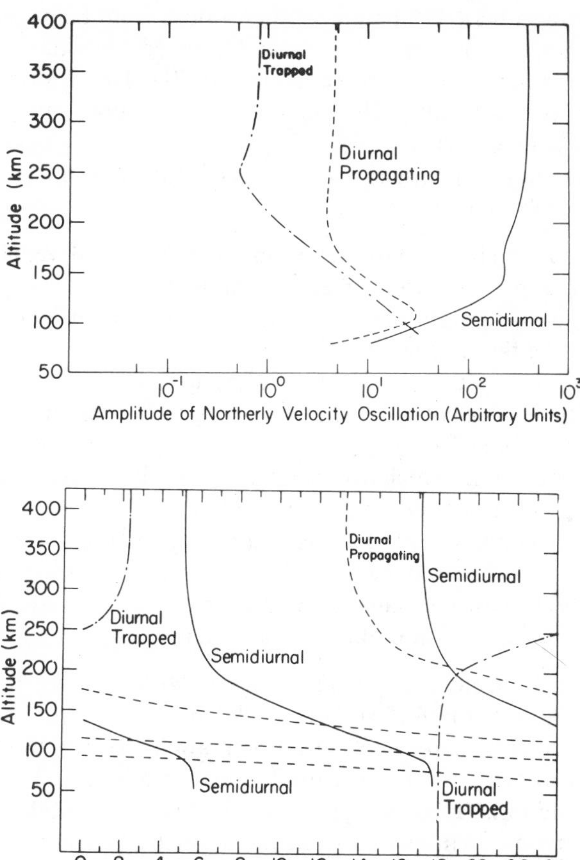

Figure 8 (*a*) Altitude structure for the amplitude of the southerly velocity for the main semidiurnal (solid line), the main propagating diurnal (dashes), and the main trapped diurnal (dots and dashes) modes when molecular viscosity and thermal conductivity are taken into account by means of the EGM formalism. (*b*) Same as (*a*), except for the phases of the various modes. (From Lindzen 1971a.)

heating. Increased ion drag inhibits this effect and thus permits a larger thermal response.

The situation with respect to the semidiurnal thermospheric tide has proven more complicated. From Figure 8 we see that the main semidiurnal mode entering the thermosphere from below continues to grow with height up to about 150 km and barely decays above that height. Calculations in Lindzen (1971b) suggested that this would lead to semidiurnal temperature oscillations in the thermosphere of magnitude 200°K at solar minimum (and magnitudes of about 80°K at solar maximum—since higher thermospheric temperatures damp tides from below and in situ forcing was somewhat less important). Observed amplitudes were only 30°K in the upper thermosphere (Harper 1971). The resolution of this gross discrepancy required the rectification of a substantial number of individually modest errors.

According to the spherical calculations of Hong & Lindzen (1976), the following resolve the discrepancy:

1. Mode coupling induced by mean winds (e.g. Section 3) reduces the 2, 2 mode entering the thermosphere by about 30%.
2. Improved calculations of ozone heating yield a similar reduction in 2, 2 entering the thermosphere.
3. Ion drag in Lindzen (1971b) was taken to act only on the east-west wind. The new calculations show ion drag to be more nearly isotropic and much more effective.
4. Mode broadening, which occurs only in the full spherical calculations, also leads to amplitude reductions of about 30%.
5. At solar maximum in situ forcing becomes comparably important with forcing from below and the two sources are not generally in phase. Consistent with earlier calculations, smaller semidiurnal amplitudes are found at solar maximum than at solar minimum.

Using results from Lindzen & Hong (1974) which are described in Section 3, Hong & Lindzen (1976) calculated thermospheric semidiurnal tides in reasonable agreement with observations. Below 130 km higher order modes propagating from below dominate, while the 2, 2 mode dominates above 150 km. Observations of the higher order modes differ substantially among themselves and hence detailed agreement is not to be expected. This variability is not surprising if the origin of the higher order modes is in mode coupling induced by mean winds. Mean winds below the thermosphere are known to vary but they are not observed globally on a regular basis.

It should finally be mentioned that in addition to the above described finite difference spherical calculations, Harris & Mayr (1975) have carried out similar calculations with a truncated spectral model. The effective resolution in these calculations was, however, much less than that employed in the finite difference calculations.

5 SEMI-EMPIRICAL APPROACHES TO THERMOSPHERIC TIDES

As has already been noted, tides in the thermosphere have been a main focus of tidal research in the last decade. There remain notable difficulties in making detailed and dependable predictions for this region. Not the least of these difficulties involves the absence of data upon which to base predictions (observations of zonally averaged winds below the thermosphere, reliable measurements of EUV and UV solar fluxes) or with which to check predictions (global observations of thermospheric tides at all levels). Moreover, we shall see in Section 6 that there are some doubts about sources of excitation in the lower atmosphere. These problems have given rise to semi-empirical approaches which attempt to achieve practical results while bypassing certain of the above difficulties.

As has been shown in Sections 2, 3, and 4, the region between 110 km and 130 km is dominated by semidiurnal tides in primarily the 2, 4 and 2, 5 modes. It is generally believed that these modes originate below the thermosphere. The theoretical work described in Section 3 suggests that they are produced by mode coupling with the 2, 2 mode induced by mean winds; in the absence of continuous observation of such winds, accurate predictions seem unlikely. At the very least, however, it might prove possible to use data from a few stations as well as theoretical knowledge of which modes should be important at various levels in order to produce a reliable global picture of the tides. This procedure is quite simple for the neighborhood of 115 km. Here it was originally believed that the 2, 4 mode was dominant since the observed vertical wavelength was on the order of 40 km (Salah & Wand 1974, Amayenc 1974). Moreover, theoretical results all suggested that diurnal tides and lowest order semidiurnal modes (2, 2 and 2, 3) should have much smaller amplitudes. However, simultaneous observations at Arecibo and Millstone Hill were inconsistent with the latitude structure of the 2, 4 mode (Salah et al 1975). Lindzen (1976) noted that the observations were consistent with the joint presence of 2, 4 and 2, 5 modes (the latter antisymmetric mode being important even during equinoxes). Data from two stations made it possible to calibrate the amplitude and phase of both modes, thus providing a global description of the tides in the neighborhood of 115 km. Such a modal decomposition was rendered simple by the theoretical fact that at 115 km 2, 4 and 2, 5 modal shapes were not significantly altered from their inviscid forms. The decomposition permitted the determination of observed vertical wavelengths for the individual modes. These were very close to theoretical values, lending credence to both the modal decomposition and to the use of observational analyses based solely on data from daylight hours.

A far more ambitious semi-empirical model has been developed by Garrett & Forbes (1978) for both diurnal and semidiurnal tides at all thermospheric levels during equinoxes. Without any appeal to theory, one could not have reasonably obtained such a model on the basis of data from only two latitudes. Their procedure ran roughly as follows:

1. Using the theoretical result that the diurnal tide in the upper thermosphere is forced primarily by EUV heating (Lindzen 1971b, Forbes & Garrett 1976), the value of the EUV flux was estimated by forcing agreement between calculations and diurnal temperature oscillations obtained from satellite drag data (Jacchia & Slowey 1968). The result was consistent with direct measurements of EUV flux (Hinteregger 1970) and the calculated diurnal temperatures and winds are in agreement with incoherent backscatter measurements (Amayenc 1974, Fontanari & Alcayde 1974, Roble et al 1974).
2. Having determined EUV heating, one calculates its semidiurnal component which, in turn, allows one to compute that portion of the semidiurnal tide forced in situ. Since the semidiurnal tide is expected to involve both in situ forcing and forcing from below, one cannot directly measure the in situ component. Garrett & Forbes (1978) note that a significant contribution to the in situ semidiurnal forcing arises from the interaction of the diurnal variations in ion drag with the diurnal tide. This component was neglected in Hong & Lindzen (1976).
3. Given the semidiurnal tide forced in situ, one uses linear regression methods to obtain the amplitudes and phases of semidiurnal tidal modes (2, 2; 2, 3; 2, 4; and 2, 5) propagating upward from 100 km which together with the tide forced in situ provide a best fit to incoherent backscatter data. Such a procedure is rendered possible (given that one has data from only two latitudes) because the 2, 4 and 2, 5 modes dominate in the lower thermosphere while the 2, 2 and 2, 3 modes assume importance primarily above 150 km. It should be noted that the calibration of Hough modes is rendered more complicated by the fact that above 115 km, Hough modes are significantly distorted from their inviscid forms due to viscosity and ion drag. Instead of inviscid Hough modes one must use Hough mode extensions of the sort described in Hong & Lindzen (1976) and extensively tabulated in Lindzen et al (1977). The latter work presents extensive descriptions of the height and latitude dependence (within the thermosphere) of semidiurnal

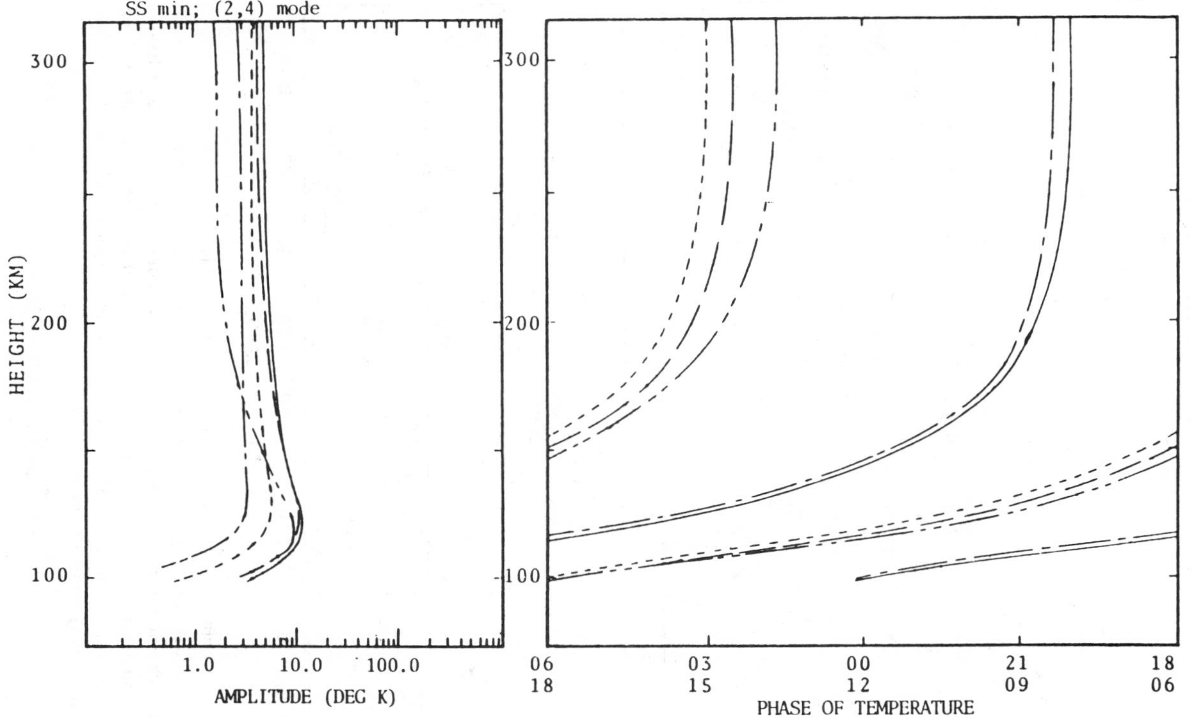

Figure 9 (*a*) Amplitude of temperature oscillation of the 2, 4 Hough mode extension (HME) as a function of altitude for various latitudes under minimum sunspot conditions. Different line conventions apply to different latitudes as follows: —, 0°; — - —, 15°; — - - —, 30°; — — —, 45°; - - - -, 60°. (*b*) Same as (*a*) but for the phase of the temperature oscillation (hour of maximum, local time) of the 2, 4 HME. (From Lindzen et al 1977.)

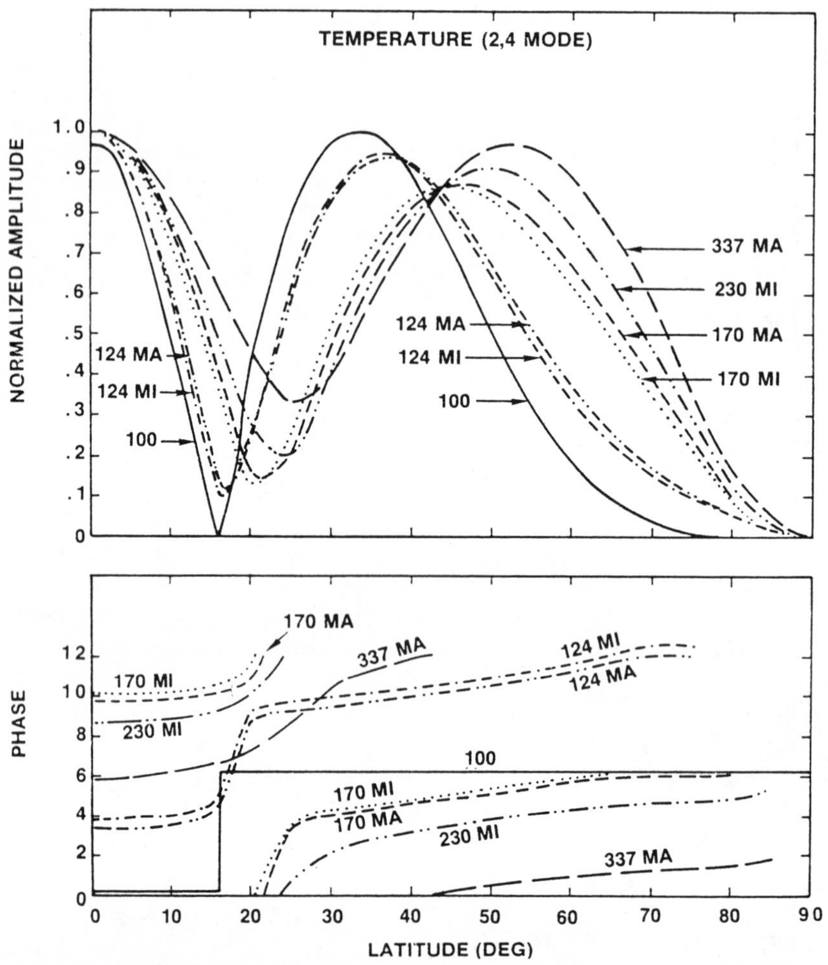

Figure 10 (*a*) Amplitude of temperature oscillation of the 2, 4 HME as a function of latitude for selected altitudes and sunspot conditions (values normalized by maximum value at each altitude). MI refers to sunspot minimum; MA refers to sunspot maximum. Values at 100 km are essentially independent of sunspot conditions. (*b*) Phase of temperature oscillation (time of maximum, local time) of the 2, 4 HME as a function of latitude for selected altitudes and sunspot conditions. (From Lindzen et al 1977.)

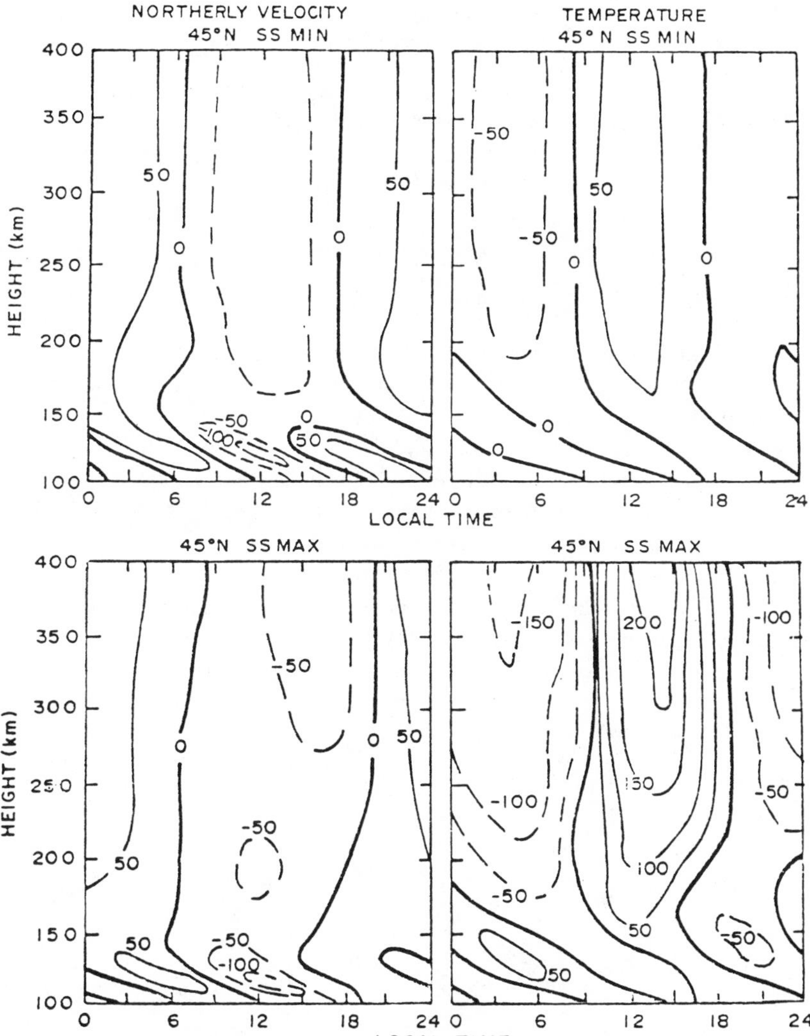

Figure 11 Altitude–local time contour plots of the northerly wind (*left*) and temperature (*right*) for sunspot minimum (*top*) and sunspot maximum (*bottom*) at 45° latitude. Contours are in 50 m s^{-1} and 50°K steps. (From Garrett & Forbes 1978.)

components of temperature and horizontal and vertical winds (under various sunspot conditions) resulting from forcing at 100 km by individual inviscid Hough modes (2, 2 through 2, 5). Examples of such results are shown in Figures 9 and 10.

In Figure 11 we see an example of the results obtained by Garrett & Forbes (1978), namely, time height cross sections of the total daily variation (diurnal plus semidiurnal) in northerly velocity and temperature at 45° latitude. The model yields results of this nature for all fields and latitudes.

6 POSSIBILITY OF AN ADDITIONAL SOURCE OF THERMAL FORCING

As noted in Section 1, existing calculations of the semidiurnal solar tide fail to correctly predict the phase of the surface pressure oscillation or to account for the absence of a dramatic 180° phase shift in horizontal wind oscillations in the neighborhood of 28 km in all seasons.

With respect to the phase discrepancy at the surface, it has been shown that neither the consideration of mean winds (Lindzen & Hong 1974) nor dissipation (Lindzen & Blake 1971) resolve the problem. It has also been found that a more accurate calculation of ozone heating fails to improve matters (D. W. Blake, personal communication). Recent analyses have noted semidiurnal oscillations in tropical rainfall (Brier & Simpson 1969) and mid-latitude rainfall (Wallace 1975). (Further references are given in

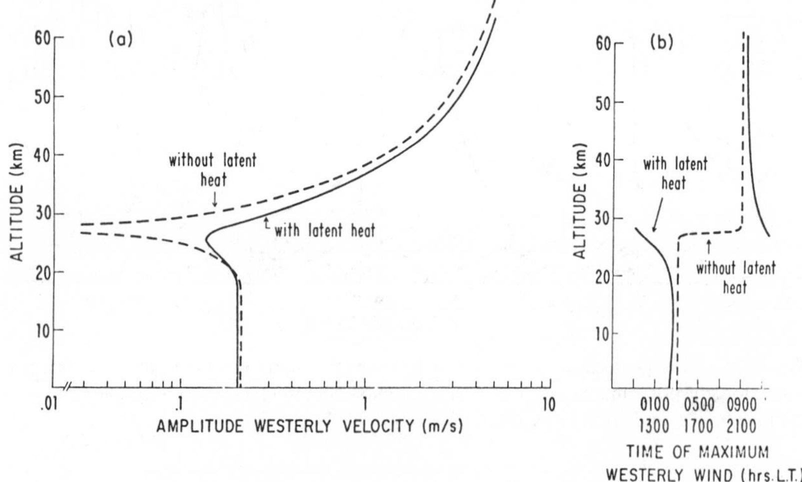

Figure 12 Calculated semidiurnal oscillation in westerly wind over the equator with and without the inclusion of latent heat forcing (ozone and water vapor forcing present in both cases): (*a*) Amplitude (m/s); (*b*) phase (hour of maximum).

Lindzen 1978.) In the former paper a correlation was noted between variations in amplitude of both rainfall and surface pressure oscillations, leading Simpson and Brier to suggest that tides produced the variation in rainfall. An alternative approach has been considered by Lindzen (1978) who determined the magnitude and phase of the semidiurnal oscillation in rainfall (and hence latent heat release) which would correct the surface phase discrepancy. He found that an oscillation of amplitude $\simeq 0.12$ cm/day at the equator with maxima occurring between 0230 and 0430, and between 1430 and 1630 LT, would suffice. Analysis of very limited available data appears to support the existence of such an oscillation. The resulting forcing is comparable in magnitude to the forcing due to ozone heating. The enhancement of tropospheric forcing leads (as anticipated by Green 1965) to the elimination of the sharp phase shift at 28 km as shown in Figure 12 (taken from Lindzen 1978). The results in Figure 12 are compatible with the observational results shown in Figure 1. To be sure, there is still a 180° phase shift, but it is now distributed over a broad range of altitudes. According to Reed (1972) even such a phase shift should occur at higher altitudes in the summer hemisphere. Such an effect, however, is still explicable in terms of the effects of mean winds as described in Section 3 of this review. It should be noted from Figure 12 that the posited latent heat forcing negligibly affects results above 40 km.

Although the existence of such an additional forcing is consistent with a variety of observations, its existence is by no means firmly established. No global analysis of daily variations in rainfall is available; neither is there an adequate theory of why such variations should exist. Lindzen (1978) has shown that tidally induced convergence of moisture is inadequate to explain matters.

7 MISCELLANEOUS TOPICS

It has long been realized that the characteristic growth with height (approximately inversely with the square root of mean pressure) of vertically propagating tidal modes would lead at some height to the importance of nonlinear terms unless damping processes curtailed growth before that height. That damping becomes important before pronounced nonlinearity occurs appears to be the case for all major tidal modes except the main propagating diurnal mode (Lindzen & Blake 1971). As shown in CL, however, the situation is not so straightforward; before nonlinear terms become very important for the 1, 1 mode, the tidal fields associated with this mode become convectively unstable (Lindzen 1967, 1968, Lindzen & Blake 1971). To be sure, the evolution of the convective instabilities will involve nonlinearity; however, such nonlinearities cannot be handled through simple amplitude expansion since the time and space scales of the convection differ profoundly from those associated with the tide. The convective instabilities should set on at about 80 km. Under the assumption that such instabilities lead to turbulent mixing, Lindzen & Blake (1971) point out that such turbulence should cease when molecular viscosity and thermal conductivity begin to damp this mode. This occurs near 108 km (see Figure 8). Whether the observed cessation of turbulence near this height (turbopause) is coincidental is currently unclear. However, the same approach applied to Mars once again predicts a reasonable turbopause (Lindzen 1970b).

A somewhat surprising application of tides has been to account for the forcing of mean zonal winds. The process is described in detail by Fels & Lindzen (1974). Briefly, Eliassen & Palm (1961) have shown that internal gravity waves (of which vertically propagating tides are a special case), away from their sources of excitation, produce a mean momentum flux which if absorbed by the atmosphere, would accelerate the mean flow toward the apparent speed of the sun. Fels & Lindzen (1974) show that this momentum comes from the mean flow in the region of thermal excitation where the mean flow is accelerated primarily in a direction opposite to that of the apparent motion of the sun. In the mesosphere the diurnal propagating and trapped modes interact to produce accelerations of ~ 1 ms/day, which may prove to be important. The same mechanism has been invoked to explain the 100 m/s circulation of the Venusian atmosphere.

8 CONCLUDING REMARKS

Overall, theory and observation of atmospheric tides have generally kept pace with each other over the last decade. Our understanding of the semidiurnal tide from the surface to 40 km rests, in part, on a conjectural daily variation in rainfall, which must be further checked in the data. If it exists, a theory will have to be developed to explain such a variation; if it does not exist, then the discrepancies described in Section 6 still remain. As concerns tides above 40 km, existing theory appears to explain virtually all observed qualitative and gross quantitative features. However, accurate predictions of detailed day-to-day variations in upper atmosphere tides seems out of the question. Moreover, data are insufficient to determine input parameters to tidal calculations (like mean winds or EUV heating) with precision; nor are there sufficient data to check the numerous theoretical predictions of global tidal behavior.

Acknowledgments

This review has been supported by the Office of Naval Research and by the National Science Foundation under Grant ATM-75-20156.

Literature Cited

Amayenc, P. 1974. Tidal oscillations of the meridional neutral wind at midlatitudes. *Radio Sci.* 9:281–93

Amayenc, P., Fontanari, J., Alcayde, D. 1973. Simultaneous neutral wind and temperature oscillations near tidal periods in the F-region over St. Santin. *J. Atmos. Terr. Phys.* 35:1499–505

Brier, G. W., Simpson, J. 1969. Tropical cloudiness and rainfall related to pressure and tidal variations. *Q. J. R. Meteorol. Soc.* 95:120–47

Butler, S. T., Small, K. A. 1963. The excitation of atmospheric oscillations. *Proc. R. Soc. London Ser. A* 274:91–121

Chapman, S., Lindzen, R. S. 1970. *Atmospheric Tides*. Dordrecht, Holland: Reidel. 200 pp.

Dickinson, R. E. 1975. Meteorology of the upper atmosphere. *Rev. Geophys. Space Phys.* 13:771–90

Eliassen, A., Palm, E. 1961. On the transfer of energy in stationary mountain waves. *Geofys. Publ.* 22:1–23

Evans, J. V. 1972. Ionospheric movements measured by incoherent scatter: a review. *J. Atmos. Terr. Phys.* 34:175–209

Evans, J. V. 1975. High-power radar studies of the ionosphere. *Proc. IEEE* 63:1636–50

Evans, J. V. 1978. Incoherent scatter contributions to studies of the dynamics of the lower thermosphere. *Rev. Geophys. Space Phys.* 16:195–216

Fels, S., Lindzen, R. S. 1974. Interaction of the thermally excited gravity waves with mean flows. *Geophys. Fl. Dyn.* 6:149–91

Fontanari, J., Alcayde, D. 1974. Observation of neutral tidal-type oscillations in the F_1-region. *Radio Sci.* 9:275–80

Forbes, J. M. 1978. Tidal variations in thermospheric O, O_2, N_2, Ar, He and H. *J. Geophys. Res.* 83:3691–98

Forbes, J. M., Garrett, H. B. 1976. Solar diurnal tides in the thermosphere. *J. Atmos. Sci.* 33:2228–41

Forbes, J. M., Garrett, H. B. 1978a. Seasonal-latitudinal structure of the diurnal thermospheric tide. *J. Atmos. Sci.* 35:148–59

Forbes, J. M., Garrett, H. B. 1978b. Solar tidal wind structures and the E-region dynamo. *J. Geomagn. Geoelectr.* In press

Forbes, J., Lindzen, R. S. 1976a. Atmospheric solar tides and their electrodynamic effects, I. The global Sq current system. *J. Atmos. Terr. Phys.* 38:897–910

Forbes, J., Lindzen, R. S. 1976b. Atmospheric solar tides and their electrodynamic effects, II. The equatorial electrojet. *J. Atmos. Terr. Phys.* 38:911–20

Forbes, J., Lindzen, R. S. 1977. Atmospheric solar tides and their electrodynamic effects, III. The polarization electric field. *J. Atmos. Terr. Phys.* 39:1369–77

Garrett, H. B., Forbes, J. M. 1978. Tidal structure of the thermosphere at equinox. *J. Atmos. Terr. Phys.* 40:657–68

Glass, M., Spizzichino, A. 1974. Waves in the lower thermosphere: recent experimental investigation. *J. Atmos. Terr. Phys.* 36:1825–39

Green, J. S. A. 1965. Atmospheric tidal oscillations: an analysis of the mechanics. *Proc. R. Soc. London Ser. A* 288:564–74

Groves, G. V. 1976. Rocket studies of atmospheric tides. *Proc. R. Soc. London Ser. A* 351:437–69

Groves, G. V. 1978. Diurnal and semidiurnal wind oscillations in the stratosphere. *J. Geophys. Res.* In press

Harper, R. M. 1971. *Dynamics of the neutral atmosphere in the 200–500 km height region at low latitudes*. PhD thesis. Rice Univ., Houston

Harris, I., Mayr, H. G. 1975. Diurnal variations in the thermosphere, 1. Theoretical formulation. *J. Geophys. Res.* 80:3925–33

Haurwitz, B., Cowley, A. 1973. The diurnal and semidiurnal barometric oscillations, global distribution and annual variation. *Pure Appl. Geophys.* 102:193–221

Hinteregger, H. E. 1970. The extreme ultraviolet solar spectrum and its variation during a solar cycle. *Ann. Geophys.* 26: 547–54

Hong, S.-s., Lindzen, R. S. 1976. Solar semidiurnal tide in the thermosphere. *J. Atmos. Sci.* 33:135–53

Ivanovsky, A., Semenovsky, I. 1971. Theory of the semidiurnal tide allowing for the effect of the zonal wind. *Izv. Atm. Ocean. Phys.* 7:159–65

Jacchia, L. G., Slowey, J. W. 1968. Diurnal and seasonal latitudinal variations in the upper atmosphere. *Planet. Space Sci.* 16: 509–24

Lindzen, R. S. 1966. On the theory of the diurnal tide. *Mon. Weather Rev.* 94:295–301

Lindzen, R. S. 1967. Thermally driven diurnal tide in the atmosphere. *Q. J. R. Meteorol. Soc.* 93:18–42

Lindzen, R. S. 1968. The application of classical atmospheric tidal theory. *Proc. R. Soc. London Ser. A* 303:299–316

Lindzen, R. S. 1970a. Internal gravity waves in atmospheres with realistic dissipation and temperature: Part I. Mathematical development and propagation of waves into the thermosphere. *Geophys. Fl. Dyn.* 1:303–55

Lindzen, R. S. 1970b. The application and

applicability of terrestrial atmospheric tidal theory to Venus and Mars. *J. Atmos. Sci.* 27:536–49

Lindzen, R. S. 1971a. Atmospheric tides. *Lect. Appl. Math.* 14:293–362

Lindzen, R. S. 1971b. Internal gravity waves in atmospheres with realistic dissipation and temperature: Part III. Daily variations in the thermosphere. *Geophys. Fl. Dyn.* 2: 89–121

Lindzen, R. S. 1972a. Atmospheric tides. In *Structure and Dynamics of the "Upper Atmosphere,"* ed. F. Verniani, pp. 21–88. New York: Elsevier

Lindzen, R. S. 1972b. Equatorial planetary waves in shear, Part II. *J. Atmos. Sci.* 29:1452–63

Lindzen, R. S. 1976. A modal decomposition of the semidiurnal tide in the lower thermosphere. *J. Geophys. Res.* 81, 1561–71

Lindzen, R. S. 1978. Effect of daily variations of cumulonimbus activity on the atmospheric semidiurnal tide. *Mon. Weather Rev.* 106:526–33

Lindzen, R. S., Blake, D. W. 1971. Internal gravity waves in atmospheres with realistic dissipation and temperature: Part II. Thermal tides excited below the mesopause. *Geophys. Fl. Dyn.* 3:31–61

Lindzen, R. S., Chapman, S. 1969. Atmospheric tides. *Space Sci. Rev.* 10:3–188

Lindzen, R. S., Hong, S.-s. 1974. Effects of mean winds and horizontal temperature gradients on solar and lunar semidiurnal tides in the atmosphere. *J. Atmos. Sci.* 31:1421–46

Lindzen, R. S., Hong, S.-s., Forbes, J. 1977. *Semidiurnal Hough mode extensions and their application. Naval Res. Lab. Memo. Report. 3442.* 65 pp.

Manson, A. H., Gregory, J. B., Stephenson, D. G. 1973. Wind and wave motions (70–100 km) as measured by a partial reflection radiowave system. *J. Atmos. Terr. Phys.* 35:2055–67

Mayr, H. G., Harris, I. 1977. Diurnal variations in the thermosphere, 2. Temperature, composition, winds. *J. Geophys. Res.* 82:2628–40

Murata, H. 1972. Atmospheric tidal oscillations in a viscid atmosphere. *J. Geomagn. Geoelectr.* 25:387–402

Nunn, D. 1967. *A theoretical study of tides in the upper atmosphere.* MSc thesis. McGill Univ., Montreal

Pitteway, M. L. V., Hines, E. O. 1963. The viscous damping of atmospheric gravity waves. *Can. J. Phys.* 41:1935–48

Reed, R. J. 1967. Semidiurnal tidal motions between 30 and 60 km. *J. Atmos. Sci.* 24:315–17

Reed, R. J. 1972. Further analysis of semidiurnal tidal motions between 30 and 60 kilometers. *Mon. Weather Rev.* 100:579–81

Reed, R. J., Oard, M. J., Sieminski, M. 1969. A comparison of observed and theoretical diurnal tidal motions between 30 and 60 km. *Mon. Weather Rev.* 17:456–59

Richmond, A. D. 1975. Energy relations of atmospheric tides and their significance to approximate methods of solution for tides with dissipative forces. *J. Atmos. Sci.* 32:980–87

Richmond, A. D., Matsushita, S., Tarpley, J. D. 1976. On the production mechanisms of electric currents and fields in the ionosphere. *J. Geophys. Res.* 81:547–55

Roble, R. G., Emery, B. A., Salah, J. E., Hays, P. B. 1974. Diurnal variation of the neutral thermospheric winds determined from incoherent scatter radar data. *J. Geophys. Res.* 79:2868–76

Salah, J. E., Wand, R. H. 1974. Tides in the temperature of the lower thermosphere at mid-latitudes. *J. Geophys. Res.* 79:4295–4304

Salah, J. E., Wand, R. H., Evans, J. V. 1975. Tidal effects in the E-region from incoherent scatter radar observations. *Radio Sci.* 10:347–55

Schmidlin, F. J., Yamasaki, Y., Motta, A., Brynztein, S. 1975. *Diurnal experiment, data report, March 19–20, 1974. NASA SP-3095.* Washington, D.C. 150 pp.

Siebert, M. 1961. Atmospheric tides. *Adv. Geophys.* 7:105–82

Smith, W. S., Theon, J. S., Swartz, P. C., Katchen, L. B., Horvath, J. J. 1968. *Temperature, pressure, and wind measurements in the stratosphere and mesosphere, 1966. NASA TR-288.* Washington, D.C. 91 pp.

Smith, W. S., Theon, J. S., Swartz, P. C., Casey, J. F., Horvath, J. J. 1969. *Temperature, pressure, density and wind measurements in the stratosphere and mesosphere, 1967. NASA TR-R-316.* Washington, D.C. 83 pp.

Smith, W. S., Theon, J. S., Casey, J. F., Horvath, J. J. 1970. *Temperature, pressure, density, and wind measurements in the stratosphere and mesosphere, 1968. NASA TR T-340.* Washington, D.C.

Smith, W. S., Theon, J. S., Wright, D. U. Jr., Ramsdale, D. J., Horvath, J. J. 1974. *Measurements of the structure and circulation of the stratosphere and mesosphere, 1971–2. NASA TR R-416.* Washington, D.C.

Volland, H., Mayr, H. G. 1972a. A three-dimensional model of thermosphere dynamics, I. Heat input and eigenfunctions. *J. Atmos. Terr. Phys.* 34:1745–68

Volland, H., Mayr, H. G. 1972b. A three-dimensional model of thermosphere dynamics, II. Tidal waves. *J. Atmos. Terr. Phys.* 34:1769–99

Volland, H., Mayr, H. G. 1974. Tidal waves within the thermosphere. *Radio Sci.* 9: 263–26

Volland, H., Mayr, H. G. 1977. Theoretical aspects of tidal and planetary wave propagation at thermospheric heights. *Rev. Geophys. Space Phys.* 15:203–26

Wallace, J. M. 1975. Diurnal variations in precipitation and thunderstorm frequency over the coterminous United States. *Mon. Weather Rev.* 103:406–19

Wallace, J. M., Hartranft, F. R. 1969. Diurnal wind variations; surface to 30 kilometers. *Mon. Weather Rev.* 97:446–55

Wallace, J. M., Tadd, R. F. 1974. Some further results concerning the vertical structure of atmospheric tidal motions within the lowest 30 kilometers. *Mon. Weather Rev.* 102:795–803

Zimmerman, S. P., Rosenberg, N. W., Faire, A. C., Golomb, D., Murphy, E. A., Vickery, W. K., Trowbridge, C. A., Rees, D. 1974. Aladdin II Experiment: Part I, Dynamics. *COSPAR: Space Research XIV*. Proceedings of Open Meetings of Working Groups of the 16th Plenary Meeting of COSPAR, 1973, pp. 81–87

Ann. Rev. Earth Planet. Sci. 1979. 7:227–48

PRIMARY SEDIMENTARY STRUCTURES

J. C. Harms
Marathon Oil Company, P.O. Box 269, Littleton, Colorado 80160

INTRODUCTION

While rivers flow, waves pound the shore, or winds shift sand and dust across the desert, particles moved and dropped by these processes build mounds, ridges, and corrugations or form layering that subtly records the history of events. We can see the most recent products of this action when we stroll across a now-dry river bed or tidal zone or over a desert dune, knowing full well that what we see will be partially destroyed and modified by what comes next in the natural course of events. Similarly, ancient sedimentary rocks (and certain extrusive igneous rocks) display layering and undulations of bedding caused by episodes of deposition and erosion. These features, created by the physical interactions of grains and fluid, are called primary sedimentary structures. They are the statements, sometimes complete but more often fragmentary, written by the rivers, shores, and dunes of the remote past.

Can we read this record? Yes, in part fairly well, in part with difficulty and a good deal of inference, and in part scarcely at all. I shall outline a system of sedimentary structures about which a good deal is known and some useful interpretations can be made, as well as other structures whose significance is elusive. In presenting such a summary, the viewpoint is inescapably incomplete (limited space prohibits a review of the vast and burgeoning literature) and subjective. My own interest and perspective is that of a geologist interpreting ancient facies, rather than that of one interested in theoretical or engineering aspects of fluid dynamics and sediment transport.

BREAKING THE CODE FOR SHALLOW WATER FLOWS

Ancient sedimentary rocks commonly display delicate textural lamination within somewhat thicker beds or have wavy parting surfaces that somehow

0084-6597/79/0515-0227$01.00

reflect fluid-particle interactions at the time of deposition. The principal scheme on which interpretation of these features is based is founded on two propositions: 1. there is an *orderly* system of shapes (called bed forms) acquired by a loose granular bed that results from flow conditions and

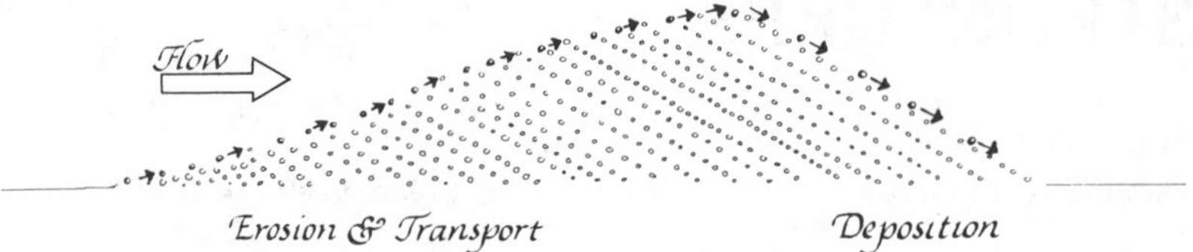

Figure 1 Sediment moved along the bed by fluid flow tends to accumulate in ridges called bed forms. Grains are rolled or bounced by fluid shear along the up-current side; this sort of movement near the sediment-water interface is designated as traction transport. When grains reach the crest of the ridge, they slide or tumble down a somewhat steeper down-current face to form inclined laminae marked by slight variations in texture. The position of the bed form moves down the flow with time because of erosion and deposition. The laminae constructed by a migrating bed form are termed a set. At high flow velocities, grains move more continuously, and bed forms change in style.

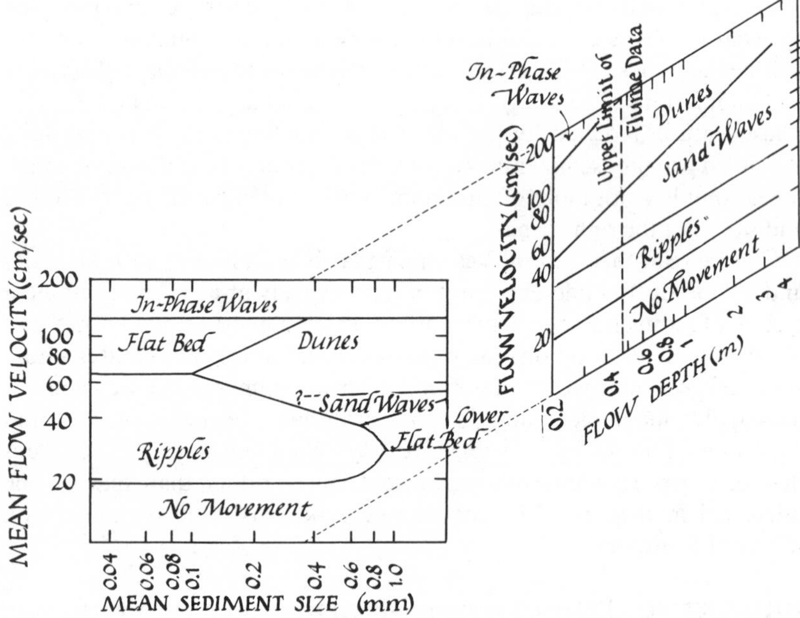

Figure 2 Relationship of bed-form types to the dimensions of mean flow velocity, grain size, and mean flow depth for simple water currents. The front panel shows the various fields for a flow depth of 20 cm, while the influence of depth is shown only for a single plane extending toward the back through a grain size of 0.4 mm. The bed forms are illustrated and described in Figures 3 through 7. Adapted from data compiled by Southard (Harms et al 1975).

fluid and particle properties, and 2. these bed forms, which migrate with prolonged flow, construct laminae that mimic important elements of their shapes. These axioms are verified by experiments and by observations of natural environments. Flows of prescribed depth and velocity are seen to consistently build bed forms of statistically similar sizes and shapes, and these shifting bed forms make recognizable sets of laminae. Figure 1 shows how a simple bed form builds a planar set of cross laminae as grains are moved along the up-current slope and tumble down the steeper down-current face.

The Rosetta stone for interpreting structures developed by unidirectional water flows was provided by flume experiments. Gilbert (1914) recognized a succession of dune to plane to antidune bed forms with increasing flow velocity; the basic succession was expanded and modified on the basis of extensive flume studies reported by Simons, Richardson & Nordin (1965). Later flume data were incorporated in a still more extensive scheme by Southard (Ch. 2 in Harms et al 1975). This latest rendition is shown in Figure 2 for one flow depth for which there are many data. Based on these experimental data, bed forms with distinctive shapes or sizes are seen to occupy fairly closely defined fields of mean flow velocity, grain size, and mean flow depth, with other variables held constant. Thus Figure 2 indicates that ripples, sand waves, dunes, and so forth, if consistently defined and recognized, have specific meaning in flow environment terms.

The bed forms of Figure 2 are perhaps most easily defined and visualized by simple diagrams. Figures 3 to 7 show typical configurations of loose sand or silt beds under unidirectional water flow, as well as the stratification

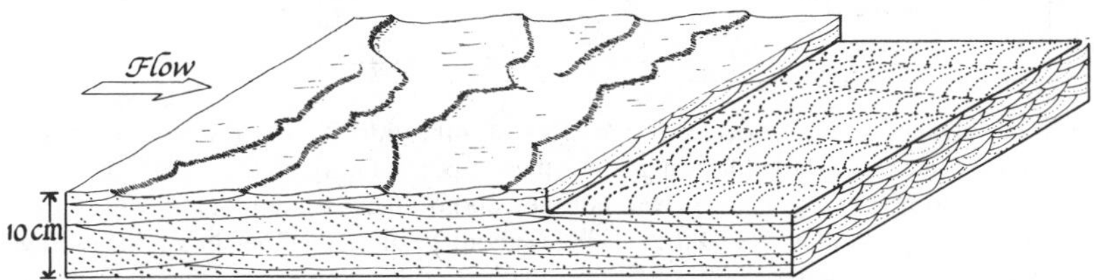

Figure 3 Current ripples and small-scale trough cross stratification. Ripples are small features with crest spacing of about 10–15 cm and heights mostly less than 3 cm. Upstream slopes are gentle, downstream faces are commonly at the angle of repose for sand, and forms are cuspate or linguoid in plan view. Stratification developed as current ripples migrate with slow deposition is invariably small trough sets, consisting of erosional hollows elongate parallel to flow filled with scoop-shaped laminae plunging down-flow. This configuration reflects infill of small, slightly deeper scour pockets that erode below the average bed level. Thus the preserved stratification reflects only certain elements of the bed configuration when sediment is deposited at a slow rate. Ripples develop in sediment sizes ranging from silt to medium-grained sand. Current ripples and associated stratification have been extensively discussed in the literature (Harms & Fahnestock 1965, Allen 1968, Harms 1969).

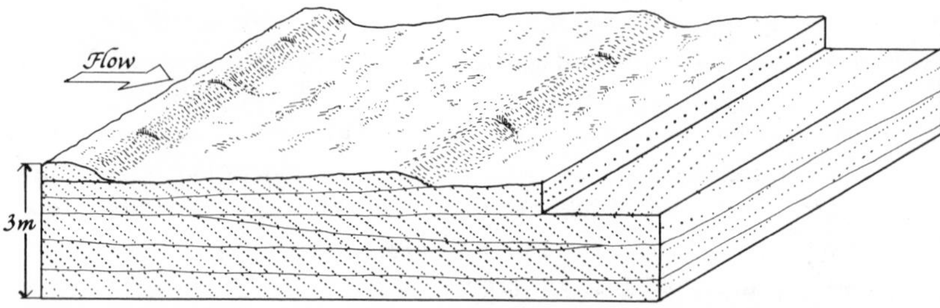

Figure 4 Sand waves and tabular cross stratification. Sand waves are relatively straight-crested features that have a large range of dimensions. In deep flows, they can be many meters high and have crest spacings of hundreds of meters. In shallow flows, sand waves can be very low but have characteristically straighter crests and broader spacing than would current ripples or dunes of similar heights. A key feature is the downstream faces that slope at the angle of repose of sand. This and the lateral continuity of straight or slightly sinuous crests yield tabular sets of planar cross laminae as waves migrate and deposit. Sediment is moved along the upstream sides by ripples or dunes superimposed on the larger sand wave. Sand waves are not known to develop in sediment finer than 0.1 mm in flume experiments; similarly, tabular cross stratification in ancient very-fine-grained sandstone or siltstone is unknown to me. Additional discussion of sand waves can be found in Harms et al (1975, pp. 24–49).

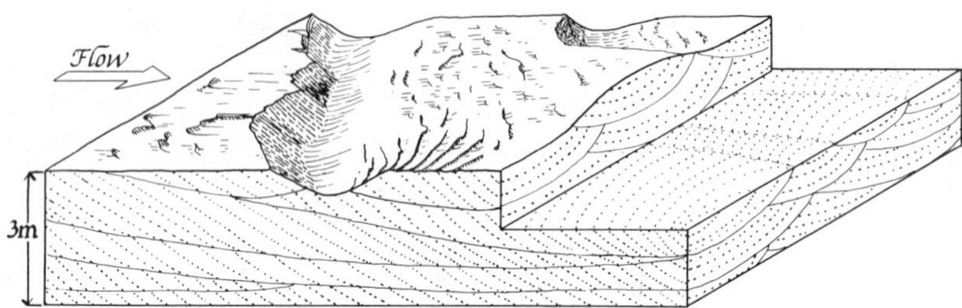

Figure 5 Dunes and large-scale trough cross stratification. Dunes resemble ripples or sand waves in profile, but differ in many other important ways. Their height is several centimeters to perhaps a meter or two, but they do not increase additionally with greater flow depths as can sand waves. Dune spacing is about 10 times their height, on the average. The forms are less regular than sand waves, with shorter and less continuous crests and deep scour holes. Toward the upper range of flow velocities, dunes appear to be lower and more rhomboid in the transition to flat bed. Downstream faces can be as steep as the angle of sand repose, but are commonly less inclined because sediment is distributed across these faces by powerful eddies. As a result, cross laminae are curved and tangential to their lower boundary; laminae dips are in the range of 18 to 30 degrees in uncompacted sediment. Trough cross strata are the product of dune migration, with elongate scours filled with curved sets of laminae. Like ripples, the form of stratification is dominated and controlled by deep scours. The trough shapes and the curved laminae allow distinction between stratification formed by dunes and sand waves in ancient rocks. Dunes do not appear to form in sediments finer than about 0.1 mm. Additional discussions of dune forms and stratification appear in Harms et al (1975) and Harms & Fahnestock (1965).

formed within a limited thickness of sediment by the migration of these features. A brief description of distinguishing characteristics and dimensions for each bed form is given in the captions of these figures.

The relationships described by Figure 2 are relatively simple and useful. With given grain size, flow depth, and flow velocity, quite distinct bed forms develop which deposit recognizable styles of stratification. In

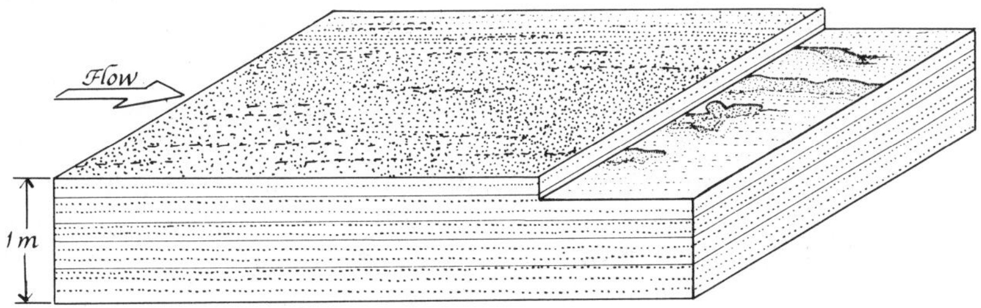

Figure 6 Flat bed and horizontal stratification. A nearly featureless flat bed (sometimes called a plane bed) is formed under two conditions. In one case, and certainly the one most readily attained in nature, energetic flows moving at velocities in excess of those for ripples or dunes cause a flattening of all transverse irregularities. In the other case, slow flows over coarser beds cause flat bed transport in a small range of velocities below the advent of dunes. For both, turbulence takes the form of small helical vortices with axes parallel to current which sort grains of differing sizes or densities into subtle windrows that give the bed a streaked appearance. The lamination produced is horizontal and has lineation parallel to flow. In ancient rocks, this fabric on bedding is called parting lineation.

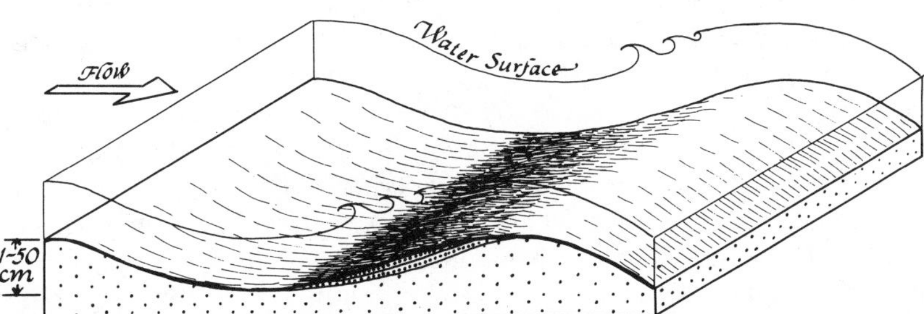

Figure 7 In-phase waves and anti-dune bedding. At high flow velocities, transverse features form on the bed, which are also strongly reflected by the water surface. These water-surface waves are in phase with the bed undulations, hence the name. The bed forms can move downstream, remain stationary, or migrate upstream. In the latter case, called anti-dunes, flow deepens over the crest of beds wave until the water surface becomes steep and unstable and forms a breaking roller, as illustrated. Sediment eroded and carried in temporary suspension drops to the bed on the up-current side of the transverse ridge, causing the seemingly unlikely upstream migration. The bedding formed by this process dips upstream but is generally rather indistinct. Antidune bedding is relatively rarely preserved in ancient rocks but can be found in some turbidity current and alluvial fan deposits (an example is described by Hand, Wessel & Hayes 1969).

applying the diagram to interpret ancient rocks, it is quite logical to reverse the procedure and use the stratification to infer bed form (or the bed form itself may be preserved) and, with the observed grain size of the sediment, to imply a range of flow depths and velocities. A unique flow velocity cannot be deduced unless depth is somehow independently determined, for example by the cross-sectional outline of a preserved channel.

The scheme just described has evolved as a major interpretive approach over the past 15 years. The relationship of bed forms to flow conditions and to stratification is based on many flume experiments and observations of natural flows and stratification in modern sediments. Yet the approach does have distinct limitations. First, the depth range of most flume experiments is only 15–30 cm or rarely reaches 1 m, and detailed observations of deeper natural steady flows are still very limited. The same bed forms developed under shallow flows are seen with deeper currents, and we infer that the boundaries between fields can be extrapolated to significantly greater depths as shown for medium-grained sand in Figure 2. But we also know that larger-scale bed irregularities develop with deeper flows, commonly in combination with ripples or dunes, and we know very little of the character and significance of these features when observed in modern sediments or ancient rocks. Additionally, some bed forms may not develop in very deep currents in the appropriate range of velocities because the scale of turbulence is large.

A second limitation of the flume-based scheme of interpretation of primary structures is inherent in the simplicity of the experimental approach. Experiments are carefully designed to be steady and uniform, that is, invariant in time and space, so that bed forms are known to be in equilibrium with flow. In nature, depth, velocity, or sediment supply commonly change and impose variations on bed forms and stratification that are not entirely decipherable. Yet experiments to investigate the many plausible changes are so difficult and myriad as, so far, to defy serious and comprehensive effort. Viewing this limitation in the reverse, the very lack of resemblance of some structures to equilibrium forms may imply transient conditions and be of some interpretive value.

A third simplification implicit in Figure 2 is that velocity, depth, and grain size are the dominantly important variables controlling bed form. Flume data are well organized using these variables, but the experiments are generally conducted with fresh, clear, warm water and fairly well-sorted, pure quartz sand. It is known from experiments and natural flows that cold or turbid water or grains of lower density cause a shift of bed form boundaries to lower velocities. In natural channels, such a shift can cause dramatic changes in flow depth and velocity, even though slope and discharge remain constant (an example was given by Harms & Fahnestock

1965). Because a larger number of variables can be inextricably involved, some geologists have preferred to use the more qualitative concept of flow regime, wherein ripples through dunes define the lower-flow regime and flat bed through antidunes the upper-flow regime. This terminology was developed by hydraulic engineers to group flow phenomena in open channels in useful ways (Simons, Richardson & Nordin 1965). The concept and terminology of flow regime is now well rooted in the literature, and I see no reason for totally abandoning it. For example, it is concise to state that channels of deep, low gradient streams flow dominantly in the lower regime, implying bed configurations including ripples, sand waves, and dunes, whereas shallow channels on steep alluvial fans commonly flow in the upper regime with flat bed, standing wave, or antidune configurations. So flow regime is useful for general designations where control by specific variables is not or cannot be designated. However, relationships between fields of bed forms are more clearly defined using depth, velocity, and grain size, provided we recall that the secondary influences of fluid density and viscosity, grain density, and so forth have been carefully balanced out.

Two other details of Figure 2 deserve brief note. Many past studies have paid special attention to the boundary between no transport of sediment and the formation and migration of ripples. Because many flume experiments begin with a flattened bed, it is well known that higher velocities are required to initiate ripples than to maintain their active migration, simply because of the turbulence and local shear generated by the ripple crests reaching upward into the flow. Because initially flat fine-grained sand or silt beds are probably uncommon in natural environments, I prefer to simplify the diagram by using a line representing an originally rough bed. Coarser grain sizes, which do not develop ripples, have a lower flat bed phase as an equilibrium form, a fact that has only been recognized in the past few years (Southard, in Harms et al 1975, Ch. 2). This dual position of flat beds on Figure 2 leads to the only apparent ambiguity of interpretation of velocity from the flat bed form or horizontal stratification (Figure 6). Thus it would be possible to interpret horizontal stratification in sands coarser than 0.5 mm as formed at fairly low velocities or quite high velocities (in mean flow depths somewhat deeper than 20 cm shown as the front panel of Figure 2). In my opinion, it is unlikely that substantial thicknesses of horizontal lamination would be deposited in natural environments by lower flat beds because transport capacity is relatively low, and flow conditions would likely change before a very thick set developed. However, additional evidence should be sought to aid interpretation, such as the presence of larger pebbles or cobbles in the horizontal strata that would imply high velocities for transport.

MORE COMPLEX CONDITIONS

The preceding discussion shows that several very common stratification types can be related to bed forms and flow conditions through flume experiments or observations of natural flows. The conditions under which these relationships are defined are of necessity somewhat simple, in that flows are shallow, unidirectional, and steady, and deposition rates are low or nil. Of course, natural conditions are more variable, and currents can quickly vary in depth, velocity, and direction, or sediment can be supplied at low or high rates. The following examples attempt to illustrate how certain structures reflect more complex circumstances. Although interpretations are mostly qualitative and less secure than for equilibrium forms, these examples indicate the kinds of details that may be preserved in ancient rocks and the sorts of inferences that may be drawn to assist in reconstruction of ancient depositional environments.

Ripples

Ripples are the smallest in scale within the hierarchy of bed forms, as intimated in the description of Figure 3. Because of their small size, their shapes and patterns are easily surveyed in a glance at a loose sand bed or an ancient bedding surface, and ripples in great variety have been recorded in modern and old sediments. A second advantage of their small size is the quick adjustment that ripples make to changing flow conditions. Their individually small masses are readily reshaped by ephemeral influences, whereas much larger bed forms can contain so much sediment in relation to transport rate that shape adjustments lag far behind hydrodynamic shifts. As a result, the ripples we see generally reflect transport conditions of the preceding few minutes or hours.

Ripples display an amazing variety of shapes and patterns. Our eye, powerful integrator of pattern that it is, readily recognizes and distinguishes these varieties, although they are commonly difficult to define with simple and quantitative criteria. Figure 8 is an attempt to illustrate major ripple types that I believe are especially useful for interpretation and to indicate the influences that caused the patterns and stratification to be what they are.

Current ripples on sand-sized material, which have already been discussed and illustrated in Figure 3, provide a good starting point for comparison with other ripples. The pattern, typical for such ripples and shown in the left center of Figure 8, is composed of mixtures of cuspate and linguoid forms that are strongly three dimensional. Experiments with fine-to-medium-grained sand (Harms 1969) show that ripples have longer, more continuous crestlines at flow velocities in the lower range for ripples and more broken patterns at higher velocities. Small-scale sets of trough cross strata are the dominant stratification type deposited throughout the velocity range of ripples (Figure 3).

Finer grain sizes in the coarse- or medium-silt range cause patterns to change dramatically through the velocity range for ripples. At lower velocities, current ripples on silt resemble closely those formed with coarser sand (lower left block of Figure 8). At higher velocities, but below speeds required for flat beds, silt ripples become lower, more rounded in profile, and longer crested and sinuous in plan view. The physical cause for this change is stronger turbulent eddies which lift increasing volumes of silt into temporary suspension. This suspended material falls to the bed somewhat indiscriminantly and in some places mantles the up-current sides of ripples or causes lee faces to have low-angle, concave-upward profiles. In contrast, grains on lower velocity silt ripples or sand ripples rarely move much above the bed, but rather roll or bounce along the profile and commonly tumble or slide down the lee face to form planar laminae that dip 30 to 35 degrees, in the manner illustrated in Figure 1. As a result, relative flow velocity of silt ripples is expressed by changes in both pattern and lamination, and these character changes can be used to suggest velocity ranges from Figure 2.

Ripple patterns can also reflect shallow flows that inhibit fully developed vertical profiles. The height of current ripples rarely exceeds three centimeters in any flow over a few centimeters deep; with very shallow flows, the profiles must of necessity become truncated and flattened, and the pattern takes on a braided or ropey appearance (top left block of Figure 8).

A rapid supply of sediment to a rippled bed causes conspicuous changes in shapes and stratification. Two cases of lower and higher rates of aggradation, called climbing ripples, are illustrated in the lower right blocks of Figure 8. With lower rates, sediment is deposited and preserved on lee faces, but largely eroded off of up-current sides, giving the net effect of a series of ripple profiles migrating downstream but climbing upward. Because such rapid supply requires that much of the sediment arrive in suspension, the faces of the ripples have low-angle, concave-upward profiles and internal lamination. The pattern becomes long crested in comparison to ripples formed at near-equilibrium rates of deposition, taking on an elongate cellular plan. At still higher rates of supply, the bed is undulatory but sediment is added to both up-current and lee faces at nearly the same rate. The crest lines may migrate down-current, up-current, or not at all. The ripple pattern is subparallel or elongate cells. Additional details or illustrations of aggradational ripples are shown by Jopling & Walker (1968) and McKee (1965).

Recognition of patterns or laminations that indicate significant aggradation rates has value for interpretation of ancient sediments. There are relatively few environments where rapid supply of sand, which must be partly in suspension, is coupled with ripple-scale features that indicate relatively low flow velocities. Such conditions can be attained where a confined current jets into a more extensive water body, for example at the

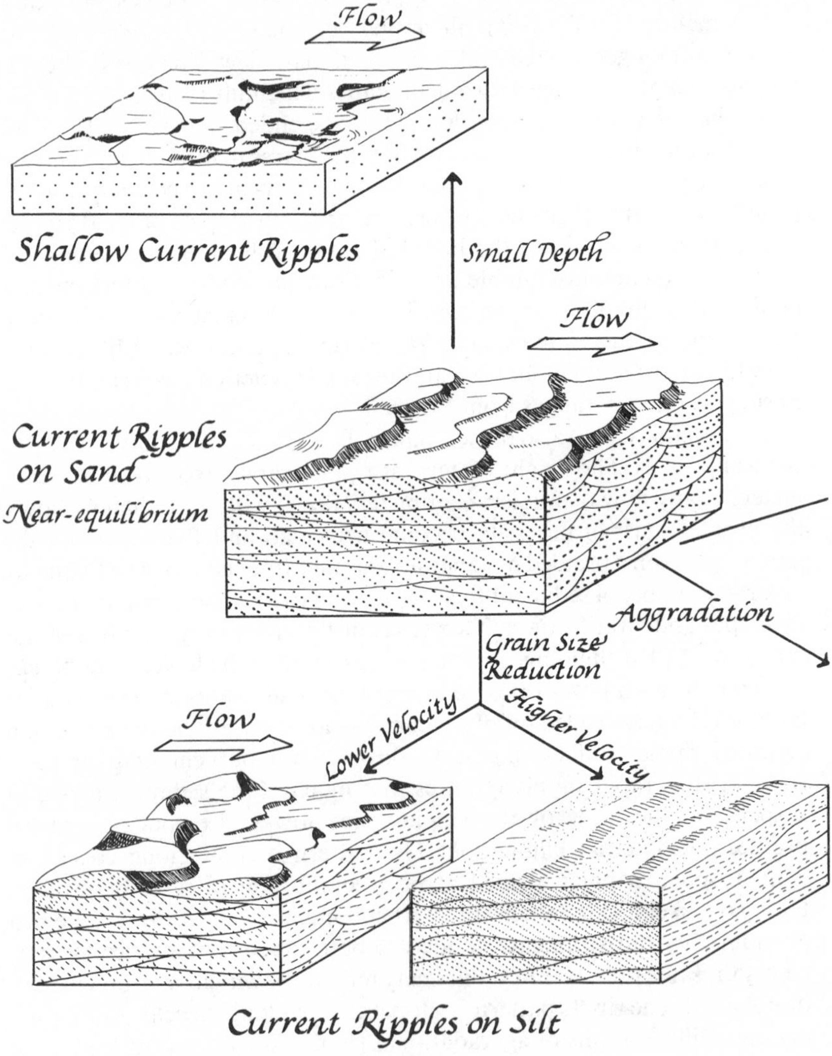

Figure 8 Variations in ripple forms and stratification caused by changes in velocity,

stream mouth of a delta or a levee break inundating a flood plain, or where an energetic flow slows, such as decelerating turbidity currents or rapidly waning flood surges in ephemeral streams. Cognizance of aggradation certainly can add useful constraints in reconstructing depositional environments from primary structures.

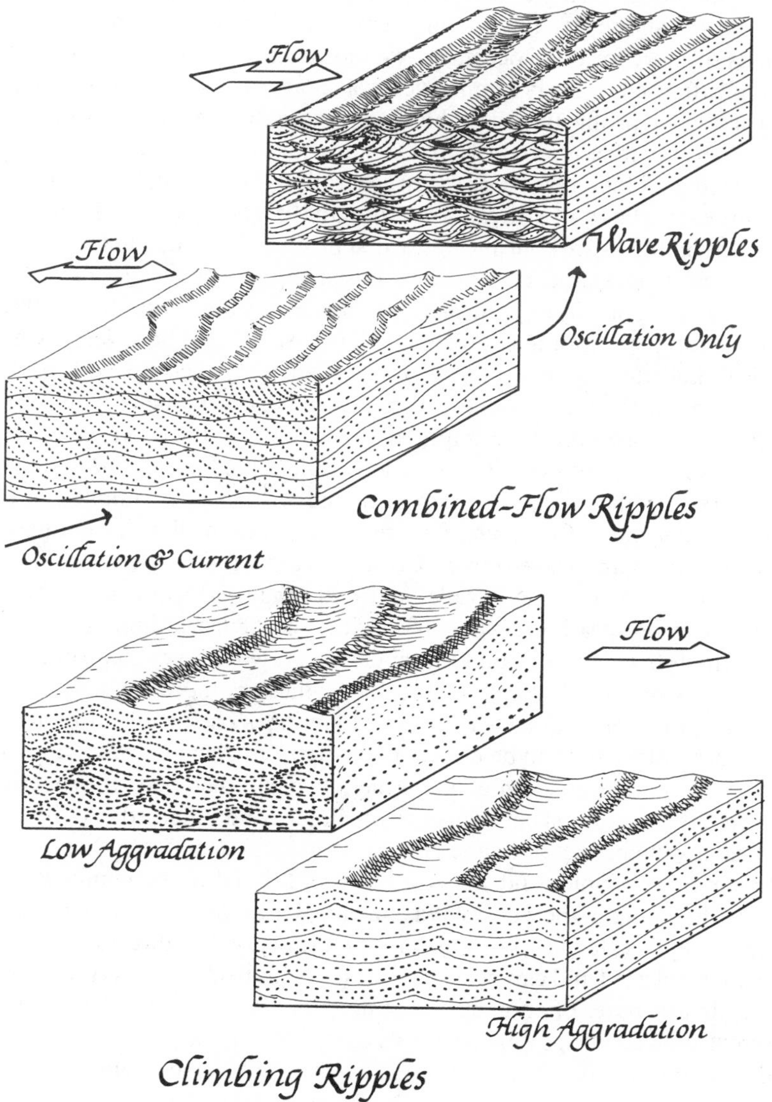

grain size, depth, rate of sediment supply, or flow direction. See text for discussion.

All of the preceding cases are based on currents that are fairly steady in direction and speed, but this is not always the case. Waves, with which we are all well acquainted, are a common example of water motion with varying direction and velocity. As a wave passes, each bit of water traces an orbit which is more or less circular near the surface and more elliptical toward the bottom. At the bed, motion is to and fro with acceleration, deceleration, and reversal in the time represented by the wave period, ten seconds or more for large ocean waves and much less for smaller waves.

Ripples formed by fairly symmetrical waves are themselves symmetrical and rounded in profile (top right block of Figure 8). The pattern is one of parallel crests with scattered tapered terminations or Y-shaped junctions. The spacing of the crests correlates generally with the scale of orbital motion near the bed and with grain size (Inman 1957, Harms 1969). Laminae deposited by waves are concave upward in profile, are grouped in sets that parallel crest lines, and can show opposed dip directions because the ripples can temporarily migrate in either direction as wave conditions vary.

The oscillatory motion associated with waves may not be strictly symmetrical, and the distance moved or the velocity associated with each wave pulse can be greater in one direction than the other. Such asymmetry occurs where waves are superimposed on currents or where waves move across a shoaling bottom. This situation is illustrated as combined flow on Figure 8, with a reversal of stronger and weaker flows implied in a time span of seconds. Combined-flow ripples are intermediate in character between current and wave ripples. Crest lines are straighter than those of current ripples, but not so parallel and well organized as those of wave ripples. The profiles are asymmetric, with steeper faces dipping in the direction of strongest flow, and the ripples migrate that way to form consistently oriented cross laminae. But the profiles are more rounded than current ripples, yielding concave upward laminae. The details of pattern and internal structure can be used to compare to current and wave ripples and infer a relative position between these two end members. If properly interpreted, the contrast between current- and wave-dominated conditions can add a valuable detail to ancient environmental analysis.

In summary, ripples and the stratification formed by their migration display the stamp of their flow environment. Variations in grain size, flow velocity, depth, rate of deposition, and complexity of flow direction are all somehow reflected and can be used at least qualitatively for interpretation of certain attributes of an ancient setting. Not all possibilities have been covered, and knowledge is by no means complete, but the general concepts and potential are apparent. Additional examples of ripples from a variety of environments are well illustrated by Reineck & Singh (1973, pp. 14–47 and 95–103).

Beaches

To extend the concept of dramatically changing flow conditions beyond the example of wave ripples, consider water motion on a beach. Waves are steepened by frictional drag as they approach the shoreline and finally break with great turbulence at the beach. The mass of water that plunges at the foot of the beach runs up the slope, stops, partly filters into the beach, and runs back toward the sea, forming the familiar smooth swash zone. Over this zone, flow depth, velocity, and direction change in periods of seconds, and the position of the zone can shift in hours if tides are substantial. As a consequence, conditions are never steady and uniform in the sense of flume experiments previously discussed.

Beach stratification appears relatively simple, even though flow conditions are complex. Figure 9 shows the sort of lamination that would be seen in a shallow trench dug in the swash zone. Slightly divergent sets of planar laminae dip seaward at angles typical of beach inclinations, generally a few to 15 degrees (beaches are steeper in coarser material and with larger waves). The dominance of planar laminae lying parallel to slightly sloping depositional surfaces is the result of the very shallow and energetic flow of the swash zone. Ephemeral ripples, antidunes, or rill marks may be formed with each pulse, but all such features are obliterated by the thin flow of the succeeding down- or up-beach swash.

Sets of laminae, which appear to have nearly parallel boundaries in small-scale exposures like Figure 9 but are probably broadly lenticular, are separated by erosion surfaces that also dip gently seaward. Erosive events are caused by changes in wave size and direction or in water level, which in turn alter beach slope or cause cuspate irregularities to migrate.

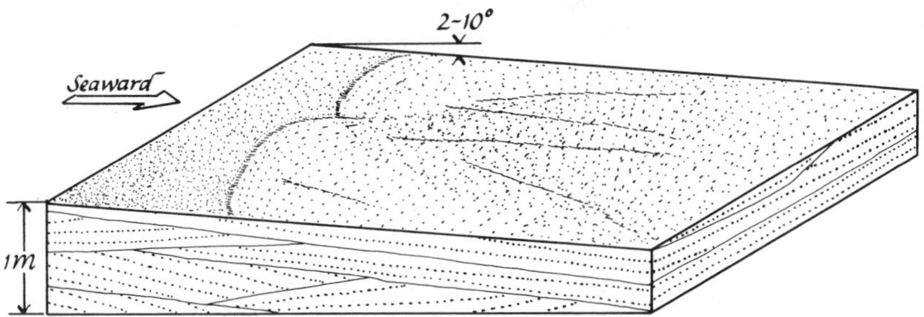

Figure 9 Beach stratification formed in the swash zone of shoreline waves. Fine, even laminae are deposited as thin flows run up the beach and return as waves break at the beach. The laminae dip seaward at the angle of the beach slope, which for sand beaches ranges commonly between 2 and 10 degrees. Slight discordances between sets of laminae reflect changes in wave size or direction or tide level. Additional illustrations, discussion, and references are given by Reineck & Singh (1973).

Thin planar sets of laminae separated by smooth erosion surfaces inclined at low angles are recognized in the ancient record and are commonly inferred to be beach deposits. This interpretation is strengthened when additional evidence is supportive, such as dip toward the likely direction of the ancient sea, heavy mineral concentrations like those noted along some modern beaches, and position in a stratigraphic sequence that suggests a shoreline environment (Harms et al 1975, Ch. 5).

Hummocky Cross Strata

Ancient sandstones, deduced to be of shallow marine origin from fossils and facies relationships, commonly contain cross strata of the distinctive style illustrated in Figure 10. I have used the term "hummocky" to describe the mound and hollow form of the bed evident from bedding surfaces (Harms et al 1975, p. 87). Similar bed forms have not been noted in flume experiments, nor in modern environments. Indeed, for very fine-grained sand or coarse silt which commonly compose hummocky cross-stratified beds, flume experiments (Figure 2) suggest that flat bed succeeds ripples and that no larger-scale bed form depositing thicker sets even exists.

The interesting problem presented by the hummocky type of cross strata is that we know it from the geologic record, not from experiments or the modern record, so that inferences about its significance are largely guided by observations of the ancient. In this respect, the hummocky style is common in shallow marine sediments which, where water depth can be inferred from position in prograding shoreline sequences, were deposited

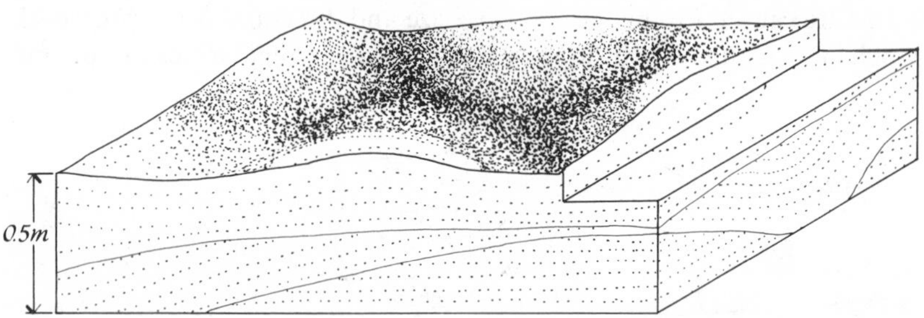

Figure 10 Hummocks and hummocky cross stratification. Low mounds and hollows form at shallow depths under strong wave action, creating an irregular bed topography without strongly aligned features. Sediment moved by strong wave surges is temporarily suspended and spread over the uneven surface as thin laminae. The resulting cross stratification has many gentle erosional surfaces overlain by parallel or nearly parallel laminae. The stratification has a similar appearance on vertical faces cut at any orientation, unlike trough cross stratification (Figure 5) which has elongate scours and well-oriented cross laminae. Hummocky sets are most common in marine sequences deposited at depths of a few to 30 meters or more and in silt or very-fine-grained sand (adapted from Harms et al 1975).

5–30 m or more below sea level. Beds composed of hummocky sets commonly have sharp bases (in many cases a muddy substrate was grooved or indented by dragged objects), burrowed tops, and are underlain or overlain by units of laminated mudstone and wave-rippled sandstone. This assemblage of features suggests pulses of rapid deposition in fairly shallow water that was wave agitated at least part of the time, and periods of slow sedimentation when animals could burrow and partly rework the upper parts of sandy beds or mud layers were deposited. The implied intermittence of deposition in turn suggests the influence of storms so as to radically change the supply of sand.

Observations of the sea floor at depths of 5–15 m made possible by diving during strong storm wave activity indicate how sediment is moved if the above inferences about hummocky cross strata are correct. Storm waves in relatively shallow water cause large, high velocity oscillations at the bed. Displacement of each wave surge can reach several meters and peak velocity of a meter or more per second. With a bed of finer sand sizes, these surges lift grains into temporary suspension at highest velocities, but allow the grains to drop back to the bed as each pulse slows and reverses direction. It is not difficult to visualize how grains moved in this way would be spread as thin laminae over an undulatory bed in the manner illustrated in Figure 10. Also, since shoreline profiles are known to be steepened during periods of storm waves and sediment is moved offshore, it is possible to imagine rapid influx of finer sand or silt into somewhat deeper water.

This somewhat tenuous web of inference suggests that hummocky sets are formed in shallow seas by storm waves. Obviously additional and well-quantified observations should be made in appropriate settings, and other likely hypotheses should be proposed and explored. However, this example illustrates the value of careful discrimination of forms of primary structures and the interpretive leverage gained through consideration of stratigraphic setting.

LARGER FORMS WITH COMPOUND BEDDING

The existence of larger bed features than those discussed in preceding sections is well known in many environments. For example, some large constructional forms are known in river channels (Coleman 1969, Collinson 1970), shallow seas (Houbolt 1968, Stride 1970), and tidal estuaries (Knight & Dalrymple 1975, Greer 1975). In many cases, a veneer of small bed forms of the type already discussed covers the bigger features and makes up sets of the appropriate kinds. What has not been generally learned is whether, or how, internal bedding reflects the large form. Are there, for example, master bedding surfaces that mimic the major shapes of tidal ridges and would provide greater interpretive insight if recognized in ancient rocks?

On an even larger scale, major depositional surfaces are present along delta fronts, reef flanks, and, for that matter, continental slopes. Precisely how sediment arrives and is deposited on these surfaces or how the slopes are expressed in bedding is little known. I believe that careful attention to primary structures in such settings will yield important information in future research. But for the moment, I have chosen two examples of more modest dimension to illustrate interpretation of two scales of structures.

Point-bar Accretion Surfaces

A meandering channel erodes its curves on the outer bank and deposits on the inner bank. The accretional crescents left by a shifting sinuous stream are called point bars, and many river flood plains are imprinted with meander scrolls that mark the progress of the migrating channel. A cross section of a meandering channel at a bend shows that the channel is deepest close to the outer eroding bank and the profile slopes more gently on the accretional point-bar side.

Sediment is added to the point bar by dunes, sand waves, or ripples that migrate along the course of the bend, forming a sequence of cross

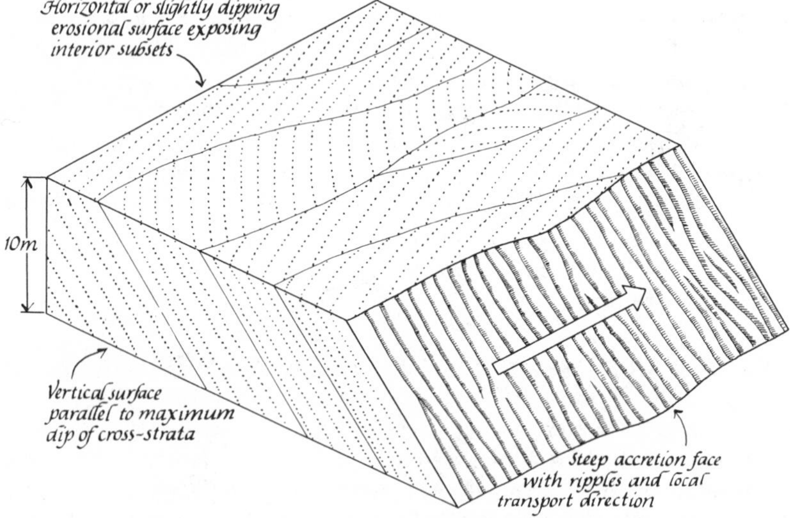

Figure 11 Composite sets of cross strata deposited on an ancient eolian dune complex. Sand was drifted by wind along dune slopes, leaving the low crests of eolian ripples on these surfaces. This example is by no means the only type deposited on eolian dunes, but it may be indicative of structures produced by forms that are subparallel to wind direction (adapted from Walker & Harms 1972).

strata that generally become smaller in scale and finer grained from the deepest part of the channel up along the point-bar profile. This organization of structures and textures has become well recognized as the "fining-upward cycle" of point-bar deposits of meandering streams.

Much of the stratification recognized in such deposits is trough or tabular sets described in Figures 3, 4, and 5. But occasional exceptional events, such as a mud drape deposited from suspension during slack flow or a broad erosion surface cut during flood stage, can leave a larger-scale surface that marks the point-bar profile. If preserved and recognized, these surfaces give detailed information on channel dimensions, especially stream width, that would otherwise be unavailable.

An illustration of how such information on channel morphology can be amplified into interpretation is provided by Cotter (1971). Using data from modern streams relating depth, width, meander length, gradient, and sinuosity to sediment and water discharge that were collected and summarized by Schumm (1972), Cotter made some estimates of sinuosity, discharge, slope, and related characteristics for a Cretaceous stream, based on larger-scale surfaces within the ancient channel. This example indicates the value garnered from observation of large structures, provided that data from modern environments are available.

Eolian Dunes

Wind heaps sand into dunes that are varied in size and shape. The large and, in some places, strikingly aligned dunes of major sand seas must rank among the most imposing bed forms on the Earth's surface. Although a detailed discussion of eolian processes and dunes is beyond the scope of this paper, I will use one ancient example of wind deposits to illustrate compound bedding of two scales related to larger and smaller bed forms and to suggest how wind processes differ from water processes.

First, air is notably less dense and viscous than water. Winds capable of moving fine-grained sand transport these grains in bouncing paths, a process called saltation. Silt or clay is lifted by turbulence and carried off in suspension by winds that move sand, and coarser material such as granules or small pebbles is slowly rolled by the impact of the bouncing sand grains. A sand bed moved by the wind develops low, beautifully parallel ripples or becomes flat at higher velocities that drive large quantities of sediment. Bagnold (1941) presented a thorough analysis of the physical processes of wind transport.

Returning to an example, Figure 11 illustrates cross stratification in a well exposed ancient dune complex described by Walker & Harms (1972). The interpretation of this Permian sandstone as eolian is based upon the large cross strata composed of well-sorted fine-grained sand, abundant ripples of wind-blown type, tracks of vertebrate animals, raindrop imprints, extensive erosion surfaces with scattered but evenly spaced coarse sand grains, and facies association with red, coarse-grained sediments of alluvial fan origin. Note in Figure 11 that there are sloping truncation surfaces that on a vertical face display the same dip as fine laminae but that on a near-horizontal plane separate discordant sets. Parallel ripple crests trend up and down the prominent slopes, indicating that sand was drifted along the flanks of dunes, adding small sets incrementally to some much larger form.

A reconstruction of dune form from Figure 11 suggests the following generalities. Dune heights were at least 10 m in many cases, perhaps substantially more, because the preserved portions are commonly this thick and obviously represent only the basal portion of some higher feature. Dune slopes were at least 20 to 26 degrees based on present flank dips, but not allowing for compaction of unconsolidated sand; slopes were less than 35 degrees because avalanche features, which should be well represented if the angle of repose of dry sand was approached, are very rare. Sand was commonly moved along dune slopes, apparently as bulges that deposited smaller sets; wind direction reversed from time to time because a few small sets show directions opposed to the majority, as viewed on a near-horizontal plane. In summary, it is likely that the dunes were more or less longitudinal forms parallel to prevailing wind directions, but the complexity of local dips viewed on a scale larger than that depicted in Figure 11 indicates that the dunes were by no means simple. In this instance, it is difficult to find precise analogs among studies of internal structures of several types of modern dunes (McKee 1966, Bigarella, Becker & Duarte 1969). Still, consideration of structures of larger scale provides valuable perspective from which to interpret this ancient example.

STRUCTURES IN FINE-GRAINED SEDIMENTS

The primary sedimentary structures described so far are common in sand and gravel. A few are formed in silt with grain diameters as small as 20 or 30 μm (Figure 2). Particles within the range of medium-grained silt to gravel generally travel close to the bed, so that they participate in tractional bed forms such as ripples or dunes in the manner illustrated in Figure 1. If flow velocity is so great that turbulence carries such grains into suspension, their settling rate is large enough to return them to the bed fairly promptly where they are reinvolved in traction processes.

Grains with diameters of a few microns or low-density particles such as clay flocs or delicate biogenic tests can be retained in suspension longer because of low fall velocities. Such grains may be lifted to the upper levels of the flow, supported by turbulent eddies, and only slowly return to the bed. As a result, structures in very fine-grained material are different and generally simpler than in coarser sediments.

Fine parallel lamination is the common structure of clay mud or its consolidated equivalent, shale. By parallel, I imply that not only are the laminae parallel one to the other, but that they drape over any irregularities in their substrate, maintaining parallelism to underlying small or large features such as ripples or channels. This parallel quality originates because sediment settles from far above the bed. A second common attribute of such laminae is textural grading, so that within the lamina thickness, finer-grained or less-dense material is progressively more concentrated toward the upper part. Such grading suggests an episodic influx of sediment, such as caused by the sudden flow of glacial melt water into a marginal lake or drift of a turbid storm-generated water mass offshore. Less pronounced grading suggests more steady supply.

Lamination in fine-grained sediments has been relatively less studied than in coarser-grained deposits. For ancient rocks, part of the reason is that shales or mudstones are commonly poorly exposed and difficult to examine in natural outcrop. But part of the reason may also be that shales are thought to be relatively simple and less interesting. Such is not always the case. Various sequences contain thin laminae or lenses of coarser grains that are poorly understood. Some may represent a residue left on a cohesive mud bed by the passage of isolated ripples, but, if this is so, the nature of the ripples or the waves or currents that formed them is conjectural. Other such coarser laminae may represent grains dropped from turbulent intrastratal density currents that flowed across a more dense layer of cold or saline water that covered the basin floor. This process was advanced as an explanation for a thick sequence of Permian laminated siltstone (Harms 1974). Additional research on stratification in fine-grained rocks should be valuable.

THE QUESTION OF MASSIVE SEDIMENTS

The examples described in preceding sections intimate that deposition creates lamination by one means or another. What of sediments that appear unlaminated and massive? In my opinion, massive units require special explanation and, therefore, careful inspection. Sometimes the cause is totally secondary, and burrowing by animals, growth of roots, or slumping and liquefaction of sediment masses can obliterate original stratification.

Other processes, by their very nature, deposit unlaminated material. Mud flows carry dense and poorly sorted slurries of sediment and deposit en masse in such a way as to prevent development of lamination (Johnson 1970). Glacial tills lack fine layering because of the way material is dragged and shoved by ice. And, finally, some sandstone beds in turbidity current deposits are massive, presumably because masses of grains settled quickly from concentrated suspensions while undergoing shearing as well as churning by expelled water (Middleton & Hampton 1973). Identification of the cause for massive character of sediments can aid in environmental reconstruction, especially for some uncommon environments.

CONCLUDING REMARKS

In this review of primary sedimentary structures, I chose to begin with those that are most ordinary and are best related to flow conditions. Then I attempted to illustrate how common variations in environmental factors or depositional rates or scale are reflected in what we observe. This approach is intended to emphasize cause and effect, not exhaustively, but as a method that may appeal to those who are unacquainted with and wish to learn something of this topic. The failure of this sort of presentation is in not providing within this length any complete catalog of structures or adequate reference to the vast literature. But these shortcomings can be overcome by the serious student through use of several recent textbooks or manuals. I especially recommend Pettijohn, Potter & Siever (1973), Reineck & Singh (1973), Blatt, Middleton & Murray (1972), Middleton (1977), Middleton & Hampton (1973), Harms et al (1975), and Middleton & Southard (1977).

As for future directions of research, more quantified data are needed on factors importantly controlling primary structures. An appealing extension of present knowledge is to larger flow dimensions and more complex flow conditions through scaled experiments, more elaborate experimental apparatus, and extensively monitored observations of modern environments. But such methods are specialized and expensive, and results may come slowly. Additional work is needed on sequences and associations of primary structures; the tactic is simple, well tried, and useful, as demonstrated by results on fluvial, turbidity current, and shoreline examples found in some of the references noted at the end of the preceding paragraph. These kinds of data can be obtained by conventional geologic approaches, requiring more a pinpointing of objective than a change in technique. Through such efforts, we can expect a clearer grasp of associations of structures in important environments and perhaps some insight into just how some structures are formed.

Over the past two decades, sedimentology has shifted in emphasis and undergone a renaissance. This has been due in part to the improved understanding of primary sedimentary structures. The impact on geology may be somewhat broader than one might at first imagine, since the reconstruction of depositional systems and ancient environments has application to topics as diverse as regional tectonics and mineral exploration. We can hope for at least incremental gains in knowledge and abilities as a result of continuing study of sedimentary structures.

ACKNOWLEDGMENTS

Figures were drawn by Alison C. Richards, to whom I am indebted.

Literature Cited

Allen, J. R. L. 1968. *Current Ripples.* Amsterdam: North-Holland. 433 pp.

Bagnold, R. A. 1941. *The Physics of Blown Sands and Desert Dunes.* London: Methuen. 165 pp.

Bigarella, J. J., Becker, R. D., Duarte, G. M. 1969. Coastal dune structures from Parana (Brazil). *Mar. Geol.* 7:5–55

Blatt, H., Middleton, G., Murray, R. 1972. *Origin of Sedimentary Rocks.* Englewood Cliffs, NJ: Prentice-Hall. 634 pp.

Coleman, J. M. 1969. Brahmaputra river: channel processes and sedimentation. *Sediment. Geol.* 3:129–239

Collinson, J. D. 1970. Bedforms of the Tana River, Norway. *Geogr. Annaler* 52:31–55

Cotter, E. 1971. Paleoflow characteristics of a Late Cretaceous river in Utah from analysis of sedimentary structures in the Ferron Sandstone. *J. Sediment. Petrol.* 41:129–38

Gilbert, G. K. 1914. The transportation of debris by running water. *US Geol. Surv. Prof. Pap.* 86. 263 pp.

Greer, S. A. 1975. Sandbody geometry and sedimentary facies at the estuary-marine transition zone, Ossabaw Sound, Georgia: a stratigraphic model. *Senckenbergiana Marit.* 7:105–35

Hand, B. M., Wessel, J. M., Hayes, M. O. 1969. Antidunes in the Mount Troy Conglomerate (Triassic), Massachusetts. *J. Sediment. Petrol.* 39:1310–16

Harms, J. C. 1969. Hydraulic significance of some sand ripples. *Geol. Soc. Am. Bull.* 80:363–96

Harms, J. C. 1974. Brushy Canyon Formation, Texas: a deep-water density current deposit. *Geol. Soc. Am. Bull.* 85:1763–84

Harms, J. C., Fahnestock, R. K. 1965. Stratification, bed forms, and flow phenomena (with an example from the Rio Grande). In *Primary Sedimentary Structures and Their Hydrodynamic Interpretation,* ed. G. V. Middleton. *Soc. Econ. Paleontol. Mineral. Spec. Publ.* 12:84–115

Harms, J. C., Southard, J. B., Spearing, D. R., Walker, R. G. 1975. *Depositional Environments as Interpreted from Primary Sedimentary Structures and Stratification Sequences.* Dallas: Soc. Econ. Paleontol. Mineral. 161 pp.

Houbolt, J. J. H. C. 1968. Recent sediments in the southern bight of the North Sea. *Geol. Mijnbouw* 47:245–73

Inman, D. L. 1957. Wave-generated ripples in nearshore sands. *US Army Corps Eng. Tech. Mem.* 100:1–42

Johnson, A. M. 1970. *Physical Processes in Geology.* San Francisco: Freeman, Cooper. 577 pp.

Jopling, A. V., Walker, R. G. 1968. Morphology and origin of ripple-drift cross lamination, with examples from the Pleistocene of Massachusetts. *J. Sediment. Petrol.* 38:971–84

Knight, R. J., Dalrymple, R. W. 1975. Intertidal sediments from the south shore of Cobequid Bay, Bay of Fundy, Nova Scotia, Canada. In *Tidal Deposits,* ed. R. N. Ginsburg. New York: Springer. 428 pp.

McKee, E. D. 1965. Experiments on ripple lamination. See Harms & Fahnestock 1965, pp. 66–83

McKee, E. D. 1966. Structures of sand dunes at White Sands National Monument, New Mexico (and a comparison with structures of dunes from other selected areas). *Sedimentology* 7:1–69

Middleton, G. V., ed. 1977. Sedimentary processes: hydraulic interpretation of primary sedimentary structures. *Soc. Econ. Paleontol. Mineral. Reprint Ser.* 3. 285 pp.

Middleton, G. V., Hampton, M. A. 1973. Sediment gravity flows: mechanics of flow and deposition. In *Turbidites and Deep Water Sedimentation,* ed. G. V. Middleton, A. H. Bouma. Anaheim: Pacific Sec. Soc. Econ. Paleontol. Mineral. 157 pp.

Middleton, G. V., Southard, J. B. 1977.

Mechanics of Sediment Movement. Binghamton: Soc. Econ. Paleontol. Mineral. 241 pp.

Pettijohn, F. J., Potter, P. E., Siever, R. 1973. *Sand and Sandstone*. New York: Springer. 618 pp.

Reineck, H.-E., Singh, I. B. 1973. *Depositional Sedimentary Environments*. New York: Springer. 439 pp.

Schumm, S. A. 1972. Fluvial paleochannels. In *Recognition of Ancient Sedimentary Environments*, ed. J. K. Rigby, W. K. Hamblin. *Soc. Econ. Paleontol. Mineral. Spec. Publ.* 16:98–107

Simons, D. B., Richardson, E. V., Nordin, C. F. Jr. 1965. Sedimentary structures generated by flow in alluvial channels. See Harms & Fahnestock 1965, pp. 34–52

Stride, A. H. 1970. Shape and size trends for sand waves in a depositional zone of the North Sea. *Geol. Mag.* 469–77

Walker, T. R., Harms, J. C. 1972. Eolian origin of flagstone beds, Lyons Sandstone (Permian), type area, Boulder County, Colorado. *Mt. Geol.* 9:279–88

Ann. Rev. Earth Planet. Sci. 1979. 7: 249–88

THE MAGNETIC FIELDS OF MERCURY, MARS, AND MOON

Norman F. Ness
Laboratory for Extraterrestrial Physics, NASA/Goddard Space Flight Center, Greenbelt, Maryland 20771

INTRODUCTION

As a result of US and USSR space research with artificial satellites, there now exists a considerable body of experimental data on the magnetic fields of the planets and the Moon. Surprisingly, Mercury has been found to possess an intrinsic global magnetic field, sufficient to deflect the flow of solar wind around it and thus forming a magnetosphere and magnetic tail surrounded by a well developed, detached bow shock wave. Equally surprising is the Moon, which, although lacking a global field, is found to possess local surface magnetic fields, which range from 3 to 300 nanoteslas (nT), with ubiquitous remanent magnetization found in the returned lunar samples. Our knowledge about Mars and Venus is less complete, primarily because of the failure of the US to include magnetometer instrumentation in its vigorous Mars exploration program of the Mariner orbiter and Viking orbiter-lander projects. However, the USSR has pursued an active exploration program for both Venus and Mars, and has contributed significantly to the study of the magnetic field and solar wind interaction at both planets.

It is the purpose of this review to survey the experimental observations and interpretations of data obtained by spacecraft or instruments which have landed or impacted on the surfaces of or been placed in orbit about Mercury, Mars, and Moon or flown past them in heliocentric orbit. Table 1 presents a list of all spacecraft missions to date which have returned information on the magnetic fields of Mercury, Mars, and Moon. The contributions of the US Apollo program are extremely significant in the study of lunar magnetism. This paper is concerned primarily with the global or large scale characteristics of these magnetic fields and so does not discuss the detailed results which have been obtained in the lunar

0084-6597/79/0515-0249$01.00

program by the analysis of the magnetic properties of the returned lunar material. A review by Fuller (1974) and the annual proceedings of the Lunar Science Conferences should be consulted for reports on that topic.

The phenomenon of magnetism in the solar system covers an extremely wide spectrum of physical problems. It touches on the origin of the Universe as well as the origin of the solar system, its evolution and present state. A common characteristic of all large celestial objects, in addition to their rotation, is an intrinsic, global magnetic field, which extends into the immediate environment. In many instances, it is principally responsible for controlling the characteristics of the environment. The origin of this magnetism is often ascribed to a self-regenerative dynamo process, acting in the interior of the object. This hypothesis is universally accepted for the origin of the terrestrial magnetic field and by inference, those of certain of the other planets, the Sun, stars and possibly other astrophysical objects such as pulsars. Although there is as yet no complete theory for such a dynamo process, advances continue to be made in studying the magneto-fluid dynamics of self-gravitating, electrically conducting matter so that observations of the external magnetic fields of an object provide indirect information on the characteristics of the interior of the object.

Table 1 Summary of space missions to study the magnetic fields at Mercury, Mars, and Moon

Target	Spacecraft	Date of initial data	Additional information
Mercury	Mariner 10	29 Mar. 1974	Flyby No. 1 at 723 km altitude
(radius =	Mariner 10	16 Mar. 1975	Flyby No. 3 at 327 km altitude
2439 km)			
Mars	Mariner 4	15 July 1965	Flyby at 9820 km altitude
(radius =	Mars 2	Dec. 1971	Orbiter (periapsis = 1150 km)
3380 km)	Mars 3	Dec. 1971	Orbiter (periapsis = 1150 km)
	Mars 5	Feb. 1974	Orbiter (periapsis = 1150 km)
Moon	Luna 2	13 Sept. 1959	Impacted
(radius =	Luna 10	Apr. 1966	Orbiter (altitude = 350–1017 km)
1738 km)	Explorer 35	July 1967	Orbiter (altitude = 730–7650 km)
	Apollo 12	Nov. 1969	Surface—astronaut deployed station
	Lunokhod 1	Nov. 1970	Lunar Rover on surface (Luna 16)
	Apollo 14	Feb. 1971	Surface—portable magnetometer
	Apollo 15	Aug. 1971	Surface—astronaut deployed station
	Apollo 15	Aug. 1971	Orbiter (altitude = 60–170 km)
	Apollo 16	Apr. 1972	Orbiter (altitude = 0–205 km)
	Apollo 16	Apr. 1972	Surface—astronaut deployed station and portable magnetometer
	Lunokhod 2	Jan. 1973	Lunar Rover on surface (Luna 21)

Before 1973–1974, when the Pioneer spacecraft explored in situ the magnetosphere of Jupiter, all our information about its magnetic field, as well as that of other astrophysical objects, had come from analyses of observational data on electromagnetic radiation emitted by these objects. This included careful study of the spectral lines emitted (or absorbed) by hot ionized gases. Simply because of the statistics involved, our knowledge of stellar magnetism is, in a certain sense, much more advanced than that of planetary magnetism, for which there are only a few observed examples. The most distinguishing physical difference between stars and planets is that the latter are primarily composed of material in a solid phase, much of which is below the Curie point temperature. As a result, that material carries, in its remanent magnetic properties, a record of the magnetic field existing at the time of cooling of the material.

One of the great revolutions in the Earth sciences was brought about in the 1960s by the concept of plate tectonics. The interpretive basis for that dynamic process shaping the Earth's crust and much of the mantle was primarily the historical record of the terrestrial magnetic field, which was being deciphered from the mid-ocean ridges and elsewhere. It is reasonable to expect that when equivalent descriptions of the global and local magnetic fields of the terrestrial planets are available, we will be able to interpret the evolutionary history of formation of the crustal layers of the individual planets themselves.

This review is intended to summarize the present status of our knowledge regarding the magnetic fields of Mercury, Mars, and Moon. A considerable time interval is required for the conception, development, and implementation of any spacecraft mission to these bodies, approximately three to eight years depending upon the target planet. Presently, the known launch schedule and technical capabilities of the US and USSR are such that until well into the next decade we do not expect any missions to Mercury, Mars, or the Moon, which could provide additional data on their magnetic fields. As a result, the body of experimental facts available for analysis and interpretation will not be changed for some time. Therefore, an effort has been made in this review to survey all experimental observations and their interpretations.

MERCURY

In the exploration of the planets, one of the more startling surprises has been the discovery by the US spacecraft Mariner 10 in 1974 (Ness et al 1975) of an intrinsic planetary magnetic field at Mercury with an intensity of approximately 1% of Earth's. The interaction with the solar wind forms a magnetosphere which is similar in many characteristics to the terrestrial one, but much smaller, relative to the planetary radius, by a factor of approximately 7.5. Since the planet then occupies such a large fractional volume of the magnetosphere, no permanently trapped radiation belts develop, although a magnetic tail and imbedded neutral sheet do exist, similar to Earth's (Ogilvie et al 1977). Detection of a magnetic field and magnetosphere at Mercury, a planet lacking radiation belts (Simpson et al 1974) or a substantive atmosphere (Hartle, Curtis & Thomas 1975, and Kumar 1976), provides another sample for comparative studies of magnetospheres (Siscoe 1978). It is the purpose of this section to review the experimental evidence for and the interpretation of the global magnetic field of Mercury.

Three close flybys of Mercury by Mariner 10 occurred on 29 March and 21 September 1974 and 16 March 1975 (see Figure 1). Multiple encounters with Mercury were achieved because of the nearly exact commensurability of the Mariner 10 heliocentric orbital period, 175.95 days, with the heliocentric orbital period of Mercury, 87.97 days. The relative

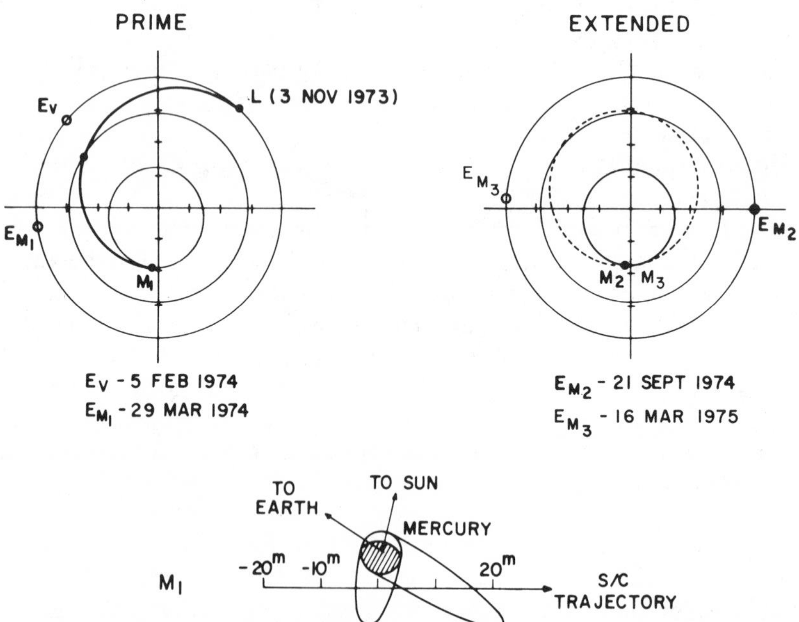

Figure 1 Trajectory of Mariner 10 and orbits of Mercury, Venus, and Earth projected onto ecliptic plane. Encounters with planets are labelled $E_{M,V}$ and show relative position of planets. The bottom panel illustrates the Mercury flyby trajectories for the first and third encounters. Ness (1978).

orientation of the planet, with respect to the Sun, was identical for all three encounters because of the resonant coupling of Mercury's orbital period with its spin period, 58.70 days, in the ratio of 3 to 2. Unfortunately, no further close encounters occurred due to a limited supply of spacecraft attitude and trajectory control gas. This mission is unique not only for its triple encounter of a target planet, while in heliocentric orbit, but also because it was the first gravity assist mission of the space age. A close encounter with the planet Venus, on 5 February 1974, provided the necessary orbit modification to accomplish the multiple Mercury encounters.

Each of the flyby trajectories was carefully selected to enhance the scientific return from the complement of experiments on the spacecraft. The first encounter was a near equatorial darkside pass, with closest approach distance of 723 km from the surface. The spacecraft passed rapidly by the planet, at approximately 11 km/sec, and almost in a straight line, since the gravitational mass of Mercury is not large enough to significantly affect the heliocentric orbital motion of the spacecraft. Improved values of the radius, 2439 ± 1 km, and density, 5.44 gm cm^{-3}, have been derived from these encounter data (Howard et al 1974).

The second encounter was selected to provide imaging picture coverage of the south polar region and passed too far from the planet, 50,000 km, to provide direct data on its magnetic field or the interaction with the solar wind. The third encounter was precisely chosen to enhance the results of the first encounter, by occurring closer to the planet, 327 km, and at a higher latitude in the north polar region. Data from the first and third encounters from the solar wind electron spectrometer, charged particle telescope, and magnetometer provide unique and complementary data regarding the nature of the magnetic field of Mercury and its magnetosphere as formed by the solar wind interaction (Ness 1978). The magnetometer instrumentation on Mariner 10 was unique in that it was the initial flight of a dual magnetometer system, permitting inflight determination and elimination of spacecraft magnetic field contamination (Ness et al 1971, 1974a).

The Sun is a continuous source of rarified, magnetized plasma, called the solar wind, which has been observed to flow out approximately radially at very high velocities, 300–1000 km/sec, at heliocentric distances greater than 0.2 AU. As this solar wind interacts with any planetary magnetic field, the electrically conducting plasma envelops the planet and confines the field to a region of space called the magnetosphere. The boundary of the magnetosphere, the magnetopause, is well distinguished by an abrupt directional and/or magnitude change of the magnetic field on opposite sides, since electric currents flow within the magnetopause, having been induced by the solar wind flow. For magnetized planets as

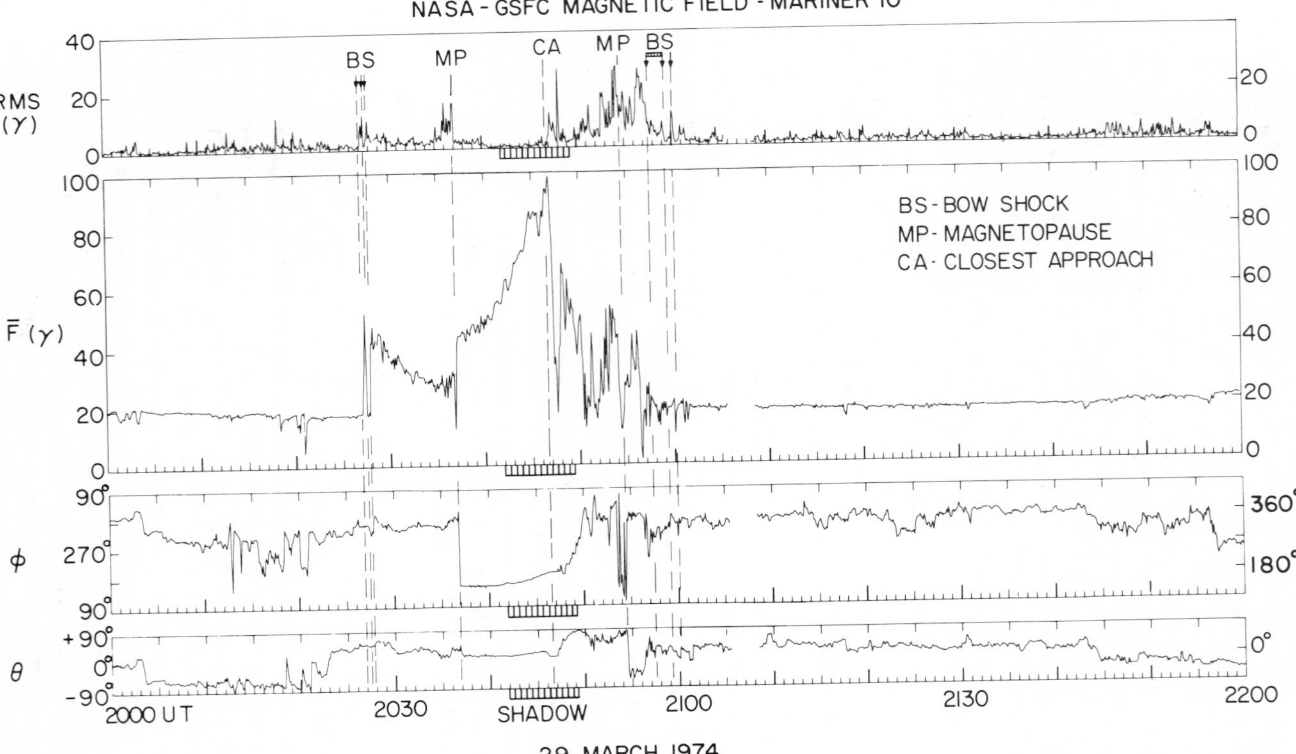

Figure 2 Magnetic data obtained by the GSFC instrument during first encounter with Mercury. The data points are six-second averages or RMS pythagorean means of 25-Hz vector measurements and are presented in solar ecliptic coordinates with θ = latitude and ϕ = longitude ($\theta = 0°$, $\phi = 0°$ points to sun, $\theta = 0°$, $\phi = 180°$ away from sun, and $\theta = 90°$ points to ecliptic pole). The period of solar occultation by the planet is indicated as a shadow period. Ness et al (1975). (Note: $1\ \gamma = 1$ nT.)

well as those possessing substantial atmospheres and ionospheres, a detached bow shock wave also develops, due to the deflection of the super Alfvénic solar wind flow around these large obstacles.

Mariner 10—First Encounter

The magnetic field data from the first encounter are shown in Figure 2, where the total time of immersion in the magnetosphere and magnetosheath of Mercury is seen to be only 32 minutes. The abrupt increase in both the magnetic field magnitude and the fluctuations, as measured by the RMS parameter, are indicative of clearly defined, multiple bow shock crossings between 2027–2028 UT. They are similar to multiple observations of the Earth's bow shock, except for the expected differences in the average magnitudes, due to the smaller distance from the Sun. The magnetopause is well distinguished by the abrupt change in direction of the magnetic field at 2037, as well as by the termination of the enhanced fluctuations of the magnetic field in the turbulent boundary layer, the magnetosheath.

Near closest approach, the magnetic field increased to a maximum of 98 nT. The configuration of the magnetic field and the variations in magnitude and fluctuations are identical to what would be observed at Earth for a spacecraft similarly transiting the terrestrial magnetosphere on the nightside of Earth, having entered the magnetosphere in the southern lobe of the geomagnetic tail. This immediately suggests that the polarity of an assumed magnetic dipole at Mercury is similar to that at Earth, roughly parallel to the axis of rotation and with a magnetic south pole in the geographic north polar region.

Immediately after closest approach, rapid and large changes in the character of the magnetic field occurred, coincident with similar changes in both the energetic particle flux and the electron plasma characteristics. These have been interpreted by Siscoe, Ness & Yeates (1975) as characteristic of a magnetospheric substorm in the Hermean magnetosphere. Following this disturbance, as the spacecraft exits from the magnetosphere, the identification of the magnetopause and bow shock traversals is less clear. In direct contrast to the inbound traversal, where the magnetic field was roughly parallel to the shock surface, the interplanetary field was approximately perpendicular to the bow shock surface, leading to the well known and well studied phenomenon of upstream waves for certain wave modes (Fairfield & Behannon 1976).

Analysis of Data from Mercury I

Quantitative analysis of planetary magnetic field data has for years followed the prescription laid down by Gauss in his pioneering studies of terrestrial magnetism. The magnetic field, **B**, is assumed to be derivable from a scalar potential, V, which is represented by a series of spherical harmonic functions:

$$V = a \sum_{n=1}^{\infty} \sum_{m=0}^{n} \left[(g_n^m \cos m\phi + h_n^m \sin m\phi) \left(\frac{a}{r}\right)^{n+1} + (G_n^m \cos m\phi + H_n^m \sin m\phi) \left(\frac{r}{a}\right)^{n} \right] P_n^m (\cos \theta)$$

$$\mathbf{B} = -\text{grad } V$$

By least squares fitting a series of observations within the magnetosphere ($\mathbf{B}_{obs}$) to an assumed finite series representation [$\mathbf{B}_{th}(n < \infty)$], it is possible to determine the harmonic coefficients g_n^m, h_n^m, G_n^m, H_n^m, or multipole parameters. For Mercury, analyses of these data have been restricted to use of an equivalent centered dipole term (g_n^m, h_n^m for $n = 1$) for the internal source and an external current system described by harmonics of degrees 1 and 2 (G_n^m, H_n^m for $n = 1,2$).

In the analyses performed (Ness et al 1975) the dipole term was found to have an equivalent equatorial surface intensity of approximately 350 nT with the dipole axis tilted at an angle of approximately 10° from the orbit plane normal (see Table 2). This value yields a moment of 5.1×10^{19} Am^2. An earlier analysis by Ness et al (1974b) neglected the external terms and assumed only an offset tilted dipole, which yielded a moment of 3.3×10^{19} Am^2 tilted at 20° from the orbit normal but with a large offset, 0.47 R_M.

Mariner 10—Third Encounter

To enhance the third encounter data and its interpretation, a more pole-ward pass at a lower altitude was selected, which led to a higher maximum field intensity observed. Data from the third encounter are shown in Figure 3, where it is seen that the maximum is 400 nT, 4 times that of Mercury I encounter and more than 20 times that of the interplanetary field of 18–20 nT. As in the first encounter, the bow shock and magnetopause were observed both inbound and outbound, although a distinguishing feature of the third encounter data is the broad extent of waves upstream from the bow shock. As previously noted, this is associated with the orientation of the interplanetary magnetic field relative to the surface of the bow shock. Throughout the entire magnetosphere transit, the magnetic field was observed to be steady, and the directional variations were small and slow, without the disturbances noted during the first encounter.

Analysis of Data from Mercury III

Computations deriving the harmonic coefficients (Ness et al 1976) yield a centered dipole term having a moment $4.9 \pm 0.2 \times 10^{19}$ Am^2, tilted at an angle of $11 \pm 1°$ to the orbit normal. The angular difference between the equivalent centered dipole moments derived from Mercury I and Mercury III encounters was 24°.

It is also possible to utilize the relative geometrical positions of the bow shock and magnetopause surfaces observed at Mercury and Earth to compute the equivalent dipole magnetic moment responsible for deflection of solar wind flow. Extensive observations of the Earth's magnetopause and bow shock during the past two decades provide a definitive

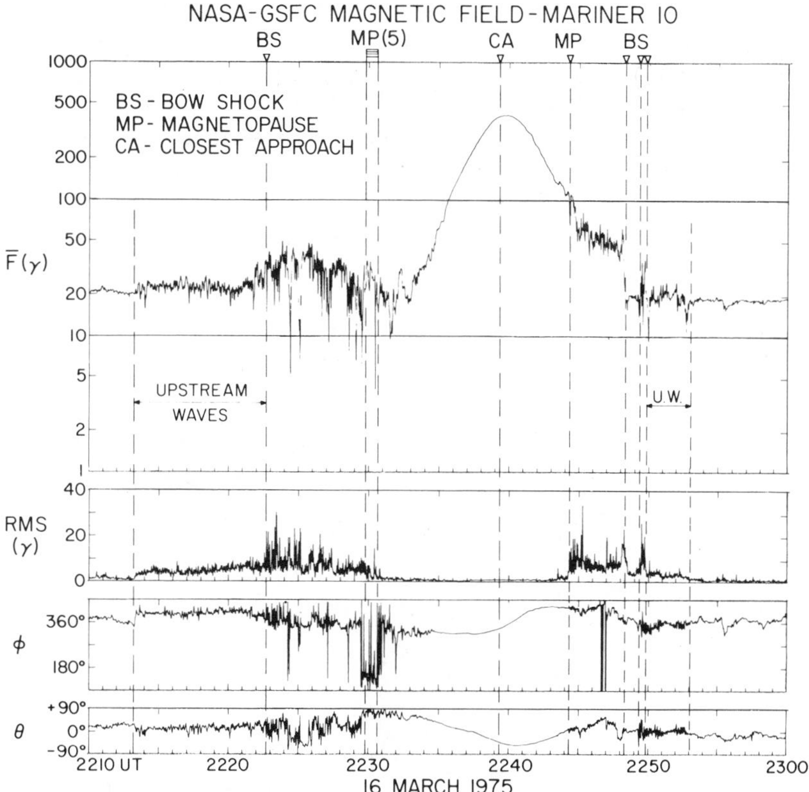

Figure 3 Magnetic field data obtained by the GSFC instrument during the third encounter with Mercury. The data points are 1.2-second averages or RMS pythagorean means of 25-Hz vector measurements. See Figure 2 caption for definition of coordinate system. There was no optical shadow. Ness et al (1976). (Note: $1\ \gamma = 1$ nT.)

Earth reference, and Figure 4 shows the results of this comparison. Note that there are regions rather than points along the trajectory for some boundary crossings. This is due to the motion of the boundary surface during the period of observation, a result of the variability in solar wind characteristics, or possibly of waves associated with intrinsic instabilities of the boundary surface itself.

The planetocentric distance to the magnetopause, at the stagnation point of solar wind flow, is found to be $1.45 \pm 0.15\ R_M$. Thus, Mercury, at the time of the Mariner 10 encounters, possessed a much smaller magnetosphere than at Earth ($11.0\ R_E$), when scaled by the planetary radius by a factor of 7.5. Consideration of the planetary radius, 6378 km vs 2439 km, leads to a smallness factor of 20. The plasma electron density and velocity measurements outside the bow shock region provide estimates of the solar wind momentum flux. Combined with the positions of the bow shock and magnetopause, and knowledge of the same parameters at Earth, this data permits a computation of the equivalent dipole moment responsible for deflection of the solar wind flow. This yields a value of $3\text{–}7 \times 10^{19}$ Am^2, in good agreement with the directly computed value obtained from the first (and third) encounters.

To illustrate the relatively large amplitude of the perturbation fields due to solar wind interaction with the Hermean magnetic field, Figure 5

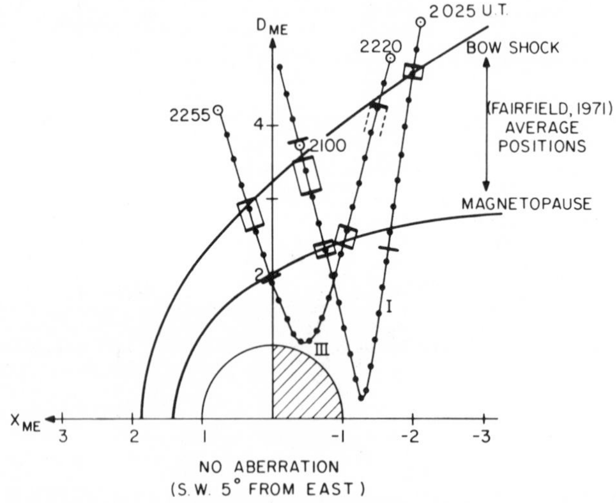

Figure 4 Positions of bow shock and magnetopause surfaces observed at Mercury compared to scaled down version of terrestrial case. The figure assumes cylindrical symmetry of the surfaces about the solar wind flow direction, which is itself assumed to be coming from 5° East of the planet-Sun line so that there is no net deviation of the flow due to the heliocentric orbital motion of the planet. Ness (1978).

presents the external magnetic field derived from the encounter I data. These perturbation fields at Mercury show the characteristics expected from the formation of a confined magnetosphere on the dayside and an extended magnetic tail on the nightside of the planet. The reversal of the field component in the X direction is readily evident in the projection on the X-Y plane, and is due to the crossing from the southern to the northern lobe of the magnetic tail. In general, our understanding of the Mariner 10

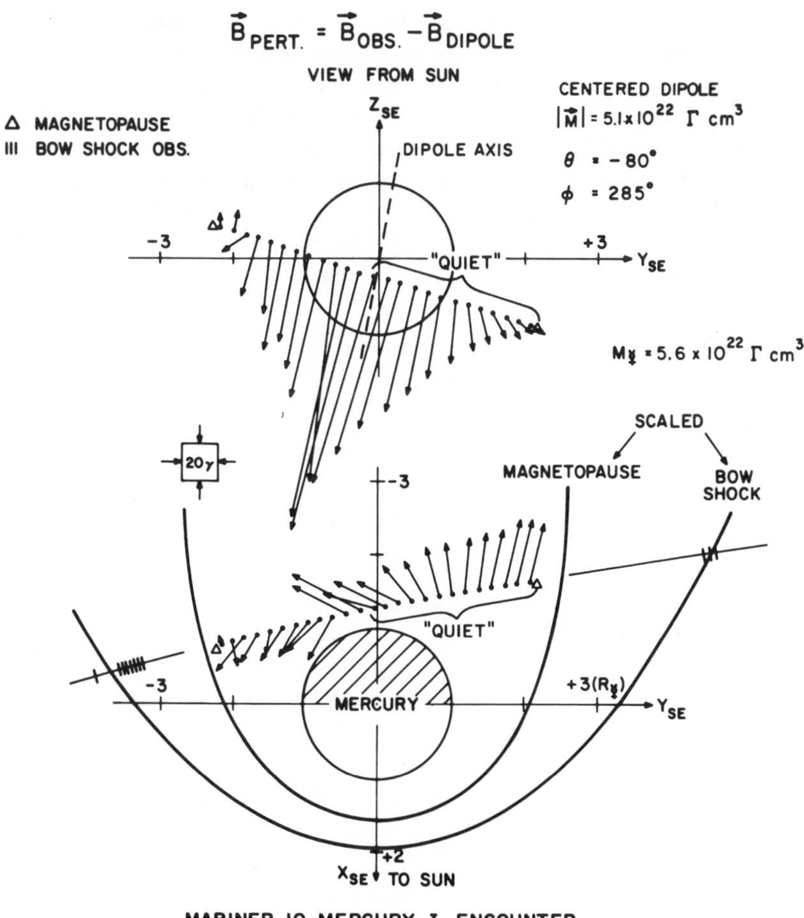

Figure 5 Perturbation magnetic field due to external electrical currents induced in solar wind flow, observed during Mariner 10 first encounter with Mercury. The dipole representing the planetary field is in the same sense as Earth's but tilted 20° from the ecliptic pole. Vectors projected on X-Y and Y-Z planes represent 48-second averages.

data depends heavily upon a comparison with the extensive suite of data on the terrestrial magnetosphere.

The results of the quantitative analysis of the third encounter data are illustrated in Figure 6, where individual orthogonal magnetic field components of observations and of a modelled magnetosphere are directly compared. The theoretical model assumed a centered, tilted dipole and a uniform external field. The departures of the observed field from this simple theoretical model could have several explanations: the internal or the external magnetic field may be more complex spatially than has been assumed in the theoretical model, there may be temporal variations of the external magnetic field, or the region of the magnetosphere in which the analysis is being conducted is not free from electrical currents so that the assumption of a magnetic field derivable from a scalar potential is violated. This data plot shows how well such a simple dipole model with uniform external field represents the observations, where the individual components range from 0 to 280 nanoteslas. The principal factor contributing to the enhanced RMS values is the spatial gradient of the magnetic

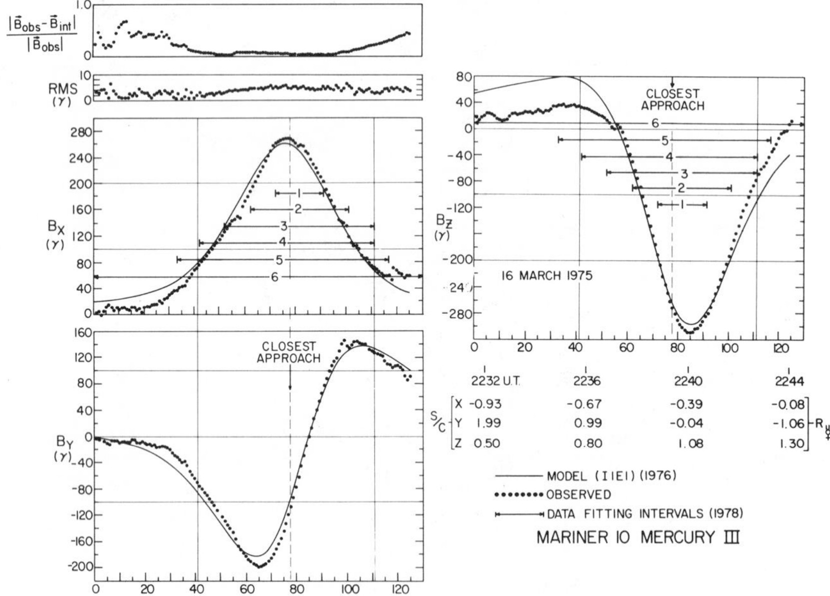

Figure 6 Comparison of observed and modelled magnetosphere field of Mercury during third encounter by Mariner 10. Data points represent six-second averages. Position of spacecraft (S/C) is given in planetocentric coordinates with X-Y plane defined by orbital plane of Mercury. Successive data sets used in the analysis are indicated by 1–6. Ness et al (1976).

field during the six-second interval when the measurements are being made.

The comparison of terrestrial and Hermean bow shock and magnetopause locations combined with the spherical harmonic analyses of the magnetic field data clearly indicates that Mercury possesses a magnetic field which, on a global scale, is well represented by a tilted, centered dipole with a polarity identical to Earth. The differences in magnetic field derived from Mercury I and Mercury III observations probably do not represent a secular change in the planetary field itself because the uncertainties of the derived terms are as large as the differences.

Latest Analyses and Magnetosphere Models

Recently, Ness (1978) has performed a sequential analysis of subsets of data taken near closest approach from the third encounter. The purpose was to investigate the variations in derived values of the tilted dipole from data centered about closest approach, when the relative contributions of the magnetic field from the external currents would be minimized. By using external harmonic representations with $n = 0$, 1, and 2, we find the different orientations of the dipole shown in Figure 7 for the series of six data sets. As is readily evident, the data subsets of increasingly larger size (from 1 to 6) approach limits which are quite close. The average of those limits is assumed to best represent the equivalent centered, tilted dipole for Mercury: 330 ± 18 nT equivalent dipole equatorial field strength and oriented at an angle of $14 \pm 5°$ with respect to the orbit plane and longitude $147 \pm 15°$. It should be noted that no claim is made for the uniqueness of the analyses reported herein, since an incomplete data set

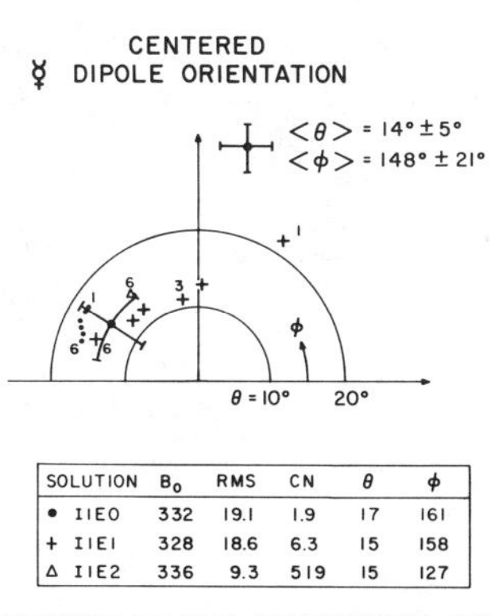

SOLUTION	B_0	RMS	CN	θ	φ
• I1E0	332	19.1	1.9	17	161
+ I1E1	328	18.6	6.3	15	158
Δ I1E2	336	9.3	519	15	127

MERCURY III DATA SOLUTIONS (N=125)

Figure 7 Summary of detailed studies of orientation of equivalent dipole magnetic moment orientation for Mercury from third encounter data set. The notation IjEk represents an expansion with internal terms up to degree j and external terms up to degree k.

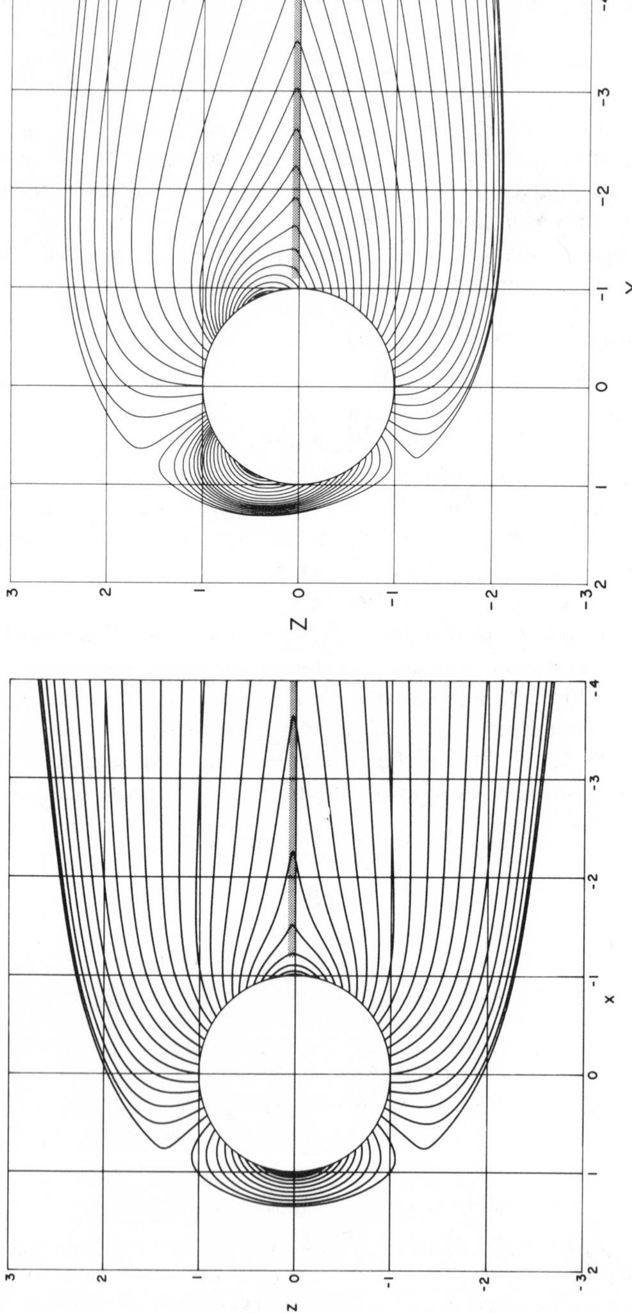

Figure 8 Moon midnight meridian plane projection of field lines in Mercury magnetosphere for different models of internal and external field. The models are a centered dipole (*a*) and a centered dipole + quadrupole + octupole (*b*) (Whang 1977).

is available for study. It is well known, in potential theory, that vector measurements over a closed surface surrounding a source are required, in order to deduce a complete and unique multipole description of the source.

Two other quantitative approaches have been used to model the magnetosphere of Mercury. Whang (1977) used a simple model of the planetary magnetic field and represented the external currents by an image dipole upstream from the planet and the magnetic tail neutral sheet by a cross tail two-dimensional current sheet. A two-dimensional view of this three-dimensional model in the noon-midnight meridian plane, including the magnetic axis of the planet, is shown in Figure 8*a*. The asymmetry of the magnetosphere of the planet is clearly indicated by the field compression on the dayside and the extension of the planetary field on the nightside, forming the magnetic tail.

The models of the magnetic field used for the internal field of Mercury employed the multipole coefficients shown in Table 2 and assumed axial symmetry of the planetary field. As the order of the harmonic representation increased, the magnitude of the dipole term decreased. This is possibly due to spatial aliasing, where the selection of only that component of the quadrupole and octupole moments, which has the required axial symmetry, masquerades as the dipole term because of the small radial extent of the data set. There is no evidence from the Earth or Jupiter that the assumption of axial symmetry is justified (Acuna & Ness 1976).

Table 2 Mercury's magnetic multipole moments[a]

Source	Data	g_1^0	g_2^0	g_3^0	External terms	Offset	Tilt[b]
Ness et al (1974a)	I	227	0	0	0	0.47 R_M	30°_E
Ness et al (1975)	I	350	0	0	$n = 2$	0	10°_E
Ness et al (1976)	III	342 ± 15	0	0	$n = 1$	0	$11^\circ \pm 1_O$
Whang (1977)	I & III	266	0	0	2-dimensional tail sheet + image dipole	0	
		165	117	0		0	2.3°_O
		166	75	48		0	
Jackson & Beard (1977)	I & III	257	0	0	scaled terrestrial analogue	0	10°–17°_O
		170	114	0		0	
Ng & Beard (1978)	I & III	190	0	0	scaled terrestrial analogue	(0.033, 0.026, 0.189)	1.2°_O
Ness (1978)	III	330 ± 18	0	0	$n = 2$	0	$14^\circ \pm 5^\circ_O$

[a] Using dipole aligned coordinates. Polarity sense is same as at Earth.

[b] Relative to normal to ecliptic or orbital plane, indicated by subscript E or O.

Jackson & Beard (1977) and Ng & Beard (1978) have scaled a representation of the terrestrial magnetosphere, using simple models of the planetary field but rather complex models of the external fields with terms up to degree $n = 8$. This leads to a large number of coefficients, whose values are varied, all in the same proportion, so as to provide for a linear scaling of the terrestrial magnetosphere to that at Mercury. The results obtained are shown in Table 2. Again, note that the addition of the g_2^0 quadrupole term to the dipole (Jackson & Beard 1977) leads to a significant decrease in the estimated dipole moment. Whang (1977) observes that the inclusion of the axially symmetric quadrupole and octupole moments corresponds to an equivalent axial offset of the dipole. This result is verified in the work of Ng & Beard, where an axial offset of 0.19–0.24 R_M was found. The topology of the magnetic field (Whang 1977), including both quadrupole and octupole moments, is shown in Figure 8*b* and clearly illustrates the axial offset of the dipole.

These studies and the earlier spherical harmonic studies indicate a global dipole field of Mercury with an equivalent dipole moment between 2.8 and 4.9×10^{19} Am^2. There is no question about the existence of the global magnetic field nor its general characteristics. Unfortunately, there are no plans by the US to return to the planet Mercury until at least the late 1980s so as to determine the higher order planetary harmonics or measure any secular variation.

Origin of Mercury's Field

Once a magnetic field at the planet is identified, the next problem is to determine its origin. The magnitude of the magnetic field observed during the third encounter is so large that it precludes any plausible mechanism of induction by the solar wind (Herbert et al 1976, Ness et al 1976). Unless there is an as yet unknown, exotic process of interaction whereby the solar wind can generate a larger field than required to stand-off the solar wind momentum flux at the stagnation point, we can conclude that Mercury possesses an intrinsic magnetic field of modest but significant magnitude. The planetary orbit is highly eccentric (perihelion, 0.307 AU, and aphelion, 0.467 AU), so that on the average the stagnation point distance is expected to be 1.85 ± 0.15 R_M (Siscoe & Christopher 1975). There is a small but finite probability that the solar wind could directly impact the sunlit hemisphere when its intensity, i.e. density and velocity, is especially high. At the present epoch, this is expected to occur for less than 1% of the time.

There are two possible sources of the observed magnetic field at Mercury: an active dynamo, or a remanent magnetization acquired either as the result of an ancient dynamo or an external magnetic field, such as that of the Sun. It is also possible that both mechanisms are present, and indeed it would be most surprising if at least locally, i.e. near and on the surface of the planet, the magnetic field were not influenced strongly by local magnetization because of the relatively high iron content deduced for the planet. Conventional wisdom has been that rapid rotation of a planet as well as an electrically conducting interior is necessary for generation of a dynamo. Upon closer examination, the appropriate and important parameter is the ratio of the coriolis forces to the other forces involved in the momentum balance (Gubbins 1977). For a general review of the dynamo problem see Gubbins (1974), Levy (1976), and Busse (1978a).

Even if a complete harmonic representation of the magnetic field of Mercury were available, there would be no way to determine uniquely the source of the magnetization responsible for it. Only if extended observations were available, permitting identification of the presence of a secular change, could we accumulate enough evidence to distinguish between an internal dynamo and a remanent magnetization source. As previously noted, the uncertainties in the magnetic field parameters of Mercury are larger than the differences observed during the one-year interval between observations.

The source of energy driving the fluid motions is an important consideration in the planetary interior. Busse (1978b) has given a plausible argument for a common origin of planetary magnetic fields in their electrically conducting cores. Either thermal convection or precession-induced

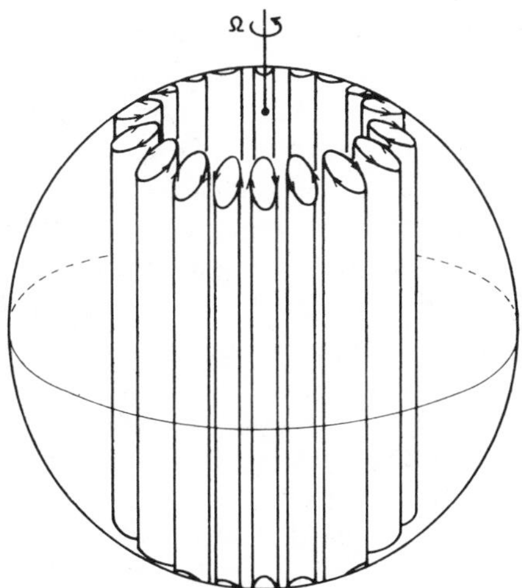

Figure 9 Sketch of the convection flow in an internally heated self-gravitating sphere. The individual cells of the convection pattern are parallel cylinders arranged circumferentially around the rotation axis of the planet (Busse 1978a).

turbulence can give rise to such motions. Precession can be eliminated as a possible source of energy in the case of Mercury, because it is probably too weak. Busse's model of dynamo generation does not require that the entire core be fluid but only a shell. Figure 9 illustrates the geometry of the convection pattern which Busse proposes as responsible for self-regenerative planetary dynamos. As long as the outer part of the Hermean core remains fluid, the observed magnetic field can be maintained by a convective dynamo driven by internal heat dissipation.

The alternative explanation of the origin of the Hermean magnetic field as being due to remanent magnetization has been considered by many authors. Ness et al (1975) have shown that a uniformly magnetized but thin outer shell of several hundred kilometers thickness can possibly be made consistent with the observed magnetic moment of the planet. The level of magnetization required is equivalent to that of the returned lunar samples which were most magnetized (3–6 $\times$ 10^{-4} Am^2/kg). The difficulty in this model is in finding some mechanism to impose a uniform magnetic field on the planet while the outer layers cool below the Curie point isotherm. The solar field is known to change polarity periodically with time and thus unless in the past such changes did not occur, during the long time required for cooling the outer shell below the Curie point, the external field would average approximately zero.

In the study of lunar magnetism, Runcorn (1975) postulates that any spherical shell cooling below the Curie point in the presence of an internally generated magnetic field will not retain any measurable external magnetic field. While this is mathematically correct for instantaneously cooling spherical planets, for real planetary materials a finite cooling time is required, and such a configuration as proposed by Runcorn would never occur. This has been shown both by Goldstein (1975) and Srnka (1976), Srnka & Mendenhall (1978). Stephensen (1976b) has proposed that the present dipole of Mercury may be explained on the basis of remanent magnetization due to an ancient dipole and permeability contrasts between mantle and core so that the core is either ferromagnetic if below the Curie point or paramagnetic if it is more than 100°C above the Curie point. These results are particularly interesting because as the core cools it goes from paramagnetic to ferromagnetic status so that the planetary dipole moment reverses orientation from negative to positive. In all the current thermal models (Solomon 1978), the core is paramagnetic and hence, with an ancient surface polar field of less than 10 gauss, the amount of free iron required in a mantle of several hundred kilometers thickness is 5% or greater. This is more than 100 times the typical values found in lunar basalts and thus seems unrealistic.

There is no question that the abundance of iron on Mercury must be high in order to explain its high density. However, whether or not the mantle will contain the necessary high percentage of iron and in a free state is quite uncertain. While remanence remains one of the possibilities for the origin of the planetary magnetic field, the most likely candidate presently appears to be an active dynamo (Srnka & Mendenhall 1978).

Whether or not this is true, the final conclusion reached from these studies is that Mercury's interior must be differentiated into a core and mantle and that quite probably the outer portion of the core, at least, is molten and can sustain a presently active hydromagnetic dynamo. Unfortunately, a lack of knowledge of the obliquity of Mercury's rotation axis precludes a determination of the moments of inertia, which would give valuable information about the internal mass distribution. In the absence of such data, and without the more explicit density distribution data obtainable by seismic studies of the planetary interior, the main information on the internal state of the planet Mercury comes from the interpretation of the data revealing a planetary magnetic field.

MARS

The first and only attempt to study the magnetic field of the planet Mars by the US was on 15 July 1965, when the Mariner 4 spacecraft passed at an altitude of 9820 km on a flyby trajectory. This corresponds to a scaled distance of 2.9 planetary radii and occurred at a position not well situated to detect any small planetary magnetic field or magnetosphere. The initial report (Smith et al 1965) did not make an interpretation of a bow shock although subsequently such an interpretation was made (Dryer & Heckman 1967). Based upon these and other data, it was concluded that Mars did not possess a dipole magnetic moment larger than 3×10^{-4} of Earth's, equivalent to an equatorial field intensity of 64 nT.

Subsequently, the USSR series of Mars spacecraft performed measurements of the magnetic field from orbits about the planet. The first were performed by Mars 2 and 3 in 1971 and were followed by additional measurements by the Mars 5 spacecraft in 1974 (see Table 1). From the plasma and magnetic field data obtained by these spacecraft, a detached bow shock was readily identified (Vaisberg et al 1975) and its position, close to the planet, showed that if any magnetic field existed at Mars, it was extremely small. In Figure 10, a summary of the bow shock positions observed at Mars is compared with those observed at Venus, a planet known to be devoid of a significant global magnetic field. Unfortunately, there are no direct measurements at Mars near the stagnation point of solar wind flow, so that the estimates of that critical position of the bow shock are based upon extrapolation of measurements made well away from the planet sun line. Without accurate knowledge of the direction and momentum flux of solar wind flow and a proper three-dimensional analysis of boundary positions, it is not possible to precisely determine the altitude of the equivalent obstacle height responsible for generation of the bow shock. Nonetheless, it is quite clear that the bow shock at Mars is substantially higher than that observed at Venus, when both are scaled by the planetary radius. Indeed, the height is consistent with a planetary magnetic field being the obstacle to solar wind flow and causing its deflection since the atmosphere-ionosphere of Mars does not extend that high.

The orbits of the Mars 2, 3, and 5 spacecraft were not well situated to study the planetary magnetic field since periapsis did not occur on the nightside of the planet where a well developed magnetic tail and/or less confined or distorted magnetosphere could be observed. Indeed, on some orbits, the magnetosphere was not identifiable in the data. One of the best sets of magnetometer data is shown in Figure 11, taken by Mars 3 on 21 January 1972. The identification of the bow shock surface at points 1 and 4 accompanies the identification of the magnetopause surface at points 2 and 3. The large quantization step size of the instrument contributes to the scatter of the individual data points. Dolginov and coworkers (Dolginov et al 1973, 1975a, 1976, Dolginov 1978a,b) have interpreted such observations of the occurrence of the magnetopause and the

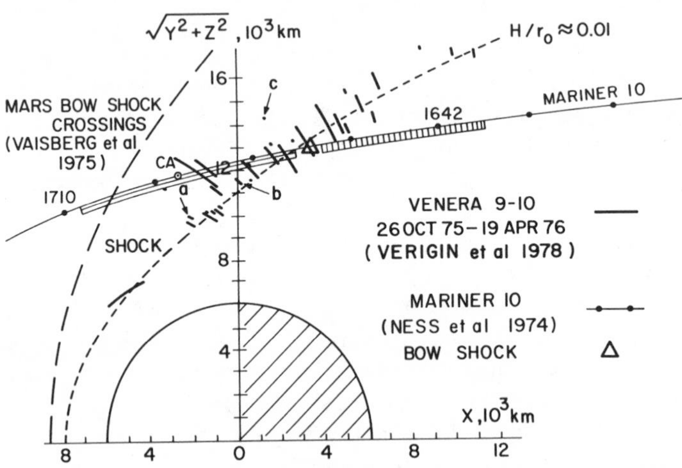

Figure 10 Comparison of the positions of the bow shocks observed at Venus and Mars, relative to the planetary surface, as reported by USSR investigations on the Mars 2, 3, 5, and the Venera 9, 10 spacecraft. Also included is the US Mariner 10 bow shock determination, as shown by the triangle (Ness et al 1974a). Cylindrical symmetry of the surfaces about the planet-Sun line has been assumed. The short line segments for Venus represent the regions of bow shock position, which is often in motion.

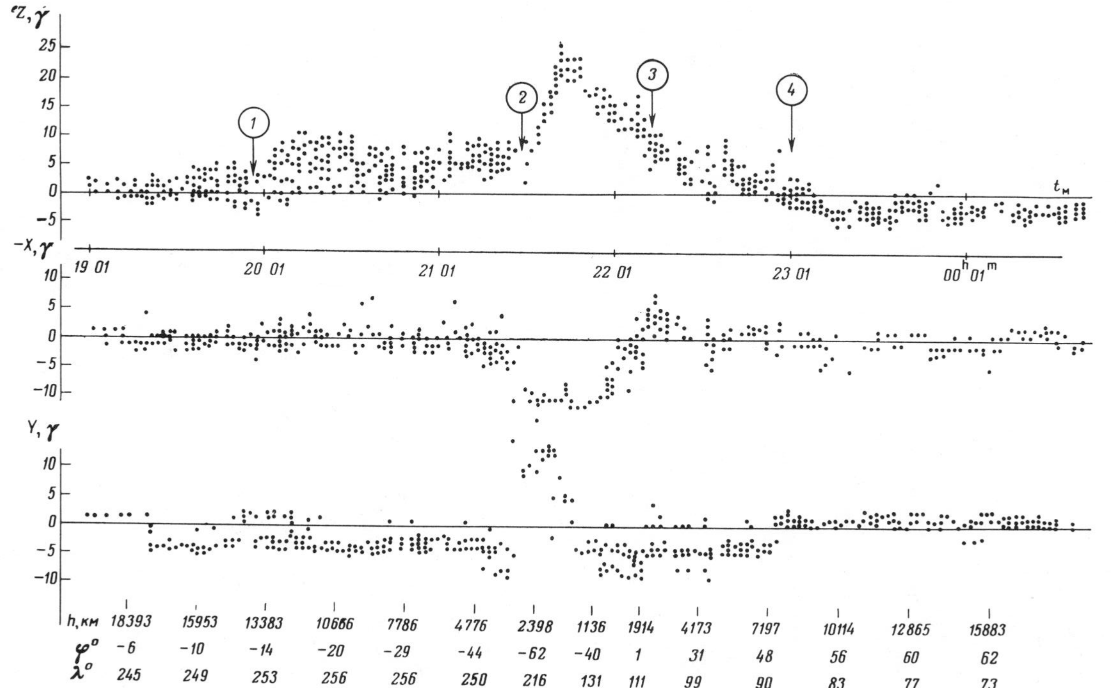

Figure 11 Magnetic field data obtained by the USSR Mars 3 spacecraft on Jan. 21, 1972. The $+X$ component points toward the Sun, the $+Z$ component to the North ecliptic pole, and the Y axis completes the right-handed solar ecliptic system. The location of the spacecraft is given in terms of altitude, h, solar ecliptic latitude, φ, and longitude, λ (Dolginov et al 1975a). (Note: $1\ \gamma = 1$ nT.)

bow shock, and the accompanying quantitative analysis of the field observations, in terms of a centered, tilted dipole representing the magnetic field of the planet. Note that the maximum field measured is approximately 30 nT at a height of 1100 km.

When the spacecraft was close to the planet, the equivalent centered tilted magnetic dipole responsible for each vector observation was computed, neglecting any contribution from external sources. Dolginov et al (1975a) obtained a dipole moment of 2.4×10^{19} Am^2, but tilted at an angle of 72° to the axis of rotation. There is no justification given for the neglect of the magnetic field of external sources and, in view of the close proximity of these sources to the planet, it would appear very plausible that any quantitative value obtained by such an analysis is subject to large error.

Dolginov (1977) has summarized the Mars 3 and 5 magnetic field (Figure 12) measurements to schematically argue for a distorted dipole field, but with only a 15° tilt of the dipole axis from the rotation axis. This is a purely qualitative sketch of the Martian magnetosphere field and although

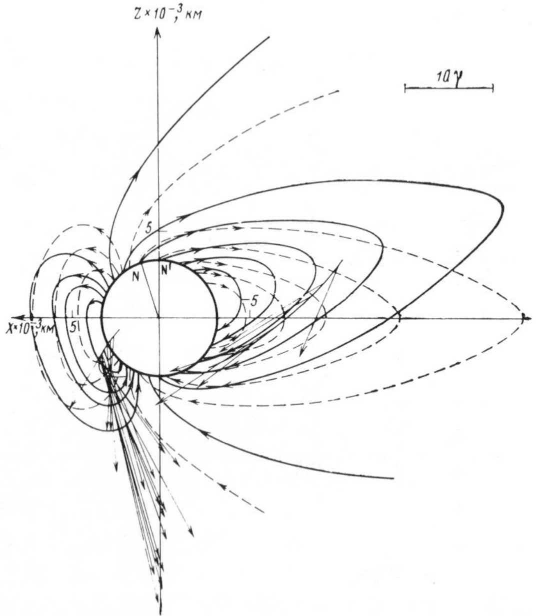

Figure 12 Projection of Mars 3 and 5 magnetic field data onto a plane perpendicular to the ecliptic when the spacecraft were inside the magnetosphere of Mars. The sketch of a distorted dipole field (dashed and solid lines) is meant to illustrate the topology for a 15° tilt of the dipole axis from the normal (Dolginov 1977).

the result appears consistent with the magnetic field measurements, the tilt is grossly inconsistent with the quantitative value earlier derived.

Russell (1978a,b) has attempted to reinterpret the USSR Mars spacecraft data in terms of a completely induced magnetosphere of the planet associated with the solar wind interaction. Dolginov (1978a,b) has refuted these and other criticisms and maintains his interpretation that the planet possesses an intrinsic magnetic field similar to that which he had earlier reported. The crux of the problem is that the data are inadequate to permit more than somewhat speculative interpretations, based in part upon qualitative models of solar wind interaction with planets lacking a significant large scale magnetic field. It is extremely unfortunate that no magnetometers were carried on the highly successful orbiting US Mariner 9 and Viking 1 and 2 spacecraft missions, which would have been able to definitely resolve this important question. Moreover, there appears to be little hope of resolving this matter in the future since there are no known plans for returning to the planet Mars. The present status of our knowledge, based upon the position of the bow shock at Mars, is that a small planetary magnetic field may exist, but its orientation is unknown.

Estimates of the moment of inertia ratio, C/MR^2, for Mars range from 0.365 to 0.377. This does not constrain sufficiently precisely an interpretation of the core size or its characteristics. Since the core composition is not known, the density of the core is uncertain and, as a result, estimates of Martian core size range from 1400 to 2400 km. Thus, in contrast to Mercury, where the definitive observation of a global magnetic field leads to a strong inference regarding the interior structure, the knowledge that Mars in all probability possesses a core (Solomon 1978) provides no further insight into the reality of any planetary magnetic field.

THE MOON

The first experimental observations related to the magnetism of the Moon were performed by the USSR probe Luna 2 in 1959. According to the Luna 2 magnetic field studies based on measurements made up to a distance of 50 km from its surface (Dolginov et al 1961), the Moon possesses a magnetic field no larger than 100 nT. The error associated with the instrumentation on the spacecraft limited the magnetic field accuracy to approximately 50–100 nT. This result indicates that the Moon clearly possesses a much weaker magnetic field than the Earth.

The next study of the lunar magnetic field was from the Luna 10 space probe. The results from the magnetic field experiment (Dolginov et al 1966) were interpreted in terms of a permanent pseudo magnetosphere which was carried along with the Moon throughout its entire orbit. In spite of uncertainties in the measurements of 10 nT, the results were believed to indicate the presence of a relatively constant magnetic field component both parallel and perpendicular to the spin axis of the spacecraft several times larger than this value. A study of the orientation of the observed field was not possible since no onboard aspect system provided directional data. Ness (1967) questioned the pseudo magnetosphere interpretation because of the failure of the magnetic field experiment on Luna 10 to observe the geomagnetic tail. The measurement period of Luna 10 extended through a full Moon period when the Moon would have been immersed in the well developed geomagnetic tail of the Earth. These matters were shortly resolved with the launching of the first US spacecraft to study lunar magnetism.

Lunar Explorer 35 was placed into orbit July 1967 and provided the first accurate measurements on the nature of the solar wind interaction with the Moon and thereby on characteristics of its global magnetic field. The results obtained showed that there was an absence of a lunar magnetic field at satellite periapsis, when the Moon was in the geomagnetic tail, at least none greater than 2 nT. More importantly, it showed the absence of a bow shock wave or magnetosheath, similar to Earth's, surrounding the Moon when it was in the interplanetary medium, clearly indicating that the Moon did not possess any global magnetic field that could deflect the solar wind flow around it. The first results indicated that the Moon did not possess a dipole moment greater than 4×10^{19} Am^2. Representative data obtained by Explorer 35 are shown in Figure 13 with field magnitude increases in the region corresponding to the plasma umbra and the decreases on either side in the plasma penumbra interpreted in terms of diamagnetic properties of the solar wind plasma.

Continued analysis of Explorer 35 data showed that the solar wind interaction with the Moon is generally very weak, when compared to its interaction with the Earth. The Moon appears to behave like a spherical obstacle in the solar wind flow, absorbing the major fraction of plasma flux incident on its surface. Studies of the magnetic field perturbations observed in the lunar wake identified several possible source mechanisms. The presence and characteristics of the perturbations were found to be related to the plasma characteristics of the solar wind, i.e. its number density, temperature, magnetic field strength. The rate of occurrence of penumbral increases depended upon certain selenographic locations being present at the limbs of solar wind flow. These locations were interpreted as representing possible sites of localized magnetic fields on the lunar surface. The study of the solar wind interaction with the Moon and the

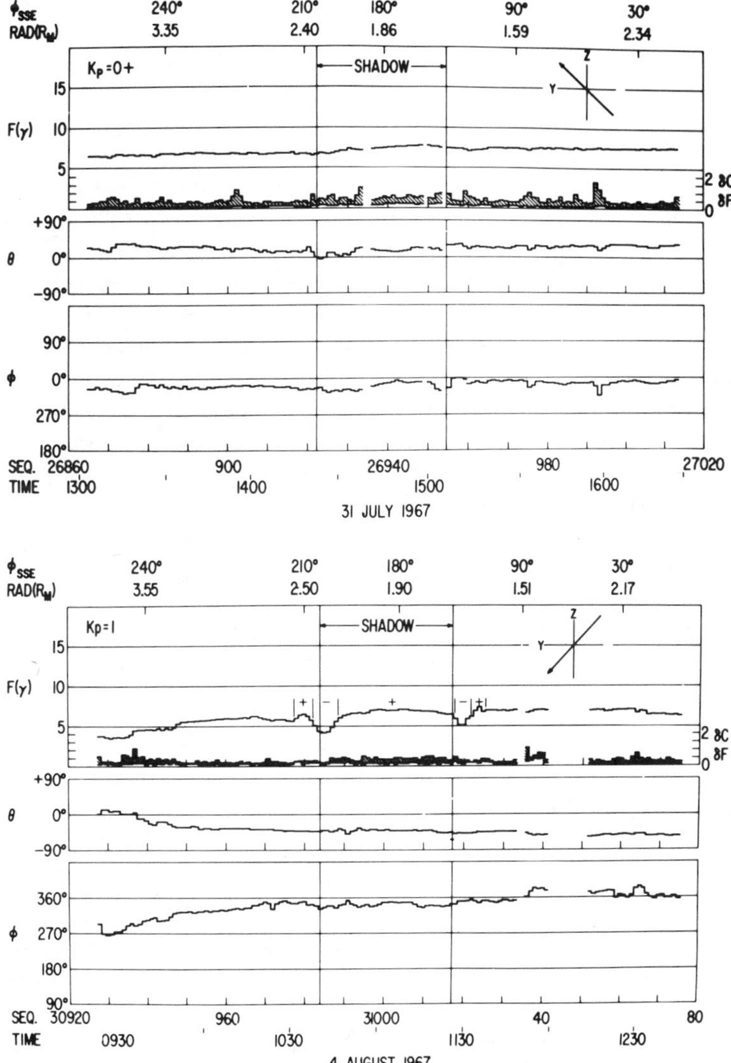

Figure 13 Explorer 35 studies of the magnetic field in the vicinity of the Moon as the solar wind flows past. There is no evidence for a bow shock, a magnetopause, or a magnetosphere. On occasion, there are slight perturbations (<20–30%) of the magnetic field (+ − + − +) in the leeward penumbral and umbral region. The selenocentric solar ecliptic position of the spacecraft at the time of measurement is given as a longitude (ϕ_{SSE}) and radius (units are lunar radii). The data points are 80.2-second averages or RMS pythagorean means of 0.2-Hz vector samples (Ness 1971). (Note: 1 γ = 1 nT.)

results from Explorer 35 have been reviewed by Ness (1971) and Schubert & Lichtenstein (1974). A review of theoretical models of solar wind interaction in the steady state was prepared by Catto (1974).

The Apollo Era

The US, in its Apollo Program, planned the placement of a suite of scientific instruments, including magnetometers, on the surface of the Moon during three of the Apollo landing missions. These were known as ALSEPs 1–3, the Apollo Lunar Science Experiment Packages. Apollo 12 in 1969, Apollo 15 in 1971, and Apollo 16 in 1972 represented three magnetic observatories set up on the lunar surface, recording the magnetic fields observed over extended periods of time with considerable inter-station overlap. In addition, Lunar Explorer 35, previously launched in 1967, continued to operate and provided extremely valuable, simultaneous observations of the magnetic field in the vicinity of the Moon. An additional feature of lunar surface exploration was the inclusion of portable magnetometers on the Apollo 14 and 16 missions. Finally, small particles and fields subsatellites carrying magnetometers and charged particle detectors were ejected into lunar orbit from Apollos 15 and 16 in 1971 and 1972 (see Table 3).

The interpretations of data from these experiments have been discussed in the literature. Reviews of the results obtained have been presented by Dyal, Parkin & Daily (1974) and Schubert & Lichtenstein (1974). Measurable magnetic fields were detected on all the lunar landing missions. In the first, the Apollo 12 mission, a lunar surface magnetic field of 38 ± 2 nT was observed. On Apollo 14, the magnetic field was 103 ± 5 and 43 ± 6 nT at two different locations, separated by 1.1 km. On Apollo 16, a series of measurements were made with the magnitude ranging from 112 to 327 nT, the length of the traverse being approximately 7 km. The orientation of the magnetic field at these different sites was not correlated in any global fashion. Measurements of the gradient of the magnetic field were possible, so that the scale sizes of the source regions were found to be on the order of hundreds of kilometers.

These magnetic fields represent localized phenomena, since, as discussed earlier, there is no evidence for a detached bow shock surrounding the Moon when it is located within the interplanetary medium. Throughout its orbit, the Moon is periodically immersed in the interplanetary medium, the magnetosheath of the Earth, and the Earth's magnetic tail where it resides approximately four days each lunar synodic month. The existence of magnetometers on the lunar surface, in concert with Lunar Explorer 35, also allowed investigation of the electrical conductivity of the lunar interior by studies of the induced magnetic field driven by fluctuations of

Table 3 Summary of spacecraft studies of lunar permeability and dipole magnetic moment

Authors	μ_{moon}/μ_o	$\mathbf{M}_{perm}$	$\mathbf{M}_{ind}$	Spacecraft
		Am^2	$Am^2/tesla$	
Dolginov et al 1961	—	$<6 \times 10^{18}$	—	Luna 2
Dolginov et al 1966	—	—	—	Luna 10
Ness et al 1967	—	$<6 \times 10^{17}$	—	Exp 35
Sonett, Colburn & Corme 1967	—	$<6 \times 10^{17}$	—	Exp 35
Behannon 1968	<1.8	$<1 \times 10^{17}$	$<1 \times 10^{25}$	Exp 35
Dyal & Parkin 1971	1.03 ± 0.13	—	—	Exp 35
Dyal, Parkin & Cassen 1972	1.01 ± 0.06	—	—	Apollo 12 Exp 35
Russell et al 1973	—	$<2.1 \times 10^{15}$	—	Apollo 12 Apollo 15 SS
Parkin, Dyal & Daily 1973	$1.029^{+0.024}_{-0.019}$	—	—	
Russell et al 1974	$1.008 - 1.03$	$<1.3 \times 10^{15}$	$-6.25 \pm 2.4 \times 10^{23}$	15SS, Exp 35
Parkin, Daily & Dyal 1974	1.012 ± 0.006	—	$+2.1 \times 10^{23}$	Apollo 12, Exp 35
Russell et al 1975	—	$<1.3 \times 10^{15}$ $<6.5 \pm 5.4 \times 10^{15}$	-5×10^{23}	15SS 16SS
Dyal et al 1975	$1.012^{+0.011}_{-0.008}$	—	—	Apollo 15 & 16

the interplanetary magnetic field. Studies of the electrical conductivity of the lunar interior have been exhaustively pursued by several research groups and are discussed in numerous publications (see, for example, Schubert & Lichtenstein 1974 and the proceedings of the Lunar Science Conferences).

Lunar Surface Magnetic Anomaly Maps

Studies of the lunar surface magnetic field were conducted from orbit by the Apollo 15 and 16 subsatellites. Launched from the lunar module, these spacecraft orbited the Moon at altitudes of several hundred kilometers or less for periods ranging from 35 days for Apollo 16 to 7 months for Apollo 15. The inclination of the orbital plane of Apollo 15 was 29° with respect to the lunar equator while that of Apollo 16 was 10°. Due to the near equality of orbital and rotational periods of the Moon, the relative location of the subsatellite track on the lunar surface, when in the geomagnetic tail, was approximately the same for each full Moon period. Thus, in spite of the 29° inclination of the Apollo 15 subsatellite, and its long life, mapping of the lunar magnetic field was accomplished over only 14% of the lunar surface.

Observations from these satellites confirmed the earlier discoveries by Explorer 35 of the absence of a bow shock wave and the intermittent presence of the penumbral increases in the interplanetary field, which because of the orbit were observed only near the limbs of the Moon (Schubert & Lichtenstein 1974). These perturbations were several nanotesla or less relative to the average fields of 5–10 nT in the interplanetary medium.

When the Moon is in the geomagnetic tail, the magnetic field magnitude is approximately 10–15 nT. Against this relatively steady and moderate field intensity, small perturbations were detected in the subsatellite data and identified as being lunar surface associated. The largest of these was in the vicinity of the Van de Graaff Crater and showed an amplitude perturbation of several nanotesla, depending upon the altitude of the spacecraft at the time of the observations. Because of the natural fluctuations of the geomagnetic tail field intensity and direction, the small amplitude of the lunar magnetic fields at subsatellite altitude and the large quantization step size of the experiments (0.2 or 0.4 nT), some difficulty was encountered in the analysis and interpretation of the data.

Anomaly maps of the lunar magnetic field observed by the subsatellites were constructed by various methods. In some cases, the measurements were reduced to a common altitude reference level (h_{ref}) and then the orbital average was subtracted from each orbit's data. In the absence of a more comprehensive extrapolation methodology, the data were simply corrected for altitude by use of one of the following algorithms

$$\mathbf{B}_{\mathrm{corr}} = \mathbf{B}_{\mathrm{obs}} \cdot \left(\frac{\mathrm{alt}}{h_{\mathrm{ref}}}\right)^{2.5}$$

$$\mathbf{B}_{\mathrm{corr}} = \mathbf{B}_{\mathrm{obs}} \cdot \left(\frac{\mathrm{alt}}{h_{\mathrm{ref}}}\right)^{1.5},$$

where h_{ref} ranged from 70 to 100 km (Russell et al 1975).

An early anomaly map obtained from Apollo 15 is shown in the upper panel of Figure 14. The large anomaly near crater Van de Graaff is readily evident in the radial component of the magnetic field, with a peak-to-peak value of 1.6 nT. An important feature, seen in these results, is the elongation of the contours parallel to the subsatellite track. Variations of the tail field were not always eliminated in these early maps and as a result led to the artifact of such an anomaly field with contours parallel to the satellite orbit.

Subsequently, these problems in analyzing the subsatellite data were partially overcome by the use of numerical filters so constructed that for each orbit they eliminated all fluctuations of the magnetic field along the orbit whose wavelength corresponded to 20° in lunar longitude or greater (Russell et al 1975). The results of this processing scheme are shown for comparison in the bottom panel of Figure 14. Many of the apparent anomalies earlier identified in the magnetometer data were thus eliminated. The anomaly near Van de Graaff remained although its characteristics were changed considerably. Strangway, Rylaarsdam & Annan (1975) have attempted to model the anomaly with a horizontal disc and find that a thickness of 1 km and a magnetization of 10^{-4} $\mathrm{Am^2/kg}$ are consistent with the data. Hood, Russell & Smith (1978) have modelled the anomalous lunar field by using a cratered, magnetized shell to investigate quantitatively a proposed hypothesis of Runcorn (1975).

One of the principal difficulties in interpreting the subsatellite data is the very poor signal-to-noise ratio at the altitude of the satellite orbit. While the lunar signal is found to be greater than 1 nT or more at altitudes of 20 km or less, from data of the Apollo 16 subsatellite, the major fraction of data used in the analysis comes from Apollo 15 and its higher altitude leads to a considerably reduced magnitude. The distribution of the magnetic field intensities used in construction of the final magnetic anomaly maps from Apollo 15 and 16 data at altitudes between 60–170 km is shown in Figure 15. Less than 1% of the entire data set has an amplitude greater than 1 nT, and less than 10% of the data has a value greater than 0.5 nT. With such small values of signal and such limited spatial coverage, it continues to be a challenge to interpret the

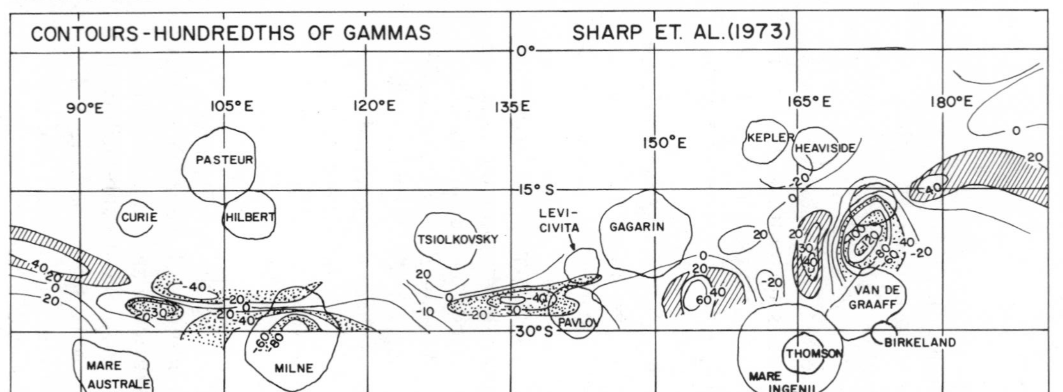

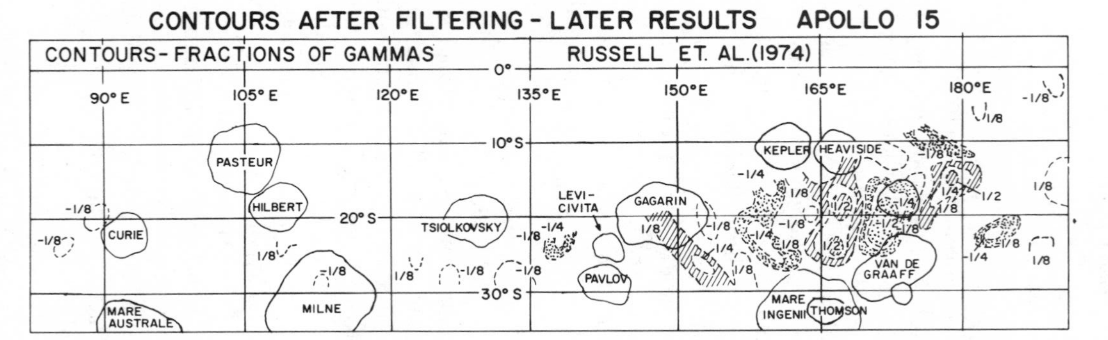

Figure 14 Comparison of two different contour maps of the radial component of the lunear magnetic field for the same Van de Graaff quadrant derived from Apollo 15 subsatellite. Striped areas represent positive, directed outward, while dotted areas indicate negative, directed inward. The upper panel, from Sharpe et al (1973), uses units of 0.01 nT while the lower panel, from Russell et al (1975) uses units of nT. As is readily evident, the major anomaly near Van de Graaff Crater changes in magnitude by more than a factor of two, reverses sign, and has its shape changed considerably.

results they have obtained. Some attempts have been made to correlate local geological characteristics with the magnetic field anomaly map (Russell et al 1977).

Lunar Permeability and Dipole Moment

Progress has been made in studying both the permeability of the Moon and the dipole moment by use of data from both orbiting spacecraft and surface measurements. Studies of the permanent magnetic moment are restricted to estimates of the component of the lunar dipole moment in the orbital plane of the subsatellites. Determinations of the lunar magnetic moment have been based upon the harmonic analysis of the radial and tangential components of the magnetic field throughout a satellite orbit, by using the orbital period as the fundamental period. Results of such an analysis are shown in Figure 16. The maximum amplitude of any of the harmonics is 0.06 nT and the $n = 1$ term, the dipole term, is $\approx$0.05 nT. This limits the magnitude of the dipole moment of the Moon

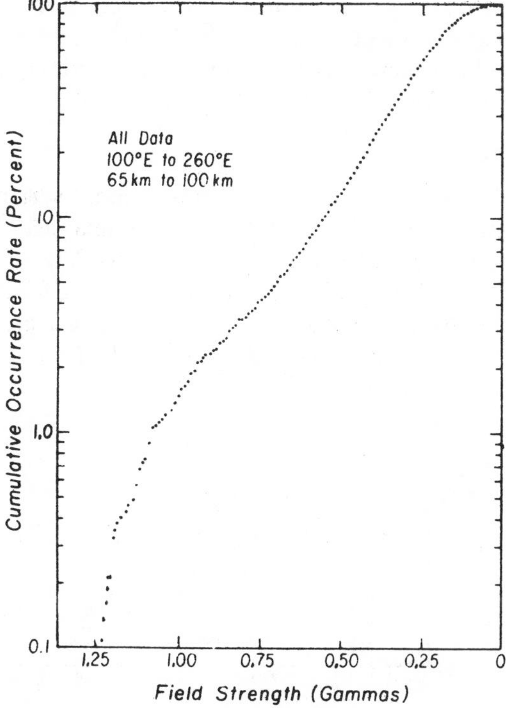

Figure 15 Cumulative statistical distribution of magnetic field strength of final filtered Apollo 15 and 16 subsatellite data (Russell et al 1977).

to less than 2.1 × 10^{15} Am^2 in the orbital plane of the subsatellites (Russell et al 1973). There is no measurable dipole moment $>1.3 \times 10^{15}$ Am^2 (Russell et al 1974, 1975; see Table 3).

Complicating the interpretation of such data is the fact that in addition to any permanent dipole moment contribution to the observation, there can be an induced dipole moment, associated with the magnetically permeable lunar interior and also a possible weak lunar ionosphere. The induced moment will be paramagnetic but the ionosphere moment will be diamagnetic. Finally, an electrically conducting core can affect the measurement of net permeability so that it masquerades as one of these effects. These several issues have been studied in the literature (Goldstein & Russell 1975). The results of all the studies of the lunar permeability and permanent and induced lunar magnetic moments are shown in Table 3.

The global permeability of the Moon is found to be very small, $\mu/\mu_0 = 1.012 \pm 0.006$. Indeed, the determination of the permeability depends critically upon the accuracy of the intercalibration of magnetic field data obtained by separate experiments on the lunar surface and in lunar orbit. King & Ness (1977) have determined that the accuracy of the data sets employed in these analyses is not sufficient to warrant their use in physical interpretation of the characteristics of the lunar interior. It appears that, at best, we can conclude that $\mu_{moon} \leqq 1.02\mu_0$.

Lunokhod Studies

In addition to the US Apollo investigations of lunar surface magnetism, the USSR carried out investigations from unmanned lunar rovers, Lunokhods 1 and 2, in the Luna 16 and 21 projects. These devices carried magnetometers which measured the magnetic field a few meters above the lunar surface, while they traversed the lunar surface along with other instruments examining the characteristics of the Moon. A sample of

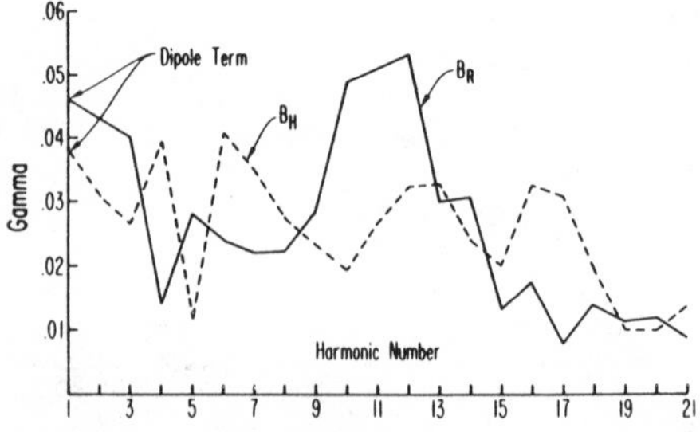

Figure 16 Results of a Fourier harmonic analysis of the magnetic field in the orbit plane. The two spectra, $\mathbf{B}_R$ and $\mathbf{B}_H$, refer to the radial component and the horizontal component, parallel to the lunar surface (Russell et al 1973).

data obtained by Lunokhod 2 is shown in Figure 17. Considerable difficulty was encountered in the analysis and interpretation of these data, due to the presence of appreciable ferromagnetic material on the Lunokhod. Attempts were made to remove these contaminating effects by having the Lunokhod vehicle travel along the same traverse but in opposite directions so that the contribution from the magnetic field of the Lunokhod itself was reversed. However, this only permits a calibration of the horizontal field, in the best of circumstances. As shown in Figure 17, there appear to exist appreciable localized magnetic fields, on the order of 10–30 nT, on the lunar surface, which are not completely random when viewed on a scale of several kilometers. Some attempt has been made to interpret these Lunokhod data in terms of the geology of the regions through which they passed (Dolginov et al 1975b).

Electron Reflectance Magnetometry

A new method of investigating the characteristics of lunar surface magnetic fields was developed from the energetic electron experiment carried onboard the Apollo 15 and 16 subsatellites (Howe et al 1974). The basis for this method is illustrated in Figure 18. A directionally sensitive detector onboard the spin-stabilized subsatellites measured electron fluxes in

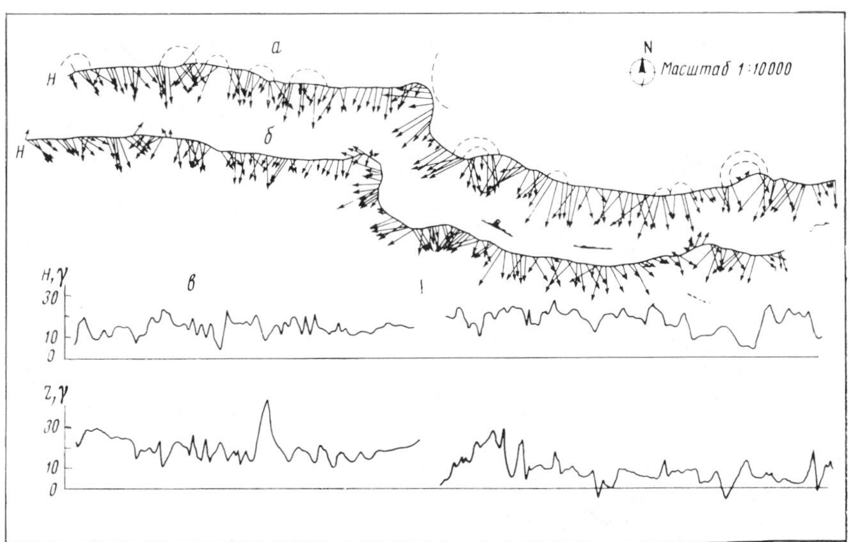

Figure 17 Results from Lunokhod 2 of a traverse of the lunar surface and the magnetic field measurements obtained (Dolginov et al 1975b). Units are $1\gamma = 1$ nT. The horizontal (H) components at the top differ by having the lunokhod field subtracted from the horizontal data curve δ. The bottom curve is the vertical component Z.

eight different angular sectors (1–4). In the absence of any lunar surface magnetic fields, electrons traveling along field lines that intersect the Moon are absorbed and no reflected electrons would be observed for detector positions sensitive to the direction leading from the Moon to the satellite, along the magnetic field direction (sectors 1–2). However, when the lunar surface magnetic field varies, regions will develop in which the local magnetic field may be strong enough to reflect a fraction of the incident particle flux, which can then be detected.

Fortunately, there exists an abundant natural supply of low energy electrons, which are especially important because the diameter of their helical trajectory about the magnetic field represents the smallest magnetic anomaly scale size which this method can probe. Electrons of 200 eV have a helix radius of 48 sin $\alpha/\mathbf{B}_0$ km, where $\mathbf{B}_0$ is the magnitude of the ambient field, as measured in nanotesla and α is the pitch angle of the particle helical trajectory. In the magnetotail, the helix diameter for these electrons is on the order of 10 km or less. In the solar wind, the flux of electrons in this energy range is typically about 3×10^6/cm^2/sec/steradian/

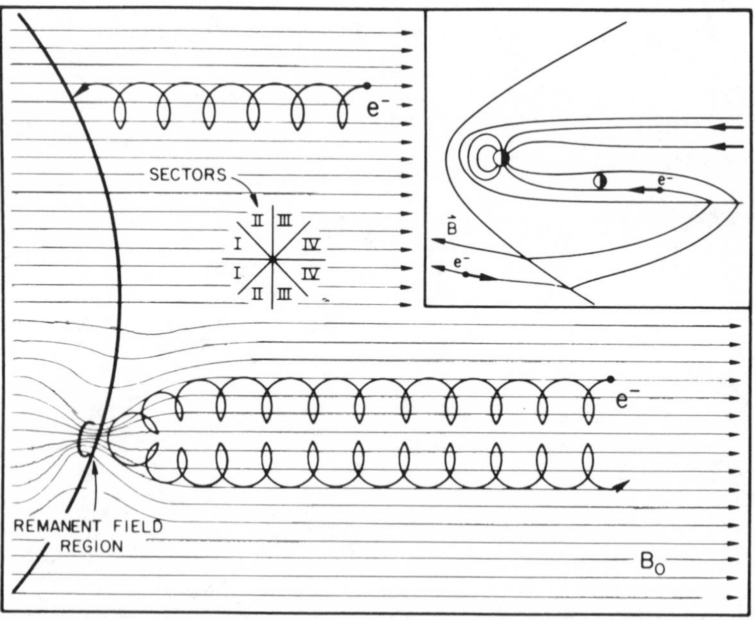

Figure 18 Illustration of electron detector angular sectoring on the Apollo 15 and 16 subsatellites and sketch of trajectory of electron reflected by localized lunar magnetic field, when Moon is in geomagnetic tail. Insert sketches manner of access of solar electrons to geomagnetic tail (Howe et al 1974).

kev. The source of such electrons is the Sun and since the geomagnetic tail is often connected to the interplanetary magnetic field, there is a copious supply of such electrons for study of the reflectance characteristics of the lunar surface magnetic field.

In this method the important parameter is the ratio of the electron flux reaching the spacecraft (due to reflection by the lunar surface magnetic field) to the incoming electron flux. As this reflection coefficient varies along the satellite track, it becomes a direct measure of the variation of the local lunar surface magnetic field. The reflection coefficient depends upon the particle pitch angle distribution, the magnetic field strength, and its orientation. Encouraging initial results have been obtained by this technique, especially in certain areas along the lunar surface. A sample of the results obtained is shown in Figure 19. This approach cannot provide a unique determination of the magnetic field on the lunar surface but only indicates regions with variations in the lunar surface magnetic

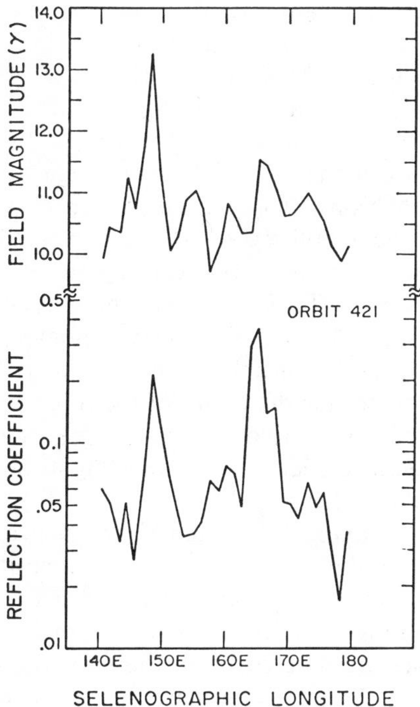

Figure 19 Observed electron reflection coefficient variation and signal observed near Crater Mandelshtam (169E) at a satellite altitude of 50 km. A second signal was detected at 149° E at an altitude of 30 km, simultaneous with an increase in the magnetic field. (Anderson et al 1975.)

field. The reason for this is that the topology of the magnetic field, as shown in Figure 18, would be altered if the orientation of the reference field, $\mathbf{B}_0$, were changed. For example, if the field polarity were simply reversed, instead of a single, high field region as shown, there would be a field minimum surrounded by an enhanced field so that the reflection coefficient would show a different variation. The method is particularly interesting for its ability to extend the study of lunar surface magnetism to small scale size from satellite orbit (Anderson et al 1977).

Origin of Lunar Magnetism

One of the more challenging problems and the most enigmatic in planetary magnetism is the origin of the lunar magnetic field responsible for the magnetization of the returned lunar samples and the localized magnetic fields measured on the Apollo and Lunokhod missions. In the absence of evidence for a global magnetic field but in the presence of substantive evidence for local magnetic fields, Runcorn (1975) hypothesized the existence of an internal dynamo, active in the past, which magnetized the lunar mantle but for which no external signature would be present. This hypothesis was challenged by Goldstein (1975), who showed that only in mathematically limiting and physically unrealistic conditions would such a situation occur.

Additional studies of the origin of lunar magnetic fields have been conducted by Stephenson (1976a) and by Srnka & Mendenhall (1978). One problem with the hypothesis of an active dynamo in the lunar past is the necessity that there be a convecting core, if a self-regenerative dynamo analogous to Earth is assumed. There is no firm evidence to support the existence of a lunar core today (Solomon 1978), only upper limits of 300–400 km if it is iron and 500–600 km if it is iron sulphide. Less conventional hypotheses for the origin of the lunar magnetic field have invoked cometary impact (Gold & Soter 1976), or electrical currents driven by thermoelectric effects in the differential heating and cooling of surficial material (Dyal et al 1975).

Investigators in studying the remanent magnetization of the lunar samples have found it difficult to agree on the intensity of the magnetic field at the time when the samples were magnetized. One reason for this is that the mechanisms whereby the remanent magnetization is carried in the lunar samples are different from those operating on Earth. Here the magnetic metallic oxides are primarily responsible for the remanent properties. On Moon, remanent properties have been found to be due primarily to free iron or iron nickel alloys in the lunar samples. Thus, such parameters as the shape factor of the grains become important in deducing the magnetizing field at time of formation. An additional complication is that lunar remanent magnetism has probably been modified due to shock impact (Cisowski et al 1975). Thus, at present, the source of the magnetic field responsible for magnetizing the lunar samples is by no means clear.

Unfortunately, there are no plans by the US to return to the Moon before at least the late 1980s. While the USSR plans are uncertain, continued problems with the accuracy of their measurements due to contamination by spacecraft magnetic fields and the inherent inaccuracy of the magnetometer sensing units, combined with the small lunar signals, are difficulties which must be overcome before adequate data can be collected from the Moon, as well as from Mars.

SUMMARY

Spacecraft measurements by the US Mariner 10 in 1974 and 1975 have shown the existence of a significant global magnetic field at Mercury, with a dipole moment of 4.8×10^{19} Am^2 at a tilt angle of $14° \pm 5$ relative to the normal to the orbit plane. This corresponds to an equatorial field intensity of 330 nT, 1% of Earth's field. While there is no question about the existence of the intrinsic planetary field and a magnetosphere at Mercury, the origin of the magnetic field is uncertain. Either remanent magnetization or an active dynamo has been postulated. A survey of all available evidence suggests that remanent magnetization is less plausible than a presently active dynamo.

The magnetic field at Mars has been studied essentially only by the USSR in its Mars spacecraft program in 1971 and 1974. A detached, bow shock wave was observed, but there is no unequivocal evidence for the existence of a planetary magnetic field. The principal problem was that the Mars spacecraft orbits were not optimumly designed to investigate the weak magnetic field of the planet, since periapsis was too high and not located near the midnight meridian plane in the equatorial region of the planet. Limited evidence suggests the possibility of an intrinsic magnetic field with dipole moment $\approx 2.4 \times 10^{19}$ Am^2, corresponding to an equatorial field intensity of 64 nT. If such a magnetic field exists, it may result from either remanent magnetism or an active dynamo.

The Moon does not possess a measurable global dipole magnetic field, the upper limit being 1.3×10^{15} Am^2, corresponding to an equatorial field intensity of 0.05 nT. There is some evidence for a magnetically permeable Moon slightly different from vacuum, but uncertainties in the data limit the interpretations to an upper limit of $\mu/\mu_0 \leqq 1.02$. There is good evidence for local magnetic fields, both directly from the Apollo and Lunokhod missions and indirectly from observations obtained from the Apollo subsatellites' magnetometer and electron reflectance spectrometer experiments. The direct surface measurements show magnetic fields ranging from 3 to 330 nT but with no global correlation of the steady state values. Fluctuating magnetic fields have been observed and interpreted in the framework of induction magnetism due to an electrically conducting lunar interior and a temporally varying external interplanetary magnetic field. One of the major limitations to progress in understanding the origin of lunar remanent magnetism is the small signal-to-noise ratio in much of the data collected thus far.

Unfortunately, there are no firm plans by either the US or USSR to return to any of these bodies in the next decade. Thus, those issues which are unresolved today shall probably persist for some time, namely:

1. Are there higher order multipole moments on Mercury?
2. What is the origin of the magnetic field at Mercury?
3. Is there a Martian magnetic field or not?
4. What was the source of the fields responsible for magnetization of the lunar surface material?

Literature Cited

Acuna, M. H., Ness, N. F. 1976. The main magnetic field of Jupiter. *J. Geophys. Res.* 81:2917–22

Anderson, K. A., Lin, R. P., McCoy, J. E., McGuire, R. E. 1975. Measurements of lunar and planetary magnetic fields by reflection of low energy electrons. *Space Sci. Inst.* 1:439–70

Anderson, K. A., Lin, R. P., McGuire, R. E., McCoy, J. E., Russell, C. T., Coleman, P. J. Jr. 1977. Linear magnetization feature associated with Rima Sirsalis. *Earth Planet. Sci. Lett.* 34:141–51

Behannon, K. W. 1968. Intrinsic magnetic properties of the lunar body. *J. Geophys. Res.* 73:7257–68

Busse, F. H. 1978a. Magnetohydrodynamics of the Earth's dynamo. *Ann. Rev. Fluid Mech.* 10:435–62

Busse, F. 1978b. Theory of planetary dynamos. In *Solar System Plasma Physics—A 20th Anniversary Review*, ed. C. F. Kennel, L. J. Lanzerotti, E. N. Parker. Amsterdam: North-Holland

Catto, P. J. 1974. A model for the steady interaction of the solar wind with the Moon. *Astrophys. Space Sci.* 26:47–94

Cisowski, S. M., Fuller, M. D., Wu, Y.-M., Rose, M. F., Wasilewski, P. J. 1975. Magnetic effects of shock and their implications for lunar magnetism. *Proc. Lunar Sci. Conf. 6th, Geochim. Cosmochim. Acta Suppl. 6*, pp. 3123–41

Dolginov, Sh. Sh. 1977. Planetary magnetism. *Geomag. Aeron.* 17:569–95

Dolginov, Sh. Sh. 1978a. On the magnetic field of Mars: Mars 2 and 3 evidence. *Geophys. Res. Lett.* 5:89–92

Dolginov, Sh. Sh. 1978b. On the magnetic field of Mars: Mars 5 evidence. *Geophys. Res. Lett.* 5:93–95

Dolginov, Sh. Sh., Yeroshenko, Ye. G., Zhuzgov, L. N. 1973. Magnetic field in the very close neighborhood of Mars according to data from the Mars 2 and Mars 3 spacecraft. *J. Geophys. Res.* 78: 4779–86

Dolginov, Sh. Sh., Yeroshenko, Ye. G., Zhuzgov, L. N. 1975a. Magnetic field of Mars from data of Mars 3 and 5. *Kosm. Issled.* 13:108–22

Dolginov, Sh. Sh., Yeroshenko, Ye. G., Zhuzgov, L. N., Pushkov, N. V. 1961. Investigation of the magnetic field of the moon. *Geomagnet. Aeron.* 1:18–25

Dolginov, Sh. Sh., Yeroshenko, Ye. G., Zhuzgov, L. N., Pushkov, N. V. 1966. Measurements of the magnetic field in the vicinity of the Moon in the AMS Luna 10. *Dokl. ANUSSR* 170:574–77

Dolginov, Sh. Sh., Yeroshenko, Ye. G., Zhuzgov, L. N., Sharova, V. A., Gringauz, K. I., Bezrukikh, V. V., Breus, T. K., Verigin, M. I., Remizov, A. P. 1976. Magnetic field and plasma inside and outside of the Martian magnetosphere,

NASA SP-397. In *Solar-Wind Interaction with the Planets Mercury, Venus and Mars*, ed. N. F. Ness, pp. 1–20. NASA, Langley, Va. 170 pp.

Dolginov, Sh. Sh., Yeroshenko, Ye. G., Zhuzgov, L. N., Sharova, V. A., Vnuchkov, C. A., Okulesski, B. A. Bazileuski, A. T., Vanyan, L. L., Egorov, I. V., Fainberg, E. B. 1975b. Magnetism and electrical conductivity of the Moon from Lunokhod 2 data. In *Cosmochemistry of Moon and Planets*, pp. 314–22. Science Publ. Moscow

Dryer, M., Heckman, G. R. 1967. Application of the hypersonic analogue to the standing shock of Mars. *Solar Phys.* 2: 112–24

Dyal, P., Parkin, C. W. 1971. The Apollo 12 magnetometer experiment: Interval lunar properties from transient and steady magnetic field measurement. *Proc. Lunar Sci. Conf. 2nd*, pp. 2391–2413

Dyal, P., Parkin, C. W., Cassen, P. 1972. Surface magnetometer experiments: Interval lunar properties and lunar surface interactions with the solar plasma. *Proc. Lunar Sci. Conf. 3rd*, pp. 2287–2307

Dyal, P., Parkin, C. W., Daily, W. D. 1974. Magnetism and the interior of the Moon. *Rev. Geophys. Space Phys.* 12: 568–91

Dyal, P., Parkin, C. W., Daily, W. D. 1975. Lunar electrical conductivity and magnetic permeability. *Proc. Lunar Sci. Conf. 6th*, pp. 2909–25

Fairfield, D. H. 1971. Average and unusual locations of the earth's magnetopause and bow shock. *J. Geophys. Res.* 76: 6700–16

Fairfield, D. H., Behannon, K. W. 1976. Bow shock and magnetosheath waves at Mercury. *J. Geophys. Res.* 81: 3897–3906

Fuller, M. 1974. Lunar magnetism. *Rev. Geophys. Space Phys.* 12: 23–70

Gold, T., Soter, S. 1976. Cometary impact and the magnetization of the Moon. *Planet. Space Sci.* 24: 45–54

Goldstein, B., Russell, C. T. 1975. On the apparent diamagnetism of the lunar environment in the geomagnetic tail lobes. *Proc. Lunar Sci. Conf. 6th*, pp. 2999–3012

Goldstein, M. L. 1975. Lunar magnetism. *Nature* 258: 175

Gubbins, D. 1974. Theories of the geomagnetic and solar dynamos. *Rev. Geophys. Space Phys.* 12: 137–54

Gubbins, D. 1977. Magnetic field of Mercury. *Icarus* 30: 186–91

Hartle, R. E., Curtis, S. A., Thomas, G. E. 1975. Mercury's helium exosphere. *J. Geophys. Res.* 80: 3689–92

Herbert, F., Wiskershen, M., Sonett, C. P., Chao, J. K. 1976. Solar wind induction in Mercury: Constraints on the formation of a magnetosphere. *Icarus* 28: 489–500

Hood, L., Russell, C. T., Coleman, P. J. Jr. 1978. Evidence for a non random magnetization of the Moon. *Geophys. Res. Lett.* 5: 305–8

Howard, H. T., et al. 1974. Mercury: Results on mass, radius, ionosphere, and atmosphere from Mariner 10 dual-frequency radio signals. *Science* 185: 179–80

Howe, H. C., Lin, R. P., McGuire, R. E., Anderson, K. A. 1974. Energetic electron scattering from the lunar remanent magnetic field. *Geophys. Res. Lett.* 1: 101–4

Jackson, D. J., Beard, D. B. 1977. The magnetic field of Mercury. *J. Geophys. Res.* 82: 2828–36

King, J. H., Ness, N. F. 1977. Lunar magnetic permeability studies and magnetometer sensitivity. *Geophys. Res. Lett.* 4: 129–32

Kumar, S. 1976. Mercury's atmosphere: A perspective after Mariner 10. *Icarus* 28: 579–91

Levy, E. H. 1976. Generation of planetary magnetic fields. *Ann. Rev. Earth Planet. Sci.* 4: 159–83

Ness, N. F. 1967. Remarks on the interpretation of Lunik 10 magnetometer results. *Geomagn. Aeron.* 3: 431–35

Ness, N. F. 1971. Interaction of the solar wind with the Moon. In *Solar Terrest. Phys. 1970 II*, pp. 159–205, ed. E. R. Dyer. Dordrecht, Holland: Reidel

Ness, N. F. 1978. The magnetosphere of Mercury. In *Solar System Plasma Physics*, ed. C. F. Kennel, L. J. Lanzerotti, E. N. Parker. Amsterdam: North-Holland. In press

Ness, N. F., Behannon, K. W., Lepping, R. P., Schatten, K. H. 1971. Use of two magnetometers for magnetic field measurements on a spacecraft. *J. Geophys. Res.* 76: 3564–72

Ness, N. F., Behannon, K. W., Lepping, R. P., Whang, Y. C., Schatten, K. H. 1974a. Magnetic field observations near Venus: Preliminary results from Mariner 10. *Science* 183: 1301–6.

Ness, N. F., Behannon, K. W., Lepping, R. P., Whang, Y. C., Schatten, K. H. 1974b. Magnetic field observations near Mercury: Preliminary results from Mariner 10. *Science* 185: 151–60

Ness, N. F., Behannon, K. W., Lepping, R. P., Whang, Y. C. 1975. The magnetic field of Mercury I. *J. Geophys. Res.* 80: 2708–16

Ness, N. F., Behannon, K. W., Lepping, R. P., Whang, Y. C. 1976. Observations of Mercury's magnetic field. *Icarus* 28: 479–88

Ness, N. F., Behannon, K. W., Scearce, C. S., Cantarano, S. C. 1967. Early results from the magnetic field experiment on Lunar Explorer 35. *J. Geophys. Res.* 72:5769–78

Ng, K. H., Beard, D. B. 1978. Possible displacement of mercury dipole. *J. Geophys. Res.* 83. In press

Ogilvie, K. W., Scudder, J. D., Vasyliunas, V. M., Hartle, R. E., Siscoe, G. L. 1977. Observations at the planet Mercury by the plasma electron experiment: Mariner 10. *J. Geophys. Res.* 82:1807–24

Parkin, C. W., Dyal, P., Daily, W. D. 1973. Iron abundance in the Moon from magnetometer measurements. *Proc. Lunar Sci. Conf. 4th*, pp. 2947–61

Parkin, C. W., Daily, W. D., Dyal, P. 1974. Iron abundance and magnetic permeability of the Moon. *Proc. Lunar Sci. Conf. 5th*, pp. 2761–78

Runcorn, S. K. 1975. On the interpretation of lunar magnetism. *Phys. Earth Planet. Inter.* 10:327–35

Russell, C. T. 1978a. The magnetic field of Mars: Mars 3 evidence reexamined. *Geophys. Res. Lett.* 5:81–84

Russell, C. T. 1978b. The magnetic field of Mars: Mars 5 evidence reexamined. *Geophys. Res. Lett.* 5:85–88

Russell, C. T., Coleman, P. J., Lichtenstein, B. R., Schubert, G., Sharp, L. R. 1973. Subsatellite measurements of the lunar magnetic field. *Proc. Lunar Sci. Conf. 4th*, pp. 2833–45

Russell, C. T., Coleman, P. J., Lichtenstein, B. R., Schubert, G. 1974. The permanent and induced magnetic dipole moment of the Moon. *Proc. Lunar Sci. Conf. 5th*, pp. 2747–60

Russell, C. T., Coleman, P. J., Fleming, B. K., Hilburn, L., Ioannides, G., Lichtenstein, B. R., Schubert, G. 1975. The fine-scale lunar magnetic field. *Proc. Lunar Sci. Conf. 6th*, pp. 2955–69

Russell, C. T., Weiss, H., Coleman, P. J. Jr., Soderblom, L. A., Stuart-Alexander, D. E., Wilhelms, D. C. 1977. Geologic-magnetic correlations on the Moon: Apollo subsatellite results. *Proc. Lunar Sci. Conf. 8th*, pp. 1171–85

Schubert, G., Lichtenstein, B. R. 1974. Observations of Moon plasma interactions by orbital and surface experiments. *Rev. Geophys. Space Phys.* 12:592–626

Sharp, L. R., Coleman, P. J. Jr., Lichtenstein, B. R., Russell, C. T., Schubert, G. 1973. Orbital mapping of the lunar magnetic field. *The Moon* 7:322–41

Simpson, J. A., Eraker, J. H., Lamport, J. E., Walpole, P. H. 1974. Electrons and protons accelerated in Mercury's magnetic field. *Science* 185:160–66

Siscoe, G. L. 1978. Towards a comparative theory of magnetospheres. In *Solar System Plasma Physics*, ed. C. F. Kennel, L. J. Lanzerotti, E. N. Parker. Amsterdam: North-Holland. In press

Siscoe, G. L., Christopher, L. 1975. Variations in the solar wind stand-off distance at Mercury. *Geophys. Res. Lett.* 2:158–60

Siscoe, G. L., Ness, N. F., Yeates, C. M. 1975. Substorms on Mercury? *J. Geophys. Res.* 80:4359–63

Smith, E. J., Davis, L. Jr., Coleman, P. J. Jr., Jones, D. E. 1965. Magnetic field measurements near Mars. *Science* 149:1241–45

Solomon, S. C. 1978. Formation, history and energetics of cones in the terrestrial planets. *Phys. Earth Planet. Inter.* In press

Sonett, C. P., Colburn, D. S., Currie, R. G. 1967. The intrinsic magnetic field of the Moon. *J. Geophys. Res.* 72:5503–7

Srnka, L. 1976. On the global TRM of the lunar lithosphere. *Proc. Lunar Sci. Conf. 7th*, pp. 3357–72

Srnka, L. J., Mendenhall, M. H. 1978. Theory of global thermoremanent magnetization of planetary lithospheres in dipole fields of internal origin. *J. Geophys. Res.* 83. In press

Stephenson, A. 1976a. The residual permanent magnetic dipole moment of the Moon. *The Moon* 15:67–81

Stephenson, A. 1976b. Crustal remanence and the magnetic moment of Mercury. *Earth Planet Sci. Lett.* 28:454–58

Strangway, D. W., Rylaarsdam, J. C., Annan, A. P. 1975. Magnetic anomalies near Van de Graaff Crater. *Proc. Lunar Sci. Conf. 6th*, pp. 2975–84

Vaisberg, O. L., Bogdanov, A. V., Smirnov, V. N., Romanov, S. A. 1975. On the nature of the solar wind-Mars interaction. In *Solar Wind Interaction with the Planets Mercury, Venus and Mars*, ed. N. F. Ness, pp. 21–40. *NASA Spec. Publ.* 397. Washington, D.C. 170 pp.

Verigin, M. I., Gringauz, K. I., Gombosi, T., Breus, T. K., Bezrukikh, V. V., Remizov, A. P., Volkov, C. I. 1978. Plasma near Venus from the Venera 9 and 10 wide angle analyzer data. *J. Geophys. Res.* 83:3721–28

Whang, Y. C. 1977. Magnetosphere magnetic field of Mercury. *J. Geophys. Res.* 82:1024–30

Ann. Rev. Earth Planet. Sci. 1979. 7: 289–342

SUBSOLIDUS CONVECTION IN THE MANTLES OF TERRESTRIAL PLANETS

Gerald Schubert
Department of Earth and Space Sciences, University of California, Los Angeles, California 90024

INTRODUCTION

Each of the terrestrial planets, Mercury, Venus, Earth, Moon, and Mars, is undoubtedly evolving toward a state of eventual quiescence, when the gravitational potential energy made available by accretion and differentiation (principally core formation), and the heat supplied by the decay of radioactives (and the energy from other possible sources), have been lost from the interior (Kaula 1975). The rate at which a planet evolves toward its inevitable fate is determined largely by its original allocation of energy sources and the efficiency of the mechanisms which transfer heat from its interior to its surface. The most efficient of these processes may be heat transport by solid state mantle convection. Our main purpose in this review is to discuss the possible dominant role of this cooling mechanism in the past and present thermal states of the terrestrial planets.

There is, we should point out, no observational evidence that absolutely requires subsolidus convection in the interiors of any of the terrestrial planets except the Earth (Phillips & Ivins 1979). In the case of our own planet, a solid mantle and a plate tectonic structure together provide incontrovertible evidence of past and present solid state mantle convection. The moving plates themselves are an integral part of the mantle convection system (Turcotte & Oxburgh 1967). However, plate tectonics is irrefutable proof of subsolidus convection only in the uppermost mantle of the Earth, and just as we cannot prove the occurrence of convection in the interiors of the other terrestrial planets, we cannot demonstrate that convection is taking place or has occurred throughout the entirety of the Earth's mantle. In fact, the issue of shallow vs deep mantle convection in the Earth is much debated in the current literature (see, for example,

0084-6597/79/0515-0289$01.00

Tozer 1972a and McKenzie & Richter 1976, who argue that convection associated with plate motions does not penetrate below 700 km, and Sammis et al 1977, O'Connell 1977, and Davies 1977, who take the opposite view).

Because the surfaces of Mercury, Moon, and Mars show no evidence of plate tectonics (Solomon 1978), convection in their interiors is largely inferred from their geometrical and dynamical figures. In a long series of papers, Runcorn (1962, 1967a, 1975) repeatedly points out that the nonhydrostatic geometrical and dynamical ellipticities of the Moon, together with the larger value of the former, can be viewed as evidence of present day convection in the lunar interior. The density beneath the bulge toward the Earth would have to be smaller than the density beneath the lunar limb to reconcile the larger geometrical ellipticity with the smaller dynamical value. Such density differences could readily be associated with temperature differences between rising and descending regions of lunar convection. However, recent numerical calculations of the dynamical ellipticity of the Moon due to finite amplitude solid state convection (Cassen, Young & Schubert 1978) show that convection could be the cause of the nonhydrostatic dynamical figure only if the lunar lithosphere were capable of resisting global scale deformation. Lithospheric inhomogeneities and surface loads then could contribute substantially to the disequilibrium of the dynamical figure (Melosh 1975). In addition, Kuckes (1977) showed that an elastic lithosphere as thin as 100 km could support the stresses associated with the disequilibrium, regardless of the state of the deeper lunar interior. Runcorn used closely related arguments to support convection in the other terrestrial planets. He suggested that the long wavelength undulations in the Earth's geoid reflect density variations associated with convection (Runcorn 1967b). Most recently, he proposed that convection in Mercury could lead to the departures from hydrostatic equilibrium apparently required by Mercury's resonant state of rotation (Runcorn 1977).

Another geometrical indication of interior convection comes from the offset of the center of figure from the center of mass in planets; Venus, Earth, Moon, and Mars are known to possess such offsets. The most straightforward and reasonable interpretation of these offsets is that they arise from a first harmonic variation in crustal thickness (Kaula et al 1972). Thus the offsets imply global differentiation of a crust, and, according to Lingenfelter & Schubert (1973), a mantle convective system at the time of crustal formation to accumulate crust preferentially in the hemisphere above the region of descending flow.

In our view, the global characteristics of internal mass distributions, as reflected in geometrical and dynamical figures and gravitational fields, argue convincingly for past or present convection in the interiors of the other terrestrial planets. Other more debatable lines of evidence are described in detail by Phillips & Ivins (1979).

Aside from the occurrence of plate tectonics on Earth, the laboratory demonstration that rocks creep while solid is the strongest reason we have for believing in the relevance of solid state convection to the interiors of the planets. Under certain conditions of temperature T, pressure p, stress, and strain rate, we know quantitatively how strain rate depends on T, p, and stress (see Heard 1976 and Carter 1976 for recent summaries of the laboratory data). Unfortunately, all the conditions in planetary interiors, especially the very low strain rates characteristic of mantle convection, cannot be reproduced in the laboratory. Under conditions relevant to the deep interior of the Earth and the interiors of the other planets, rocks may be sufficiently resistant to deformation that subsolidus creep cannot occur.

In terms of the extrapolation of the laboratory creep behavior of rocks to the temperatures and pressures encountered in planetary interiors, the deep mantles of Earth and Venus present greater uncertainties than do the interiors of Mercury, Moon, and Mars, because of the major silicate structural changes that occur only in the mantles of the larger planets. As an example, Tozer (1972a) and McKenzie & Weiss (1975) have asserted that the breakdown of spinel to a denser assemblage at a depth of about 650 km in the Earth provides an effective rheological barrier which prevents upper mantle convection from extending to much greater depth. On the other hand, Sammis et al (1977) argued that the change in effective viscosity across this phase change must be so small that it could not seriously impede whole mantle convection. Inferences of a nearly uniform viscosity for the Earth's mantle from glacial rebound observations (Cathles 1975, Peltier 1976) support the latter view.

While the ability of rocks to creep is a necessary prerequisite for convection in the planets, it does not guarantee that the process will actually occur. Other crucial factors include the amount and distribution of energy sources as reflected in the superadiabatic temperature gradient, the direct driver of convective circulation. Buoyancy forces tending to drive convection must be able to overcome the viscous forces tending to resist motion. We need, finally, a theoretical statement for the onset of instability, i.e. a way of quantitatively evaluating the net outcome of the force competition. Stability criteria are available for fluids confined to a plane layer or spherical shell, heated from below or from within containing phase changes, viscosity variations, and chemical compositional gradients. In general, when reasonable estimates of thermal, mechanical, and rheological properties are used to evaluate these stability criteria, it is found that solid state convection should occur in the interiors of the terrestrial planets. This exercise was first carried out for the Earth's upper mantle by Pekeris (1935), Knopoff (1964), and Tozer (1965a) using the stability criterion for a constant viscosity plane fluid layer heated from below. Schubert, Turcotte & Oxburgh (1969) developed the stability criterion for a viscously stratified fluid layer heated from below and applied it to the entire mantles of Earth, Venus, Mars, and Moon. This stability calculation and others we discuss later support vigorous whole mantle convection in the planets. On the other hand, an adverse chemical compositional stratification in any of the terrestrial planets would be a strong deterrent against convection in that planet (Richter & Johnson 1974). Also, since adiabatic temperature gradients are especially uncertain in the lower mantles of the large planets Earth and Venus, whole mantle convection in these bodies could have been or could now be retarded by subadiabatic deep mantle conditions (Sharpe & Peltier 1979).

In summary, there are three major reasons for believing that the thermal and dynamical states of all the terrestrial planets are controlled by the solid state convective transport of heat. First and most important is the proof of convection in the Earth's upper mantle provided by plate tectonics. Second is the demonstration of the subsolidus creep deformation of rocks in the laboratory. Third are the theories which predict vigorous whole mantle convection based on reasonable and generally accepted values of material properties. Since the last two reasons are based on extrapolation and theory, and therefore subject to obvious weaknesses, we should recognize that solid state convection could be a phenomenon unique to the Earth's upper mantle, its occurrence therein entirely dependent on the presence of volatiles (water in particular), for example.

We do not intend to make the question of the existence of whole mantle convection in the Earth and the other terrestrial planets the central theme of this review. Indeed, we have long advocated the importance of whole mantle convection in planetary interiors (Schubert, Turcotte & Oxburgh 1969). Our main purpose here is to discuss the physical aspects of whole mantle convection, assuming that it occurs, and to describe how it will control the thermal evolution of a planet. Further, we do not intend to produce a comprehensive review of the many specific models of convection in the Earth's upper mantle. Recent reviews by Oxburgh & Turcotte (1978) and Richter (1978) have concentrated on shallow mantle convection in the Earth; we emphasize the role subsolidus whole mantle convection could play in the terrestrial planets.

RELEVANT PHYSICAL PROPERTIES OF PLANETARY MANTLES

Before beginning our discussion of convection in the planets we give a brief summary of what we know about the relevant physical parameter values. Thermal expansivity α, thermal diffusivity κ, thermal conductivity k, density ρ, and specific heat at constant pressure c_p are all known probably to within factors of two or three for the Earth's upper mantle (Turcotte & Oxburgh 1972, McKenzie, Roberts & Weiss 1974). We do not expect the values of these parameters to be very different for the mantles of the other planets or for the lower mantle of the Earth and therefore we adopt $\alpha = 3 \times 10^{-5}\mathrm{K}^{-1}$, $\kappa = 0.01\ \mathrm{cm}^2/\mathrm{s}$, $k = 0.01$ cal/cm s K, $\rho = 3.3\ \mathrm{g/cm}^3$, $c_p = 0.25$ cal/g K as representative for general discussions of convection in any planetary mantle.

Certain situations require more careful specification of one or more of these parameters. For example, due to phase changes and compressibility, the average density of the Earth's mantle increases with depth to a value of about 5.5 $\mathrm{g/cm}^3$ at the core-mantle interface. Such a density stratification in itself may not significantly influence mantle convection, but an associated effect of compression, namely the adiabatic increase of temperature with depth, is of prime importance. The mantle temperature gradient must exceed the adiabatic temperature gradient over at least a portion of the mantle if convection is to occur. The magnitude of the adiabatic temperature gradient $\alpha g T/c_p$ (g is the acceleration of gravity) increases with depth due to the increase of T with depth, an effect which would tend to concentrate convection in the upper mantle if the increase were to become significant. Changes in α with depth, e.g. an increase in α by a factor of two or three, would directly influence the adiabatic temperature gradient. Whereas α tends to increase with T, it probably tends to decrease with p, leaving its depth dependence in a mantle uncertain due to these competing effects.

In regions of a planetary mantle where there are phase transitions, the effective c_p may increase due to the latent heat of the transformations. The phase change contribution to c_p in these regions is probably no larger than the specific heats of the individual phases (Schubert, Yuen & Turcotte 1975). The changes in density associated with phase transitions can result in effective thermal expansivities in two-phase regions one to two orders of magnitude larger than the ordinary values of α of individual phases (Schubert, Yuen & Turcotte 1975).

Table 1 gives values of g, the acceleration of gravity at the surface of a

Table 1 Characteristic parameter values for the planets

Planet	$g(\mathrm{cm/s^2})$	$T_s(°\mathrm{C})$	$D(\mathrm{km})$	Ra	Nu	Di
☿	370	100	640	10^4	2	10^{-1}
♀	890	500	3000	10^7	20	1
⊕	980	0	3000	10^7	20	1
☾	160	0	1740	10^5	5	10^{-1}
♂	375	0	2030	10^6	10	10^{-1}

planet, T_s the surface temperature, and D the thickness of the mantle, for Mercury ☿, Venus ♀, Earth ⊕, Moon ☾, and Mars ♂. Surface gravity values are well known. Surface temperatures are 0°C except for ☿ and ♀, for which the appropriate values are about 100°C for ☿ (Cuzzi 1974) and about 500°C for ♀ (Marov 1972). The value of D for the planets is somewhat uncertain, except of course for ⊕. The moment of inertia of the Moon is so close to that of a homogeneous sphere (Williams et al 1974, Blackshear & Gapcynski 1977) that if the Moon has an iron core its radius is limited to about 500 km (Kaula et al 1974, Dainty et al 1974). We used the entire radius of the Moon for its value of D. The observed mass and radius of Mars, and the value of its moment of inertia inferred from observations of J_2 and the assumption of hydrostatic equilibrium (Reasenberg 1977) constrain models of its internal density structure. The value of D for Mars assumes that the planet has a core whose radius is 0.4 times the planetary radius (Reynolds & Summers 1969, Anderson 1972, Binder & Davis 1973, Johnston & Toksöz 1977). Observations of the mass and radius of Mercury and Venus constrain models of their internal density structures. The value of D used for ☿ is based on the assumption of a core with radius 0.75 times the planetary radius (Siegfried & Solomon 1974, Gault et al 1977). The value of D for ♀ is based on the planet's similarity to Earth (Toksöz & Johnston 1977, Ringwood & Anderson 1977). The uncertainties in the values of D for most of the terrestrial planets are not significant for our purposes. Table 1 also contains approximate values of several dimensionless parameters which are important in assessing the vigor of convection. These will be discussed later.

The most uncertain and yet most important of the mantle properties relevant to convection are the rheological properties and the radiogenic heat source concentrations. One must admit to order of magnitude(s) uncertainties in these properties, especially when the terrestrial planets other than Earth are considered. The uncertainties and the overriding importance of mantle rheology and energy sources are discussed in the following two sections.

Mantle Rheology

Of all the thermodynamic and mechanical parameters influencing mantle convection, the rheology of the mantle is perhaps the most important. There are a number of excellent and relatively recent papers on the rheological behavior of rocks (Stocker & Ashby 1973, Weertman & Weertman 1975, Heard 1976, Carter 1976).

Subsolidus deformation of rocks is due to the motions of either point defects, such as vacancies and interstitial atoms, or line defects such as screw or edge dislocations. The volumetric diffusion of point defects through mineral grains results in flow called Nabarro-Herring creep. The surface diffusion of point defects in grain boundaries produces deformation known as Coble creep. The glide motion of dislocations yields a deformation referred to by that name, while the ability of dislocations to both climb and glide results in deformation generally referred to as dislocation creep.

Nabarro-Herring and Coble creep give linear or Newtonian constitutive equations (stress $\tau \propto$ strain rate e). Since diffusion is a thermally activated process the viscosity μ for Nabarro-Herring or Coble creep is temperature and pressure dependent according to

$$\mu \propto \exp\left(\frac{E^* + pV^*}{\mathrm{R}T}\right), \tag{1}$$

where R is the universal gas constant and E^* and V^* are the activation energy and activation volume, respectively, for the relevant diffusion process (Nabarro 1948, Herring 1950, Coble 1963). It is uncertain whether Nabarro-Herring or Coble creep can occur in a planetary mantle. Indeed, these creep mechanisms have never been identified in laboratory deformations of rocks. If they occur at all, they might govern deformation in a mantle at low stresses. It is possible to choose reasonable values of E^*, V^*, and other microscopic parameters which influence diffusion to yield reasonable estimates of mantle viscosity (Gordon 1965, Tozer 1965a, Turcotte & Oxburgh 1969a).

In Nabarro-Herring creep, grains change their shape but not their nearest neighbors. Thus the mechanism is fundamentally non-steady, since the grains would continue to elongate with sustained deformation (Weertman 1968, Green 1970). It is possible that grains can also slide past one another, changing their neighbors and altering their shapes (by diffusion) only insofar as it is necessary to maintain continuity (Green 1970, Ashby & Verrall 1973). This grain boundary sliding mechanism is believed responsible for the phenomenon of superplastic flow in metals. Superplastic creep can be linear or nonlinear according to whether the grain size is independent of or dependent on the stress. Twiss (1976) has discussed the possibility that superplastic creep might occur in the mantle.

While a form of diffusion creep may govern deformation in the mantle at low stresses, the motion of dislocations by glide and climb may control creep in the mantle at intermediate stresses and over a wide range of stress. Dislocation creep is known to govern the steady state deformation of olivine at high temperature and pressure and at laboratory strain rates. The experimental data on the creep of olivine below about 2 kbar differential stress are consistent with the power law

$$e = \frac{B_n}{T} \exp\left\{\frac{-(E^* + pV^*)}{RT}\right\} \tau^n. \tag{2}$$

Kohlstedt & Goetze (1974) and Kohlstedt, Goetze & Durham (1976) have determined the values of the power law exponent n and the activation energy E^*; they find $n = 3$ and $E^* = 125$ kcal/mol. The T^{-1} dependence multiplying the exponential has not been resolved experimentally; it has only been inferred on theoretical grounds (Weertman 1970; this results in some uncertainty in determining a value of B_3 from the data). The activation volume V^* for the creep of olivine has recently been determined by Ross, Avé Lallement & Carter (1978). They give V^* values between 10.6 and 15.4 cm^3/mol with a mean value of 13.4 cm^3/mol. This is near the theoretical value of about 11 cm^3/mol appropriate if O^{2-} ion diffusion is rate-controlling.

A number of theoretical arguments have indicated that V^* should have such a low value. Sammis et al (1977) estimated V^* using Keyes' (1963) relation between activation energy and activation volume, together with the measured value of E^* for olivine and values of elastic constants either determined in the laboratory or inferred from seismic data. They predicted that at low pressure, V^* would lie between about 8 and 11.5 cm^3/mole, depending on T. Another estimate of V^* can be made from the connection between E^*, V^*, and melting temperature (Weertman 1970). From estimates of melting temperature and its pressure derivative, Sammis et al (1977) derived $V^* \approx 11.5$ cm^3/mole. Models of temperature and flow in the Earth's upper mantle using the olivine flow law (2) have yielded geophysically more reasonable results when low values of V^* (about 15 cm^3/mole) were used to calculate the effective viscosity (Froidevaux & Schubert 1975, Schubert, Froidevaux & Yuen 1976, Schubert et al 1978).

If the above constitutive relation also describes the deformation of olivine at the much slower mantle strain rates, then it is probably the relevant rheological law for at least the Earth's upper mantle, since there is abundant seismologic and petrologic evidence that the upper mantle is predominantly olivine of the approximate composition $(Mg_{0.9}Fe_{0.1})_2SiO_4$. The similarities in the results of optical and electron transmission microscopy studies of both mantle-derived and laboratory-deformed olivine crystals (Raleigh 1968, Phakey, Dollinger & Christie 1972, Goetze & Kohlstedt 1973) lend support to this conclusion. Further, since pressures throughout most of the mantles of the smaller terrestrial planets are not higher than those encountered in the Earth's upper mantle, the olivine flow law may be relevant throughout these mantles rather than just in their upper portions. The major phase changes encountered in the Earth's mantle, and presumably in Venus' as well, preclude the direct applicability of olivine flow laws to the lower mantles of these planets.

It should also be noted in connection with the issue of the applicability of the above power law constitutive relation to flow in the Earth's mantle, that deviatoric stresses in the mantle are expected to be smaller than kilobars on the basis of variations in the gravity field (Kaula 1963a, Lambeck 1976).

Since laboratory measurements of creep in rocks are not directly applicable to the mantles of the terrestrial planets, it is desirable to infer mantle creep laws directly from geophysical and geological observations. The vertical motions of the Earth in response to the redistribution of surface loads (e.g. by deglaciation) offer such a possibility although the time scale of these motions (10^3–10^4 yr) is much shorter than geologic time. The observations of such movements cannot presently distinguish among the various linear and nonlinear creep laws that have been proposed for the Earth's mantle. One may also hope that characteristics of tectonic plates, e.g. their velocities, and other plate tectonic data, e.g. heat flow and topography vs age, will distinguish between linear and nonlinear mantle creep laws. However, temperature and flow models of the upper mantle show that this is not the case (Froidevaux & Schubert 1975, Schubert, Froidevaux & Yuen 1976).

Thus, at present, we must hypothesize a mantle flow law based on theoretical mechanisms of deformation or extrapolate empirical flow laws from laboratory circumstances to conditions relevant to the mantle. Mantle temperatures and pressures are accessible in the laboratory, but relevant mantle strain rates, between about 10^{-16} s^{-1} and 10^{-12} s^{-1} are not. Laboratory strain rates are in the range 10^{-8} to 10^{-3} s^{-1}.

Although we cannot establish which flow law is relevant to a planetary mantle, it may not be essential for us to do so. All mechanisms of creep deformation have a dependence of viscosity (or effective viscosity) on T and p given by (1). According to model calculations of flow and temperature in the Earth's mantle, the dependence of effective viscosity on temperature and pressure far outweighs any influence of a possible nonlinear connection between stress and strain rate (Froidevaux & Schubert 1975, Schubert, Froidevaux & Yuen 1976, Yuen & Schubert 1976, Parmentier, Turcotte & Torrance 1976). Later in this review we discuss in more detail the overwhelming influence of the temperature dependence of the viscosity on the thermal histories of planets cooling by subsolidus mantle convection. Thus we will no longer be concerned with which deformation law is applicable to the mantle. Rather, we will employ the concept of a viscosity or effective viscosity (stress/twice the strain rate) which is T and p dependent according to (1).

If activation energy and activation volume were constants, then viscosity would tend to decrease with depth due to the increase of temperature and increase with depth due to the increase of pressure. In a large planet, the temperature effect would tend to predominate at shallow depths where the temperature increase is especially pronounced and the pressure is relatively small, and the pressure effect would take over at sufficiently large depth where the temperature would tend to increase relatively slowly along an adiabat. The net result would be a viscosity which had a minimum somewhere in the upper mantle, depending on the actual values of E^* and V^*. This is undoubtedly the situation, qualitatively at least, with the Earth's asthenosphere, the region of viscosity minimum, serving to decouple the rigid surface plates from the underlying mantle (Froidevaux & Schubert 1975, Schubert, Froidevaux & Yuen 1976). The inference from glacial rebound studies is that the viscosity of the Earth's mantle does have a minimum in the upper mantle, but that there is at best only a modest increase of viscosity with depth throughout the entire mantle; a Newtonian viscosity of 10^{22} poise is consistent with the glacial rebound observations (Cathles 1975, Peltier 1976). Weertman (1978) has argued that, because of the small strains involved, glacial rebound may be a transient creep phenomenon, in which case the rheological inferences therefrom would not pertain to steady creep on geologic time scales. However, we basically adopt the view of mantle viscosity inferred from glacial rebound data.

How then are we to reconcile the inference of nearly uniform mantle viscosity for the Earth with the viscosity dependence on T and p given in (1)? We described just above how μ would tend to increase in the Earth's lower mantle if the temperature rose along an adiabat. The pressure dependence of μ would dominate in the lower mantle and viscosity would increase with depth at a rate inconsistent with glacial rebound inferences. The answer lies in the variations of E^* and V^* with depth and across phase transitions. It is possible to construct theoretical estimates of the depth dependences of E^* and V^* and the changes in these quantities across phase transitions which result in little increase in viscosity with depth in models of the Earth's mantle (Sammis et al 1977). A major factor in limiting the increase in viscosity with depth is an inferred decrease in V^* with depth to values of only 4–6 cm^3/mol in the Earth's lower mantle (Sammis et al 1977, O'Connell 1977). Viscosity changes across any of the major upper mantle phase transitions are inferred to be less than an order of magnitude (Sammis et al 1977). There is no inconsistency in an adiabatic, nearly uniform viscosity mantle even though viscosity may be strongly temperature and pressure dependent according to (1).

Mantle Heat Sources

The average heat flow through the Earth's surface, about 1.5 μcal/cm^2 s, has been generally believed to originate mainly in the mantle from the decay of radioactive elements uranium, thorium, and potassium (Turcotte & Oxburgh 1972, Oxburgh & Turcotte 1978). Further, it has been argued that it is reasonable to assume a nearly steady state balance between mantle and crustal heat production and heat flow through the Earth's surface if convection is the fundamental mode of heat transport (Tozer 1965a, Turcotte & Oxburgh 1972). If this is the case, then the average rate of heat generation in the Earth's mantle and crust is 0.084×10^{-13} cal/cm^3 s (Oxburgh & Turcotte 1978). A number of recent studies have indicated, however, that these ideas may require revision; in particular, even with convective heat transport in the Earth's mantle, the average rate of mantle and crustal heat production may be considerably less than 0.084×10^{-13} cal/cm^3 s. In fact, Sharpe & Peltier (1978) have shown that it is possible to construct a reasonable thermal history, including cooling by subsolidus mantle convection, in which the Earth model, devoid of mantle radioactivity, evolves to a state consistent with present observations of surface heat flow (and other constraints). Although the Earth is probably not completely devoid of mantle radioactivity, it is highly possible that the average mantle concentration of radioactives is smaller, by as much as a factor of two, than the value indicated by the present surface heat flow.

The Moon is the only other terrestrial body for which we have measurements of surface heat flux. The two values of heat flow reported by Langseth, Keihm & Peters (1976) are 0.5 μcal/cm^2 s and 0.38 μcal/cm^2 s. Detailed analyses of topographic effects indicate that 0.33 μcal/cm^2 s, rather than 0.38 μcal/cm^2 s, might be more representative of the regional heat flux at the Taurus-Littrow site (Langseth, Keihm & Peters 1976). If a heat flux intermediate between the two in situ determinations, i.e. about 0.4 μcal/cm^2 s, is representative of the mean lunar surface heat flow (Langseth, Keihm & Peters 1976 argue that this may be the case), then the average heat source concentration throughout the Moon, assuming a steady state relation between production and loss, is about 0.07×10^{-13} cal/cm^3 s, similar to the value computed for the Earth's crust and mantle under the steady state assumption. Langseth, Keihm & Peters (1976) in fact employ the steady state approach to estimate uranium concentration in the Moon from the assumed mean surface heat flow. The steady state assumption is highly questionable, as we have noted, and the average lunar radiogenic heat source concentration can be much less than the value given above.

There are several reasons why the heat produced by radioactivity in a planet's interior may be substantially less than the heat flowing through its surface. First, primordial heat, i.e. energy of accretion and gravitational potential energy released in core formation, can contribute to the surface heat loss. Schubert, Cassen & Young (1979a) have quantitatively modelled the cooling of terrestrial planets without any sources of energy other than an initial thermal energy resulting from accretion and core formation. Even with heat removal from the mantle by vigorous convection, there is enough thermal energy remaining after 4.5 billion years of cooling to supply as much as a third of the presently observed surface heat flows on Earth and the Moon. Similar conclusions have been reached in recent studies by Cassen et al (1979), Stevenson (1979), and Stevenson & Turner (1979). Second, there is a fundamental lag between the decay of the surface heat flux on a cooling terrestrial planet and the decay of the heat flux into the base of the lithosphere (Schubert, Cassen & Young 1979a). This is due to the formation and thickening of the lithosphere. At any time in the cooling history of a planet there is more heat flowing through the surface than flowing into the base of the lithosphere. Even if convection in the mantle is in a quasi-steady state, the heat flow from the mantle is not in balance with the heat flow through the surface. The physical mechanism which accounts for the lag between "internal heat production" (heat from the mantle) and surface heat flux is the thickening of a rigid lithosphere, i.e. the tendency of the lithosphere to supplement the decaying mantle heat flux by feeding on the internal thermal energy of the mantle. Third, according to Daly & Richter (1978), reasonably vigorous convection may not be able to remove heat sufficiently rapidly from the central region of a system with decaying heat sources for surface heat flow to be in equilibrium with the instantaneous heat production. For all these reasons, the use of present day surface heat flux observations to infer the total concentration of radiogenic heat sources in a planet on the basis of a presumed steady state thermal balance must be viewed with caution.

Our discussion so far has emphasized the difficulty in using surface heat flow data to estimate the internal radiogenic heat content of a planet. Although these difficulties indicate that surface heat flow derived estimates of radioactive concentrations may be overly generous, there is no question that radioactive decay contributes substantially to planetary heat flow. Abundances of radioactive elements have been measured in rocks from the surfaces of Earth, Venus (Vinogradov, Surkov & Kirnozov 1973, Surkov 1977, Keldysh 1977), and the Moon (LSPET 1972, 1973). For Mercury and Mars we have only equilibrium condensation models of the solar nebula (see, for example, Lewis 1972, Grossman & Larimer 1974) with which to guess the radioactive elemental abundances.

In general, it is clear that radioactives have been concentrated in surface rocks by the processes which led to their differentiation; radioactive elements are concentrated in the partial melts of silicates which migrate toward the surface to form crustal rocks. From the known exponential decrease of radioactives with depth in the Earth's continental crust (Birch, Roy & Decker 1968, Lachenbruch 1968), Oxburgh & Turcotte (1978) have estimated that 9% of the Earth's heat flow originates there, leaving 0.076×10^{-13} cal/cm^3 s to be accounted for by radioactives in the mantle if the steady state production vs loss model is assumed. Oxburgh & Turcotte (1978) compare this latter value of mantle heat production with values that can be inferred from measurements of radioactivity in lavas and estimates of the percentage of melting of upper mantle source rocks, which presumably produced the lavas. The upper mantle sources of the lavas are estimated to have heat production rates between 2 and 4 times smaller than the surface heat flow derived average. Oxburgh & Turcotte (1978) also consider radiogenic heat production rates based on measurements of element abundances in xenoliths presumably originating in the Earth's upper mantle. The heat production rates in xenoliths are also generally less than the surface heat flow derived average. Oxburgh & Turcotte (1978) call upon prior episodes of partial melting to deplete the radioactives in the source rocks of abyssal tholeiites, thereby reconciling the low values of heat production in these lavas with the "expected" mantle-wide average. Perhaps these lower values of radiogenic heat production are more representative of the Earth's mantle than the surface heat flux derived value for the reasons stated above.

The distributions of radioactives in the mantles of the terrestrial planets are completely unknown. The processes, which concentrate radioactive elements in crustal rocks and lead to a stratified distribution of radioactives in the continental crust of the Earth, might be expected to produce similar upward concentrations of heat sources in a planetary mantle; in the extreme case, these processes could completely deplete a mantle in heat producing elements. Whole mantle convection, on the other hand, would tend to homogenize the heat sources if a quasi-steady circulation had sufficient time to be established.

We have already mentioned that the gravitational potential energy made available during accretion (Hanks & Anderson 1969, Mizutani, Matsui & Takeuchi 1972, Wetherill 1976, Weidenschilling 1976, Safronov 1979, Kaula 1979) and core formation are important energy sources for terrestrial planets. We are reasonably certain that all the terrestrial planets with the possible exception of the Moon have metallic cores. Core formation probably occurred very early in the histories of Earth and Mercury (Solomon 1979). By analogy with Earth, core formation would also be expected to occur early in Venus' evolution. Because of the overwhelming energy release during core formation, the Earth's core must have formed prior to the oldest known rocks presently at the surface; this would require core formation within the first 750 million years of the Earth's history (Moorbath, O'Nions & Pankhurst 1975). Other lines of evidence which suggest early core formation in the Earth include remanent magnetism in 2.7 billion year old rocks with paleointensities comparable to the Earth's present field intensity (Hanks & Anderson 1969) and the radiogenic nature of Pb in the average crust and mantle (Vollmer 1977). Core formation in the Earth is likely to have been a catastrophic or runaway process (Ringwood 1960, Tozer 1965b, Ringwood 1975) and the energy released could easily overshadow the contributions of the other sources mentioned above. For the Earth, core differentiation is estimated to have raised the temperature of the planet 2000°C (Birch 1965, Tozer 1965b).

Solomon (1977, 1979) argues that core formation in Mercury is likely to have occurred prior to 4 billion years ago because of the absence of tensional tectonic features in the old cratered terrain. The heavily cratered terrain on Mercury is estimated to be at least 4 billion years old by analogy with the probable age of the lunar highlands and the end of heavy bombardment in the inner solar system. Core formation in Mercury would lead to such a large increase in the planetary radius that tensional cracks would be expected to occur in the ancient cratered terrain if it existed prior to core formation. A global system of lobate scarps on Mercury indicates that the surface has been subjected to horizontal compressive stress throughout much of its history, consistent with a thermal evolution dominated by cooling from a hot initial state. On the other hand, the dominance of tensile features in the martian surface constrains the amount that Mars may have cooled over geologic time and suggests a relatively late core formation in that planet (Solomon & Chaiken 1976). Core segregation in Mercury could have raised the temperature by about 700°C (Solomon 1979, Toksöz, Hsui & Johnston 1978); the average temperature rise due to core formation in Mars is estimated to be about 200°C to 300°C (Solomon 1979, Toksöz & Hsui 1978).

The Moon shows neither the compressional features displayed by Mercury nor the tensional ones characteristic of Mars. However, the absence of these features in itself constrains the lunar thermal history; Solomon & Chaiken (1976) and Solomon (1977, 1979) have argued, on the basis of this evidence, that the early formation of a lunar core is not likely. However, there are other persuasive arguments indicating that the Moon may have formed a core very early in its history. These include the presence of remanent magnetization in a 4 billion year old lunar sample and the apparently global remanent magnetization of the entire lunar crust and the ancient farside highlands in particular (Runcorn 1976). In any case, the temperature increment associated with the formation of a small lunar core is only about 10°C (Solomon 1979), negligibly small for it to have any direct impact on thermal history. It is generally agreed however that the Moon was hot early in its history, at least in its outer several hundred kilometers because of the requirement of early differentiation of its crust (Toksöz & Johnston 1977). In the case of the Moon, accretional heating is the dominant early source of energy (Kaula 1979), not core formation, but the net result is that the Moon also undergoes a cooling history, at least in its outer regions. Solomon & Chaiken (1976) would argue that the lunar interior on the other hand has had to heat with geologic time by radioactive decay to offset the cooling of the outer parts so that no net lithospheric compression or tension would result.

Thus, for the Earth, Mercury, and perhaps Venus, core formation may have been the single event controlling early thermal evolution. Core formation in Mars may not have played such an important role early in martian history, while core formation in the Moon probably had negligible thermal consequences. For the Moon, accretional heating is probably the important early energy source. Thermal history calculations (Schubert, Cassen & Young 1979a, Cassen et al 1979, Stevenson 1979, Sharpe & Peltier 1979, Stevenson & Turner 1979) show that the energy released in core differentiation or accretion could still be escaping through the surfaces of the Earth and Moon (and perhaps those of the other terrestrial planets as well) in amounts competitive with heat originating from radioactive decay. Other energy sources which are generally believed to be of minor importance are tidal dissipation (Kaula 1963b, 1964, Burns 1976, Kaula & Yoder 1976, Peale & Cassen 1978) (however, see Turcotte, Cisne & Nordmann 1977 for a contrary view), short-lived radionuclides such as Al^{26} (however Runcorn 1976 has suggested that Al^{26} may be an important heat source for the Moon), and joule heating by solar wind driven planetary electrical induction currents (Sonett, Colburn & Schwartz 1975, Herbert, Sonett & Wiskerchen 1977).

DIMENSIONLESS PARAMETERS AND THEIR SIGNIFICANCE

The equations governing mantle convection are the equations of conservation of mass, momentum, and energy, the equation of state, and constitutive equations for the rheological and thermal parameters. Standard nondimensionalization procedures applied to these equations show that the behavior of a convecting system is governed by relatively few dimensionless combinations of parameters. The number of these dimensionless ratios (and their specific forms) is not unique, but depends, in particular, on the degree to which one simplifies the state and constitutive equations. If, as is often done, one assumes that the mantle is a Boussinesq fluid (one whose density can be assumed constant for all purposes except the calculation of buoyancy forces) with a constant Newtonian viscosity (and constant values of other relevant thermal and mechanical parameters) then there are only two dimensionless ratios that determine the form of mantle convection; these are the Rayleigh number Ra

$$\mathrm{Ra} = \frac{\alpha g \Delta T D^3}{\kappa \nu}, \tag{3}$$

and the Prandtl number

$$\mathrm{Pr} = \frac{\nu}{\kappa}. \tag{4}$$

With ν, the kinematic viscosity, equal to 10^{21} cm^2/s and $\kappa = 10^{-2}$ cm^2/s, we get $\mathrm{Pr} = 10^{23}$. The effectively infinite value of the Prandtl number implies that inertial forces are unimportant in planetary mantles, and that only pressure forces, buoyancy forces, and viscous forces need be included in the equations of motion. Once such a simplification is made, the Prandtl number no longer explicitly appears in the equations and the Rayleigh number is the single parameter governing the nature of convection. Ra is the ratio of the rate at which buoyancy forces do work on the flow to the rate at which energy dissipation occurs. The form of the Rayleigh number given in (3) is for a fluid layer with no heat sources across which a temperature difference ΔT is applied. Actually, only the temperature difference in excess of the value associated with adiabatic compression of the fluid can drive convection, and ΔT should be so interpreted. In classical fluid dynamical situations, the adiabatic increase of temperature is usually unimportant, but for planetary mantles it can be as large as 1000°C to 2000°C. In addition to heating from below, internal heat sources can drive convection in a fluid layer. In this case, Ra takes the form

$$\mathrm{Ra} = \frac{\alpha g Q D^5}{k\kappa\nu}, \tag{5}$$

where Q is the constant rate of internal heat production per unit volume.

The Rayleigh number must exceed a critical value Ra_{cr} before convection can occur in a fluid layer. Ra_{cr} can be determined by a linearized stability analysis; its exact value depends on particular forms of boundary conditions (e.g. isothermal vs adiabatic boundaries, rigid vs stress-free boundaries), the geometry of the convecting region (e.g. spherical vs plane), and whether heating is from below, or internal, or a combination of both. For convection in spherical shells, the ratio of outer to inner radius also influences the exact value of Ra_{cr} (Chandrasekhar 1961). Calculations have shown that Ra_{cr} is $O(10^3)$ for all the many different circumstances under which convection may occur (see, for example, Chandrasekhar 1961). Convection becomes more vigorous as the Rayleigh number increases beyond the critical value. Ra is the essential dimensionless ratio which describes the character of a convecting system, even when more complicated equation of state and constitutive relationships introduce additional parameters.

Table 1 includes estimates of the Rayleigh numbers for the mantles of the terrestrial planets. These values of Ra were calculated using $\nu = 10^{21}$ cm^2/s and a ΔT based on a 0.1 K/km superadiabatic gradient across the entire mantle. Other parameter values have been given in the previous section. The Rayleigh numbers for all the terrestrial planets, except perhaps Mercury, are so large compared with the critical value of $O(10^3)$, that based on linear stability theory one would expect convection in their mantles. Rayleigh numbers for the Earth and Venus are so large compared to Ra_{cr} that one would expect quite vigorous present day mantle convection. Convection would be expected to be less vigorous for Mars and still less vigorous for the Moon. For Mercury, the estimate of Ra is sufficiently small that convection if it is taking place at all in that planet's mantle would be expected to be rather weak (Cassen et al 1976). Such simple evaluations of Ra and comparisons with Ra_{cr} led Pekeris (1935), Knopoff (1964), and Tozer (1965a) to conclude that convection could be expected for the Earth's upper mantle. Schubert, Turcotte & Oxburgh (1969) determined the critical Rayleigh number for a viscously stratified fluid and concluded that convection was likely in all the terrestrial planets.

Cassen & Reynolds (1973, 1974) analyzed the stability of lunar temperature profiles established by radioactive and accretional heating and concluded that solid state convection would play an important role in lunar thermal history (see also Turcotte & Oxburgh 1969b).

Estimates of the Nusselt numbers Nu for the planetary mantles are also given in Table 1. Nu is the ratio of the heat flux carried by convection and conduction to the heat flux that would be carried by conduction alone if the system were subjected to the same temperature difference across its boundaries. At the onset of convection Nu = 1. Nu is much greater than unity for a vigorously convecting system. The values of Nu in the table were based on the values of Rayleigh number and a relation between Nu and Ra which has support from experimental, theoretical, and numerical studies. The form of this relation is

$$\mathrm{Nu} = b\,\mathrm{Ra}^{\beta}, \tag{6}$$

where b and β are constants. If one argues, following Priestley (1954), that the heat transport through a fluid layer heated from below at high Rayleigh number is independent of overall layer thickness, because such heat transport is controlled by conduction through thin thermal boundary layers, then it is straightforward to show from the definitions of Nu and Ra that $\beta = 1/3$. Slightly different values of β are predicted by experiments and numerical studies. We will discuss this in more detail later. Here we simply note that the approximate Nusselt number estimates given in Table 1 were obtained assuming

$$\mathrm{Nu} = \left(\frac{\mathrm{Ra}}{\mathrm{Ra}_{\mathrm{cr}}}\right)^{1/3}, \tag{7}$$

with $\mathrm{Ra}_{\mathrm{cr}} = 10^3$. These estimates of Nu indicate that about 10 times as much heat is being transported to the surface by convection as compared with conduction in the mantles of Earth, Venus, and Mars. Convective heat transport in the Moon may be several times the conductive heat transport, while in Mercury, convective heat transport, if it is occurring at all, should be at best comparable to conductive heat flow.

The final dimensionless ratio presented in Table 1 is the Dissipation number Di

$$\mathrm{Di} = \frac{\alpha g D}{c_p}. \tag{8}$$

Di is the dimensionless ratio which determines the influence of viscous dissipation and adiabatic compression on convection, effects not explicitly incorporated in the framework of the standard Boussinesq approximation

(Peltier 1972, Turcotte et al 1974, Hewitt, McKenzie & Weiss 1975, Jarvis & McKenzie 1979). When Di is much less than unity, viscous dissipation and adiabatic compression have little influence on convection. However, increasing Di has a stabilizing effect on the flow; for sufficiently large Di the flow becomes penetrative. From the order of magnitude values of Di listed in Table 1, it is obvious that viscous dissipation should be relatively unimportant in the mantles of the smaller terrestrial planets ☿, ☾, ♂, but it may be of importance for whole mantle convection in the large planets ⊕ and ♀.

APPROACHES TO THE STUDY OF MANTLE CONVECTION

A number of different approaches are available for the study of mantle convection. One can study convection experimentally in the laboratory (Whitehead 1976) or theoretically, using various schemes to integrate the equations. Boundary layer techniques and other semi-rigorous scaling simplifications can be used to obtain solutions for convection at high Rayleigh number. Laboratory experiments are limited in their ability to simulate convection in the mantle essentially because one cannot carry out the laboratory experiments on mantle materials under mantle conditions of temperature, pressure, etc. Even if convection in ordinary viscous fluids is relevant to the mantle, laboratory experiments cannot be performed under circumstances wherein values of both Ra and Pr are comparable to those in the mantle. Many laboratory experiments that purport to be relevant to the Earth's mantle (Richter & Parsons 1975) are carried out at high Ra for large Prandtl number fluids [$\mathrm{Pr} = O(10^4)$]. However, we do not have laboratory fluids with Pr as large as 10^{23}, a value which insures the unimportance of inertial forces in the mantle. Inertial forces may not be negligible in high Rayleigh number laboratory experiments even with large Prandtl number fluids if the relative importance of these forces scales according to the ratio Ra/Pr (Corcos, private communication, Peltier 1972) or $\mathrm{Ra}^{2/3}/\mathrm{Pr}$ (Elsasser, Olson & Marsh 1979) instead of 1/Pr (Oxburgh & Turcotte 1978). In the experiments of Richter & Parsons (1975), Ra/Pr varies between 1 and 10, while $\mathrm{Ra}^{2/3}/\mathrm{Pr}$ varies between 1 and 5.

Rigorous numerical modelling of highly nonlinear mantle convection is probably not feasible at present, because such convection is likely to be fully three-dimensional and perhaps time-dependent. Nevertheless, there has been a great deal of effort expended in constructing two-dimensional numerical models of convection in the Earth's mantle (see e.g. Torrance & Turcotte 1971a,b, Richter 1973a, Turcotte, Torrance & Hsui 1973,

McKenzie, Roberts & Weiss 1974, Parmentier, Turcotte & Torrance 1976). The emphasis on two-dimensional numerical models must largely be attributed to the relative ease of carrying out such calculations. Only if convection in the Earth were confined to its upper mantle would two-dimensional models of convection have potential relevance. If convection extends throughout the Earth's mantle it should be represented by spherical shell models. Even if convection were restricted to the Earth's upper mantle, strictly two-dimensional models of it have questionable relevance since such solutions may be unstable to perturbations in the third dimension, i.e. even shallow upper mantle convection may be three-dimensional. Busse (1967) has shown that two-dimensional convection in a layer of infinite Prandtl number fluid becomes unstable when the Rayleigh number exceeds 2.26×10^4. Most estimates of Ra even for the upper mantle exceed this value. The authors of two-dimensional numerical models of mantle convection have been among the strongest advocates of shallow mantle convection in the Earth (McKenzie, Roberts & Weiss 1974, Richter 1978). The depth of convection in the Earth's mantle is a topic of much current debate and we will discuss it in more detail in a later section.

Basic fluid dynamical calculations of finite-amplitude thermal convection in spherical geometry have been carried out by Hsui, Turcotte & Torrance (1972) and Young (1974). Numerical models of the thermal states of planetary interiors based on computations of convection in spheres and spherical shells have been constructed by Turcotte et al (1972), Young & Schubert (1974), Cassen & Young (1975), Schubert & Young (1976), and Schubert, Young & Cassen (1977). These computations have been restricted to axisymmetric solutions and their relevance to convection in planetary interiors can only be determined by testing their stability to nonaxisymmetric perturbations. Recent studies indicate that axisymmetric modes of convection may in fact be unstable to nonaxisymmetric perturbations. Busse's (1975) stability analysis of axisymmetric convection in spherical geometry shows that axisymmetric modes of convection which are symmetric about an equatorial plane are generally unstable to nonaxisymmetric perturbations for Rayleigh numbers near the critical value. The one exception to this conclusion is the lowest order even axisymmetric mode, which is stable to nonaxisymmetric perturbations. Zebib, Schubert & Straus (1979) have studied heated from below convection in a spherical shell the size of the Earth's mantle for Rayleigh numbers up to ten times critical. The critical motion may be axisymmetric but it is not symmetric about an equatorial plane. At a given supercritical Rayleigh number, it is possible to calculate axisymmetric solutions that are symmetric about an equatorial plane, and ones that do not possess such symmetry. Zebib, Schubert & Straus (1979) have shown that the former are unstable to the latter and, further, that the latter are unstable to nonaxisymmetric perturbations at Rayleigh numbers near the critical value. However, these axisymmetric solutions that are not symmetric about the equator can be stable to azimuthal perturbations for Rayleigh numbers which are not too close to the critical value.

The pattern of axisymmetric convection at the onset of instability is often compared with physical observations which suggest convection in the planets (see, for example, Runcorn 1977, Elsasser, Olson & Marsh 1979). However, these studies of finite amplitude convection in spherical geometry show that the form of convection at the onset of instability may have little relevance to the patterns of vigorous convection in planetary mantles. Not only must we question the relevance of the linearized axisymmetric flows to actual motions in the planets, but instability of axisymmetric finite amplitude convection to general three-dimensional perturbations (Busse 1979) would obviously also preclude the relevance of detailed characteristics of even these nonlinear solutions. The same criticism also applies, of course, to nonlinear two-dimensional models of convection. It is possible however if interest is confined to the average characteristics of convection, e.g. mean heat flux or temperature, that the axisymmetric (or two-dimensional) solutions could give results similar to those of the more complex fully three-dimensional motions. It would be fortunate if this were the case since numerical modelling of three-dimensional convection at very high Rayleigh number is a formidable task. If one adds to the burden of carrying out high Rayleigh number three-dimensional convection calculations by incorporating realistic equation of state and rheological behavior, e.g. a nonlinear stress-rate of strain connection, a temperature- and pressure-dependent effective viscosity, etc, then the effort will be totally beyond our capabilities for some years to come. These difficulties argue for the development of simplified theoretical approaches to the description of mantle convection, a point forcefully argued by Tozer (1972a). The very large uncertainties in rheological properties and heat source content of the mantles of the terrestrial planets makes a simplified treatment of convection necessary for the systematic investigation of parameter variations.

Boundary layer theories of high Rayleigh number convection represent one important type of simplified theoretical description of the phenomenon. They are based on observations of the form of steady two-dimensional convection of a constant viscosity Boussinesq fluid layer heated from below at high Rayleigh number. As Ra increases, convection tends to take the form of a nearly isothermal core region (or an adiabatic one if the adiabatic temperature gradient is not negligible) with horizontal thermal boundary layers at the upper and lower boundaries and vertical plumes adjacent to the lateral boundaries. The core region of the convection cell is at the average temperature of the upper and lower boundaries and the temperature differences between the boundaries and the fluid interior occur across the horizontal boundary layers. Heat transport across these boundary layers is by the process of conduction. Buoyancy forces in the vertical plumes drive the circulation of the convection cell. Turcotte & Oxburgh (1967) developed this two-dimensional boundary layer model of mantle convection and showed, if the upper and lower boundaries are isothermal, stress-free surfaces, that the maximum thickness of the thermal boundary layer δ is

$$\delta = 7.38 \text{ Ra}^{-1/3} D, \tag{9}$$

and the overall heat transport q is

$$q = 0.167 \text{ Ra}^{1/3} \left(\frac{k\Delta T}{D}\right), \tag{10}$$

where D is the thickness of the fluid layer and ΔT is the temperature difference across the layer. Equation (10) shows that the Nusselt number is given by

$$\text{Nu} = 0.167 \text{ Ra}^{1/3}. \tag{11}$$

Thus as the Rayleigh number increases, the boundary layers become thinner and harder to resolve by direct numerical solution techniques. Corcos (private communication) has found that the numerical factors in (9)–(11) need revision to correct an error in the details of the original boundary layer solution by Turcotte & Oxburgh (1967).

The scaling and planform assumptions that enter the development of a boundary layer theory place potentially serious limitations on the ability of the theory to properly describe high Rayleigh number mantle convection. Assumptions such as two-dimensionality and steady state can only be verified by direct numerical or experimental tests; the difficulties involved in performing these checks motivate the formulation of a boundary layer theory in the first place. Boundary layer theories which can cope with three-dimensional, time-dependent flow, or nonlinear, temperature- and pressure-dependent creep behavior are not presently available.

In certain instances, it may suffice to know some average characteristic of a convecting system, e.g. the mean heat flux. It is then possible to use a relation such as (6) to infer the heat flux from the average properties of a convecting region. This power law relation between Nusselt number and Rayleigh number is known from laboratory experiments and theoretical and numerical calculations to characterize the heat flux from a vigorously convecting system ($\mathrm{Ra} \gg \mathrm{Ra}_{cr}$) in a variety of circumstances. Tozer (1967, 1972a,b, 1974) made extensive use of it throughout his papers discussing the importance of subsolidus creep in regulating the temperatures of the terrestrial planets. Sharpe & Peltier (1979) strongly advocated its utility for studying the thermal histories of the planets. The power law relation was used by Kaula (1979) and Stevenson & Turner (1979) to investigate thermal history models of the Earth and by Cassen et al (1979) to model the thermal evolution of the Moon. Schubert, Cassen & Young (1979a,b) have used the relation to develop cooling history models of the terrestrial planets. Most of the evidence supporting relation (6) applies to a fluid layer heated from below which is transporting a steady quantity of heat. The data from laboratory experiments fit (6) quite well (Rossby 1969, Chu & Goldstein 1973, Garon & Goldstein 1973). Values of b and the power law exponent β depend somewhat on the particular fluid used in the experiment (i.e. on the Prandtl number of the fluid) and on the Rayleigh number range studied; b is generally $O(10^{-1})$ and β is about 0.3. This value of β is consistent with the value 1/3 which comes from various scaling arguments and boundary layer theories (Priestley 1954, Kraichnan 1962, Howard 1966, Turcotte & Oxburgh 1967, Long 1976), and the $O(10^{-1})$ value of b is consistent with the form of the power law relation given in (7) since Ra_{cr} is $O(10^3)$. There are numerous other experimental, theoretical, and numerical studies which support relation (6) for the fluid layer heated from below; Busse's recent review (1978a) provides a detailed discussion of many of these. Relation (6) appears to adequately describe the steady heat transport through a vigorously convecting layer heated from below. However, the values of b and β depend on boundary conditions, on Prandtl number, and weakly on the Rayleigh number itself.

Because the mantles of the planets are not simply fluid layers heated from below in steady state, it is important to expand the justification of relation (6) to more general situations, including internally heated convection and transient convection. Schubert, Cassen & Young (1979a) show that the experimental data of Kulacki & Nagle (1975) and Kulacki & Emara (1977) for steady convection in an internally heated layer insulated at the bottom are in good agreement with the power law (6) (with Ra based on the temperature difference across the layer); these experiments imply $b = 0.23$ and $\beta = 0.29$ (Schubert, Cassen & Young 1979a). Convection in spherical shells with internal heating and insulated lower boundaries was studied numerically by Young & Schubert (1974), Schubert & Young (1976), and Schubert, Young & Cassen (1977). These papers presented thermal models of Mars, Earth, and the Moon from which one can calculate the amounts by which the average temperatures of the convecting mantles exceed the base temperatures of the lithospheres. Log-log plots of these excess temperatures ΔT against $\mathrm{Ra}/\mathrm{Ra}_{cr}$ reveal approximate power law dependences of ΔT on $\mathrm{Ra}/\mathrm{Ra}_{cr}$ which are consistent with (6) and imply $\beta \approx 0.3$.

Cassen & Young (1975) analyzed the situation in which steady convection is driven both by internal energy sources and an imposed constant temperature differential across the boundaries. They found that the heat flux through the bottom boundary is approximately a linear function of the strength of the internal energy sources, with negative slope. This is the expected result for steady state, as long as the flux through the top of the layer is independent of heat source content, as is assumed in (6).

It is also possible to test the applicability of (6) to transient convection. Kulacki & Nagle (1975) and Kulacki & Emara (1977) also studied the response of the internal temperature of their convection cell (again, volumetrically heated and insulated at the bottom) to step changes in the heating rate. They measured the time required for a system, initially with a uniform temperature, to come to steady convective equilibrium after a step increase in heating and the time for the system in equilibrium with a given heat source concentration to return to a constant temperature after the heat was turned off. Schubert, Cassen & Young (1979a) compared these data with an analytic solution of the heat conservation equation made possible by relation (6); they found good agreement between the theoretical prediction and the experimental results.

The experiments of Booker (1976) and Booker & Stengel (1978) with Bénard convection in a variable viscosity fluid show that temperature-dependent viscosity has a relatively minor effect on the Nu-Ra relation when the Rayleigh number is based on the viscosity corresponding to the mean temperature of the layer. Even this minor effect can be accounted for by writing

$$\mathrm{Nu} = b' \left(\frac{\mathrm{Ra}}{\mathrm{Ra}_{cr}} \right)^{\beta} \tag{12}$$

and using the actual value of the critical Rayleigh number for the variable viscosity situation. Booker & Stengel (1978) suggest $b' = 1.49$. Finally, numerical calculations of convection in non-Newtonian fluids (Parmentier, Turcotte & Torrance 1976, Parmentier 1978) suggest that relation (6) holds for even more complicated rheologies.

Thus the power law Nusselt number–Rayleigh number relation (6) can be used with some confidence to study the thermal balance of planetary interiors. However, although the experimental and theoretical justifications for such a law are numerous, there are presently no data available to rigorously validate its use under the circumstances required for general studies of convection in planetary mantles. The application of the power law to studies of mantle convection involves the extension of a principle beyond its rigorously defendable region of validity with the expectation of at least qualitatively correct results.

DEPTH OF MANTLE CONVECTION

One of the major issues facing mantle convection theorists is whether convection is confined only to the upper regions of the Earth's mantle or extends throughout. Proponents of shallow mantle convection in the Earth argue that the major upper mantle structural transformations should restrict convection to the upper mantle. Since similar phase transformations could occur in the mantles of Venus and perhaps Mars, the issue of shallow vs whole mantle convection applies to these planets as well. However, pressures in the mantle of Mercury and throughout the Moon are sufficiently low that if convection occurs in these bodies it should be of the whole mantle type.

McKenzie and Richter have been among the strongest advocates of shallow mantle convection. Throughout their papers (Richter 1973a,b, 1977, 1978, McKenzie, Roberts & Weiss 1974, McKenzie & Weiss 1975, McKenzie & Richter 1976, McKenzie 1977, Richter & McKenzie 1978) they argue that the spinel-postspinel phase change should act as a barrier, restricting convection associated with plate motions to the upper 650–700 km of the mantle (they do not preclude a separate convective circulation in the lower mantle). Their numerical models describe two-dimensional flows in plane fluid layers. These authors are not alone in constructing such models of convection in the Earth's mantle; in fact, the major effort to model convection in the Earth's mantle has been the development of two-dimensional, plane layer models (see, for example, Torrance & Turcotte 1971a,b, Houston & De Bremaecker 1975, Parmentier, Turcotte & Torrance 1976). McKenzie and Richter point out that the predominance of compressional focal mechanisms in deep earthquakes between depths of 500 and 700 km and the absence of earthquakes at depths greater than 700 km (Isacks & Molnar 1971) support the view that the spinel-oxide phase change is a barrier to convection.

Although the compressional nature of deep earthquakes indicates that descending slabs meet some resistance at the 650-km phase change, this does not necessarily imply the inability of a slab to penetrate the phase transition. Schubert, Yuen & Turcotte (1975) have shown that, while the spinel-oxide transition may exert an upward body force on a descending slab due to the depression of the phase boundary within the slab, the downward body forces, due to the negative buoyancy of the cold slab and the upward distortion of the olivine-spinel phase boundary within the slab, are overwhelming and readily drive the slab through the 650-km phase transition. The absence of earthquakes below 700-km depth may indicate only that we have not detected earthquakes any deeper. Also cessation of earthquake activity within the slab below a certain depth does not imply that the slab itself ceases to exist below that depth. An alternative explanation for this behavior is that the upper layers of the slab have become sufficiently heated by the time they reach depths in excess of about 700 km that earthquakes can no longer occur. There is, in addition, a growing body of seismic travel time data (Julian & Sengupta 1973, Jordan & Lynn 1974, Engdahl 1975, Dziewonski, Hager & O'Connell 1977) providing evidence for lateral heterogeneity in the Earth's lower mantle; such heterogeneity may be associated with the temperature differences of a deep mantle circulation.

At one time it was argued that the thermodynamic properties of a phase change would inhibit convective motions from occurring across the major phase transitions of the Earth's upper mantle (Knopoff 1964, Verhoogen 1965). The stability analysis of a fluid layer with a univariant phase transition heated from below (Schubert, Turcotte & Oxburgh 1970, Busse & Schubert 1971, Schubert & Turcotte 1971, Peltier 1972) clarified the physics of the stabilizing (latent heat release) and destabilizing (phase boundary distortion caused by advection of ambient temperature) effects of an exothermic phase change and showed that the olivine-spinel phase change in the presence of a negative temperature gradient could enhance deep mantle convection. Richter's (1973c) finite-amplitude numerical calculations of convection with a univariant phase change supported the conclusions of the stability analysis. The destabilizing character of the olivine-spinel phase change is dramatically illustrated by the elevation of the phase change in the descending slab. This phase boundary elevation in the descending slab provides an important driving force for mantle convection (Schubert & Turcotte 1971, Turcotte & Schubert 1971, Griggs 1972, Schubert, Yuen & Turcotte 1975).

Solid-solid phase transitions in the Earth are divariant in nature and a linear stability analysis of such an exothermic phase transformation (Schubert, Yuen & Turcotte 1975) shows that the destabilizing effect of an enhanced effective coefficient of volume expansion in the two-phase region can dominate the stabilizing effect of an enhanced adiabatic temperature gradient in the transition region (Ringwood 1972, Tozer 1972a). The destabilizing effect of phase boundary distortion also occurs for finite amplitude motions through exothermic divariant phase changes.

It is presently not clear whether the spinel-oxide phase transition is exothermic or endothermic (Liebermann, Jackson & Ringwood 1977). Schubert, Yuen & Turcotte (1975) discuss the consequences of an endothermic spinel-oxide phase change, since this possibility distinguishes the behavior of the 650-km phase transition from that of the 400-km phase change. An endothermic spinel-oxide phase change would offer some resistance to mantle convection (this would help explain the compressional nature of deep earthquakes, while an exothermic behavior could not), but not enough to terminate the descent of lithospheric plates into the deep mantle. It seems reasonable to conclude, on the basis of the above discussion, that thermodynamically, upper mantle phase changes do not necessarily confine convection to the upper mantle. On the contrary, they may promote whole mantle convection and provide an important driving force for plate motions.

Another way of discussing phase changes in the Earth's upper mantle as barriers to convection is by hypothesizing a dramatic increase in viscosity across them. McKenzie & Weiss (1975) assert that an increase in activation energy of 2 eV/mol ($\approx$46 kcal/mol) across the 650-km phase change leads to such a large increase in viscosity below 650 km as to confine convection to the upper mantle. Such a large change in activation energy across the spinel-oxide phase transition could increase the viscosity by a factor of 10^5 across it. Tozer (1972a) also suggests an increase in viscosity by a factor of 10^5 or 10^6 across the spinel-oxide phase transition which could confine convection to the upper mantles of all terrestrial bodies with radii larger than some value between about 3000 and 6000 km. Such enormous increases in viscosity across the spinel-oxide phase change, however, seem in conflict with the inference of a uniform mantle viscosity from glacial rebound data (Cathles 1975, Peltier 1976). Also, a systematic relation betwen activation energy and oxygen ion packing predicts an increase in activation energy across the spinel-oxide phase change of only several kcal/mol implying a viscosity increase of no more than one order of magnitude (Sammis et al 1977). The upper mantle phase transitions, in particular the spinel-oxide phase change, should not necessarily act as rheological barriers to whole mantle convection.

Instead of relying on sudden changes in viscosity across phase transitions to limit convection to the upper mantles of large terrestrial planets similar to Earth and Venus, one could hypothesize a gradual increase in viscosity with depth associated mainly with the large increase in pressure in the lower mantle. The inference of uniform viscosity in the Earth's mantle from glacial rebound observations also argues against this possibility. In addition, a decrease of activation volume with increasing depth in the Earth's mantle would make it unlikely that the pressure effect would increase viscosity substantially (Sammis et al 1977, O'Connell 1977).

Two theoretical studies emphasize that only unreasonably large mantle viscosity stratifications could preclude whole mantle convection. Schubert, Turcotte & Oxburgh (1969) considered the stability of the mantles of the terrestrial planets assuming a viscosity that increased exponentially with depth. If the depth of the convecting layer D is large compared with the scale height of the viscosity increase h, then the Rayleigh number, based on the entire thickness of the mantle and the minimum viscosity ν_s,

$$\mathrm{Ra} = \frac{\alpha g}{\kappa \nu_s}\left(\frac{\Delta T}{D}\right) D^4, \tag{13}$$

would have to exceed either $23(D/h)^4$ (for a stress-free upper surface) or $30(D/h)^4$ (for a rigid upper surface) for whole mantle convection to occur (Schubert, Turcotte & Oxburgh 1969). This condition for the onset of convection in the entire viscously stratified mantle can be compared with the criterion for onset of instability in a shallow, constant viscosity upper mantle of thickness D_u and viscosity ν_s across which the destabilizing temperature rise is ΔT_u; that criterion is

$$\mathrm{Ra} = \frac{\alpha g}{\kappa \nu_s}\left(\frac{\Delta T_u}{D_u}\right) D_u^4 > \begin{array}{l} 1707 \text{ (rigid upper surface)} \\ 1101 \text{ (stress-free upper surface).} \end{array} \tag{14}$$

For the constant viscosity shallow upper mantle layer to be more unstable than the entire viscously stratified mantle (assuming the same destabilizing temperature gradient for both configurations, i.e. $\Delta T/D = \Delta T_u/D_u$) equations (13) and (14) show that

$$\begin{aligned} \frac{D}{h} \geqq & \left(\frac{1101}{23}\right)^{1/4} \frac{D}{D_u} \quad \text{(stress-free upper surface)} \\ & \left(\frac{1708}{30}\right)^{1/4} \frac{D}{D_u} \quad \text{(rigid upper surface).} \end{aligned} \tag{15}$$

Thus, before shallow mantle convection would be more likely than whole mantle convection, at least on the basis of linear stability theory, the viscosity increase across the mantle would have to exceed the value given by

$$\begin{aligned} \frac{\nu}{\nu_s} &= \exp\left(2.63\,\frac{D}{D_u}\right) \quad \text{(free-surface upper boundary)} \\ &= \exp\left(2.74\,\frac{D}{D_u}\right) \quad \text{(rigid upper boundary).} \end{aligned} \tag{16}$$

If we use $D/D_u = 30/7$ for the Earth, then according to equation (16), the viscosity increase across the mantle would have to exceed about 10^5 before shallow mantle convection would be the preferred mode. Davies (1977) recently arrived at the same conclusion from a linear stability analysis of a mantle separated into two constant viscosity layers. A viscosity increase of 10^5 across the Earth's mantle is wholly inconsistent with glacial rebound data.

Other physical properties of a mantle that could conceivably restrict the depth of convection are chemical compositional stratification and an enhanced lower mantle adiabatic temperature gradient. Both these possibilities are rather speculative. Even a very small chemical compositional stratification would be very effective in limiting convection to the upper mantle of a planet (Richter & Johnson 1974). Although Anderson (1977) argues that it is likely for the Earth's mantle to be compositionally stratified, our knowledge of the mantle is consistent with its being compositionally homogeneous (Wang & Simmons 1972, Liebermann & Ringwood 1973, Davies 1974, Watt, Shankland & Mao 1975).

If the scaling and boundary layer arguments already discussed indeed characterize high Rayleigh number convection, then they can be used, together with measured characteristics of the Earth's convection, e.g. plate velocities, surface heat flow, etc, to shed light on whether shallow or deep convection in the Earth's mantle is more likely. One possibility is to use Equation (7) or Equation (11) [they are essentially identical for approximate calculations if $\mathrm{Ra}_{\mathrm{cr}}$ is $O(10^3)$] to estimate the heat flux from convecting layers of different depth and compare these estimates with the observed average heat flow at the Earth's surface. With $D = 3000$ km, a temperature gradient of 0.1 K/km and other parameter values given previously, Equation (7) predicts a heat flux of about 0.3 μcal/cm^2 s. This estimate is only about a factor of 5 less than the Earth's mean heat flux; it can be adjusted upward to better agree with observation by reasonable changes in the parameter values entering Equation (7), e.g. one could have taken a somewhat larger value of temperature gradient. With $D = 700$ km, on the other hand, the estimated heat flux would be about an order of magnitude smaller (for the same temperature gradient), and it would be more difficult to reconcile such a reduced estimate with observations.

An estimate of the boundary layer thickness from Equation (9), or the essentially equivalent form

$$\delta = \left(\frac{\mathrm{Ra}_{\mathrm{cr}}}{\mathrm{Ra}}\right)^{1/3} D, \tag{17}$$

can be compared with an oceanic lithosphere thickness of about 100 km. For $D = 3000$ km, and a temperature gradient of 0.1 K/km, Equation (17) gives $\delta = 100$ km, while for $D = 700$ km and the same temperature gradient, one gets $\delta = 170$ km, a somewhat poorer estimate of average oceanic lithosphere thickness.

The scaling arguments can actually be used to derive an estimate of D from observations alone (Elsasser, Olson & Marsh 1979). This requires the introduction of a velocity scale which can be obtained by equating heat transport by horizontal advection with heat transfer by vertical conduction in the thermal boundary layer at the top of the convecting region. If u is the velocity scale, then this procedure gives

$$u = \frac{\kappa}{D}\frac{D^2}{\delta^2} = \frac{\kappa}{D}\frac{\delta}{D}\frac{\text{Ra}}{\text{Ra}_{\text{cr}}}, \tag{18}$$

where we have used Equation (17) to introduce the Rayleigh number. To solve for D, eliminate δ using Equation (17), ΔT using Equation (7), and find

$$D = u\left(\frac{\text{Ra}_{\text{cr}}\, k\nu}{\alpha g \kappa q}\right)^{1/2} \tag{19}$$

With $u = 4$ cm/yr (the average plate speed), $\text{Ra}_{\text{cr}} = 10^3$, $k = 0.01$ cal/cm s K, $\nu = 10^{21}$ cm^2/s, $\alpha = 3 \times 10^{-5}$ K^{-1}, $g = 10^3$ cm/s^2, $\kappa = 10^{-2}$ cm^2/s, and $q = 1.5 \times 10^{-6}$ cal/cm^2 s, Equation (19) gives $D = 6000$ km, an estimate that is more consistent with whole mantle convection than it is with shallow mantle convection.

Another observation consistent with deep mantle convection is the length scale of the largest tectonic plates on Earth. It is not immediately obvious how the sizes of plates should relate to an underlying convective pattern that may be three-dimensional and time-dependent, but conventional wisdom would suggest that the length scales of the plates should be comparable to the depth of the convecting system. It is, of course, much easier to reconcile the size of the Pacific plate, for example, with whole mantle convection than with shallow upper mantle convection.

An alternative explanation of the relation between plate scale and mode of convection relies on the strong temperature dependence of mantle viscosity. The thermal boundary layer of a mantle convection system is also a rheological boundary layer because viscosity is exponentially dependent on the inverse absolute temperature. The rheological and thermal boundary layer at the Earth's surface, the lithosphere, is effectively rigid on geologic time scales and this rigidity may tend to prevent subduction at the relatively young ages that would be expected on the basis of convection in ordinary viscous fluids. The numerical experiments of Parmentier & Turcotte (1978) indicate that this effect results in two-dimensional convection cells with large aspect ratios.

Related to the problem of understanding the sizes of the large plates is the difficulty of explaining the range of different plate sizes in a single convecting system, shallow or deep (Richter 1978).

The observed angles of subducting slabs also seem to be more readily understandable by the deep mantle circulation models of Hager & O'Connell (1978). These authors used observed plate motions and geometries to calculate the viscously induced flow driven in spherical shell models of the Earth's mantle by the prescribed surface velocities. Dip angles inferred from streamline patterns could be readily reconciled with the observed subduction angles of the plates only for deep circulation models.

Schubert & Turcotte (1971) argued against a shallow mantle circulation by suggesting that an excessively large pressure gradient might be required to drive the return flow beneath the plates. Limits on the magnitude of such a pressure gradient could be set by the slope of the ocean floor. More recent two-dimensional boundary layer calculations of shallow mantle return flow beneath oceanic plates (Schubert et al 1978) indicate that shallow return flows can be driven by much smaller pressure gradients than previously thought, thus weakening the constraint that ocean floor topography places on models of mantle convection.

There are strong arguments in favor of whole mantle convection in the Earth. If there is convection in the interiors of the smaller planets, ☿, ☾, ♂, then it should be of the whole mantle variety. Thus more emphasis needs to be placed on the construction of models of convection in spherical shells, rather than simply on modelling two-dimensional convection in plane layers.

EFFECTS OF MANTLE CONVECTION ON THE THERMAL AND MECHANICAL STATE OF A PLANETARY INTERIOR

In this section we discuss, in general terms, the thermal and mechanical state of the interior of a planet in which the mantle is in a vigorous state of convection. We describe the planet's internal structure at one instant, leaving to the following section a discussion of how the interior evolves with time. Since vigorous mantle convection is likely to be fully three-dimensional and perhaps time-dependent, we cannot describe the interior temperature and velocity fields in detail. However, calculations (see e.g. Schubert, Young & Cassen 1977) show that the average properties of a convecting mantle, in particular the spherically averaged temperature profile, may be quasi-steady even when convection is basically unsteady. We may also hope that the spherically averaged temperature profile, for example, is reasonably well-determined by axisymmetric models even if convection is nonaxisymmetric.

As repeatedly emphasized by Tozer (e.g. 1967, 1972a, 1974), the temperature-dependence of the viscosity of rocks [see Equation (1)] is the single most important factor controlling the thermal and mechanical state of a planetary interior. The main variation of temperature and viscosity in a planet at a given time in its evolution is between the cold rigid lithosphere and the underlying mantle. This thermomechanical state is illustrated by the model temperature vs depth profiles shown in Figure 1 for the present day Moon (Turcotte et al 1972, Schubert, Young & Cassen 1977, Cassen et al 1979). These average lunar temperature profiles are representative of the average thermal structures of the mantles of the smaller terrestrial planets. The average temperature profiles show that the lunar interior is divided into essentially two regions, one just below the surface in which the temperature increases from its value at the surface to its interior value, and the other, the underlying interior in which the mean temperature is essentially constant. The outer region wherein most of the temperature rise occurs is a thermal boundary layer, the lithosphere. Convection in the deep interior maintains the average temperature constant with depth. The lithosphere or thermal boundary layer forms because the average temperature in the deep interior cannot be maintained throughout the planet; the interior temperature must decrease with proximity to the surface to match the low value of surface temperature. Heat transfer through the lithosphere is by conduction. On a one

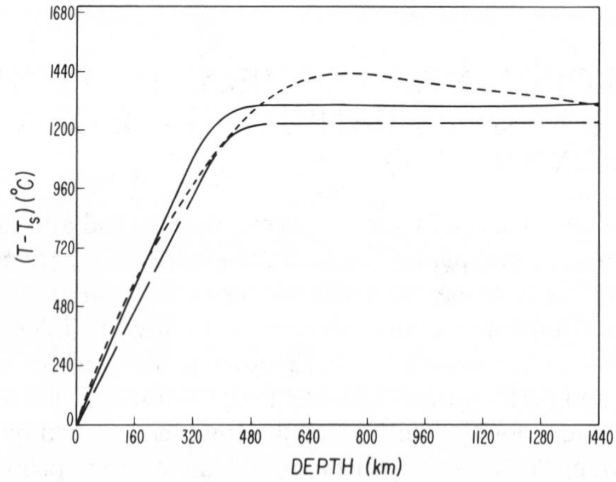

Figure 1 Average lunar temperature profiles based on calculations of finite amplitude convection in spherical geometry. Solid curve (Schubert, Young & Cassen 1977), short-dashed curve (Turcotte et al 1972), long-dashed curve (Cassen et al 1979).

plate planet like the Moon, all the internal heat must be conducted through a relatively thick lithosphere to the surface. The lithospheres in the models of Figure 1 are about 300-km thick. Superimposed on the spherically averaged quasi-steady temperature profiles, such as those shown in Figure 1, are temperature variations due to the spatial dependence of the convective state. A number of numerical calculations show that such temperature differences between hot ascending regions and cold descending ones are of the order of hundreds of degrees (Hsui, Turcotte & Torrance 1972, Turcotte et al 1972, Young & Schubert 1974).

The average viscosity in the Moon is essentially determined by the mean temperature profile because the increase of pressure with depth is too small to have much of an influence on viscosity. Also, no major silicate phase transformations can occur under the low pressures in the lunar interior. In the lithosphere, the temperature is so low that the material is effectively rigid on a geologic time scale. (It is generally accepted that subsolidus creep is negligible at temperatures below about 800°C). Thus the lithosphere is also a rheological boundary layer. In the deeper interior, the average viscosity is essentially constant, reflecting the uniformity in mean temperature. The viscosities corresponding to the temperature profiles shown in Figure 1 are about 10^{21} cm^2/s. The deep interior is convecting rather vigorously, for such a viscosity corresponds to a Rayleigh number about five-hundred times critical.

There is a relatively thin transition region just below the base of the lithosphere in which the viscosity decreases rapidly with depth from its effectively infinite value in the lithosphere to its constant value in the deep interior. The temperature profile also adjusts in this region, changing its character from a profile with a monotonic increase with depth to one which is constant with depth. This transition region could be referred to as the lunar asthenosphere, although there is no pronounced viscosity minimum in the lunar models.

Although the temperature-dependence of the viscosity is responsible for the basic thermomechanical structure just described, the very nature of this structure allows one to construct relevant models of the mean temperature profile in a planet's interior using calculations of convection in fluid spheres and spherical shells with constant viscosity. It is only necessary to combine a thermal conduction calculation in an outer, rigid, spherical shell with a temperature calculation in an inner convecting, constant viscosity, fluid spherical shell. However, one must be careful to maintain an internal consistency in the model; the temperature in the deep interior must correspond to a viscosity value (on the basis of an acceptable rheological law) similar to the one used in the constant viscosity convection calculation, and the temperature at the base of the lithosphere must not allow significant creep to occur, also on the basis of the same rheological law (Schubert, Young & Cassen 1977). When modelling the thermal evolutions of the planets, lithosphere thickness and deep temperature are functions of time, and adapting constant viscosity convection calculations to a basically time-dependent situation would be more complex, if possible at all.

The lunar temperature profiles shown in Figure 1 extend all the way to the center of the Moon because the models do not include a core. The effect of a core on the mean temperature profile in a mantle would depend on the heat flux through the core-mantle boundary as illustrated by the temperature profiles in Figure 2. The figure shows spherically averaged quasi-steady temperature profiles from Schubert & Young (1976) for models of constant viscosity, internally heated convecting fluid shells with rigid, conducting, internally heated outer shells. The dimensions of the shells are those of the Earth's lithosphere and underlying mantle and the total heat through the surfaces of the models matches that through the Earth's surface. The viscosity of the convecting interiors is 10^{25} cm^2/s, probably too large to be representative of the Earth's mantle, so the actual temperatures are not expected to represent those in the Earth's interior. In one case, the heat flux entering the mantle from the core is zero and the temperature profile is flat near the core-mantle boundary. In the other case, about 13% of the surface heat flux emanates from the core, and there is a large rise in temperature near the core-mantle boundary

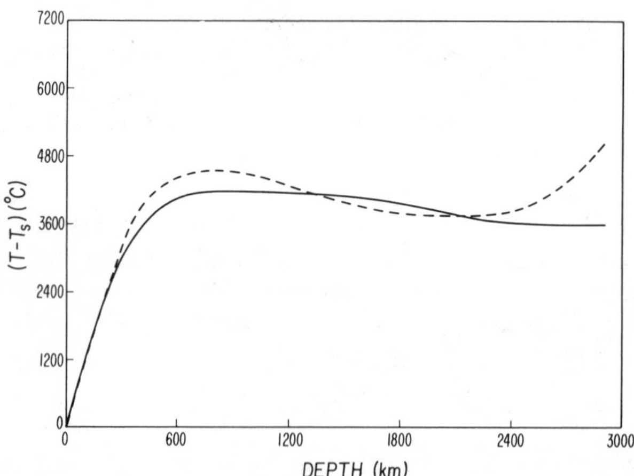

Figure 2 Average temperature profiles in an internally heated, constant viscosity fluid undergoing convection in a spherical shell the size of the Earth's mantle (Schubert & Young 1976). Solid curve (adiabatic lower boundary), dashed curve (heat flux into the lower boundary).

through a lower mantle thermal boundary layer. In this model, the mean temperature profile consists of three regions, two boundary layers, one near the surface and the other near the core-mantle interface, and an interior region in between with reasonably uniform temperature. The distinctive character of the thermal boundary layers and the interior isothermal region would presumably be more readily apparent if it were possible to carry out the numerical calculations with much smaller viscosities, more representative of the Earth's.

None of the mean thermal profiles shown in Figures 1 and 2 include the adiabatic increase of temperature with depth. For the smaller terrestrial planets like the Moon, the temperature rise due to adiabatic compression is small, and a convecting interior would be nearly isothermal, with upper and lower boundary layers as appropriate. For the larger planets, Earth and Venus, the adiabatic temperature increase with depth is substantial. Mean temperature profiles in these planets would consist of a nearly adiabatic interior region with surface and core-mantle thermal boundary layers. It is possible that the effects of adiabatic compression in the large planets could modify the thermal profiles even more substantially by leading to penetrative convection (Peltier 1972, Turcotte et al 1974, Jarvis & McKenzie 1979).

It is usually assumed that the mean temperature in the interior of a convecting mantle would lie along an adiabat, since such is the case for a constant viscosity fluid. However, it can also be argued that a convecting mantle should have nearly uniform viscosity because enhanced convection in regions of relatively low viscosity, for example, would tend to remove heat more efficiently from the region thereby reducing its temperature and raising the viscosity. Tozer (1967) asserts that viscosity will be constant along an adiabat, in which case there is no difficulty in reconciling an adiabatic temperature distribution with a constant viscosity. However, since viscosity is determined by the rheological parameters E^* and V^*, together with T and p, and the adiabatic temperature gradient depends on α, g, T, and c_p, it is not clear that there should be a connection between these thermodynamic and rheological parameters which insures that μ remains constant on an adiabat. Sammis et al (1977) and O'Connell (1977) have shown that it is possible for an adiabatic mantle to have nearly uniform viscosity because of the likely decrease of V^* with pressure.

We have mentioned several times that high Rayleigh number convection in the planets may be basically time-dependent. The numerical calculations of finite amplitude axisymmetric convection in spherical shells with internal heating by Schubert & Young (1976) and Schubert, Young & Cassen (1977) show that convection is unsteady even at modestly supercritical Rayleigh numbers. If convection in the planets is indeed unsteady, it may have important geophysical consequences. Jones (1977) argued that the possible intermittency of high Rayleigh number convection in the Earth's mantle may explain certain temporal variations in the geomagnetic field by modulating conditions at the core-mantle boundary Recent papers by Busse (1978b) and Walzer (1978) emphasize the potential tectonic importance of time-dependent models of mantle convection.

Effects of Mantle Convection on Thermal History

The profound effect that solid state convection can have on the thermal evolution of a planet is most dramatically illustrated by thermal histories which involve cooling from initial high temperature states. Cooling is the ultimate destiny of any planet, but it may have dominated the entire thermal histories of the larger terrestrial bodies, Earth and Venus, following early core formation. The same may be true of Mercury, as we have already discussed. The gravitational potential energy released upon core formation in the Earth, and perhaps Venus, could have been sufficient to melt these planets. Efficient convection in the molten state would remove some of this energy quite rapidly, but at some point in its cooling, the mantle would solidify, and subsequent cooling would be controlled mainly by subsolidus convection. Thermal history models in which subsolidus convective cooling from a hot initial state governs the evolution of a planet have recently been studied by Schubert, Cassen & Young (1979a), Sharpe & Peltier (1978), and Stevenson & Turner (1979).

The temperature-dependence of mantle viscosity is the single most important factor controlling the thermal evolution of a cooling planet (Tozer 1967, 1972a, 1974). It acts as a thermostat to regulate the mantle temperature. Initially, when the planet is hot, mantle viscosity is low, and extremely vigorous convection rapidly cools the planet. Later in its history, when the planet is relatively cool, its viscosity is higher and more modest convection cools the planet at a reduced rate. This is illustrated in Figure 3, which shows average mantle temperatures as functions of time for Earth models cooling from initial temperatures between 1500 and 3000°C (Schubert, Cassen & Young 1979a). When the planet is initially hot, there is an extremely rapid reduction in mantle temperature very early in the cooling history. This is followed by a much more gradual decrease in temperature over most of the lifetime of the planet. An initial temperature of 3000°C is reduced to 2690°C after only 10 Myr of vigorous convective cooling. The temperature is only 2060°C after 100 Myr and by 500 Myr it has fallen to 1670°C. Cooling between 500 Myr and 4.5 Gyr reduces the temperature by only an additional 360°C. Cooling is self-regulated through the dependence of μ on T.

The temperature after 4.5 Gyr of convective cooling is extremely insensitive to the initial temperature; for a starting temperature of 1500°C, the temperature after 4.5 Gyr is 1275°C, while for an initial temperature of 3000°C, T after 4.5 Gyr is 1307°C. Convection reduces the initial 1500°C temperature difference between the models to only 32°C after 4.5 Gyr. Thus, subsolidus convection rapidly cools a hot planet to a temperature determined essentially by the rheology of the mantle alone; the temperature in the interior of a planet has no memory of its initial value after convective cooling over geologic time.

Once the interior temperature of a planet reaches the value determined by its rheology, there is very little further change in temperature even after billions of years of convective cooling. This is shown by the Earth models in Figure 3, especially the one with an initial temperature of 1500°C. It is even more dramatically illustrated by the evolution of average temperature in the lunar thermal history model also shown in Figure 3 (Schubert, Cassen & Young 1979a). The initial temperature of 1300°C is reduced by only 96°C after 4.5 Gyr of convective cooling. Except for a relatively short period of time when the temperature of a planet's mantle may decrease substantially during the early stages of cooling, a planet cools mainly by thickening its lithosphere; the underlying mantle temperature decreases relatively slowly. For the Moon model of Figure 3, cooling produces a thick lithosphere; it reduces the temperature beneath this lithosphere only slightly. The temperatures of the Moon and Earth models are nearly the same after 4.5 Gyr despite the difference in radii of these planets. This is because mantle temperature is determined mainly

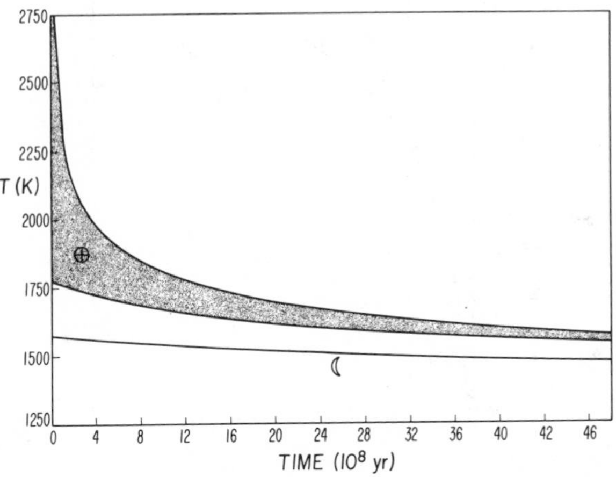

Figure 3 Average mantle temperature vs time for cooling history models of Earth and Moon (Schubert, Cassen & Young 1979a). The shaded region includes Earth models with initial temperatures between 1500 and 3000°C.

by rheology, independent of planetary size. The size of a planet influences the cooling history mainly through its effect on lithosphere thickness. Smaller planets generally have thicker lithospheres. Because average mantle temperature is principally fixed by rheology, the temperatures in planetary mantles are essentially independent of lithosphere thickness. If fact, calculations show that mantle temperatures for models that have no lithospheres at all are the same as temperatures for models that include lithospheric growth (Schubert, Cassen & Young 1979a).

Mantle viscosity vs time for the cooling history models of Figure 3 are shown in Figure 4. The dramatic increase in viscosity early in the cooling of the Earth models reflects the rapid decrease in T. The viscosity for the Moon model increases only by about one order of magnitude during 4.5 Gyr because of the small decrease in T over geologic time. At $t = 4.5$ Gyr, ν is substantially the same for ⊕ and ☾ (ν for the highest temperature Earth model is 4.1×10^{21} cm^2/s and ν for ☾ is 4.8×10^{22} cm^2/s) reflecting the approximate equality of temperature calculated for the mantles of these planets.

The calculations of Schubert, Cassen & Young (1979a) include the thickening of a lithosphere as a planet cools. They assumed that litho-

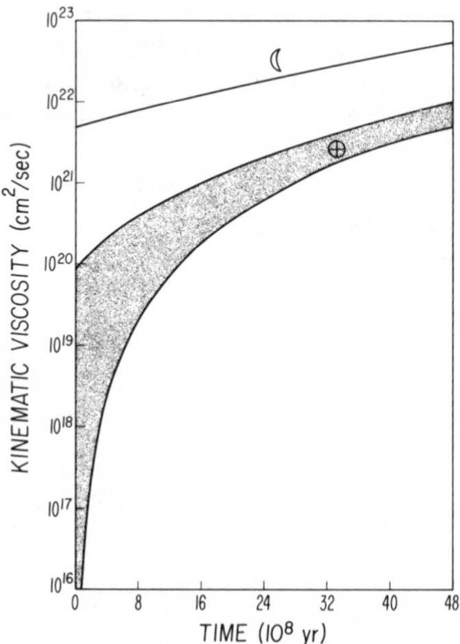

Figure 4 Mantle viscosity vs time for the cooling history models of Earth and Moon shown in Figure 1 (Schubert, Cassen & Young 1979a). Viscosity curves for Earth models with initial temperatures between 1500 and 3000°C lie in the shaded region.

spheres grow from an initial thickness of only 100 m; the growth is extremely rapid during the very earliest stages of cooling, as shown in Figure 5. The growth rate is more modest throughout most of a planet's history. The lithosphere thickness vs time curves of Figure 5 correspond to the cooling histories of Figure 3. Lithospheres thicken to 1 km at $t = 10^4$ yr for the Moon model and 35 Myr for the highest temperature Earth model. Ten-kilometer-thick lithospheres are formed after only 1 Myr for ☾ and 280 Myr for ⊕. At $t = 4.5$ Gyr lithospheres have thickened to 225 km for ⊕ and 550 km for ☾. The growth curves in Figure 5 assume that lithospheric thickening on the planets is unimpeded by other processes. During the initial growth, vigorous convection may preclude the formation of a competent lithosphere. A higher rate of impacting objects during this time than at present may also slow the accumulation of a competent lithosphere. Thus, while the figure shows that nearly a third of a billion years is required to form a ten-kilometer thick lithosphere for the hot Earth model, the time is probably closer to a billion years, in agreement with other lines of evidence suggesting a very thin terrestrial lithosphere during its first billion years of evolution (Wetherill 1972). Except for the hot Earth model, lithosphere growth is so rapid during these early cooling stages that the impediments to lithosphere formation probably make

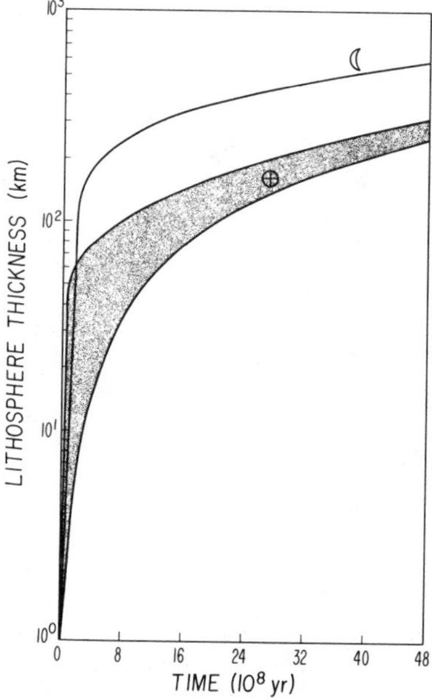

Figure 5 Lithospheric growth in the Earth and Moon models of Figure 1 (Schubert, Cassen & Young 1979a).

little difference to its thickness after 4.5 Gyr. On Earth, constraints on the growth of a lithosphere continue through the present. Plate tectonics severely limits the lithosphere thickness beneath the ocean basins. However, the estimates of lithosphere thickness in Figure 5 should be relevant to the lithosphere beneath continental shields. Since the Moon shows no evidence of plate tectonics the estimates of lithosphere thickness should be directly applicable.

Why we have plate tectonics on Earth with the continual creation and destruction of oceanic lithosphere is still an open question. When we are asked to explain why there is no plate tectonics on Mars, Mercury, or the Moon we are tempted to say that these planets have thicker lithospheres which are more resistant to breakup in the style of Earth tectonics. However, one may wonder if such tectonics occurred on these smaller terrestrial planets when their lithospheres were thinner only to have the evidence obliterated by meteoritic bombardment. We have no evidence from the present geologic surface features of these planets that plate tectonics ever occurred on them while very ancient cratered surfaces have survived. Kaula (1975) suggested that all the terrestrial planets did in fact pass through a stage of plate tectonics in their evolutions, but that

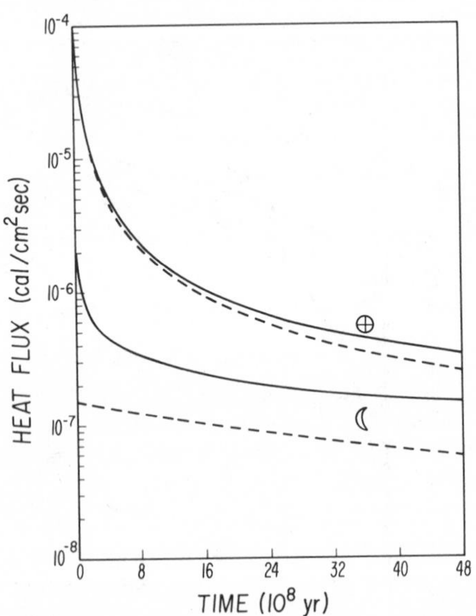

Figure 6 Temporal decay of surface heat flow (solid curves) and heat flow into the base of the lithosphere (dashed curves) for the cooling history models of the Moon and the Earth (for a 3000°C initial temperature) shown in Figure 1 (Schubert, Cassen & Young 1979a).

this stage occurred too early and too rapidly on Mercury and the Moon for any traces of it to remain.

The temporal behavior of surface heat flux q_s and heat flux into the base of the lithosphere q_c can be deduced from the cooling history calculations of Schubert, Cassen & Young (1979a). These quantities are shown in Figure 6 for the hot Earth model and the lunar model discussed in previous figures. The values q_s and q_c for the Earth model undergo dramatic early decreases due to the rapid decay of extremely vigorous convection followed by more gradual decreases throughout most of geologic time. Because of the rapid early thickening of the lunar lithosphere, q_s decreases markedly for the Moon model; q_c for the Moon model decreases slowly over the entire thermal evolution because convection is not very vigorous at the relatively low initial temperature of the model.

A significant result of these calculations is the lag between the decay of the surface heat flux and the decay of the heat flux into the base of the lithosphere. This is entirely due to the formation and thickening of the lithosphere. At any time in the cooling history of a planet there is more heat flowing through the surface than is flowing into the base of its lithosphere (assuming a competent lithosphere that is not a part of the mantle convection system). Even if convection in the mantle is in a quasi-steady state, the heat flow from the mantle is not in balance with the heat flow through the surface. The physical mechanism which accounts for the lag between "internal heat production" (heat from the mantle) and surface heat flux is the tendency of the lithosphere to supplement the decaying heat flux from the mantle by feeding on the internal energy of the mantle.

Daly & Richter (1978) recently addressed the issue of whether there is a balance between surface heat flux and instantaneous internal heat sources. On the basis of numerical calculations of convection in a two-dimensional box with decaying radiogenic heat sources, they concluded that the surface heat flux exceeds the instantaneous internal heat production even for convection with initial Rayleigh number as large as 10^6. They attribute this to the fact that conduction across closed streamlines must still play a role in removing heat from the core region of a convecting system with decaying internal heat sources. This source of nonequilibrium between internal heat production and surface heat flux is distinct from the one associated with lithospheric thickening. It is also not clear how relevant Daly & Richter's (1978) result is for the real Earth, since Rayleigh numbers much larger than 10^6 have probably characterized the Earth's mantle throughout most of its evolution.

Thus the use of present day surface heat flux observations to infer the total concentration of radiogenic heat sources in a planet on the basis of a presumed steady state thermal balance must be viewed with caution.

The surface heat flux after 4.5 Gyr of cooling without internal heat sources calculated for the Earth is still 0.35 μcal/cm^2 s (Schubert, Cassen & Young 1979a), a significant fraction of the actual present day mean surface heat flux of 1.5 μcal/cm^2 s (Oxburgh & Turcotte 1978). The Moon's surface heat flux after 4.5 Gyr of convective cooling from a modest initial temperature with no internal heat sources is calculated to be 0.15 μcal/cm^2 s (Schubert, Cassen & Young 1979a), nearly 1/3 to 1/2 of the two measured lunar surface heat flow values (Langseth, Keihm & Peters 1976). Thus, a rather substantial fraction of the present day heat flow from a planet could be attributed to primordial heat, still another reason to exercise caution in estimating the present day concentration of radiogenic heat sources in a planet from surface heat flux observations. Sharpe & Peltier (1978) and Stevenson & Turner (1979) have also recently concluded that whole mantle convection in the Earth driven solely by primordial heat content could persist for the age of the Earth and that cooling could be a significant contribution to the present day terrestrial surface heat flux. The convective lunar thermal history models of Cassen et al (1979) also exhibit a surface heat flux in excess of that attributable to radioactive heat sources.

Figures 7 and 8 show Ra and Nu vs time for the cooling history models of Schubert, Cassen & Young (1979a) already discussed in the previous

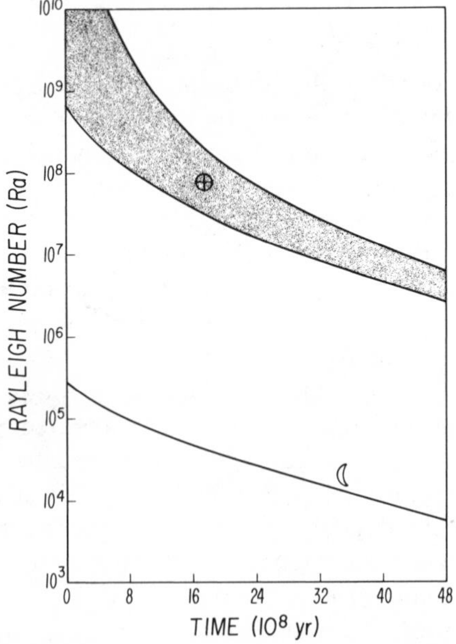

Figure 7 Decrease of Rayleigh number with time in the cooling history models of Earth and Moon shown in Figure 1.

figures. During their early histories, the Earth and Moon models are highly supercritical; even the present day values of Ra for ⊕ and ☾ indicate a vigorously convecting mantle for ⊕ and modest convection in the Moon. The present day estimate of Nu in ⊕ indicates that about 10 times as much heat is being transported to the surface by convection as compared with conduction; for the Moon model the factor is only about two.

The thermal history calculations already referenced in this section were all carried out using the Nusselt number – Rayleigh number power law relation given in Equation (6). Recently, Hsui & Toksöz (1978), Toksöz & Hsui (1978), and Toksöz, Hsui & Johnston (1978) reported thermal evolution computations for all the terrestrial planets except Earth in which they incorporated solid state convective heat transport by numerically solving the equations of motion and heat transfer for axisymmetric convection in spherical geometry of a Newtonian fluid with temperature-dependent viscosity. Their models evolve from arbitrary temperature profiles determined mainly by accretional heating and they simulate core formation, melting, and upward differentiation of radioactives as well as solid state convection. For the Moon, Toksöz, Hsui & Johnston (1978) predict a present day thermal state which involves solid state convection

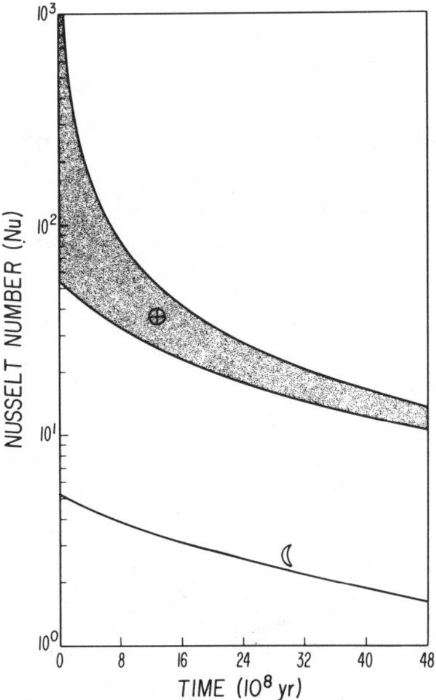

Figure 8 Decrease of Nusselt number during the evolution of the Earth and Moon for the thermal history models of Figure 1.

below a depth of 800 km. However, their mean temperature profile does not look very much like the typical convection profiles shown in Figure 1. This may result from their assumption of rather extensive differentiation of the outer volume of the Moon early in its history with accompanying efficient upward differentiation of radioactives, leading to a present day model which has a rather thick lithosphere highly depleted in radioactives and a central convecting region undepleted in radioactives. While the deep interior temperature is probably still regulated mainly by the rheology of the convecting material, the thermal profile in the outer part of the Moon has been strongly influenced by processes other than convection, e.g. initial conditions and upward differentiation of radioactives. Cassen et al (1979) also concluded that the thermal state of the lunar lithosphere is sensitive to the efficiency of heat source redistribution while that of the deep interior depends primarily on rheology.

For Mercury, Toksöz, Hsui & Johnston (1978) concluded that solid state mantle convection would cease about 2 billion years after formation of the planet. While this is in agreement with the cooling history calculation of Schubert, Cassen & Young (1979a), our previous discussion and other models of Mercury's internal thermal state show that convection in Mercury's mantle at present cannot be ruled out. However, present day convection in Mercury's mantle should at best be rather weak (Cassen et al 1976). Toksöz & Hsui (1978) and Toksöz, Hsui & Johnston (1978) calculate present day Mars models which involve convection beneath a lithosphere two hundred kilometers thick. The simple cooling history model of Schubert, Cassen & Young (1979a) produces a martian lithosphere about 300 km thick at present.

Hsui & Toksöz (1978) asserted that the size of a planet is more important than any other factor in controlling thermal evolution, whereas we, and Tozer (e.g. 1972a), have emphasized the importance of mantle rheology. We would agree with Hsui & Toksöz (1978) that the size of a planet is important, to the extent that planetary size limits the occurrence of convection; objects which are too small may not be convecting. For planetary bodies sufficiently large to be convecting, rheology, not size, will control the deep temperature. Size, however, will determine the thickness of the lithosphere.

Mantle Convection and Core Freezing

Mantle convection is so efficient at cooling a planet that it can readily lead to core freezing. This was first demonstrated quantitatively by Young & Schubert (1974) and Schubert & Young (1976), who calculated temperatures in convecting, constant viscosity, internally heated fluid models of the mantles of Mars and Earth. Schubert & Young (1976) showed that the temperature at the core-mantle boundary would lie

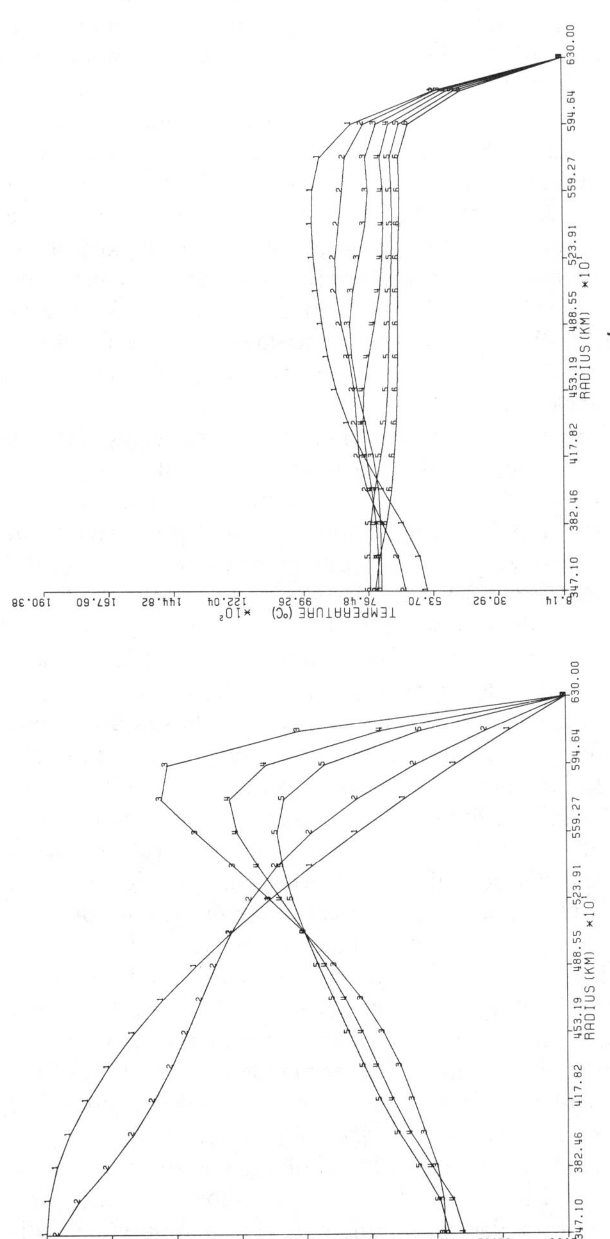

Figure 9 (*a*) Rapid decay of temperature in a model of the Earth's mantle (Schubert & Young 1976, Schubert, Cassen & Young 1979b) which cools from a hot initial conduction profile (curve 1). The profiles with successively higher numbers correspond to later times. The time between successive profiles is only 5×10^{-3} times the conduction time across the Earth's mantle (about 3×10^{11} yr). The large temperature drop early in the evolution of the model requires only 3×10^9 yr. The numerical calculations are for $Ra/Ra_{cr} = 100$. (*b*) The continued evolution of the temperature profiles of Figure 9(*a*). Curve (1) on this figure is the temperature profile 5×10^{-3} conduction times after curve (5) on Figure 9(*a*).

significantly below the iron melting point if the Earth's mantle viscosity were less than 10^{24} cm^2/s. Cassen et al (1976) showed that convection in the relatively thin mantle of Mercury could freeze its core in a billion years or less.

If the viscosity of the earth's mantle is indeed nearly uniform with the value of 10^{22} cm^2/s, as inferred from glacial rebound data (Cathles 1975, Peltier 1976), how could the outer core still be liquid? The existence of a liquid outer core in the Earth places a significant constraint on the efficacy of subsolidus convective cooling during the thermal history of our planet. While we must find a reason why overly efficient mantle convection has not frozen the Earth's core, this problem may not exist for one or more of the other terrestrial planets if future seismic observations should reveal a solid core or if future magnetic observations should confirm the absence of a planetary field.

Schubert & Young (1976) reported that the quasi-steady, average mantle temperatures in their Earth models were established on a time scale no larger than a tenth of a conduction time across the mantle (a conduction time for the Earth's mantle is about 3×10^{11} yr). In fact, upon examining the transient development of mean temperature profiles for the Earth models in more detail, Schubert, Cassen & Young (1979b) found that only a few hundredths of a mantle conduction time is required to establish the low quasi-steady average temperatures at a Rayleigh number only 100 times the critical value. Figure 9, from Schubert, Cassen & Young (1979b), shows how quickly a hot initial conduction temperature profile is reduced by vigorous convection to relatively low temperatures throughout the mantle and in particular at the core-mantle interface for $\text{Ra} = 100\ \text{Ra}_{\text{cr}}$. Since we estimated the Rayleigh number for the present Earth to be about 10^7, and the Rayleigh number is likely to have been many orders of magnitude larger just after core formation (see Figure 7), the ease with which whole mantle convection could freeze the core on time scales very much less than the age of the Earth is clear.

For the Earth, there are several ways in which core solidification by subsolidus convective cooling can be prevented. One way is to have a significant source of radioactive heating in the core. Another way is to prevent convection from reaching the lower mantle for a portion of the Earth's history, particularly during the initial period of cooling after core formation when convection should be especially vigorous. This may be accomplished by chemically or viscously stratifying the lower mantle or by hypothesizing that the lower mantle geotherm is subadiabatic. In view of both the inference from glacial rebound data that mantle viscosity is essentially uniform and the argument of Sammis et al (1977) against large viscosity jumps across the major mantle phase transitions, a subadiabatic lower mantle accessible only by some form of weak penetrative convection may be the more likely explanation. The models of Sharpe & Peltier (1978) rely on an assumed subadiabaticity of the lower mantle to prevent core solidification by solid state convective cooling.

CONCLUDING REMARKS

Whereas a decade ago subsolidus convection even in the Earth's upper mantle was a minority view, today this is no longer true. The debate usually centers on the characteristics of convection in the terrestrial planets rather than on its existence. We have made considerable progress in our understanding of mantle convection. Two-dimensional and axisymmetric three-dimensional numerical calculations incorporating the temperature- and pressure-dependence of mantle rheology and its non-Newtonian character have allowed us to gain an appreciation for the relative importance of these properties. Theoretical scaling arguments and boundary layer theories have extended our knowledge of convection to higher Rayleigh numbers than direct numerical calculations can deal with. Even so, we are far from being able to model the extremely high Rayleigh number convection likely to have occurred in all the terrestrial planets at some time in their evolutions and likely to be occurring in the larger terrestrial planets at present. Very high Rayleigh number convection in the planets is undoubtedly fully three-dimensional and probably time-dependent as well; present computing limitations probably preclude the direct numerical modelling of such convection. Our understanding of high Rayleigh number mantle convection is made difficult by both these computational barriers and the uncertainties in thermodynamic and rheologic properties of mantle materials, especially for the constituents of the lower mantles of the large planets.

Nevertheless, we need to persevere in our attempts to study mantle convection using all the approaches available to us whether theoretical, numerical, or experimental. Each of these approaches has its own set of advantages and disadvantages and all of them are worth pursuing for the different insights they provide. Rigorous fluid dynamic studies of convection should be carried out even for parameter values not directly applicable to planetary interiors because of the fundamental knowledge we gain of the convective process. Reasonable though nonrigorous modelling of mantle convection for parameter values directly relevant to the planets is also worthwhile as a way of providing some quantitative assessments of our ideas about the way convection might actually work in the planets.

Much emphasis has been placed on the building of two-dimensional models of mantle convection by the community of modellers primarily interested in the Earth. However, the evidence for shallow mantle convection vs whole mantle convection in the Earth necessitates efforts in directions other than two-dimensional numerical computations. Since whole mantle convection in the Earth and the other terrestrial planets is a highly likely form of convection, we should direct attention toward constructing models of convection in spherical geometry. Even models of axisymmetric convection in spheres and spherical shells would be a welcome addition to the literature, although their stability to nonaxisymmetric disturbances should always be determined.

Future planetary exploration will provide crucial data against which to test our ideas about mantle convection. We will learn a great deal when radar observations of Venus allow us to determine whether the surface contains any record of plate tectonic activity (Weertman 1979). Earth-based radar observations (Malin & Saunders 1977) and Pioneer Venus radar measurements (Masursky et al 1977) will contribute significantly toward this end, but definitive conclusions will require the global coverage and resolution of a Venus Orbiting Imaging Radar. Our discussion of convection in terrestrial planets has concentrated on the planets of the inner solar system. Yet the outer solar system contains bodies whose global properties, density in particular, would place them in the category of terrestrial planets. Io and Europa, two of the Galilean satellites, are examples of such bodies (Johnson 1978) and future spacecraft reconnaisance of their surfaces, shapes, gravitational and magnetic fields, etc will provide still additional experimental tests of our theories of convection. The Galilean satellites Ganymede and Callisto, which by virtue of their low densities must have water as a major constituent (Johnson 1978), may allow us to view the consequences of solid state convection in a planet with a rheology quite different from that of the silicate-dominated bodies of the solar system (Reynolds & Cassen 1979). The imminent Voyager exploration of the Galilean satellites and the planned Galileo observations of these objects may reveal new worlds whose evolutions have also been influenced, if not dominated, by solid state convection.

Acknowledgments

I would like to thank my colleagues P. Cassen, R. E. Young, and D. A. Yuen for stimulating conversations and ideas throughout the course of our continued collaboration. This research was supported in part by the Planetology Program, Office of Space Science, NASA grant NGR 05-007-317, and by NSF grant EAR 77-15198.

Literature Cited

Anderson, D. L. 1972. Internal constitution of Mars. *J. Geophys. Res.* 77:789–95

Anderson, D. L. 1977. Composition of the mantle and core. *Ann. Rev. Earth Planet. Sci.* 5:179–202

Ashby, M. F., Verall, R. A. 1973. Diffusion-accommodated flow and superplasticity. *Acta Metall.* 21:149–63

Binder, A. B., Davis, D. R. 1973. Internal structure of Mars. *Phys. Earth Planet. Inter.* 7:477–85.

Birch, F. 1965. Energetics of core formation. *J. Geophys. Res.* 70:6217–21

Birch, F., Roy, R. F., Decker, E. R. 1968. Heat flow and thermal history in New England and New York. In *Studies of Appalachian Geology: Northern and Maritime*, ed. E. Zen, W. S. White, J. B. Hadley, J. B. Thompson, Jr., pp. 437–51. New York: Interscience

Blackshear, W. T., Gapcynski, J. P. 1977. An improved value of the lunar moment of inertia. *J. Geophys. Res.* 82:1699–701

Booker, J. R. 1976. Thermal convection with strongly temperature-dependent viscosity. *J. Fluid Mech.* 76:741–54

Booker, J. R., Stengel, K. C. 1978. Further thoughts on convective heat transport in a variable viscosity fluid. *J. Fluid Mech.* 86:289–91

Burns, J. A. 1976. Consequences of the tidal slowing of Mercury. *Icarus* 28:453–58

Busse, F. H. 1967. On the stability of two-dimensional convection in a layer heated from below. *J. Math. Phys.* 46:140–50

Busse, F. H. 1975. Patterns of convection in spherical shells. *J. Fluid Mech.* 72:67–85

Busse, F. H. 1978a. Nonlinear properties of thermal convection. *Rep. Prog. Phys.* 41:1929–67

Busse, F. H. 1978b. A model of time-periodic mantle flow. *Geophys. J. R. Astron. Soc.* 52:1–12

Busse, F. H. 1979. High Prandtl number convection. *Phys. Earth Planet. Inter.* In press

Busse, F. H., Schubert, G. 1971. Convection in a fluid with two phases. *J. Fluid Mech.* 46:801–12

Carter, N. L. 1976. Steady state flow of rocks. *Rev. Geophys. Space Phys.* 14:301–60

Cassen, P., Reynolds, R. T. 1973. The role of convection in the Moon. *J. Geophys. Res.* 78:3203–15

Cassen, P., Reynolds, R. T. 1974. Convection in the Moon: Effect of variable viscosity. *J. Geophys. Res.* 79:2937–44

Cassen, P., Reynolds, R. T., Graziani, F., Summers, A., McNellis, J., Blalock, L. 1979. Convection and lunar thermal history. *Phys. Earth Planet. Inter.* In press

Cassen, P., Young, R. E. 1975. On the cooling of the Moon by solid convection. *The Moon* 12:361–68

Cassen, P., Young, R. E., Schubert, G. 1978. The distortion of the Moon due to convection. *Geophys. Res. Lett.* 5:294–96

Cassen, P., Young, R. E., Schubert, G., Reynolds, R. T. 1976. Implications of an internal dynamo for the thermal history of Mercury. *Icarus* 28:501–8

Cathles, L. M., III. 1975. *The Viscosity of the Earth's Mantle.* Princeton: Univ. Press. 386 pp.

Chandrasekhar, S. 1961. *Hydrodynamic and Hydromagnetic Stability*, Chaps. II and VI. Oxford: Clarendon. 652 pp.

Chu, T. Y., Goldstein, R. J. 1973. Turbulent convection in a horizontal layer of water. *J. Fluid Mech.* 60:141–59

Coble, R. L. 1963. A model for boundary diffusion controlled creep in polycrystalline materials. *J. Appl. Phys.* 34:1679–82

Cuzzi, J. N. 1974. The nature of the surface of Mercury from microwave observations at several wavelengths. *Astrophys. J.* 189:577–86

Dainty, A. M., Toksöz, M. N., Solomon, S. C., Anderson, K. R., Goins, N. R. 1974. Constraints on lunar structure. *Proc. Lunar Sci. Conf. 5th*, pp. 3091–3114

Daly, S. F., Richter, F. M. 1978. Convection with decaying heat sources: A simple thermal evolution model. *Lunar and Planetary Science IX*, pp. 213–14 (Abstr.)

Davies, G. F. 1974. Limits on the constitution of the lower mantle. *Geophys. J. R. Astron. Soc.* 38:479–503

Davies, G. F. 1977. Whole mantle convection and plate tectonics. *Geophys. J. R. Astron. Soc.* 49:459–86

Dziewonski, A. M., Hager, B. H., O'Connell, R. J. 1977. Large-scale heterogeneities in the lower mantle. *J. Geophys. Res.* 82:239–55

Elsasser, W. M., Olson, P., Marsh, B. D. 1979. The depth of mantle convection. *J. Geophys. Res.* In press

Engdahl, E. R. 1975. Effects of plate structure and dilatancy on relative teleseismic P-wave residuals. *Geophys. Res. Lett.* 2:420–22

Froidevaux, C., Schubert, G. 1975. Plate motion and structure of the continental asthenosphere: A realistic model of the upper mantle. *J. Geophys. Res.* 80:2553–64

Garon, A. M., Goldstein, R. J. 1973. Velocity and heat transfer measurements in thermal convection. *Phys. Fluids* 16: 1818–25

Gault, D. E., Burns, J. A., Cassen, P., Strom, R. G. 1977. Mercury. *Ann. Rev. Astron. Astrophys.* 15:97–126

Goetze, C., Kohlstedt, D. L. 1973. Laboratory study of dislocation climb and diffusion in olivine. *J. Geophys. Res.* 78: 5961–71

Gordon, R. B. 1965. Diffusion creep in the Earth's mantle. *J. Geophys. Res.* 70: 2413–18

Green, H. W., II 1970. Diffusional flow in polycrystalline materials. *J. Appl. Phys.* 41:3899–902

Griggs, D. T. 1972. The sinking lithosphere and the focal mechanism of deep earthquakes. In *The Nature of the Solid Earth*, ed. E. C. Robertson, pp. 361–84. New York: McGraw-Hill

Grossman, L., Larimer, J. W. 1974. Early chemical history of the solar system. *Rev. Geophys. Space Phys.* 12:71–101

Hager, B. H., O'Connell, R. J. 1978. Subduction zone dip angles and flow driven by plate motion. *Tectonophysics* 50:111–33

Hanks, T. C., Anderson, D. L. 1969. The early thermal history of the Earth. *Phys. Earth Planet. Inter.* 2: 19–29

Heard, H. C. 1976. Comparison of the flow properties of rocks at crustal conditions. *Philos. Trans. R. Soc. London Ser. A* 283: 173–86

Herbert, F., Sonett, C. P., Wiskerchen, M. J. 1977. Model 'zero-age' lunar thermal profiles resulting from electrical induction. *J. Geophys. Res.* 82:2054–60

Herring, C. 1950. Diffusional viscosity of a polycrystalline solid. *J. Appl. Phys.* 21: 437–45

Hewitt, J. M., McKenzie, D. P., Weiss, N. O. 1975. Dissipative heating in convective flows. *J. Fluid Mech.* 68:721–38

Houston, M. H., Jr., De Bremaecker, J. C. 1975. Numerical models of convection in the upper mantle. *J. Geophys. Res.* 80: 742–51

Howard, L. N. 1966. Convection at high Rayleigh number. In *Proc. 11th Cong. Appl. Mech.*, ed. H. Görtler, pp. 1109–15. Berlin: Springer

Hsui, A. T., Toksöz, M. N. 1978. Thermal evolution of planetary size bodies. *Proc. Lunar Sci. Conf. 8th*, pp. 447–61

Hsui, A. T., Turcotte, D. L., Torrance, K. E. 1972. Finite amplitude thermal convection within a self-gravitating fluid sphere. *Geophys. Fluid Dyn.* 3:35–44

Isacks, B., Molnar, P. 1971. Distribution of stresses in the descending lithosphere from a global survey of focal-mechanism solutions of mantle earthquakes. *Rev. Geophys. Space Phys.* 9:103–74

Jarvis, G. T., McKenzie, D. P. 1979. Infinite Prandtl number compressible convection. *J. Fluid Mech.* Submitted

Johnson, T. V. 1978. The Galilean satellites of Jupiter: Four worlds. *Ann. Rev. Earth Planet. Sci.* 6:93–125

Johnston, D. H., Toksöz, M. N. 1977. Internal structure and properties of Mars. *Icarus* 32:73–84

Jones, G. M. 1977. Thermal interaction of the core and the mantle and long-term behavior of the geomagnetic field. *J. Geophys. Res.* 82:1703–09

Jordon, T. H., Lynn, W. S. 1974. A velocity anomaly in the lower mantle. *J. Geophys. Res.* 79:2679–85

Julian, B. R., Sengupta, M. K. 1973. Seismic travel time evidence for lateral inhomogeneity in the deep mantle. *Nature* 242: 443–47

Kaula, W. M. 1963a. Elastic models of the mantle corresponding to variations in the external gravity field. *J. Geophys. Res.* 68:4967–78

Kaula, W. M. 1963b. Tidal dissipation in the moon. *J. Geophys. Res.* 68: 4959–65

Kaula, W. M. 1964. Tidal dissipation by tidal friction and resulting orbital evolution. *Rev. Geophys.* 2:661–85

Kaula, W. M. 1975. The seven ages of a planet. *Icarus* 26: 1–15

Kaula, W. M. 1979 Thermal evolution of Earth and Moon growing by planetesimal impacts. *J. Geophys. Res.* In press

Kaula, W. M., Schubert, G., Lingenfelter, R. E., Sjogren, W. L., Wollenhaupt, W. R. 1972. Analysis and interpretation of lunar laser altimetry. *Proc. Lunar Sci. Conf. 3rd*, pp. 2189–204

Kaula, W. M., Schubert, G., Lingenfelter, R. E., Sjogren, W. L., Wollenhaupt, W. R. 1974. Apollo laser altimetry and inferences as to lunar structure. *Proc. Lunar Sci. Conf. 5th*, pp. 3049–58

Kaula, W. M., Yoder, C. F. 1976. Lunar orbit evolution and tidal heating of the moon. *Lunar Science VII*, pp. 440–42. (Abstr.)

Keldysh, M. W. 1977. Venus exploration with the Venera 9 and Venera 10 Spacecraft. *Icarus* 30:605–25

Keyes, R. W. 1963. Continuum models of the effect of pressure on activated processes. In *Solids Under Pressure*, ed. W. Paul, D. M. Warshauer, pp. 71–99. New York: McGraw-Hill

Knopoff, L. 1964. The convection current hypothesis. *Rev. Geophys.* 2:89–123

Kohlstedt, D. L., Goetze, C. 1974. Low stress and high temperature creep in olivine single crystals. *J. Geophys. Res.* 79:2045–51

Kohlstedt, D. L., Goetze, C., Durham, W. B. 1976. Experimental deformation of single crystal olivine with application to flow in the mantle. In *Petrophysics: The Physics and Chemistry of Minerals and Rocks*, ed. S. K. Runcorn. London: Wiley

Kraichnan, R. H. 1962. Mixing-length analysis of turbulent thermal convection at arbitrary Prandtl numbers. *Phys. Fluids* 5:1374–89

Kuckes, A. F. 1977. Strength and rigidity of the elastic lunar lithosphere and implications for present-day mantle convection in the Moon. *Phys. Earth Planet. Inter.* 14:1–12

Kulacki, F. A., Emara, A. A. 1977. Steady and transient thermal convection in a fluid layer with uniform volumetric energy sources. *J. Fluid Mech.* 83:375–95

Kulacki, F. A., Nagle, M. E. 1975. Natural convection in horizontal fluid layers with volumetric energy sources. *J. Heat Transfer* 97:204–11

Lachenbruch, A. 1968. Preliminary geothermal model of the Sierra Nevada. *J. Geophys. Res.* 73:6977–89

Lambeck, K. 1976. Lateral density anomalies in the upper mantle. *J. Geophys. Res.* 81:6333–40

Langseth, M. G., Keihm, S. J., Peters, K. 1976. Revised lunar heat-flow values. *Proc. Lunar Sci. Conf. 7th*, pp. 3143–71

Lewis, J. S. 1972. Metal/silicate fractionation in the solar system. *Earth Planet. Sci. Lett.* 15:286–90

Liebermann, R. C., Jackson, I., Ringwood, A. E. 1977. Elasticity and phase equilibria of spinel disproportionation reactions. *Geophys. J. R. Astron. Soc.* 50:553–86

Liebermann, R. C., Ringwood, A. E. 1973. Birch's law and polymorphic phase transformations. *J. Geophys. Res.* 78:6926–32

Lingenfelter, R. E., Schubert, G. 1973. Evidence for convection in planetary interiors from first order topography. *The Moon* 7:172–80

Long, R. R. 1976. Relation between Nusselt number and Rayleigh number in turbulent thermal convection. *J. Fluid Mech.* 73:445–51

LSPET. 1972. The Apollo 15 lunar samples: A preliminary description. *Science* 175:363–75

LSPET. 1973. The Apollo 16 lunar samples: Petrographic and chemical description. *Science* 179:23–34

Malin, M. C., Saunders, R. S. 1977. Surface of Venus: Evidence of diverse landforms from radar observations. *Science* 196: 987–90

Marov, M. Ya. 1972. Venus: A perspective at the beginning of planetary exploration. *Icarus* 16:415–61

Masursky, H., Kaula, W. M., McGill, G. E., Pettengill, G. H., Phillips, R. J., Russell, C. T., Schubert, G., Shapiro, I. I. 1977. The surface and interior of Venus. *Space Sci. Rev.* 20:431–49

McKenzie, D. P. 1977. Surface deformation, gravity anomalies and convection. *Geophys. J. R. Astron. Soc.* 48:211–38

McKenzie, D. P., Richter, F. 1976. Convection currents in the Earth's mantle. *Sci. Am.* 235:72–89

McKenzie, D. P., Roberts, J. M., Weiss, N. O. 1974. Convection in the Earth's mantle: Towards a numerical solution. *J. Fluid Mech.* 62:465–538

McKenzie, D., Weiss, N. 1975. Speculation on the thermal and tectonic history of the Earth. *Geophys. J. R. Astron. Soc.* 42: 131–74

Melosh, H. J. 1975. Mascons and the Moon's orientation. *Earth Planet. Sci. Lett.* 25:322–26

Mizutani, H., Matsui, T., Takeuchi, J. 1972. Accretion process of the Moon. *The Moon* 4:476–89

Moorbath, S., O'Nions, R. K., Pankhurst, R. J. 1975. The evolution of early Precambrian crustal rocks at Isua, West Greenland-geochemical and isotopic evidence. *Earth Planet. Sci. Lett.* 27: 229–39

Nabarro, F. R. N. 1948. Deformation of crystals by the motion of single ions. In *Strength of Solids*. The Physical Society of London. 175 pp.

O'Connell, R. J. 1977. On the scale of mantle convection. *Tectonophysics* 38: 119–36

Oxburgh, E. R., Turcotte, D. L. 1978. Mechanisms of continental drift. *Rep. Prog. Phys.* 41:1249–312

Parmentier, E. M. 1978. A study of thermal convection in non-Newtonian fluids. *J. Fluid Mech.* 84:1–11

Parmentier, E. M., Turcotte, D. L. 1978. Two-dimensional mantle flow beneath a rigid, accreting lithosphere. *Phys. Earth Planet. Inter.* 17:281–89

Parmentier, E. M., Turcotte, D. L., Torrance, K. E. 1976. Studies of finite amplitude non-Newtonian thermal convection with application to convection in the Earth's mantle. *J. Geophys. Res.* 81: 1839–46

Peale, S. J., Cassen, P. 1978. Contribution

of tidal dissipation to lunar thermal history. *Icarus*. 36:245–69

Pekeris, C. L. 1935. Thermal convection in the interior of the Earth. *Mon. Not. R. Astron. Soc., Geophys. Suppl.* 3:343–67

Peltier, W. R. 1972. Penetrative convection in the planetary mantle. *Geophys. Fluid Dyn.* 5:47–88

Peltier, W. R. 1976. Glacial-isostatic adjustment, II, The inverse problem. *Geophys. J. R. Astron. Soc.* 46:669–705

Phakey, P., Dollinger, G., Christie, J. 1972. Transmission electron microscopy of experimentally deformed olivine crystals. In *Flow and Fracture of Rocks, Geophys. Monogr. Ser.*, ed. H. C. Heard, I. Y. Borg, N. L. Carter, C. B. Raleigh, 16:117–38. Washington, DC: AGU

Phillips, R. J., Ivins, E. R. 1979. Geophysical observations pertaining to solid state convection in the terrestrial planets. *Phys. Earth Planet. Inter.* In press

Priestley, C. H. B. 1954. Convection from a large horizontal surface. *Aust. J. Phys.* 7:176–201

Raleigh, C. B. 1968. Mechanisms of plastic deformation in olivine. *J. Geophys. Res.* 73:5391–406

Reasenberg, R. D. 1977. The moment of inertia and isostasy of Mars. *J. Geophys. Res.* 82:369–75

Reynolds, R. T., Cassen, P. 1979. On the internal structure of the major satellites of the outer planets. *Geophys. Res. Lett.* In press

Reynolds, R. T., Summers, A. L. 1969. Calculations on the composition of the terrestrial planets. *J. Geophys. Res.* 74: 2494–511

Richter, F. 1973a. Dynamical models for sea floor spreading. *Rev. Geophys. Space Phys.* 11:223–87

Richter, F. M. 1973b. Convection and the large-scale circulation of the mantle. *J. Geophys. Res.* 78:8735–745

Richter, F. M. 1973c. Finite amplitude convection through a phase boundary. *Geophys. J. R. Astron. Soc.* 35:265–76

Richter, F. M. 1977. On the driving mechanism of plate tectonics. *Tectonophysics* 38:61–88

Richter, F. M. 1978. Mantle convection models. *Ann. Rev. Earth Planet. Sci.* 6: 9–19

Richter, F. M., Johnson, C. E. 1974. Stability of a chemically layered mantle. *J. Geophys. Res.* 79:1635–39

Richter, F. M., McKenzie, D. P. 1978. Simple plate models of mantle convection. *J. Geophys.* 44:441–71

Richter, F. M., Parsons, B. 1975. On the interaction of two scales of convection in the mantle. *J. Geophys. Res.* 80:2529–41

Ringwood, A. E. 1960. On the chemical evolution and densities of the planets. *Geochim. Cosmochim. Acta* 15:257–83

Ringwood, A. E. 1972. Phase transformations and mantle dynamics. *Earth Planet. Sci. Lett.* 14:233–41

Ringwood, A. E. 1975. *Composition and Petrology of the Earth's Mantle*, pp. 573–79. New York: McGraw-Hill. 618 pp.

Ringwood, A. E., Anderson, D. L. 1977. Earth and Venus: A comparative study. *Icarus* 30:243–53

Ross, J. V., Avé Lallemant, H. G., Carter, N. L. 1978. The activation volume for creep of olivine. *EOS Trans. AGU* 59: 374–75 (Abstr.)

Rossby, H. T. 1969. A study of Bénard convection with and without rotation. *J. Fluid Mech.* 36:309–35

Runcorn, S. K. 1962. Convection in the Moon. *Nature* 195:1150–51

Runcorn, S. K. 1967a. Convection in the Moon and the existence of a lunar core. *Proc. R. Soc. London Ser. A* 296:270–84

Runcorn, S. K. 1967b. Flow in the mantle inferred from the low degree harmonics of the geopotential. *Geophys. J. R. Astron. Soc.* 14:375–84

Runcorn, S. K. 1975. Solid-state convection and the mechanics of the Moon. *Proc. Lunar Sci. Conf. 6th*, pp. 2943–53

Runcorn, S. K. 1976. Inferences concerning the early thermal history of the Moon. *Proc. Lunar Sci. Conf. 7th*, pp. 3221–28

Runcorn, S. K. 1977. Convection in Mercury. *Phys. Earth Planet. Inter.* 15: 131–34

Safronov, V. S. 1979. The heating of the Earth during its formation. *Icarus*. 33: 3–12

Sammis, C. G., Smith, J. C., Schubert, G., Yuen, D. A. 1977. Viscosity-depth profile of the Earth's mantle: effects of polymorphic phase transitions. *J. Geophys. Res.* 82:3747–61

Schubert, G., Cassen, P., Young, R. E. 1979a. Cooling histories of terrestrial planets. *Icarus*. In press

Schubert, G., Cassen, P., Young, R. E. 1979b. Core cooling by subsolidus mantle convection. *Phys. Earth Planet. Inter.* Submitted

Schubert, G., Froidevaux, C., Yuen, D. A. 1976. Oceanic lithosphere and asthenosphere: Thermal and mechanical structure. *J. Geophys. Res.* 81:3525–40

Schubert, G., Turcotte, D. L. 1971. Phase changes and mantle convection. *J. Geophys. Res.* 76:1424–32

Schubert, G., Turcotte, D. L., Oxburgh, E. R. 1969. Stability of planetary interiors. *Geophys. J. R. Astron. Soc.* 18:441–60

Schubert, G., Turcotte, D. L., Oxburgh, E.

R. 1970. Phase change instability in the mantle. *Science* 169:1075–77.

Schubert, G., Young, R. E. 1976. Cooling the Earth by whole mantle subsolidus convection: A constraint on the viscosity of the lower mantle. *Tectonophysics* 35: 201–14

Schubert, G., Young, R. E., Cassen, P. 1977. Solid state convection models of the lunar internal temperature. *Philos. Trans. R. Soc. London Ser. A* 285:523–36

Schubert, G., Yuen, D. A., Froidevaux, C., Fleitout, L., Souriau, M. 1978. Mantle circulation with partial shallow return flow: Effects on stresses in oceanic plates and topography of the sea floor. *J. Geophys. Res.* 83:745–58

Schubert, G., Yuen, D. A., Turcotte, D. L. 1975. Role of phase transitions in a dynamic mantle. *Geophys. J. R. Astron. Soc.* 42:705–35

Sharpe, H. N., Peltier, W. R. 1978. Parameterized mantle convection and the Earth's thermal history. *Geophys. Res. Lett.* 5:737–40

Sharpe, H. N., Peltier, W. R. 1979. A thermal history model for the Earth with parameterized convection. *Geophys. J. Astron. Soc.* In press

Siegfried, R. W., II, Solomon, S. C. 1974. Mercury: Internal structure and thermal evolution. *Icarus* 23:192–205

Solomon, S. C. 1977. The relationship between crustal tectonics and internal evolution in the Moon and Mercury. *Phys. Earth Planet. Inter.* 15:135–45

Solomon, S. C. 1978. On volcanism and thermal tectonics on one-plate planets. *Geophys. Res. Lett.* 5:461–64

Solomon, S. C. 1979. Formation, history, and energetics of cores in the terrestrial planets. *Phys. Earth Planet. Inter.* In press

Solomon, S. C., Chaiken, J. 1976. Thermal expansion and thermal stress in the Moon and terrestrial planets: Clues to early thermal history. *Proc. Lunar Sci. Conf. 7th*, pp. 3229–43

Sonett, C. P., Colburn, D. S., Schwartz, K. 1975. Formation of the lunar crust: An electrical source of heating. *Icarus* 24: 231–55

Stevenson, D. J. 1979. Whole Earth cooling and primordial heat. *Nature*. Submitted

Stevenson, D. J., Turner, J. S. 1979. Fluid models of mantle convection. In *The Earth, Its Origin, Evolution and Structure*, ed. M. W. McElhinney. New York: Wiley

Stocker, R. L., Ashby, M. F. 1973. On the rheology of the upper mantle. *Rev. Geophys. Space Phys.* 11:391–426

Surkov, Yu. A. 1977. Geochemical studies of Venus by Venera 9 and 10 automatic interplanetary stations. *Proc. Lunar Sci. Conf. 8th*, pp. 2665–89

Toksöz, M. N., Hsui, A. T. 1978. Thermal history and evolution of Mars. *Icarus* 34: 537–47

Toksöz, M. N., Hsui, A. T., Johnston, D. H. 1978. Thermal evolutions of the terrestrial planets. *The Moon and The Planets* 18:265–72

Toksöz, M. N., Johnston, D. H. 1977. The evolution of the Moon and the terrestrial planets. In *The Soviet-American Conference on Cosmochemistry of the Moon and Planets*, ed. J. H. Pomeroy, N. J. Hubbard, NASA SP-370, pp. 295-327. Washington, DC: GPO

Torrance, K. E., Turcotte, D. L. 1971a. Structure of convection cells in the mantle. *J. Geophys. Res.* 76:1154–61

Torrance, K. E., Turcotte, D. L. 1971b. Thermal convection with large viscosity variations. *J. Fluid Mech.* 47:113–25

Tozer, D. C. 1965a. Heat transfer and convection currents. *Philos. Trans. R. Soc. London Ser. A* 258:252–71

Tozer, D. C. 1965b. Thermal history of the Earth: 1. The formation of the core. *Geophys. J. R. Astron. Soc.* 9:95–112

Tozer, D. C. 1967. Towards a theory of thermal convection in the mantle. In *The Earth's Mantle*, ed. T. F. Gaskell, pp. 325–53. London: Academic

Tozer, D. C. 1972a. The present thermal state of the terrestrial planets. *Phys. Earth Planet. Inter.* 6:182–97

Tozer, D. C. 1972b. The Moon's thermal state and an interpretation of the lunar electrical conductivity distribution. *The Moon* 5:90–105

Tozer, D. C. 1974. The internal evolution of planetary-sized objects. *The Moon* 9:167–82

Turcotte, D. L., Cisne, J. L., Nordmann, J. C. 1977. On the evolution of the lunar orbit. *Icarus* 30:254–66

Turcotte, D. L., Hsui, A. T., Torrance, K. E., Oxburgh, E. R. 1972. Thermal structure of the Moon. *J. Geophys. Res.* 77: 6931–39

Turcotte, D. L., Hsui, A. T., Torrance, K. E., Schubert, G. 1974. Influence of viscous dissipation on Bénard convection. *J. Fluid Mech.*. 64:369–74

Turcotte, D. L., Oxburgh, E. R. 1967. Finite amplitude convective cells and continental drift. *J. Fluid Mech.* 28:29–42

Turcotte, D. L., Oxburgh, E. R. 1969a. Convection in a mantle with variable physical properties. *J. Geophys. Res.* 74: 1458–74

Turcotte, D. L., Oxburgh, E. R. 1969b. Implications of convection within the Moon. *Nature* 223:250–51

Turcotte, D. L., Oxburgh, E. R. 1972. Mantle convection and the new global tectonics. *Ann. Rev. Fluid Mech.* 4: 33–68

Turcotte, D. L., Schubert, G. 1971. Structure of the olivine-spinel phase boundary in the descending lithosphere. *J. Geophys. Res.* 76:7980–87

Turcotte, D. L., Torrance, K. E., Hsui, A. T. 1973. Convection in the Earth's mantle. *Methods Comput. Phys.* 13:431–54

Twiss, R. J. 1976. Structural superplastic creep and linear viscosity in the Earth's mantle. *Earth Planet. Sci. Lett.* 33:86–100

Verhoogen, J. 1965. Phase changes and convection in the Earth's mantle. *Philos. Trans. R. Soc. London Ser. A* 258:276–83

Vinogradov, A. P., Surkov, Yu. A., Kirnozov, F. F. 1973. The content of uranium, thorium and potassium in the rocks of Venus as measured by Venera 8. *Icarus* 20:253–59

Vollmer, R. 1977. Terrestrial lead isotopic evolution and formation of the Earth's core. *Nature* 270:144–47

Walzer, U. 1978. On non-steady mantle convection: The case of the Bénard problem with viscosity dependent on temperature and pressure. *Gerlands Beitr. Geophys.* 87:19–28

Wang, H., Simmons, G. 1972. FeO and SiO_2 in the lower mantle. *Earth Planet. Sci. Lett.* 14:83–86

Watt, J. P., Shankland, T. J., Mao, N. H. 1975. Uniformity of mantle composition. *Geology* 3:91–94

Weertman, J. 1968. Dislocation climb theory of steady-state creep. *Trans. ASME* 61:681–94

Weertman, J. 1970. The creep strength of the Earth's mantle. *Rev. Geophys. Space Phys.* 8:145–68

Weertman, J. 1978. Creep laws for the mantle of the Earth. *Phil. Trans. R. Soc. London, Ser. A* 288:9–26

Weertman, J. 1979. Height of mountains on Venus and the creep properties of rock. *Phys. Earth Planet. Inter.* In press

Weertman, J., Weertman, J. R. 1975. High temperature creep of rock and mantle viscosity. *Ann. Rev. Earth Planet. Sci.* 3:293–315

Weidenschilling, S. J. 1976. Accretion of the terrestrial planets II. *Icarus* 27:161–70

Wetherill, G. W. 1972. The beginning of continental evolution. *Tectonophysics* 13: 31–45

Wetherill, G. W. 1976. The role of large bodies in the formation of the Earth and Moon. *Proc. Lunar Sci. Conf. 7th*, pp. 3245–57

Whitehead, J. A. Jr. 1976. Convection models: Laboratory vs. mantle. *Tectonophysics* 35:215–29

Williams, J. G., Sinclair, W. S., Slade, M. A., Bender, P. L., Hauser, J. P., Mulholland, J. D., Shelus, P. J. 1974. Lunar moment of inertia constraints from lunar laser ranging *Lunar Science V*, p. 845 (Abstr.)

Young, R. E. 1974. Finite-amplitude thermal convection in a spherical shell. *J. Fluid Mech.* 63:695–721

Young, R. E., Schubert, G. 1974. Temperatures inside Mars: Is the core liquid or solid? *Geophys. Res. Lett.* 1:157–60

Yuen, D. A., Schubert, G. 1976. Mantle plumes: A boundary layer approach for Newtonian and non-Newtonian, temperature-dependent rheologies. *J. Geophys. Res.* 81:2499–510

Zebib, A., Schubert, G., Straus, J. M. 1979. Infinite Prandtl number thermal convection in a spherical shell. *J. Fluid Mech.* Submitted

Ann. Rev. Earth Planet. Sci. 1979. 7:343–55

THE NORTH ATLANTIC RIDGE: OBSERVATIONAL EVIDENCE FOR ITS GENERATION AND AGING[1]

J. R. Heirtzler

Department of Geology and Geophysics, Woods Hole Oceanographic Institution, Woods Hole, Massachusetts 02543

INTRODUCTION

During the period from 1963 to 1969 a beautifully simplistic picture of seafloor spreading and global tectonics was developed. One of the most attractive aspects of the seafloor spreading theory postulates creation of the seafloor at mid-ocean ridges, and its magnetization with a polarity determined by the direction of the earth's ambient field at the time (Vine & Matthews 1963).

It was realized early that this simplistic theory would require elaboration and possibly modification as knowledge increased. In the last decade three major types of studies of the ocean crust bear on the theory of seafloor spreading:

1. Geological and geophysical studies of the axes of the ridges with precisely located instruments and with manned submersibles, especially on the Mid-Atlantic Ridge south of the Azores (Projects FAMOUS and AMAR), in the Cayman Trough of the Caribbean, in the Galapagos Rift, and near the Tamayo Fracture Zone of the East Pacific Rise. These studies have revised our simple ideas about crustal emplacement and have provided insight into its episodic nature.
2. Deep sea drilling at a number of sites. In the North Atlantic, holes were drilled from near the ridge axis to sites in very old crust. At 21 Atlantic sites we have penetrated more than 20 m into the basaltic basement on Legs 37, 38, 45, 46, 49, 51, 52, and 53 (Figure 1). These sites range in age from 1.6 to 110 million years, and shallower penetration was made

[1] Contribution No. 4255 from the Woods Hole Oceanographic Institution.

0084-6597/79/0515-0343$01.00

in crust about 150 million years old. Recovered samples have indicated the upper part of the oceanic basement changes with age, and have provided help in the interpretation of geophysical profiles taken by research vessels operating on the ocean surface and using remote sensing instruments.

3. Geophysical studies between the mid-ocean ridge axial region and the edges of the oceans. Significant achievements include the discovery that the uppermost seismic layer (Layer 2) can be subdivided into Layers 2A, 2B, and 2C, and that these subdivisions change in thickness

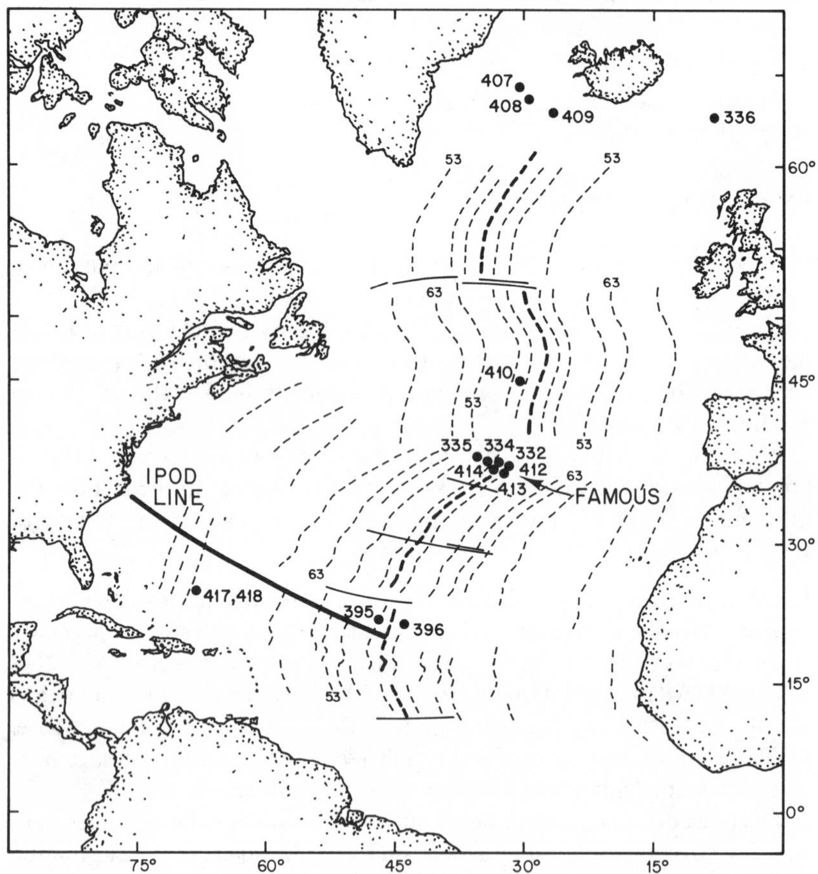

Figure 1 Map of North Atlantic showing deep sea drill sites that have penetrated more than 20 m of basaltic basement. Figures 2–4 are details of the FAMOUS area shown and Figure 5 is a profile along the IPOD line shown. Age of the seafloor from magnetic anomalies is also shown.

in a generally systematic way with age. Also, detailed mapping of magnetic anomalies, especially in the North Atlantic, has permitted us to understand how the initial opening of the ocean took place, how spreading rates have changed with time, and how fracture zones have altered their positions as the ocean opened (Schouten & Klitgord 1977).

While all of the investigations have provided fresh observational data, these data are not always in agreement with the simple seafloor spreading theory. For example, it is not yet clear how many of the observations are related to spreading rate. It is not clear how many of the major rises and non-mid-ocean (aseismic) ridges are to be reconciled with the age depth curve. We need to understand why many of the seamounts have subsided as they have and whether seamount chains are related to isolated upwellings (hot spots) of mantle material. The location of the magnetic anomaly source layer has not been clearly identified by drilling. The details of the oceanic continental crust interface at passive margins—like the margins of the Atlantic—have not yet been explored in any detail. Even in mid-plate regions the question of regional variability is not appreciated because of the thick sediment cover.

It is not possible to discuss here all of the recent studies. Most have been covered in comprehensive papers. Here we discuss those observations that will probably have the most far-reaching theoretical implications, and sets of observations that aid in the identification of new phenomena.

THE INNER RIFT VALLEY

In the North Atlantic, much of the Mid-Atlantic Ridge has a central rift valley a few tens of kilometers wide. Within this valley is centered an inner rift valley, which in the FAMOUS area south of the Azores is a few kilometers wide. Precisely navigated deep-towed instruments and manned submersibles have explored this region (Heirtzler & van Andel 1977). A very detailed profile across this inner valley has been made from four individual submersible dive profiles (Figure 2). This east-west profile passes between two of the many important elongated volcanic hills that stretch north-south along the central volcanic province. It shows other less prominent hills, which, together with the axial hills, are believed to be each created by a single volcanic event at the centerline, then to have moved to either side, a distance dependent upon their age (Ballard & van Andel 1977). The hills are the result of episodic volcanism with a periodicity believed to be of the order of 10,000 years (comparable to that of the last interglacial periods). Visual observations and photographs showed much

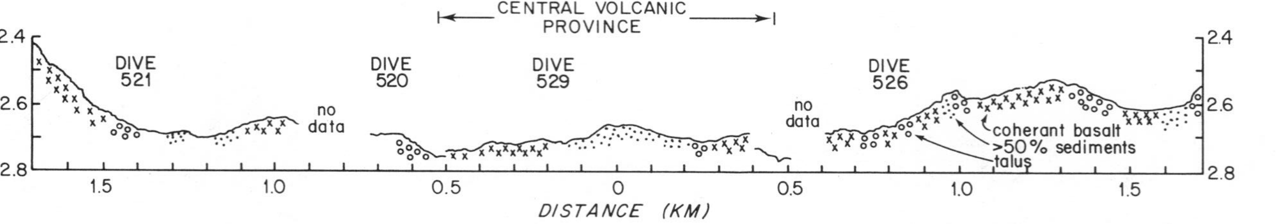

Figure 2 A detailed profile made by submersible across the inner rift valley of the FAMOUS area showing typical bathymetric relief and places with appreciable sediment, talus, and consolidated basalt. The central volcanic zone is where nearly all the new material is emplaced and frequently hills elongated along the valley axis (perpendicular to paper) are found. The west and east inner valley walls are on the left and right sides respectively.

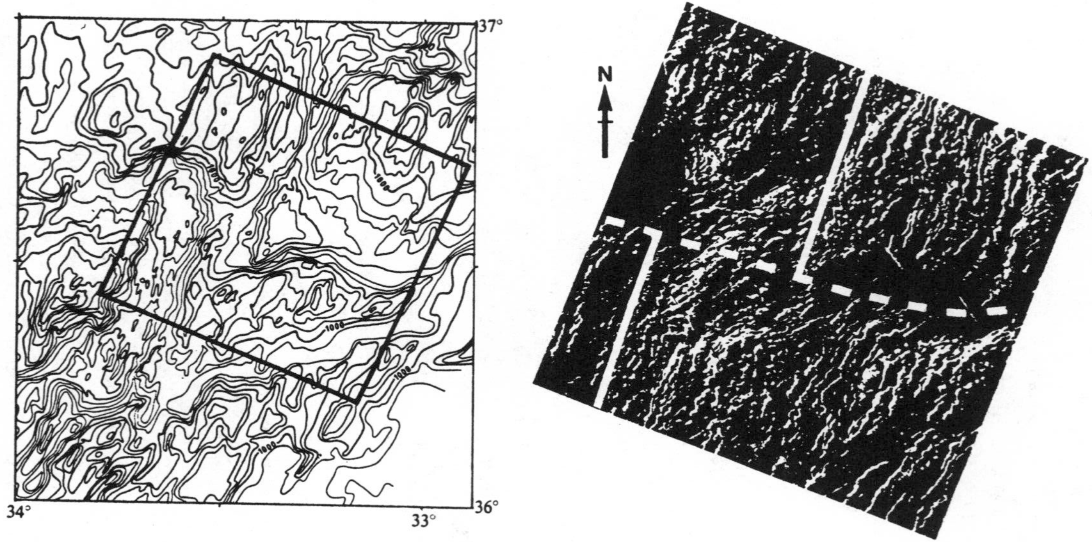

Figure 3 (*Left*) Bathymetric map of FAMOUS. Fracture Zone B area with contour interval 100 fathoms. (*Right*) GLORIA side-scan record of box on left. Median valley and fracture zone are shown by solid and broken lines respectively.

new pillow basalt in this area, with the freshest on the central volcanic hills (Ballard et al 1975, Arcyana 1975), and showed fissures a few meters wide, and up to ten meters deep paralleling the axis of the rift for a few hundred meters (Brundage & Cherkis 1975).

It is now believed that these fissures, even as they age and are buried under sediment, play a significant role in the geochemistry and metallogenesis of the seafloor. They are believed to provide pathways for the circulation of metal-rich hydrothermal seawater from hot sub-bottom regions. They may concentrate metals or otherwise fill the cracks between basaltic fragments (Kerr 1978). Layer 2A, with its lower seismic velocities, is now believed to be unconsolidated basaltic fragments. It is becoming consolidated by hydrothermal fluids with time. This would account for the disappearance of Layer 2A with age. Drilling has confirmed that the youngest oceanic rocks are less consolidated and the older ones more so. Although drilling into hydrothermal areas will be emphasized on legs in late 1979, there has already been some evidence of the drill penetrating a previously closed hydrothermal flow system on Leg 46 (Dmitriev, Heirtzler et al 1976).

Careful observations with ocean bottom seismographs have shown that the inner rift valley floor, at least in the vicinity of the younger axial hills, is underlain by a low compressional wave velocity (3.2 km/s) layer. This low velocity does not imply an intrusive zone across the entire inner rift valley, but is probably caused by the large number of cracks and voids here (Whitmarsh 1973). No elevated temperatures were found on the seafloor here, suggesting that a steady state magma chamber does not exist within a one kilometer depth of the inner valley floor.

Seismic velocities outside of the inner rift valley and near the seafloor average 4.1 to 5.3 km/s, which is more typical of upper basaltic seafloor (Layer 2) everywhere. Layer 2 is estimated to be 2 km thick immediately outside the inner valley and to be underlain by about 5 km of Layer 3, which has a seismic velocity of about 6.2 km/s (Fowler 1976).

Whereas low off-axis volcanic hills are scattered across the inner rift valley in a nearly random manner, the inner valley has distinct east and west walls that run uninterrupted between the north and south bounding fracture zones—a distance of over 30 km. Outside the inner valley, the linearity and regularity of the faults or tectonic blocks are nicely illustrated by large-area sidescan sonar records (Figure 3; Whitmarsh & Laughton 1975, 1976). Whether this linear pattern is true only of slow-spreading ridges like the Atlantic is not yet known. This grain to the Mid-Atlantic Ridge topography has been observed from about 36° to 62° N in the North Atlantic (Laughton & Searle 1979). The distance between the blocks is only a few kilometers. The same two-dimensional wavelength has been recently observed in heatflow profiles in the Galapagos Rift, the South Indian Ocean, and the South Atlantic Ocean areas. In these areas, the basement rocks were not exposed or mapped in detail. Why the heat-flow wavelength should be related to the size of tectonic blocks is not clear and there may, in fact, be no causal relationship between the two phenomena. No other observations have this two-dimensional wave-length, however, and such observations hint at a scale for regional variability for the upper part of Layer 2.

THE RIFT VALLEY

The rift mountains reach their maximum elevation some 12 km to the west and 19 km to the east of the valley axis in the FAMOUS area (Figure 4; Luyendyk & MacDonald 1976). The asymmetry in the topography reflects the generally faster spreading rates to the east in this region. On the outer limits of the region, the first sediment ponds up to 100 m thick appear. Drill site 332 was located in one of the ponds to the west, as shown in Figure 4.

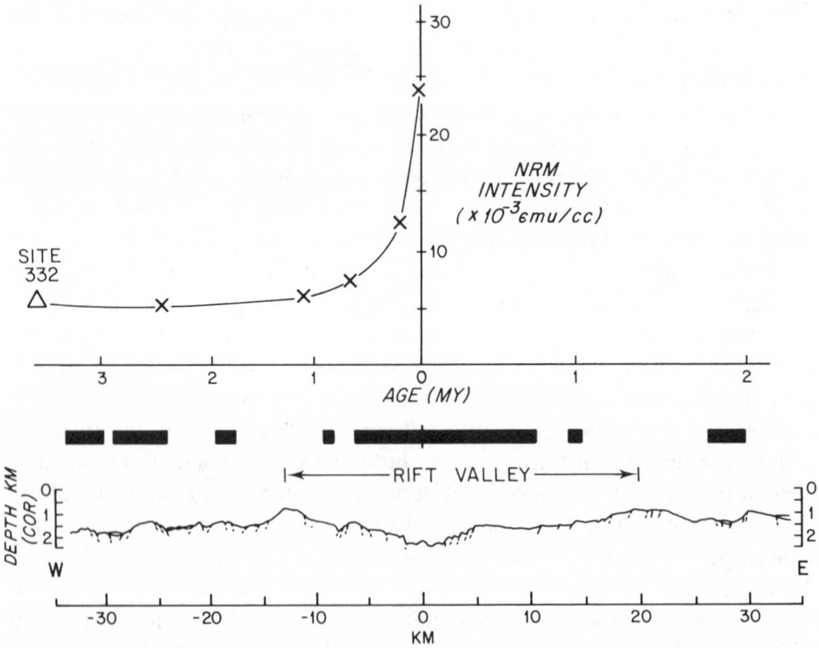

Figure 4 A detailed bathymetric profile extending to about 30 km on either side of the axis in the FAMOUS area (Luyendyk & MacDonald 1976). The age of the crust is from magnetic anomalies. Bars show areas of crust that are presumably normally magnetized. The intensity of magnetization (Johnson & Atwater 1977) drops sharply away from the axis.

The normally magnetized zone includes all of the inner rift valley, but not the entire rift valley. Oriented specimens obtained by the submersible in the inner rift valley were normally magnetized, except for one which appeared to be reversely magnetized. A return to the area by the submersible ALVIN in the summer of 1978 permitted an investigation of this apparently anomalous region. Rock specimens obtained by submersible in the inner valley, by dredging in all regions, and by drilling at site 332 show a sharp reduction in intensity of magnetization away from the axis within the rift valley (Johnson & Atwater 1977).

We think that the uplift which creates the rift mountains is caused by a cooling of the lithosphere as rising deep earth materials get tacked on the accreting edge of the two tectonic plates and move apart. In any case, the uplift begins abruptly at the inner rift valley walls and continues, though less dramatically, across the rift mountains.

It is noteworthy that the uplift at the inner valley walls and beyond does not destroy the magnetic anomaly pattern observed. The fracturing and tilting that accompanies uplift would be expected to cause the individual rock pieces or blocks to lose their coherent orientation with respect to the direction in which they were originally magnetized. The fact that the magnetic anomaly is not destroyed by such uplifts is enigmatic in terms of the seafloor spreading theory. We would not expect rock units to be uplifted without change in orientation. It is also difficult to conceive of the magnetic layer being deeper than the uppermost basaltic layer, or being so deep that it is not fractured and disoriented by uplift. It is likewise difficult to imagine the basaltic rocks acquiring appreciable magnetization outside of the inner rift valley, and therefore after uplift, because volcanic injection is much less likely.

Geologic reconnaissance in the inner rift valley shows ubiquitous pillow basalts that tumbled or were rotated after cooling and acquiring their magnetization. Also, drilling results frequently show magnetic "units" of a few meters dimension, with consistent anomalous directions of magnetization. These magnetic units, although strongly magnetized, could not produce the magnetic anomaly observed at the surface because they are so small and disoriented (Scientific Party 1975). Thus, what rocks cause the anomaly and how they become magnetized is a mystery at this time. It may be that the simple picture of rocks becoming magnetized as they rise and cool at the ridge axis is realistic only in a statistical sense.

Some off-axis volcanism is known to occur outside the inner rift valley. In the sediment pond where drill site 332 is located, small cones of basalt pillows were seen during a submersible dive there (Heirtzler & Ballard 1977). Had these been created in the inner rift valley, they would probably have been destroyed during uplift at the inner valley walls. No other manned submersible dives have been made outside the rift valley, but one would assume that many cases of off-axis volcanism would be observed if dives were made. The presence of such extrusives could modify the two-dimensional pattern of magnetized rocks if the direction of the earth's magnetic field had changed since the native rock was magnetized.

THE OLDER SEAFLOOR

At this time, no area outside the rift valley has been examined in as much detail as areas on the axis because of the blanket of sediment over the typical older basement (Figure 5), and because depths are too great for manned submersible observation. For a number of reasons, scientific interest has focused on the edges of the tectonic plates rather than on mid-plate regions.

Since seafloor spreading requires an injection of hot material at the mid-ocean ridge axis and a convective system to move the tectonic plates apart, the distribution of temperature with distance from the axis and with depth is of fundamental importance and has been studied theoretically by a number of investigators. The various thermal models, whether based on slabs or half-spaces, with narrow or more diffuse axial injection zones, all give favorable age-depth curves with observation within 750 km of the axis in the North Atlantic. The differences in these models are made most clear by a study of the variation of the depth of the seafloor as a function of age (t). A plot of observed depth vs the square root of t yields

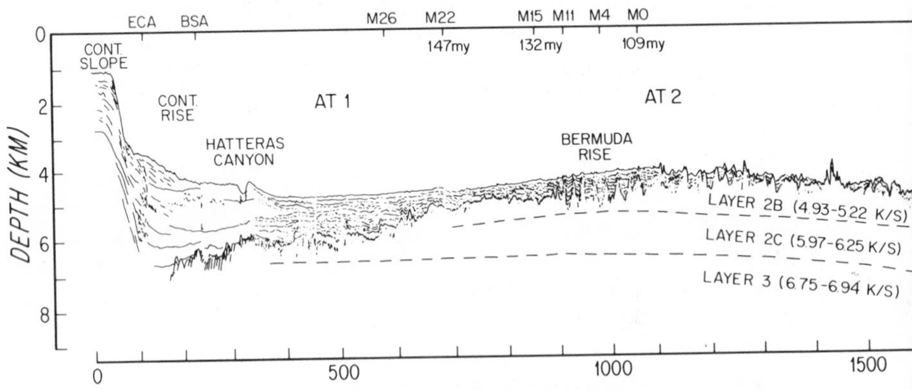

Figure 5 A bathymetric and sub-bottom profile along the "IPOD Line" extending from near Cape Hatteras to the Mid-Atlantic Rift valley at about 23°N. Locations indicated (AT1, AT2, etc) are proposed drilling sites. Site 395 on Leg 45 was drilled near AT5. Magnetic anomaly numbers and ages are indicated along the top. ECA is East Coast Anomaly and

a straight line to an age of about 60 m.y. Departure from a straight line for older crust suggests that a slab model may be more appropriate than a half-space model (Parsons & Sclater 1977).

The measurement of heat flow in the rift valley is difficult since there is little sediment and the heatflow probe requires sediment for insertion. As was mentioned previously, there are few sediment ponds within 25 km of the axis and the few ponds that exist nearer are in special places, such as rift valley fracture intersections, which would not yield typical values of heat flow. It is usually found, however, that the average heat flow from the seafloor is higher near the axis than further away. An average value of about 3 heatflow units (microcal/cm^2/s) is found for the axial area of the Atlantic (LePichon & Langseth 1969). However, isolated values as high as 15 heatflow units have been found for the intersection of Fracture Zone B and the axial valley of the FAMOUS area (Williams et al 1977). The average value is asymptotically reduced to about 1 heatflow unit halfway between the axis and the Atlantic margins. A plot of heat flow vs distance or age is far from a smooth curve, with 90% confidence limits of one-half to one-third the average value.

The depth vs age relationship has been studied in detail in the North Atlantic out to about 80 m.y. (Sclater et al 1975). There are deviations from the depth vs age curve, and correlations between these deviations (or residuals, as they are called) and gravity anomalies have been attempted. There is an average gravity anomaly maximum of about 25 mgal over the ridge axis, falling off to about zero at approximately 40 m.y., but there is

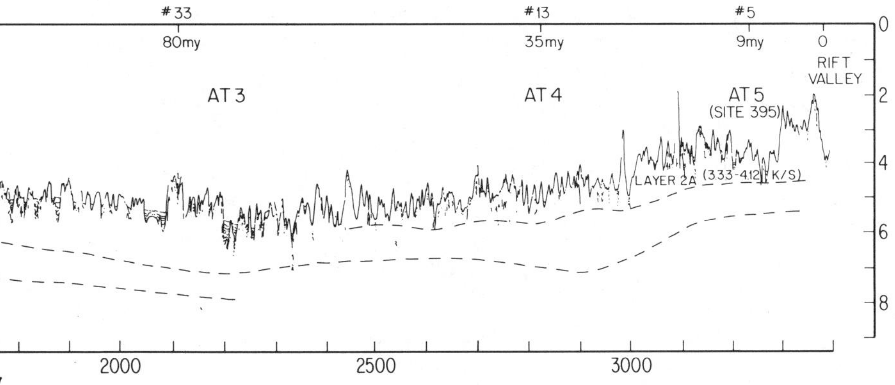

BSA is Blake Spur Anomaly. The ocean-continent boundary is thought to lie between these two anomalies. Generalized Layer 2A, 2B, and 3 velocities and thicknesses for the North Atlantic are shown. Solid lines at AT3 are divisions between layers from site survey data there.

still some uncertainty about correlation of this variation with residual depths (Cochran & Talwani 1977). Such a correlation could be evidence of flow in the upper mantle.

The seismic velocity through each layer of the older crust and upper mantle has been reviewed by Lewis (1978) and given in a very abbreviated form by Raitt (1963), shown in Table 1.

A larger standard deviation for Layer 2 velocity is believed to be caused by real variation from place to place rather than by instrumental error. The rugged topography where Layer 2A is found may account for the error in thickness. In recent years, relatively inexpensive sonobuoys have allowed the regional variability of Layer 2 to be explored. Layer 2 has been divided into three layers: 2A, 2B, and 2C with the largest regional variability in Layer 2A (Houtz & Ewing 1976). Starting with an attempt to drill in young crust on DSDP Leg 34 in the Pacific, and subsequently on other legs, it has become apparent that the upper part of young crust is fractured and broken. Houtz and Ewing hypothesized that this crust becomes more cemented and consolidated with age. Support for this idea was found later when, in drilling 110 million-year-old crust on Legs 51, 52, and 53, less fracturing and more infilling of cracks was revealed. As Layer 2A becomes cemented, its velocity increases, causing it to become unidentifiable. The expected decrease in thickness of layer 2A with age has been observed, but layers 2B and 2C do not seem to thin or thicken with age.

Several analyses of teleseismic events have yielded information on the seafloor structure in the oceans in general and in the North Atlantic in particular. Usually such signals have passed through many different regions of the seafloor before being recorded, and so lack a great deal of geographic resolution. Information has been gained from the study of high-frequency shear wave (S_n) velocity, of attenuation of shear waves, of surface wave data, and of anisotropy of seismic wave propagation.

Hart & Press (1973) have found an average velocity for S_n in the Atlantic of 4.58 ± 0.02 km/s when the seafloor is less than 50 m.y. old (about 750 km), and 4.71 ± 0.01 km/s for older seafloor. Why this change occurs is not clear, but it does show that regional changes in the lithosphere can occur.

Table 1 Seismic layers of oceanic basement

Layer	Velocity (km/s)	Thickness (km)
Layer 2	5.07 ± 0.63	1.71 ± 0.75
Layer 3	6.69 ± 0.26	4.86 ± 1.42
Mantle	8.13 ± 0.24	—

Regional changes in the lithosphere have, in fact, been found in the Pacific as well as in the Atlantic.

One of the very interesting characteristics of S_n is that it does not propagate across mid-ocean ridges (Molnar & Oliver 1969) except possibly at very shallow depths. Kausel (1972) showed that these waves were quenched in the low-Q zone that begins at about 20 km depth and goes to 60 or 70 km depth. Solomon (1973) showed that near the Gibbs Fracture Zone in the North Atlantic, the highly attenuating zone ($Q \leqq 10$) is between 50 and 100 km wide and 50 to 150 km deep. Furthermore, Solomon & Julian (1974) used ray tracing techniques to show that the top of this zone is at a depth proportional to an age of about 5 m.y. (75 km).

Studies of Rayleigh wave dispersion have shown that velocities in the 20–200 s period range increase with age of the Atlantic floor (Weidner 1974, Forsyth 1975). These results indicate a lithosphere that cools and thickens with age from 25 km or less at 5 m.y. (75 km) to 80–100 km thick at about 150 m.y. (2250 km). A seismic low-velocity zone (LVZ) is hypothesized to lie below the lithosphere. The spatial relationships of these zones has been given in a model by Forsyth (1977; Figure 6).

Although there is still insufficient data to make definitive statements, especially for the Atlantic, seismologists agree that there is some directional dependence of Rayleigh wave phase and group velocities, and possibly of Love waves as well. The cause of this dependence may be liquid-filled

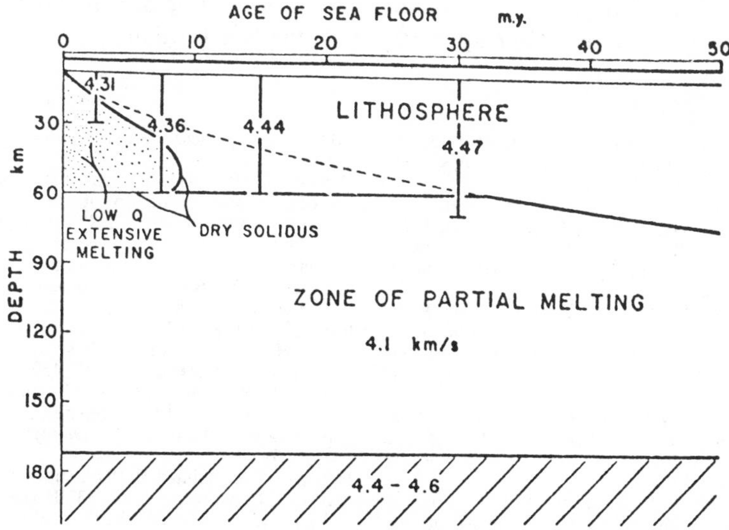

Figure 6 Crustal cross section through Low-Velocity Zone consistent with shear wave velocities from teleseismic events (Forsyth 1977).

cracks in LVZ (Mitzutani & Abe 1972, Schlue 1975). Others have suggested that olivine crystals, either during implacement or later, were preferentially aligned.

CONCLUSIONS

Some of the main conclusions that can be drawn about the Mid-Atlantic Ridge in the North Atlantic at this time are

1. New material is emplaced along a linear zone a few hundred meters wide near the center of the inner rift valley. This material is emplaced episodically about every 10,000 years. How this material is uplifted at the rift valley walls without destroying the magnetic source layer is still not explained.
2. Outside the inner rift valley walls the basaltic basement is faulted into linear blocks a few kilometers wide paralleling the rift valley axis. It is not clear if these blocks are related to the width between fissures in the inner rift valley, but they do seem to have the same wavelength as heatflow variations across the sedimentary blanket away from the axis, suggesting that the heatflow may be related to blocks.
3. Layer 2A is apparently a layer of basaltic rubble which gets cemented increasingly as it ages. As Layer 2A thins and disappears, teleseismic data indicates that the lithosphere thickens. With age, the seafloor cools and contracts so that depth age curves can be drawn with confidence for crust of age up to 80 m.y. in the North Atlantic.

Literature Cited

Arcyana, 1975. Transform fault and rift valley from bathyscaph and diving saucer. *Science* 190: 108–16

Ballard, R. D., Bryan, W. B., Heirtzler, J. R., Keller, G., Moore, J. G., van Andel, T. H. 1975. Manned submersible observations in the FAMOUS area: Mid-Atlantic Ridge. *Science* 190: 103–8

Ballard, R. D., van Andel, T. H. 1977. Morphology and tectonics of the inner rift valley at latitude 36°50′N on the Mid-Atlantic Ridge. *Geol. Soc. Am. Bull.* 88: 507–30

Brundage, W. L. Jr., Cherkis, N. Z. 1975. Preliminary LIBEC/FAMOUS cruise results. *NRL Rep. 7785*. Naval Res. Lab., Wash. DC

Cochran, J. R., Talwani, M. 1977. Free-air gravity anomalies in the world's oceans and their relationship to residual elevation. *Geophys. J. R. Astron. Soc.* 50: 495–552

Dmitriev, L. V., Heirtzler, J. R. et al. 1976. Drilling into ocean crust. *Geotimes* 21: 21–23

Forsyth, D. W. 1975. The early structural evolution and anisotropy of the oceanic upper mantle. *Geophys. J. R. Astron. Soc.* 43: 289–325

Forsyth, D. W. 1977. The evolution of the upper mantle beneath mid-ocean ridges. *Tectonophysics* 38: 89–118

Fowler, C. M. R. 1976. Crustal structure of the Mid-Atlantic Ridge Crest at 37°N. *Geophys. J. R. Astron. Soc.* 47: 459–91

Hart, R. S., Press, F. 1973. Sm velocities and the composition of the lithosphere in the regionalized Atlantic. *J. Geophys. Res.* 78: 407–11

Heirtzler, J. R., Ballard, R. 1977. Submersible observations at the Hole 332B Area. In Aumento, F., Melson, W. G. et al. *Initial Reports of the Deep Sea Drilling Project, Vol. 37*, pp. 363–66. Washington, DC: GPO

Heirtzler, J. R., van Andel, T. H. 1977. Project FAMOUS: Its origin, programs and setting. *Geol. Soc. Am. Bull.* 88:481–87

Houtz, R., Ewing, J. 1976. Upper crustal structure as a function of plate age. *J. Geophys. Res.* 81:2490–98

Johnson, P., Atwater, T. 1977. Magnetic study of basalts from the Mid-Atlantic Ridge, Lat. 37°N. *Geol. Soc. Am. Bull.* 88:637–47

Kausel, E. G. 1972. *Regionalization of the lithosphere and asthenmosphere of the Pacific Ocean.* PhD thesis. Columbia University, New York

Kerr, R. A. 1978. Seawater and the ocean crust: The hot and cold of it. *Science* 200:1138–41, 1187

Laughton, A. S., Searle, R. C. 1979. Tectonic processes on slow-spreading ridges (Abstr.). *2nd Maurice Ewing Mem. Symp.: Implications of Deep Drilling Results in the Atlantic Ocean.* In press

LePichon, X., Langseth, M. G. Jr. 1969. Heatflow from the mid-ocean ridges and sea-floor spreading. *Tectonophysics* 8: 319–44

Lewis, B. T. 1978. Evolution of ocean crust seismic velocities. *Ann. Rev. Earth Planet. Sci.* 6:377–404

Luyendyk, B. P., MacDonald, K. C. 1976. Spreading center terms and concepts. *Geology* 4:369–70

Mizutani, H., Abe, K. 1972. An Earth model consistent with free oscillation and surface wave data. *Phys. Earth Planet. Interiors* 5:345–56

Molnar, P., Oliver, J. 1969. Lateral variations of attenuation in the upper mantle and discontinuities in the lithosphere. *J. Geophys. Res.* 74:2648–82

Parsons, B., Sclater, J. G. 1977. An analysis of the variation of ocean floor bathymetry and heat flow with age. *J. Geophys. Res.* 82:803–27

Raitt, R. W. 1963. The crustal rocks. In *The Sea,* Vol. 3, pp. 85–101. New York: Interscience

Schlue, J. W. 1975. Anisotropy of the upper mantle in the Pacific Basin. PhD thesis. Univ. Calif., Los Angeles. 129 pp.

Schouten, H., Klitgord, K. 1977. Mesozoic magnetic anomalies, Western North Atlantic. Misc. Field Studies Map MF-915. US, Arlington, Va.

Scientific Party. 1975. Deep sea drilling project Leg 37, sources of magnetic anomalies on the Mid-Atlantic Ridge. *Nature* 255:389–90

Sclater, J. G., Lawrer, L. A., Parsons, B. 1975. Comparison of long-wavelength residual elevation and free air gravity anomalies in the North Atlantic and possible implications for the thickness of the lithospheric plate. *J. Geophys. Res.* 80:1031–52

Solomon, S. C. 1973. Shear-wave attenuation and melting beneath the Mid-Atlantic Ridge. *J. Geophys. Res.* 78:6044–59

Solomon, S. C., Julian, B. R. 1974. Seismic constraints on ocean ridge mantle structure: Anomalous fault-plane solutions from first motions. *Geophys. J. R. Astron. Soc.* 38:265–85

Vine, F. J., Matthews, D. H. 1963. Magnetic anomalies over oceanic ridges. *Nature* 199:947–49

Weidner, D. J. 1974. Rayleigh wave phase velocities in the Atlantic Ocean. *Geophys. J. R. Astron. Soc.* 36:105–39

Whitmarsh, R. B. 1973. Median valley refraction line, mid-Atlantic ridge at 37°N. *Nature* 246:297–99

Whitmarsh, R. B., Laughton, A. S. 1975. The fault pattern of a slow-spreading ridge near a fracture zone. *Nature* 258: 509–10

Whitmarsh, R. B., Laughton, A. S. 1976. A long-range sonar study of the Mid-Atlantic Ridge crest near 37°N (FAMOUS area) and its tectonic implications. *Deep-Sea Res.* 23:1005–23

Williams, D. L., Lee, T., von Herzen, R. P., Green, K. E., Hobart, M. A. 1977. A geothermal study of the Mid-Atlantic Ridge near 37°N. *Geol. Soc. Am. Bull.* 88:531–40

Ann. Rev. Earth Planet. Sci. 1979. 7:357–84

THE DIAMOND CELL AND THE NATURE OF THE EARTH'S MANTLE

×10117

William A. Bassett
Department of Geological Sciences, Cornell University, Ithaca, New York 14853

INTRODUCTION

Our knowledge of the Earth's interior has grown rapidly over the past several decades by careful interpretation of geophysical observations. Our very limited means of access to the Earth's interior has made the quest for knowledge about its constitution and the nature of the processes that take place there one of the most fascinating and challenging areas of geologic investigation. In spite of the limited means of access, methods of geophysical observation have improved dramatically with new technological developments in recent years, especially in the area of data reduction by computers.

Methods of geophysical observation include study of the behavior of elastic waves traversing the Earth (seismology), the dimensions of the Earth (geodesy), the motion of the Earth (celestial mechanics), and measurement of heat flow, gravity, and magnetism, as well as direct studies of materials thought to have been thrust up from the Earth's interior in the form of volcanic nodules, ophiolites, and mid-ocean ridges.

The interpretation of these observations, however, depends to a large extent on another branch of geophysics which has also undergone some remarkable advances in recent years, namely experimental geophysics. Experimental geophysics encompasses that area of endeavor devoted to the investigation of the properties of the materials believed to exist within the Earth's interior at the conditions believed to prevail within the Earth's interior. Without this information, it would have been impossible to piece together as complete a picture as we have today.

The desire to generate within the laboratory conditions which accurately simulate conditions within the Earth has been the driving force for some

0084-6597/79/0515-0357$01.00

of the most diverse and innovative research in the field of high pressure–high temperature physics. There have been two very different approaches to achieving high pressures and temperatures in the laboratory, static and dynamic. In the static approach, the sample is contained within a vessel or squeezed between anvils constructed of very strong materials where it can be heated and examined by various analytical techniques either in situ or by postmortem examination. In the dynamic approach, a sample is hit by a flying projectile or by the blast of an explosive charge. The sample is at high pressures and temperatures for only microseconds and so in situ measurements must be made very rapidly if anything is to be learned or the sample must be recovered after subjection to such a drastic event. It is worth noting that the pressures within the Earth do not derive from either of these two mechanisms but rather from self-compression of a large mass by gravitational forces, a mechanism which, due to scaling considerations, is simply unavailable in the laboratory for achieving such high pressures.

While the desire to generate high pressures and temperatures for the study of geophysically significant materials has led to important advances in all aspects of high pressure–high temperature technology, it is the purpose of this paper to examine the contributions of one particular type of device, the diamond anvil cell. This device has proved to be an unusually versatile means of studying the properties of materials to very high static pressures and temperatures. It takes advantage of some important principles of high pressure generation. The first and most important is that since pressure is defined as force divided by area, it is possible to push to high pressures either by increasing force or decreasing area. If one wishes to achieve high pressures by decreasing area, then one must be willing to study small samples. In the case of the diamond anvil cell, it has been remarkable that such a large variety of analytical techniques have been found to be applicable to the small samples required by the device.

The small size of the sample, and hence of the anvils, greatly diminishes the requirements for large forces and large supporting frames. In fact it is possible to achieve pressures over a megabar in an instrument not much larger than a pocket camera. The small anvil size (100 mg or less) has the additional advantage that it permits the choice of a rather exotic material with just the right properties for the anvils. It is also possible to exclude all flaws. Finally, the cost of such a device is quite modest when compared with other high pressure devices.

The choice of diamond is a particularly happy one since it is not only the hardest material known but is transparent to light and to X rays. The technology for the mining, cutting, and polishing of diamonds has developed over centuries to a fine art, making it possible to acquire specimens of high quality and reasonable price. This unique combination of properties has made it possible to subject specimens to 1.7 Mbar and 3000°C, conditions well within the Earth's core. Significant geophysical measurements, however, have been made up to only 1 Mbar. The pressure range above 1 Mbar has yet to be explored. The accomplishments and the potential of the diamond cell in the study of geophysics are the topics of this article.

History of the Diamond Cell

P. W. Bridgman did much of his pioneering high pressure research by squeezing samples between flat parallel faces on tungsten carbide anvils which came to be known as Bridgman anvils. In 1959 Weir and his colleagues (Weir et al 1959) recognized that the principle of the Bridgman anvils could be applied to much smaller samples and that spectroscopic studies could be made on these samples while under pressure simply by using diamond instead of tungsten carbide for the anvils. Their first design was very simple and contained the basic elements found in present day designs. It consisted of two brilliant cut gem quality single crystal diamonds mounted on the ends of pistons driven together by a lever and spring device. Faces produced by cutting and polishing the culet points provided the anvil faces. Shortly after this Piermarini & Weir (1962) developed a version of the diamond cell for making X-ray diffraction patterns of the sample under pressure, while Van Valkenburg (1963) found that the sample under pressure could be observed visually simply by looking at it through one of the diamonds using a microscope. These methods for studying the sample under pressure have continued to be basic to all others that have followed. Other discussions of the historical development of the diamond cell can be found in Bassett & Takahashi (1974) and Block & Piermarini (1976).

EXPERIMENTAL TECHNIQUES

Diamond Cell Designs

The basic principle of the diamond cell, as has been pointed out earlier, is a very simple one. A sample placed between flat parallel faces of two diamond anvils is subjected to pressure as a force pushes the diamond anvils together. However, the variations of diamond cell designs are surprisingly diverse. There are many ways to apply a force (Figure 1): first class lever (Weir et al 1959), second class lever (Bassett & Takahashi 1965), screw pushing a piston (Bassett et al 1967), screws pulling two platens (Merrill & Bassett 1974), screw and piston guided by pins (W. K. Hensley, unpublished design), or hydraulic (Webb et al 1976). The alignment of the diamonds is critical. The diamonds must be mounted in such a way as to permit both angular and translational adjustment so that the opposing faces meet properly and are parallel. This has been accomplished in various ways: by mounting the diamond anvils on plates that can be tilted by the screws that support them, by mounting the diamond anvils on hemicylinders in troughs so that each anvil can be rocked or translated individually, or by mounting one anvil on a hemisphere for angular adjustment and the other on a sliding disc for translational adjustment.

The greatest variety of diamond cell designs, however, results from the various requirements for access to the sample while it is under pressure (Figure 2). Visual observations and most X-ray observations are made by passing radiation through the diamond anvils and so the openings on the supporting members of the anvils are most important (Van Valkenburg 1963, Piermarini & Weir 1962). X rays are able to traverse beryllium metal, which is strong enough to support the diamonds, thus permitting access through supporting members as shown in Figure 2*c* (Weir et al 1965, Merrill & Bassett 1974). Some X-ray determinations are made by directing an X-ray beam across the load axis. In this geometry, the openings on the side of the cell become important (Kinsland & Bassett 1976). In some Brillouin and Raman scattering experiments, the light scattering at 90° from the incident beam must be collected. The instruments designed for this purpose have holes at 45° with respect to the load axis of the cell

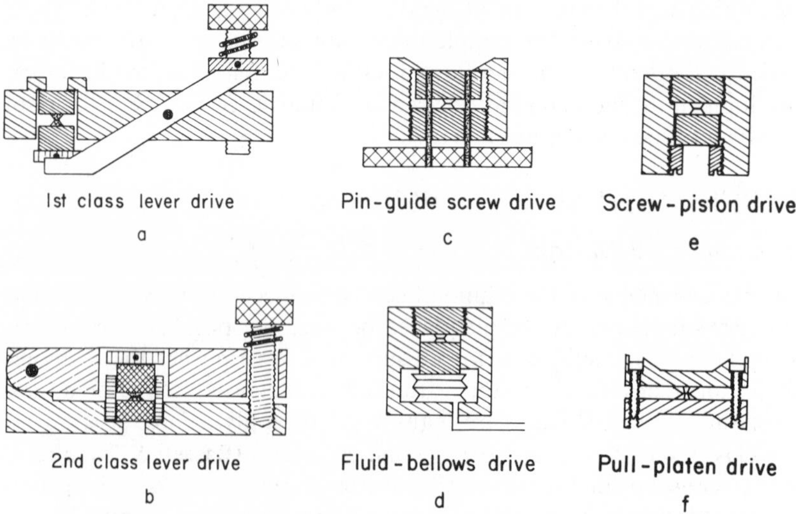

Figure 1 Six basic ways of providing force in the diamond cell.

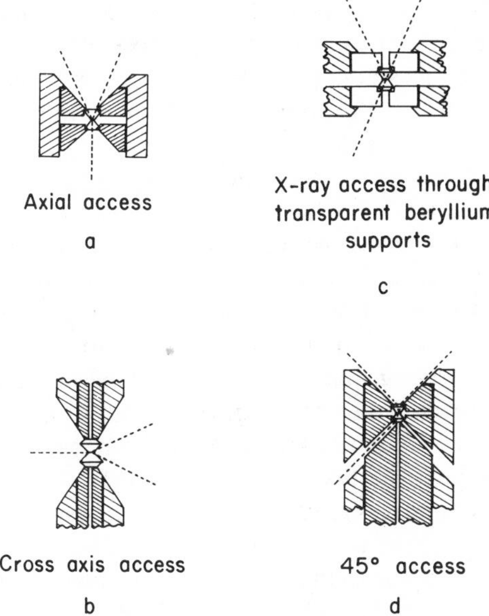

Figure 2 Various ways of providing access to the sample while under pressure in the diamond cell.

(Whitfield et al 1976, Bassett et al 1979). Finally, various combinations of these geometries are used when more than one technique is planned for a sample under pressure.

In addition to the variations described above, there are two fundamental ways of containing the sample between the diamond anvils. The first is simply to let the sample serve as its own gasket. That is, place the sample between the anvil faces and let it extrude with pressure until friction prevents further extrusion. If the sample is a hard brittle material, it is usually mixed with a pressure transmitting medium such as NaCl. A pressure gradient always forms with the highest pressure at the center of the anvil area. As pressure increases, the anvil faces become more and more cup shaped (Figure 3*a*) and the pressure distribution becomes quite flat across the center of the sample area and drops abruptly at the edges (Figure 3*b*). This occurs because the gap between the anvil faces filled with sample closes more rapidly at the edges of the sample area than in the center (Bassett & Takahashi 1974).

The second method for containing samples between the anvil faces is by the use of a metal gasket. The gasket is produced by drilling a small hole ($\sim 300\ \mu$m) in a thin metal foil, placing the sample in the hole, and trapping it there between the anvil faces. Various metal alloys have been used for this purpose. One of the most successful has been tempered 301 stainless steel. With this technique it is possible to trap liquids in the gasket, making hydrostatic pressure entirely feasible. A mixture of methanol and ethanol in a ratio of 4:1 has proved to remain liquid up to 100 kbar (Piermarini et al 1973).

The use of a gasket also serves another very important purpose. The gasket metal extrudes from between the anvils and accumulates just outside the sample area in the form of a belt around the plane of the sample (Figure 3*c*). It thus provides a vital function by acting as a supporting ring that prevents failure of the anvils due to concentration of stress at the edges of the anvil faces In order to further minimize the concentration of stress gradient at the anvil edges (Figure 3*d*), Mao & Bell (1978) have beveled the anvil edges as shown in Figure 3*c*.

It has been these principles that have made it possible to extend the pressure capability of the diamond cell to 1.7 Mbar. So successful has the minimizing of the stress gradients been that catastrophic brittle failure no longer determines the upper limit of the technique but instead slow plastic deformation of the diamond anvils takes place (Figure 4; Mao & Bell 1978).

Temperature Generation

There are two approaches to producing high temperatures in the diamond anvil cell. One is to heat the sample externally by placing a resistance heater around the cell itself, or around the members that support the diamonds, or immediately around the diamond anvils. The other is to heat the sample internally without raising the temperature of the diamond anvils to the temperature of the sample. This is accomplished either by internal resistance heating or by passing electromagnetic radiation through one of the diamonds to the sample.

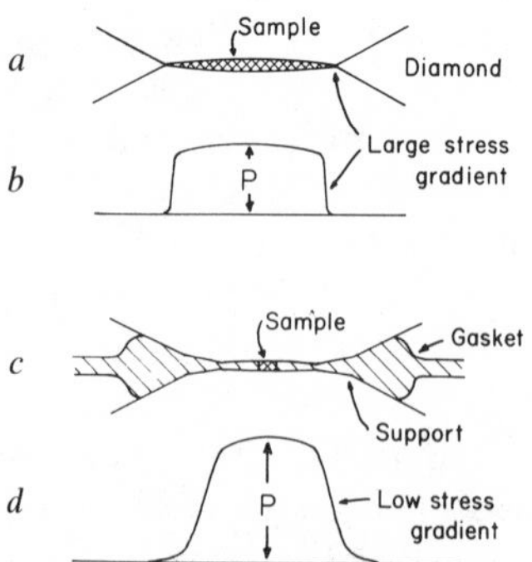

Figure 3 Diamond anvil face designs and the resulting pressure distributions.

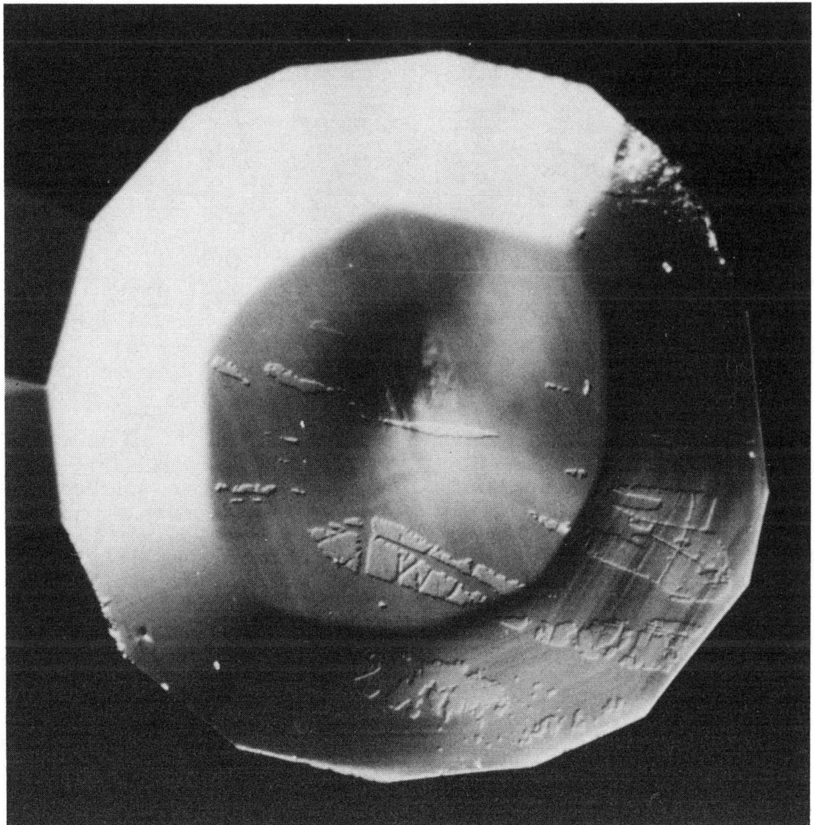

Figure 4 A diamond anvil face showing a depression resulting from plastic deformation of the diamond when subjected to a pressure of 1.7 Mbar (Mao & Bell 1978).

All forms of external heating produce within the sample very uniform temperatures which can easily be measured by placing a very fine thermocouple in the sample. No correction has to be made for the pressure effect on the thermocouple since the points at which the thermocouple wires enter the high pressure zone are at the same temperature as the junction. The closer the heater can be placed to the sample anvils, the fewer the cell parts that need to be raised to high temperatures. It has been possible to achieve close to 1000°C with each of the approaches to external heating (Bassett & Ming 1972, Sung 1976; D. Wilburn and W. Bassett, unpublished observations). In all of the diamond cells using external heating, the force was produced by a screw and spring located well outside of the heated zone so that the force on the anvils was unaffected by the temperature. Heaters close to the anvils are more difficult to design and construct but result in far less damage to the diamond cell parts.

Internal heating by electrical resistance can be used only on samples that are electrical conductors. Liu & Bassett (1975) studied the melting of iron up to 2000°C and 200 kbar by placing a very fine wire of iron 0.005 mm in diameter across a gap 0.05–0.1 mm wide in a relatively inert matrix of alumina, periclase, or forsterite between the diamond anvils. The diamonds remained close to room temperature. The temperature of the sample was measured by optical pyrometry.

Internal heating of samples has been accomplished by pulsed ruby laser and by continuous YAG (yttrium aluminum garnet) laser (Ming & Bassett 1974). With the pulsed laser it is possible to achieve temperatures as high as 3000°C for approximately a millisecond and with the continuous laser it is possible to reach 2000°C continuously. In each case a beam of light from the laser is focused onto the sample through one of the diamond anvils. The sample must be partially absorbing so that the beam penetrates and is converted to heat within the sample. The diamonds are such good thermal conductors that it is necessary to introduce on the order of 6 W into a volume of sample only 10 μm across and 5 μm thick. The diamonds rapidly conduct the heat to the metal parts of the press, and the temperatures of the anvils and press rise to only 10° or 20°C above room temperature. The portion of the sample directly in the laser beam becomes incandescent. It is possible to observe the sample during heating with a microscope and to measure its temperature by optical pyrometry, i.e. by measuring the brightness of the incandenscent light being emitted from it.

The internal methods of heating are suitable for studies of high pressure–high temperature phase relations because of the high temperature that can be achieved. Because accurate temperature measurements, however, are difficult due to the uncertainty of the emissivity of samples, the technique is used mostly as a means of making reconnaisance studies at high pressures and temperatures. Accurate mapping of stability fields will probably have to await the extension of the external heating methods to higher temperatures.

Pressure Measurement

When a new means of producing pressure has been developed, the most important task is to find a way to measure that pressure as accurately as possible.

FORCE DIVIDED BY AREA The applied pressure, defined as the force divided by the total sample area, is easily measured with considerable accuracy in the diamond anvil cell since both the force and the sample area can be measured with a high degree of accuracy. This method, however, usually fails as a means of measuring the pressure of the portion of the sample being studied (generally the central pressure) because the distribution of pressure over the whole sample area is too poorly known. The highest ratio of central pressure to applied pressure probably occurs when the pressure is a linear function of radius; this yields a calculated ratio (central pressure to applied pressure) of 3.4. Measurements of pressure distributions by Piermarini et al (1973) and Sung et al (1977) show remarkably little departure from such a linear relationship. Bassett & Takahashi (1974) give a diagram (their Figure 4) showing the central pressure versus the applied pressure in a diamond cell. From their plot it is possible to determine that the ratio of central pressure to applied pressure starts at just over 3 at low pressures and declines to 1 at about 300 kbar. They attribute this loss of pressure efficiency to the cupping of the diamond faces described in the previous section (Figure 3*a*).

In spite of the problems mentioned above the force/area method is so direct and so reliable that it should probably be utilized to a greater extent for resolving controversies concerning the other methods of pressure measurement described next.

FIXED POINT SCALE There are a number of substances which undergo phase transitions at known pressures. In many cases these transitions are visible under the microscope. These transitions can be used as a means to calibrate the load, usually in the form of a spring length, in a diamond cell. This is not very accurate, as the relationship between the load and the pressure depends to some extent on the nature of the sample. It is fast, however, and can be used to establish the approximate pressure at the start of a long run in which the pressure is to be determined more accurately by some other method such as the X-ray diffraction of NaCl.

The substance most commonly used for this purpose is AgI, which undergoes very conspicuous transitions at 4, 5, and 100 kbar (Bassett & Takahashi 1965). Liu et al (1973) studied samples in the solid solution series KCl–NaCl. These undergo a visible B1–B2 phase transition anywhere between 20 kbar for pure KCl and 300 kbar for pure NaCl. By adjusting the composition one can produce a substance of known transition pressure anywhere between 20 and 300 kbar.

EQUATION OF STATE The molar volume of NaCl, which can easily be calculated from X-ray diffraction patterns, decreases in a predictable way with increasing pressure. If one knows the compression curve or equation of state of NaCl, then the molar volume becomes a means of measuring pressure. A number of these equations of state are discussed and compared by Decker et al (1972). The one most commonly used for measuring pressure is given by Decker (1971), which he based on the Mie-Gruneisen equation. For many years the University of Rochester group used the relationship given by Weaver et al (1971) based on the Hildebrand equation. It is apparent from a comparison of these two relationships (Weaver 1971, Decker et al 1972) that the discrepancy between them is less than the experimental error in measuring molar volume by X-ray diffraction. The good agreement (Weaver 1971) of both of these pressure scales with a pressure scale based on shock compression measurements given by Fritz et al (1971) further supports their accuracy.

The NaCl scale is useful only to 300 kbar where NaCl undergoes a phase transition from the B1 structure to the B2 structure. Other substances (e.g. NaF, Ag, Mo, Cu, Pd, Fe, and MgO) have been used as pressure scales in the same way as the NaCl. In some cases their equations of state have been determined independently; in others they have been calibrated against the NaCl scale.

RUBY PRESSURE SCALE The ruby pressure scale (Barnett et al 1973, Piermarini et al 1975) has been an important innovation in the development of diamond cell technology. It takes advantage of the fact that the R_1 ruby fluorescence line shifts to longer wavelengths as pressure increases. A small chip of ruby is placed between the diamonds along with the sample to be studied. Fluorescence is excited in the ruby chip by directing light from a mercury lamp or from a laser through one of the diamond anvils to the chip. The red light emitted is analyzed by a spectrometer and the change in wavelength determined. The wavelength shift was calibrated against the NaCl pressure scale and found to be linear within experimental error to 195 kbar.

More recently the ruby scale has been extended to much higher pressures. Mao & Bell (1978) report a pressure of 1.7 Mbar. They have calibrated the ruby scale against shock compression measurements Cu, Mo, Pd, and Ag by making X-ray diffraction determinations of molar volumes of these metals at the same time that they made ruby fluorescence measurements. They obtained the following empirical equation for the ruby scale:

$$P = 3.808\ [(\Delta\lambda/694.2 + 1)^5 - 1],$$

where P is the pressure in megabars and $\Delta\lambda$ is the wavelength shift in nanometers. They estimate the error at one megabar to be $+20\%$, -10%. In addition to tieing the ruby scale to the shock data, which must be considered one of the most reliable ways of determining such high pressures, H. Mao and P. Bell (personal communication) have calculated the pressure on the basis of force divided by area and found it to be in good agreement.

Simultaneous Pressure and Temperature Measurement

Although electrical and visual observations have been made at simultaneous high pressures and temperatures in the diamond cell, there have been no in situ X-ray diffraction studies in which an attempt was made to accurately measure both pressure and temperature. Yagi (1976) has made simultaneous high pressure and temperature measurements in a large cubic press by measuring the temperature with a thermocouple and then using Decker's pressure-temperature equation of state for NaCl (Decker 1971) to calculate the pressure from molar volume measurements determined by X-ray diffraction. It should be possible to do this in the diamond cell as well, but recrystallization of the NaCl might make the diffraction patterns of so small a sample difficult to obtain because of the small number of crystallites doing the scattering. A very intense source of X rays capable of giving a diffraction pattern before recrystallization can take place may be the solution.

Some phase boundaries may be well enough known so that the pressure at a point along the boundary can be derived when the temperature is known. This is clearly an area of diamond cell technology that has yet to be explored.

GEOPHYSICAL APPLICATIONS

Geophysicists have been able to divide the Earth's interior into major zones based on the various properties they exhibit when studied by observational geophysical techniques such as seismology (Figure 5). The most prominent are the crust, which extends from the surface to a depth ranging from 5 to 50 km; the upper mantle, which extends from the bottom of the crust to a depth of approximately 1000 km and is distinguished by its rapidly increasing seismic velocities with depth; the lower mantle, which extends from 1000 to 2900 km and is characterized by seismic velocities that increase gradually with depth; the outer core, which extends from 2900 to 5000 km and is believed to be liquid because no transverse seismic waves traverse it; and the inner core, which extends from 5000 km to the center of the Earth at 6340 km and is considered to be solid.

This layering is believed to have developed over a long period of time by the slow differentiation of a once homogeneous Earth which formed by the aggregation of dust and gas from the solar nebular cloud 4.5 billion years ago. The Earth's interior is far from static today. Recent plate

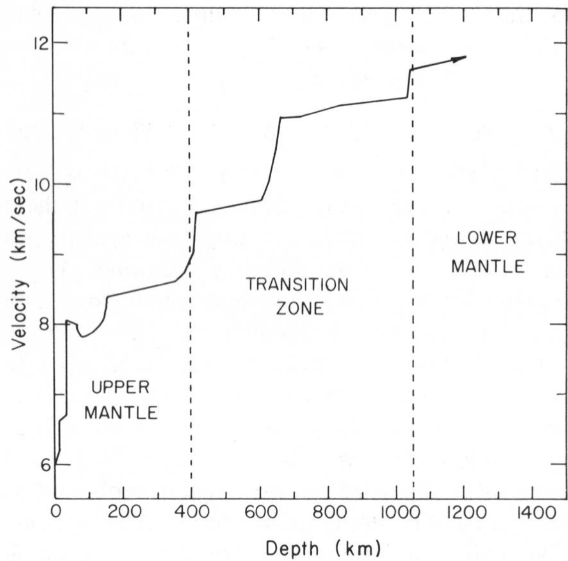

Figure 5 P-wave velocities versus depth within the Earth's mantle.

tectonic studies have provided extensive evidence that oceanic crust is being produced along ocean ridges by material rising from within the mantle, that oceanic crust is descending into the mantle at ocean margins, and that continents sometimes collide and thrust mantle material up onto the crust in the form of ophiolites. There is evidence that the motion of the crustal plates is driven by deep mantle convection cells. There is also evidence that the Earth's magnetic field is produced by convection within the liquid part of the core. Studies of the Earth's interior deal not only with the composition and properties of the various zones but with the evolution and dynamic processes as well.

Compression Measurements

One of the most fundamental properties that can be measured as a function of pressure, P, is molar volume, V. From such measurements it is possible to calculate the bulk modulus, $K_0 = -V(\partial P/\partial V)$, and its pressure derivative $K_0' = \partial K_0/\partial P$. For the interpretation of most of the diamond cell compression measurements the Birch-Murnaghan equation of state has been used for establishing K_0 and K_0':

$$P = \tfrac{3}{2}K_0\left[\left(\frac{V}{V_0}\right)^{-7/3} - \left(\frac{V}{V_0}\right)^{-5/3}\right]\left\{1 - \tfrac{3}{4}(4 - K_0')\left[\left(\frac{V}{V_0}\right)^{-2/3} - 1\right] + \cdots\right\}.$$

There are three important geophysical applications of measurements of molar volume as a function of pressure: determination of density, elastic behavior, and thermodynamic stability. Geophysicists have been able to establish the radial distribution of density within the Earth with a surprising degree of reliability. Models for the constitution of the Earth's interior must conform to the constraints placed by these geophysical observations. In other words, when a composition is proposed for a layer within the Earth, the crystalline phases making up the composition must yield a density consistent with the density based on the observations.

The behavior of seismic waves as they pass through the Earth's interior is determined by the elastic properties of the materials traversed. Measurements of isothermal bulk modulus and its pressure derivative constitute important information on elastic properties. Static compression measurements such as those made by X-ray diffraction in the diamond cell are particularly valuable since they are isothermal. The other two major experimental sources of information on elastic properties, namely sonic velocities and shock experiments, are not isothermal.

The effect of pressure on volume at constant temperature is directly related to the Gibbs free energy, G, by the equation $(\partial G/\partial P)_T = V$. Thus, the measurement of molar volume as a function of pressure is very important to the understanding of thermodynamic stabilities of geophysically significant phases.

The molar volume of a crystalline substance is easily calculated from its X-ray diffraction pattern. Early diamond cell compression measurements were made by placing a mixture of the sample and NaCl between the diamond anvils without a gasket. Bassett & Takahashi (1974) summarize compression measurements on some 23 substances studied at the University of Rochester between 1964 and 1974. Wilburn & Bassett (1976, 1977, 1978) and Wilburn et al (1978) have redetermined the compression measurements for Fe, Fe_2O_3, Fe_3O_4, and γ-Fe_2SiO_4 under hydrostatic conditions in the diamond cell by placing the sample in fluid medium (4:1 methanol-ethanol) encapsulated in a metal gasket. They found a significant difference between the recent measurements made under truly hydrostatic conditions and the earlier measurements made under quasi-hydrostatic conditions. The greater the shear strength of the sample, the greater the discrepancy. They attribute the discrepancy to the fact that the sample is subjected to an anisotropic elastic strain in the nonhydrostatic environment. Because of the geometry of the diamond cell, the lattice parameters that are calculated are based on the larger set of d-spacings of the anisotrophically compressed sample (Kinsland & Bassett 1976, 1977).

The discrepancy between molar volumes determined under hydrostatic and quasihydrostatic conditions appears to be constant above 50 kbar for most geophysically significant substances. Therefore, the higher the pressures used to measure compression curves, the smaller the resulting error in bulk modulus. Nonetheless, a downward revision of all bulk moduli based on measurements made under quasihydrostatic conditions in the diamond cell should probably be made. The correction is not large and the rather crude relationship developed by Wilburn & Bassett (1978) may be adequate. Such a correction is necessary not only for the earlier nonhydrostatic measurements but for the measurements made above the pressure region in which liquids are available for producing truly hydrostatic conditions.

Some of the most significant compression measurements made recently are those of Mao & Bell (1977c) on iron and periclase (MgO) up to 1 Mbar. These are some of the highest geophysical measurements made under static pressure. Both substances are believed to exist at the core-mantle boundary and so their densities at the pressures found there are of considerable interest. At these high pressures hydrostaticity is impossible but the discrepancy mentioned above is very small.

High Pressure Temperature Phases

Some of the most important contributions in experimental geophysics in recent years have resulted from the relatively new technique of heating samples while under pressure in the diamond cell by means of a laser beam (Ming & Bassett 1974). With this technique it is possible to routinely achieve temperatures and pressures comparable to conditions within the Earth's lower mantle. Space does not permit discussion of all the samples that have been studied by laser heating in the diamond cell. The most important results are those obtained on the very compositions which seem best to fit all the constraints placed on the Earth's interior.

Liu (1979) proposes a simple model for the composition of the mantle. It consists of a 1:1 mixture of two species $(Mg_{0.9}Fe_{0.1})_2SiO_4$ and 90% $(Mg_{0.9}Fe_{0.1})SiO_3 \cdot 10\%\ Al_2O_3$. For convenience I call these $Fo_{90}Fa_{10}$ and $En_{81}Fs_9Co_{10}$, where Fo stands for forsterite (Mg_2SiO_4), Fa for fayalite (Fe_2SiO_4), En for enstatite ($MgSiO_3$), Fs for ferrosilite ($FeSiO_3$), and Co for corundum (Al_2O_3). This mixture has the following chemical composition: SiO_2, 45.6 wt %; MgO, 42.0 wt %; FeO, 8.3 wt %; and Al_2O_3, 4.1 wt %. This is very close to the proportions of these components found in the pyrolite model for the mantle composition proposed by Ringwood (1975).

In Liu's model these two species behave independently, each going through a series of phase transitions as pressure increases. The first of

Table 1 The high pressure phases of Mg_2SiO_4, $MgSiO_3$, 90% $MgSiO_3$ 10% Al_2O_3

Starting composition	Phase(s)	Phase name or structure type	Pressure (kbar) of first appearance	Temperature (°C)	Volume change at first appearance (%)	References
Mg_2SiO_4	Mg_2SiO_4	Olivine	0	25	—	—
Mg_2SiO_4	Mg_2SiO_4	β-phase	~120	>1000	7.9	Moore & Smith (1969)
Mg_2SiO_4	Mg_2SiO_4	Spinel	~200	>1000	2.4	Suito (1972)
Mg_2SiO_4	$MgSiO_3$ MgO	Perovskite Periclase	~270	>1000	10.7	Liu (1976)
$MgSiO_3$	$MgSiO_3$	Pyroxene	0	25	—	—
$MgSiO_3$	Mg_2SiO_4 SiO_2	β-phase Stishovite	~180	>1000	15.2	Ito et al (1972)
$MgSiO_3$	Mg_2SiO_4 SiO_2	Spinel Stishovite	~200	>1000	1.8	Ito et al (1972)
$MgSiO_3$	$MgSiO_3$	Ilmenite	~220	>1000	2.0	Kawai et al (1974)
$MgSiO_3$	$MgSiO_3$	Perovskite	~270	>1000	7.0	Liu (1976)
$En_{90}Co_{10}$	$En_{90}Co_{10}$	Pyroxene	0	25	—	—
$En_{90}Co_{10}$	SS[a] SS	Pyroxene Garnet	~20	>1000	8.5	Boyd & England (1964)
$En_{90}Co_{10}$	$En_{90}Co_{10}$	Garnet	~170	>1000	—	Ringwood (1967)
$En_{90}Co_{10}$	SS SS	Garnet Ilmenite	~225	>1000	8.7	Liu (1977a)
$En_{90}Co_{10}$	$En_{90}Co_{10}$	Ilmenite	~235	>1000	—	Liu (1977a)
$En_{90}Co_{10}$	SS SS	Ilmenite Perovskite	~265	>1000	7.4	Liu (1977a)
$En_{90}Co_{10}$	$En_{90}Co_{10}$	Perovskite	~275	>1000	—	Liu (1977a)

[a] Solid solution.

the two species, $Fo_{90}Fa_{10}$, has a behavior very similar to that of pure forsterite. Because Fe^{2+} has the same charge as Mg^{2+} and an ionic radius which is only slightly larger, the presence of 10% fayalite has the effect of shifting the pressures of the phase transitions slightly and of blurring the transitions because of the two-phase regions that exist along the phase boundaries in two component systems having solid solutions. These features can be seen in Table 1 and Figure 6. The second species, $En_{81}Fs_9Co_{10}$, can be thought of as being principally enstatite with substitution of iron and aluminum. As with forsterite, the substitution of iron has only minor effects on the high pressure behavior of enstatite. The presence of alumina, however, greatly changes the high pressure relationships. This can be seen in Figure 7, the phase diagram for enstatite-corundum. The vertical line at 10% Al_2O_3 indicates the high pressure phases for $En_{90}Co_{10}$. Since the effect of iron substitution is minor, the high pressure phases of $En_{90}Co_{10}$ are probably very similar to those expected for $En_{81}Fs_9Co_{10}$.

The density changes (at 1 bar pressure) which take place at the phase

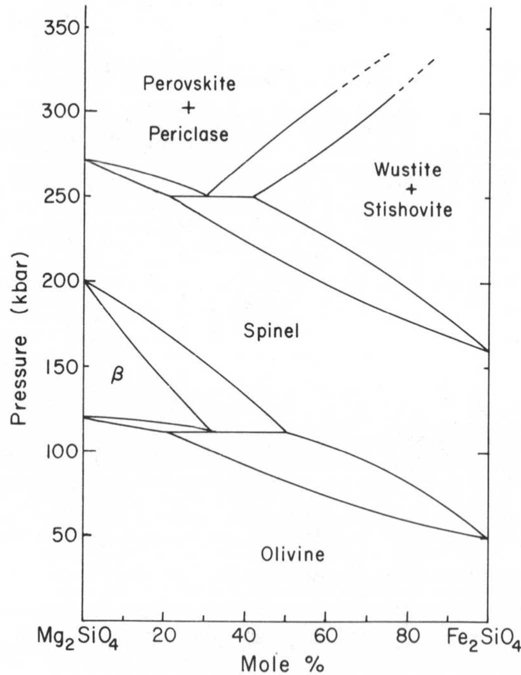

Figure 6 Proposed pressure-composition phase diagram for the forsterite-fayalite system at 1000°C.

transitions have been determined experimentally and are reported in Table 1. Using this information, Liu plots the density (at 1 bar) versus the depth in the mantle for each of the species along with the density for the 1:1 mixture (Figure 8). Thus he suggests that the observed seismic discontinuity at 420 km is due mainly to the forsterite-β-phase transition in the $Fo_{90}Fa_{10}$ species. The seismic discontinuity at 650 km is due to the transition from the spinel phase to the perovskite plus periclase phases in the $Fo_{90}Fa_{10}$ species and from the garnet phase to the ilmenite phase $En_{81}Fs_9Co_{10}$ species. A small seismic discontinuity at a depth between 420 and 650 km has sometimes been thought to exist and may be attributable to the transition from the β-phase to the spinel phase in the $Fo_{90}Fa_{10}$ species. Another small seismic discontinuity has sometimes been reported at depths greater than 650 km and may be attributable to the ilmenite-perovskite transition in the $En_{81}Fs_9Co_{10}$ species.

In the above model the correspondence between the seismic discontinuities and the phase transitions is still quite uncertain. Pressure and temperature have been difficult to measure accurately in the laser-heated diamond cell experiments. Both compression and thermal expansion have

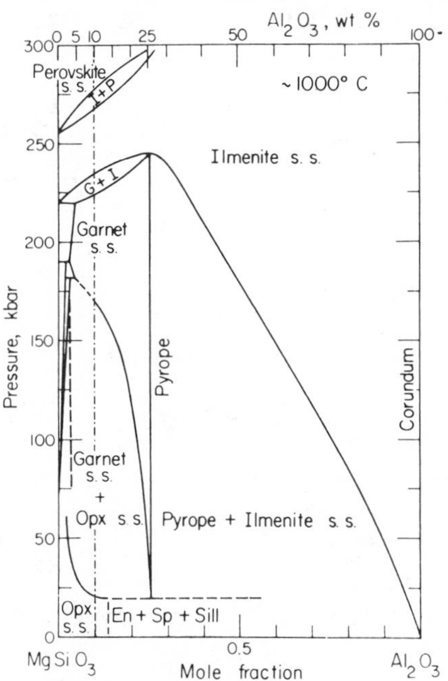

Figure 7 Proposed pressure-composition phase diagram for the enstatite-corundum system at 1000°C (Liu 1977a).

been ignored in the densities shown in Figure 8. The effect of temperature on the pressures of some of the phase transitions is poorly known or not known at all. No experiments have been done yet to determine if the two species do indeed remain independent without reacting with each other at high pressure and temperature. I have not considered the role of any other elements such as Na, K, and Ca on which there has already been a great deal of experimental work done (Liu 1977b, 1978). Clearly the details of the phases that make up the Earth's mantle, their compositions and crystal structures, require more work before we can state some of the conclusions with as much confidence as we would like. However, we should realize just how far we have come in the past few years since the time when the nature of many of the high pressure phases and their relationships to each other were a matter of speculation. At best they were based on comparisons with analogs of similar composition and therefore presumably similar properties.

Shock compression data on both olivine and pyroxene (Jeanloz & Ahrens 1977) have provided evidence that the ferromagnesian silicates

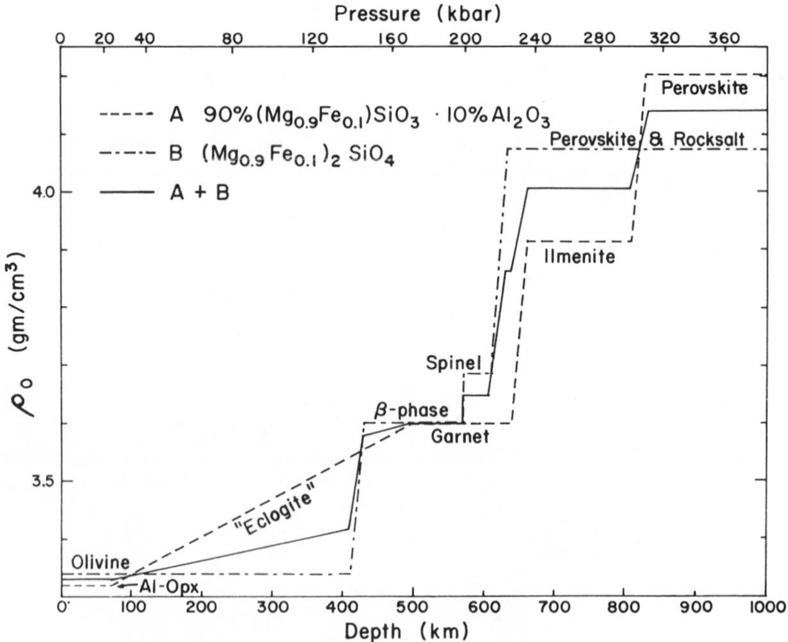

Figure 8 Effects of phase transformations upon densities of aluminous pyroxene, olivine, and their 1:1 molecule mixture. Pressure and depth scales are shown only for reference. Effects of compression and thermal expansion on each phase are not indicated (Liu 1979).

may undergo further transitions to even denser phases. These have not been observed by static high pressure–high temperature experimental techniques but may be observed once higher pressures and temperatures are achieved on a more routine basis.

Single Crystal X-Ray Diffraction

The first high pressure, single crystal, X-ray diffraction study made in the diamond cell was conducted by Weir et al (1965), who found they could place a liquid in a gasketed diamond cell and by judicious cycling of the pressure get the liquid to solidify into a single crystal that filled the entire sample area. Using a special cell constructed of beryllium they were able to direct the X-ray beam through the body of the cell to the sample and to detect scattered X rays that emerged through the beryllium body. By mounting the cell on a specially designed goniometer head they were able to collect single crystal diffraction data by the precession photographic method.

Meanwhile, Van Valkenburg (1965) found that he could visually observe phase transitions in single crystals by placing them in a liquid in a gasketed diamond cell and applying pressure. Some substances undergo displacive transitions which do not disrupt the single crystal. Thus, for instance, he found that a single crystal of the high pressure phase of calcite could be produced intact.

Merrill & Bassett (1974) developed a smaller, more compact diamond cell that required the X rays to traverse a shorter distance through the beryllium (Figure 1d). In their design, the diamonds are mounted on beryllium discs that are in turn mounted over holes in platens drawn together by three screws. The whole press is small enough to be mounted on a standard goniometer head. Merrill & Bassett (1974, 1975) used a single crystal diffractometer to collect data on the high pressure phase of calcite, $CaCO_3$ (II), that Van Valkenburg had observed visually. This phase is metastable since it would revert to aragonite if the sample were heated under pressure but is of considerable interest in understanding the mechanisms of phase transitions in carbonates and nitrates, which have similar structures. Merrill & Bassett found that the transition from $CaCO_3$ (I) to $CaCO_3$ (II) consists of the rotation of the carbonate groups with accompanying displacement of the calcium ions.

Hazen & Burnham (1974), using the diamond cell design of Merrill & Bassett, analyzed the crystal structures of gillespite I and II. Their results were important in establishing that the drastic color change from red to blue at the pressure-induced phase transition can be explained by structural change alone rather than by a high-low spin transition in the iron ion as had been suggested.

In more recent years, the high pressure single crystal X-ray diffraction technique has been applied extensively to the study of continuous crystallographic changes that take place in minerals as a function of pressure (Hazen 1976, 1977, Hazen & Prewitt 1977, Hazen & Finger 1978, Finger et al 1977). Hazen & Finger (1979) have drawn several conclusions from these studies:

1. For divalent cations there is a linear relationship between the polyhedron volume and its compressibility.
2. Oxides and silicates which have edge-shared polyhedra in a rigid array have low compressibilities, and their compressibilities are controlled by the compressibilities of the constituent polyhedra.
3. In oxides and silicates such as MgO and α-Al_2O_3 in which the polyhedra are all the same, the compressibilities are the same as the compressibilities of the polyhedra. In oxides and silicates in which there is a mixture of polyhedra such as spinel, zircon, and garnet, the compressibilities are intermediate between the compressibilities of the polyhedra.
4. In framework silicates such as α-quartz and feldspar which have corner-linked polyhedra, the compressibilities may be controlled to a large extent by changes in angles between polyhedra (polyhedral tilting). Thus substances with very incompressible polyhedra can have large compressibilities.
5. In layered structures such as SnS_2 and phlogopite the bonding within layers is usually more rigid than between layers. Thus the compressibility is usually much greater perpendicular to the layers than parallel to them.
6. In many structures including pyroxenes, olivines, and rutile, polyhedra share edges with other polyhedra but do not form a continuous three-dimensional network. In these, both metal-oxygen compression and polyhedron distortion contribute to the compressibility of the material. Thus the compression may be greater than that of the component polyhedra. For instance, the compressibility of olivine is greater than that of the spinel phase having the same composition and consisting of the same polyhedra but differently arranged.
7. The pressure derivative of bulk modulus, K', is a measure of the rate at which the structure becomes stiff as it is compressed. If the response of the structure to compression is a combination of mechanisms such as polyhedron compression, distortion, and tilting, then the rate of stiffening may be greater than in a simpler structure, i.e. the value for K' is expected to be greater.

Improvements in the design of the single crystal diamond cell and in the data handling (Finger & King 1978) have made it possible to make structure refinements with a weighted residual of only 2.0%, a precision which rivals that of many studies made at atmospheric pressure and room temperature.

Disproportionation of Iron

One of the most severe constraints placed on models for the origin of the Earth was the supposed disequilibrium among the Earth's core, mantle, and crust because of different oxidation states of iron in each (Ringwood 1966). However, experimental work of Mao & Bell (1971, 1977a) and Bell & Mao (1975) using the diamond cell has shown that the various oxidation states of iron can coexist in equilibrium under pressure-temperature conditions within the Earth. They found that the spinel phase of fayalite composition broke down to stishovite plus a nonstoichiometric form of wustite ($Fe_{0.91-0.97}O$) plus metallic iron. This was later confirmed by Bassett & Ming (1972). In the earlier experiments the evidence for metallic iron was by inference. However, later experiments by Bell & Mao (1975) using the scanning electron microscope on quenched samples and by J. Sharry and W. Bassett (unpublished) using the ion probe confirmed the presence of metallic iron. The removal of the constraint of disequilibrium permits a much simpler model for the origin of the Earth, one in which the proto-Earth formed by homogeneous aggregation of nebular dust and gas and later differentiated into core, mantle, and crust by gravitational settling of the denser phases.

Kinetics of Phase Transitions

The rate at which a subducting slab of lithosphere descends into the mantle is expected to be strongly influenced by the rate at which the olivine-spinel phase transition proceeds as the olivine in the slab passes through the conditions for the transition to take place. Sung & Burns (1976) made a systematic investigation of the rate of the olivine-spinel transition in a diamond anvil cell equipped to heat samples up to 1000°C as well as subject them to pressures up to 300 kbar. From their results they concluded that the transformation rate becomes virtually zero at temperatures below 700°C. Even if the high pressure phase is the modified spinel (β-phase) rather than the spinel phase, the transformation rate would probably be very similar since the β-phase and the spinel phase have such similar structures.

During the initial stages of descent of the slab, the rate of descent may be slow enough for the interior of the slab to warm up to temperatures high enough to permit the phase transition to proceed at close to equilibrium. This would cause the phase boundary to be quite high in the descending slab, resulting in a large quantity of the dense β-phase. This in turn would have the effect of accelerating the rate of descent. As the rate of descent increased, the interior of the slab would become cooler and the phase transition would be delayed to a greater depth in the slab. The descent of the less dense olivine deeper into the mantle would make the slab more buoyant and slow its rate of descent. Thus the rate of descent of the slab would be controlled by the rate of the phase transition. The results of Sung & Burns (1976), though preliminary, indicate that the kinetics of the olivine-spinel or olivine-β-phase transition could account for such a mechanism.

This mechanism could lead to an interesting phenomenon. The olivine would be descending metastably into the portion of the mantle where the β-phase or spinel phase should be stable. This is a situation in which the transition could be suddenly triggered producing an implosive impulse resulting in a deep-seated earthquake. The fact that such deep-seated earthquakes are observed suggests that just such a mechanism may be in effect.

Solubility of Minerals in Water

The effects of pressure on the solubility of minerals in water have been studied by Van Valkenburg et al (1971a) using a gasketed diamond cell. They found that the solubility of gypsum at room temperature is not a linear function of pressure but increases rapidly at pressures above 4 kbar. They also found that pressure greatly increased the solubility of calcite, aragonite, witherite, hydromagnesite, and oldhamite.

In studying calcite and aragonite, Van Valkenburg et al (1971b) found that a new phase formed at pressures greater than 6 to 7 kbar. They identified this new phase as calcium carbonate hexahydrate, the mineral ikaite which was previously known from the arctic waters of Ika Fjord in Greenland. These studies have shown the gasketed diamond cell to be a very promising means of investigating the solubility of various species in water at high pressures.

Elastic Properties

The study of seismology has provided most of our knowledge about the interior of the Earth. The interpretation of seismic data, however, relies heavily on the study of the elastic properties of substances which determine the velocities of seismic waves as they traverse the various zones of the Earth. Direct isothermal compression measurements of materials have provided a great deal of information as described in an earlier section. However, sonic velocity measurements have been an even more important source of information. Until recently all sound velocity measurements were made by bonding a transducer to a sample and then subjecting that sample to the desired conditions of pressure and temperature. This technique requires samples that are much too large for a study in the diamond cell. Recently, however, the technique of Brillouin scattering has been applied to samples under pressure in a diamond cell (Whitfield et al 1976). This consists of directing a laser beam through one of the diamonds onto the sample, an oriented single crystal, while it is under pressure in an encapsulated liquid. A small portion of that light is scattered by reflection off thermal phonons in the crystal and is doppler shifted as a result of the velocity of the phonons. Since the thermal phonons are simply sound waves, the doppler-shifted light can be used to calculate sound velocities and in turn the elastic moduli. Initial measurements on NaCl have shown that the technique is a very promising means of collecting data on the elastic properties of geophysically significant materials.

Strength

A modification of the diamond cell was developed by Kinsland & Bassett (1976) to permit the recording of a diffraction pattern of a polycrystalline sample as the X-ray beam traverses the sample perpendicular to the load axis. Diffraction patterns recorded on flat film have elliptical diffraction rings because of a greater compression of *d*-spacings parallel to the load axis than perpendicular to the load axis. The ellipticity can be used to calculate the elastic strain in each of the crystallites. Kinsland & Bassett (1976, 1977) found that the elastic strain which a sample could endure before yielding would increase with pressure up to a point and then remain constant as the pressure was further increased. They interpreted this to mean that the strength of the sample increased up to a certain pressure but remained essentially constant at pressures higher than that. In NaCl, they found that strength increased up to 4 kbar pressure before leveling off, and for MgO it increased up to 30 kbar pressure before leveling off. They interpret these results to indicate that brittle failure ceases to be the principal mechanism of deformation and true ductile yielding by atomic scale gliding becomes the dominant mechanism of deformation.

Sung et al (1977) used a diamond cell to make strength measurements on fayalite by a different approach. They calculated the strength from the rate of extrusion of sample from between the diamond anvils. Their measurements are in reasonable agreement with measurements made by other methods. Both of the techniques for measuring strength in the diamond cell described here show considerable promise for study of the properties of mantle materials and their ability to undergo mantle convection.

Electrical Conductivity and Optical Absorption

Mao (1973a–c) and Mao & Bell (1972a–c, 1977b) have measured the effect of pressure on the electrical conductivity and optical absorption in olivines and members of the magnesiowustite series. The electrical conductivity measurements are made by placing wires in the sample and then subjecting the sample to pressure between the diamond anvils. The optical absorption measurements are made by microspectrophotometer. Their measurements show that in both olivine and magnesiowustite the electrical conductivity and the opacity are much higher than had previously been thought. Electrical conductivity, when compared with actual measurements on the Earth's mantle, indicates a ratio of 0.15 for Fe/(Fe + Mg). The opacity is so high that it seems unlikely that radiative transfer plays the major role in heat transfer in the mantle as had been postulated. These conclusions are predicated on a model for the lower mantle of a mixture of stishovite and magnesiowustite. The electrical conductivity and optical absorption in a lower mantle consisting of predominantly perovskite mixed with some magnesiowustite is unknown. Measurements on a substance with iron in twelvefold coordination as in the perovskite structure have yet to be made.

Mossbauer Method

Huggins et al (1975) have adapted the diamond anvil cell to permit the application of the Mossbauer method to the study of iron-bearing minerals under pressure. The technique consists of placing a cobalt-57 source on the end of a palladium rod and then bringing the rod as close as possible to one of the diamond anvils. The gamma rays emitted by the source travel through the diamonds and the sample. Iron in the sample

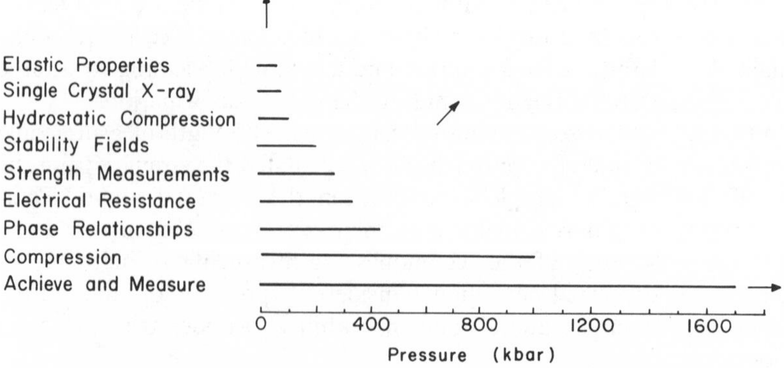

Figure 9 State of the art in high pressure experimental geophysics.

selectively absorbs the gamma rays depending on the nature of the site occupied by the iron ions. From their study of the spinel phase of Fe_2SiO_4 they concluded that within their experimental error all of the iron ions occupy the octahedral sites. Their study of gillespite generally confirmed earlier studies but indicated that the pressure-induced transition was more gradual than had been thought. Mao et al (1977) applied the Mossbauer method to a study of magnesioferrite, magnetite, and hematite and found that each of these materials became paramagnetic at high pressures. The transitions in magnetite and hematite corresponded with structural transitions that had been observed earlier.

Radioactivity

Hensley et al (1973) measured the effect of pressure on the radioactive decay rate in beryllium-7 which decays by electron capture. They found that the decay rate increases approximately 0.5% at 270 kbar. The increased decay rate was predicted because the pressure reduces the size of the electron orbitals, thus increasing the probability of capture by the nucleus. These results are of geophysical significance since one branch of the potassium-40 decay scheme takes place by the same process and is considered an important source of heat within the Earth's interior. At present the concentrations of potassium within the various portions of the Earth are less well known than the pressure effect.

STATE OF THE ART AND FUTURE RESEARCH

The application of static high pressure experimental research to problems in geophysics is advancing rapidly. It has been shown that there are many observations that can be made with the diamond cell when analytical techniques are employed that work successfully on small samples. Figure 9 is intended to show the state of the art of diamond cell research in geophysics. The arrows point in the direction of advancement: toward higher pressure, toward more analytical techniques, and finally toward the extension of the present analytical techniques to higher pressures. Add to these the parameter of temperature, perhaps as a third axis, and you see that there is much work yet to be done in the application of diamond cell technology to geophysical problems.

ACKNOWLEDGMENTS

The author would like to express his appreciation to H. K. Mao, P. M. Bell, R. M. Hazen, L. W. Finger, and L. G. Liu for providing up-to-the-minute information and giving permission to use diagrams from their papers.

Literature Cited

Barnett, J. D., Block, S., Piermarini, G. J. 1973. An optical fluorescence system for quantitative pressure measurement in the diamond-anvil cell. *Rev. Sci. Instrum.* 44:1–9

Bassett, W. A., Ming, L. C. 1972. Disproportionation of Fe_2SiO_4 to $2FeO+SiO_2$ at pressures up to 250 kbar and temperatures up to 3000°C. *Phys. Earth Planet. Inter.* 6:154–60

Bassett, W. A., Takahashi, T. 1965. Silver iodide polymorphs. *Am. Mineral.* 50:1576–94

Bassett, W. A., Takahashi, T. 1974. X-ray diffraction studies up to 300 kbar. *Adv. High-Pressure Res.* 4:165–247

Bassett, W. A., Takahashi, T., Stook, P. W. 1967. X-ray diffraction and optical observations on crystalline solids up to 300 kbar. *Rev. Sci. Instrum.* 38:37–42

Bassett, W. A., Wilburn, D. R., Hrubec, J. A., Brody, E. M. 1979. Elastic properties measured under hydrostatic pressure in the diamond anvil cell by X-ray diffraction and Brillouin scattering. *Proc. 1977 AIRAPT Int. Conf., Boulder, Colo.*

Bell P. M., Mao, H. K. 1975. Preliminary evidence of disproportionation of ferrous iron in silicates at high pressures and temperatures. *Carnegie Inst. Washington Yearb.* 74:557–59

Block, S., Piermarini, G. J. 1976. The diamond cell stimulates high-pressure research. *Phys. Today* 29:44–55

Boyd, F. R., England, J. L. 1964. The system enstatite-pyrope. *Carnegie Inst. Washington Yearb.* 63:157

Decker, D. L. 1971. High pressure equation of state for sodium chloride, potassium chloride, and cesium chloride. *J. Appl. Phys.* 42:3239–44

Decker, D. L., Bassett, W. A., Merrill, L., Hall, H. T., Barnett, J. D. 1972. High-pressure calibration, a critical review. *J. Phys. Chem. Ref. Data* 1:773–835

Finger, L. W., Hazen, R. M., Yagi, T. 1977. High-pressure crystal structures of the spinel polymorphs of Fe_2SiO_4 and Ni_2SiO_4. *Carnegie Inst. Washington Yearb.* 76:504–5

Finger, L. W., King, H. 1978. A revised method of operation of the single-crystal diamond cell and the refinement of the structure of NaCl at 32 kbar. *Am. Mineral.* 63:337–42

Fritz, J. N., Marsh, S. P., Carter, W. J., McQueen, R. G. 1971. The hugoniot equation of state of sodium chloride in the sodium chloride structure. In *Characterization of the High Pressure Environment*, ed. E. Lloyd, pp. 201–8. *Natl. Bur. Stand., Spec. Publ. 326*

Hazen, R. M. 1976. Effects of temperature and pressure on the cell dimension and X-ray temperature factors of periclase. *Am. Mineral.* 61:266–71

Hazen, R. M. 1977. Effects of temperature and pressure on the crystal structure of ferromagnesian olivine. *Am. Mineral.* 62:286–95

Hazen, R. M., Burnham, C. W. 1974. The crystal structures of gillespite I and II: a structure determination at high pressure. *Am. Mineral.* 59:1166–76

Hazen, R. M., Finger, L. W. 1978. Crystal structures and compressibilities of pyrope and grossular to 60 kbar. *Am. Mineral.* 63:297–303

Hazen, R. M., Finger, L. W. 1979. Relationship between crystal structure and compressibility in ionic compounds. *Carnegie Inst. Washington Yearb.* In press

Hazen, R. M., Prewitt, C. T. 1977. Effects of temperature and pressure on interatomic distances in oxygen-based minerals. *Am. Mineral* 62:309–15

Hensley, W. K., Bassett, W. A., Huizenga, J. R. 1973. Pressure dependence of the radioactive decay constant of beryllium-7. *Science* 181:1164–65

Huggins, F. E., Mao, H. K., Virgo, D. 1975. Mossbauer studies at high pressure using the diamond-anvil cell. *Carnegie Inst. Washington Yearb.* 74:405–10

Ito, E., Matsumoto, T., Suito, K., Kawai, N. 1972. High pressure breakdown of enstatite. *Proc. Jpn. Acad.* 48:412–15

Jeanloz, R., Ahrens, T. J. 1977. Pyroxenes and olivines: structural implications of shock-wave data for high pressure phases. In *High-Pressure Applications in Geophysics*, ed. M. H. Maghnani, S. Akimoto, pp. 439–61. New York: Academic

Kawai, N., Tachimori, M., Ito, E. 1974. A high pressure hexagonal form of $MgSiO_3$. *Proc. Jpn. Acad.* 50:378–80

Kinsland, G. L., Bassett, W. A. 1976. Modification of the diamond cell for measuring strain and the strength of materials at pressures up to 300 kilobars. *Rev. Sci. Instrum.* 47:130–33

Kinsland, G. L., Bassett, W. A. 1977. Strength of MgO and NaCl polycrystals to confining pressures of 250 kbar at 25°C. *J. Appl. Phys.* 48:978–85

Liu, L. G. 1976. Orthorhombic perovskite phases observed in olivine, pyroxene, and garnet at high pressures and temperatures. *Phys. Earth Planet. Inter.* 11:289–98

Liu, L. G. 1977a. The system enstatite-pyrope at high pressures and temperatures and the mineralogy of the Earth's mantle. *Earth Planet. Sci. Lett.* 36:237–45

Liu L. G. 1977b. High pressure $NaAlSiO_4$: The first silicate calcium ferrite isotype. *Geophys. Res. Lett.* 4:183–86

Liu, L. G. 1978. High-pressure phase transformations of albite, jadeite and nepheline. *Earth Planet. Sci. Lett.* 37:438–44

Liu, L. G. 1979. Phase transformations and the constitution of the deep mantle. In *The Earth: Its Origin, Structure, and Evolution*, ed. M. W. McElhinny. London: Academic. In press

Liu, L. G., Bassett, W. A. 1975. The melting of iron up to 200 kbar. *J. Geophys. Res.* 80:3777–82

Liu, L. G., Bassett, W. A., Liu, M. S. 1973. Polymorphism in the solid solutions, potassium chloride–sodium chloride and rubidium chloride–potassium chloride, at high pressure. *J. Phys. Chem.* 77:1695–99

Mao, H. K. 1973a. Electrical and optical properties of the olivine series at high pressure. *Carnegie Inst. Washington Yearb.* 72:552–54

Mao, H. K. 1973b. Observations of optical absorption and electrical conductivity in magnesiowustite at high pressures. *Carnegie Inst. Washington Yearb.* 72:554–57

Mao, H. K. 1973c. Thermal and electrical properties of the Earth's mantle. *Carnegie Inst. Washington Yearb.* 72:557–64

Mao, H. K., Bell, P. M. 1971. High-pressure decomposition of spinel (Fe_2SiO_4). *Carnegie Inst. Washington Yearb.* 70:176–78

Mao, H. K., Bell, P. M. 1972a. Optical and electrical behavior of olivine and spinel (Fe_2SiO_4) at high pressure. *Carnegie Inst. Washington Yearb.* 71:520–24

Mao, H. K., Bell, P. M. 1972b. Interpretation of the effect of the optical absorption bands of natural fayalite to 20 kb. *Carnegie Inst. Washington Yearb.* 71:524–27

Mao, H. K., Bell, P. M. 1972c. Crystal-field stabilization of the olivine-spinel transition. *Carnegie Inst. Washington Yearb.* 71:527–28

Mao, H. K., Bell, P. M. 1977a. Disproportionation equilibrium in iron-bearing systems at pressures above 100 kbar with applications to chemistry of the Earth's mantle. In *Energetics of Geological Processes*, ed. S. K. Saxena, S. Bhattacharji, pp. 237–49. New York: Springer-Verlag

Mao, H. K., Bell, P. M. 1977b. Techniques of electrical conductivity measurement to 300 kbar. In *High-Pressure Research Applications in Geophysics*, ed. M. H. Maghnani, S. Akimoto, pp. 493–502. New York: Academic

Mao, H. K., Bell, P. M. 1977c. Pressure-volume equations of state of MgO and Fe to 1 Mbar. *Carnegie Inst. Washington Yearb.* 76:519–22

Mao, H. K., Bell, P. M. 1978. High-pressure physics: sustained static generation of 1.36 to 1.72 megabars. *Science* 200:1145–47

Mao, H. K., Virgo, D., Bell, P. M. 1977. High-pressure ^{57}Fe Mossbauer data on the phase and magnetic transitions of magnesioferrite ($MgFe_2O_4$), magnetite (Fe_3O_4), and hematite (Fe_2O_3). *Carnegie Inst. Washington Yearb.* 76:522–25

Merrill, L., Bassett, W. A. 1974. Miniature diamond anvil pressure cell for single crystal X-ray diffraction studies. *Rev. Sci. Instrum.* 45:290–94

Merrill, L., Bassett, W. A. 1975. The crystal structure of $CaCO_3$ (II), a high-pressure metastable phase of calcium carbonate. *Acta Crystallogr. B* 31:343–49

Ming, L. C., Bassett, W. A. 1974. Laser heating in the diamond anvil press up to 2000°C sustained and 3000°C pulsed at pressures up to 260 kilobars. *Rev. Sci. Instrum.* 9:1115–18

Moore, P. B., Smith, J. V. 1969. High pressure modification of Mg_2SiO_4: crystal structure and crystallochemical and geophysical implications. *Nature* 221:653–55

Piermarini, G. J., Block, S., Barnett, J. D. 1973. Hydrostatic limits in liquids and solids to 100 kbar. *J. Appl. Phys.* 44:5377–82

Piermarini, G. J., Block, S., Barnett, J. D., Forman, R. A. 1975. Calibration of the pressure dependence of the R_1 ruby fluorescence line to 195 kbar. *J. Appl. Phys.* 46:2774–80

Piermarini, G. J., Weir, C. E. 1962. A diamond cell for X-ray diffraction studies at high pressures. *J. Res. Natl. Bur. Stand.* 66A:325–31

Ringwood, A. E. 1966. Chemical evolution of the terrestrial planets. *Geochim. Cosmochim. Acta* 30:41–104

Ringwood, A. E. 1967. The pyroxene-garnet transformation in the Earth's mantle. *Earth Planet. Sci. Lett.* 2:255–63

Ringwood, A. E. 1975. *Composition and Petrology of the Earth's Mantle*, p. 188. New York: McGraw-Hill. 618 pp.

Suito, K. 1972. Phase transformations of pure Mg_2SiO_4 into a spinel structure under high pressures and temperatures. *J. Phys. Earth* 20:225–43

Sung, C. M. 1976. New modification of the diamond anvil press: a versatile apparatus for research at high pressure and high temperature. *Rev. Sci. Instrum.* 47: 1343–46

Sung, C. M., Burns, R. G. 1976. Kinetics of high-pressure phase transformations: implications to the evolution of the olivine-spinel transition in the downgoing lithosphere and its consequences on the dynamics of the mantle. *Tectonophysics* 31: 1–32

Sung, C. M., Goetze, C., Mao, H. K. 1977. Pressure distribution in the diamond anvil press and the shear strength of fayalite. *Rev. Sci. Instrum.* 48: 1386–91

Van Valkenburg, A. 1963. High pressure microscopy. In *High Pressure Measurement*, ed. A. Giardini, E. C. Lloyd, pp. 87–94. Washington: Butterworth

Van Valkenburg, A. 1965. Visual observations of single crystal transitions under true hydrostatic pressures up to 40 kilobars. In *Conf. Int. Hautes Pressions*, Le Creusot, Saone-et-Loire, France

Van Valkenburg, A., Mao, H. K., Bell, P. M. 1971a. Solubility of minerals at high water pressures. *Carnegie Inst. Washington Yearb.* 70: 233–37

Van Valkenburg, A., Mao, H. K., Bell, P. M. 1971b. Ikaite ($CaCO_3 \cdot 6H_2O$), a phase more stable than calcite and aragonite ($CaCO_3$) at high water pressure. *Carnegie Inst. Washington Yearb.* 70: 237–38

Weaver, J. S. 1971. Comparison of four proposed P-V relations for NaCl. In *Accurate Characterization of the High Pressure Environment*, ed. E. Lloyd, pp. 325–29. *Natl. Bur. Stand., Spec. Publ. 326*

Weaver, J. S., Takahashi, T., Bassett, W. A. 1971. Calculation of the P-V relation for sodium chloride up to 300 kilobars at 25°C. In *Accurate Characterization of the High Pressure Environment*, ed. E. Lloyd, pp. 189–99. *Natl. Bur. Stand., Spec. Publ. 326*

Webb, A. W., Gubser, D. U., Towle, L. C. 1976. Cryostat for generating pressures to 100 kilobar and temperatures to 0.03 K. *Rev. Sci. Instrum.* 47: 59–62

Weir, C., Block, S., Piermarini, G. J. 1965. Single-crystal X-ray diffraction at high pressures. *J. Res. Natl. Bur. Stand. Sect. C* 69: 275–81

Weir, C. E., Lippincott, E. R., Van Valkenburg, A., Bunting, E. N. 1959. Infrared studies in the 1- to 15-micron region to 30,000 atmospheres. *J. Res. Natl. Bur. Stand. Sect. A* 63: 55–62

Whitfield, C. H., Brody, E. M., Bassett, W. A. 1976. Elastic moduli of NaCl by Brillouin scattering at high pressure in a diamond anvil cell. *Rev. Sci. Instrum.* 47: 942–47

Wilburn, D. R., Bassett, W. A. 1976. Isothermal compression of spinel (Fe_2SiO_4) up to 75 kbar under hydrostatic conditions. *High Temperature–High Pressure* 8: 343–48

Wilburn, D. R., Bassett, W. A. 1977. Isothermal compression of magnetite (Fe_3O_4) up to 70 kbar under hydrostatic conditions. *High Temperature–High Pressure* 9: 35–39

Wilburn, D. R., Bassett, W. A. 1978. Hydrostatic compression of iron and related compounds: an overview. *Am. Mineral.* 63: 591–96

Wilburn, D. R., Bassett, W. A., Sato, Y., Akimoto, S. 1978. X-ray diffraction compression studies of hematite under hydrostatic, isothermal conditions. *J. Geophys. Res.* 83: 3509–12

Yagi, T. 1976. Experimental determination of thermal expansivity of several alkali halides at high pressures. *Inst. Solid State Phys., Univ. of Tokyo, Ser. A, No. 784.* 43 pp.

Ann. Rev. Earth Planet. Sci. 1979. 7: 385–415

THE MAGNETIC FIELD OF JUPITER: A COMPARISON OF RADIO ASTRONOMY AND SPACECRAFT OBSERVATIONS

Edward J. Smith and Samuel Gulkis

Earth and Space Sciences Division, Jet Propulsion Laboratory, California Institute of Technology, Pasadena, California 91103

INTRODUCTION

The planet Jupiter is known to possess a strong magnetic field. First revealed to radio astronomers in the mid-1950s through its polarized radio emissions at decametric and decimetric wavelengths, the magnetic field was later (1973 and 1974) sampled directly by magnetometers on board the Pioneer 10 and Pioneer 11 spacecraft which passed within 2.84 and 1.59 jovian radii (r_J) respectively of Jupiter. Within about $10r_J$ of the center of Jupiter, a region we call the inner magnetosphere, the magnetic field is in the first approximation a dipole field associated with currents interior to the planet. Outside of this region, the planetary magnetic field is affected by currents external to the planet and the field departs significantly from a dipole. Recent review articles of the in situ magnetic field measurements and of the radio astronomical measurements can be found in the book *Jupiter* edited by T. Gehrels (1976).

The radio astronomical measurements of the decimetric radiation and the in situ measurements give information on the magnetic field in the overlapping spatial domain from about $1.6r_J$ to $4r_J$. The continuous decimetric emissions, which were successfully interpreted in the early 1960s as synchrotron radiation from relativistic electrons trapped in a planetary magnetic field, have allowed well-founded inferences to be drawn regarding the properties of the field and the trapped particles. In the range of overlap, good agreement has been obtained with the in situ measurements of Jupiter's vector dipole magnetic moment. The dipole moment is ~ 4.2 gauss r_J^3, is tilted $\sim 10°$ with respect to Jupiter's

0084-6597/79/0515-0385$01.00

rotation axis, and is directed out of the pole in the north (opposite to that of the Earth). (In this article, the distance corresponding to one Jupiter radius is taken to be that used by the Pioneer investigators, i.e. 71,372 km. Hence, 1 gauss $r_J^3 = 3.64 \times 10^{23}$ gauss M^3.)

Both sets of observations, however, show clear evidence that Jupiter's magnetic field is more complex than that of a simple dipole. The spacecraft data record the nondipolar variations directly while the radio astronomical data reflect nondipole structure through the asymmetric variations with rotation of both the decimeter and decameter burst emissions. Thus far, no detailed comparison has been made between the higher order magnetic moments derived from the in situ measurements and the remote data. It is the intent of this review article to lay the groundwork for such a comparison. We restrict ourselves in this review to a discussion of the decimetric emission and to the spacecraft measurements of the inner magnetosphere.

Studies of Jupiter's magnetic field are important because they have a bearing on the internal structure of Jupiter. The origin of Jupiter's main magnetic field is believed to be a hydromagnetic dynamo operating in the electrically conducting fluid part of the planet interior (e.g. Hide & Stannard 1976). Presumably, the dynamo is driven by convective motions generated by internal heat released, perhaps, as a result of gravitational contraction or material differentiation in Jupiter's interior. Whether the main dynamo is located at great depth where Jupiter's interior is metallic or in an outer nonmetallic region is not known at this time. Furthermore, there is a possibility that relatively intense but smaller scale sources also exist at, or near, the surface of Jupiter.

It is important to bring together the spacecraft and radio astronomy data for a variety of reasons. Foremost of these is the recognition that the data are complimentary and taken together can lead to a better description of Jupiter's magnetic field than either one separately. In the region of overlap, the in situ data apply to two threadlike paths through the magnetosphere whereas the radio astronomy measurements yield data on the properties of the magnetic field integrated over the volume occupied by the radiating particles. The combined set applies stronger constraints on the models than either data set used alone. Other considerations are that the magnetic field description inside $1.6r_J$ can be improved using the radio astronomy data, especially the decametric observations, and secular variations of the magnetic field and relativistic electron distributions can be monitored from the ground once the measurements are fully understood.

This review begins in the next section with a discussion of the different ways of representing a planetary magnetic field. A brief discussion of the jovian rotation period and a definition of longitudes is then given. In the subsequent section, the dipole component of Jupiter's magnetic field is discussed and the remote and in situ measurements are compared. Evidence of high order moments and magnetic anomalies are then discussed in a separate section. The evidence for and against secular variations of the jovian magnetic field is presented in the next-to-final section. The article closes with our general conclusions and suggestions for further work.

REPRESENTATION OF THE MAGNETIC FIELD

At least four different models have been used to specify Jupiter's magnetic field. These range in complexity from a three parameter model (dipole) up to a model containing fifteen spherical harmonic coefficients (octupole). The representations are described below for later use throughout this paper.

Spherical Harmonic Representation

In the region external to the sources, the field can be derived from a scalar potential function, Φ, which satisfies Laplace's equation. The most general representation of Φ is given by a superposition of the eigenfunction solutions of this differential equation (e.g. Sommerfeld 1964). In spherical coordinates (Figure 1),

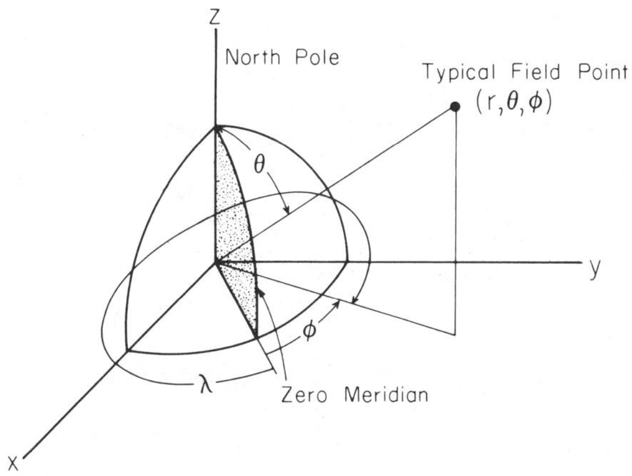

Figure 1 Coordinate system used to describe Jupiter's magnetic field. The Z axis is aligned with Jupiter's spin axis and is directed toward the north. The zero meridian coincides with System III (1957.0) zero meridian at a particular epoch. The magnetic potential is based on conventional right hand coordinate system with longitudes, ϕ, increasing to the East. Longitudes, λ, increasing westward are conventionally used by astronomers.

$$\Phi(r, \theta, \phi) = a \sum_{\ell=1}^{\infty} \sum_{m=0}^{\ell} \left(\frac{r}{a}\right)^{-\ell-1} P_\ell^m(\cos\theta)(g_\ell^m \cos m\phi + h_\ell^m \sin m\phi), \quad (1)$$

where a is the planetary radius and g_ℓ^m, h_ℓ^m are often called the Gauss coefficients. When analyzing planetary magnetic fields, the associated Legendre functions, $P_\ell^m(\cos\theta)$, are normalized in a particular way, called the Schmidt normalization (see Chapman & Bartels 1940). The Schmidt normalized associated Legendre polynomials are defined by $P_\ell^m(x) = N_{\ell m}(1-x^2)^{m/2}\, \mathrm{d}^m P_\ell(x)/\mathrm{d}x^m$ where $P_\ell(x)$ is the Legendre polynomial and $N_{\ell m} = 1$, if $m = 0$, or $[2(\ell - m)!/(\ell + m)!]^{1/2}$, if $m \neq 0$. The vector field is derived from the gradient of the potential: $\mathbf{B} = -\nabla\Phi$. In this representation, Jupiter's magnetic field is specified by the Gauss coefficients. When analyzing spacecraft data, the Gauss coefficients are determined from a least squares fit of the model to the measurements.

Dipole

To a first approximation, Jupiter's magnetic field resembles that of a dipole. Denoting the magnetic moment of the dipole by the vector **M**, the magnetic potential at **r** (r, θ, ϕ) due to a dipole located at the center of Jupiter is

$$\Phi = \frac{\mathbf{M}\cdot\mathbf{r}}{r^3}. \quad (2)$$

Dipole models of Jupiter's magnetic field have been obtained from spacecraft data both by finding the equivalent magnetic moment that would produce the observed field and by identifying the dipole terms in the spherical harmonic representations.

The former method makes use of the inversion equation (Smith et al 1974b):

$$\mathbf{M} = R^3(3\hat{R}\hat{R} - 2\hat{I})\mathbf{B}/2. \quad (3)$$

In this expression, **B** is the field, **M** the dipole moment, $\hat{R}$ is a unit vector equal to $\mathbf{R}/R$, where **R** is the radial position from the planet's center to the point of observation and $\hat{I}$ is a unit dyadic. The inversion matrix depends only on the trajectory parameters and, in rectangular coordinates, has elements $3X^2R/2 - R^3$, $3XYR/2 - R^3$, etc.

In terms of spherical harmonic coefficients, the magnitude of the dipole moment, $|\mathbf{M}|$, is given by

$$|\mathbf{M}| = [(g_1^0)^2 + (g_1^1)^2 + (h_1^1)^2]^{1/2} r_J^3. \quad (4)$$

The colatitude of the dipole (the tilt angle relative to the polar axis) is

$$\beta = \cos^{-1}(g_1^0/|\mathbf{M}|) \quad (5)$$

and the longitude of the magnetic pole in the northern hemisphere is

$$\phi = \tan^{-1}(h_1^1/g_1^1). \tag{6}$$

Thus, the dipole representation of the magnetic field requires three parameters for a complete description, the three components of the dipole moment or the first degree Gauss coefficients (g_1^0, g_1^1, h_1^1).

Offset Dipole

A magnetic configuration slightly more complicated than the centered dipole is the field produced by a dipole displaced from a planet's mass center. This solution requires a six parameter fit to the data (the three components of the dipole and the three components of the offset). The parameters can be derived using the inversion technique (given above) for a dipole but allowing the location of the dipole to be specified. In practice, the dipole offset is first guessed at and then the field vectors are inverted to find the equivalent dipole that would reproduce the observations. The offsets are then iterated successively to find the offset that minimizes the variances of the three components of **M**.

Another method that has been used involves the dipole and quadrupole coefficients of the spherical harmonic expansion. The dipole is shifted without changing its parameters and the offset is inferred from the spherical harmonic coefficients representing the five quadrupole terms, i.e. g_2^m and h_2^m. A theorem originally developed by Bartels (1936) gives algebraic expressions for the three cartesian components of the offset in terms of these coefficients:

$$\begin{aligned} X_0 &= a(L_1 - g_1^1 E)/3H_0^2 \\ Y_0 &= a(L_2 - h_1^1 E)/3H_0^2 \\ Z_0 &= a(L_0 - g_1^0 E)/3H_0^2 \\ H_0^2 &= (g_1^0)^2 + (g_1^1)^2 + (h_1^1)^2 \end{aligned} \tag{7}$$

where

$$\begin{aligned} L_0 &= 2g_1^0 g_2^0 + (g_1^1 g_2^1 + h_1^1 h_2^1)\sqrt{(3)} \\ L_1 &= -g_1^1 g_2^0 + (g_1^0 g_2^1 + g_1^1 g_2^2 + h_1^1 h_2^2)\sqrt{(3)} \\ L_2 &= -h_1^1 g_2^0 + (g_1^0 h_2^1 - h_1^1 g_2^2 + g_1^1 h_2^2)\sqrt{(3)} \\ E &= (L_0 g_1^0 + L_1 g_1^1 + L_2 h_1^1)/4H_0^2 \end{aligned}$$

The derivation is based on finding a new origin that reduces the resultant quadrupole to a minimum. Referred to this origin, the axial quadrupole and one of the nonaxial quadrupoles is reduced to zero, while the other nonaxial quadrupole is unchanged. The dipole moment is unchanged. Thus, the relation between the new and old coefficients is $\tilde{g}_1^0 = g_1^0$, $\tilde{g}_1^1 = g_1^1$, $\tilde{h}_1^1 = h_1^1$; $\tilde{g}_2^0 = \tilde{g}_2^1 = \tilde{h}_2^1 = 0$, and $\tilde{g}_2^2 = g_2^2$, $\tilde{h}_2^2 = h_2^2$.

Dual Dipole

An extension of the offset dipole representation is a model which consists of two dipoles. This model was originally suggested as a means of representing both the large scale features of the magnetic field and a field anomaly near the surface of Jupiter. It leads to a twelve parameter fit to the observations, i.e. the six parameters appropriate to each of the offset dipoles. An iterative least squares technique has been used to determine independently the two dipole moments and their offsets.

Coordinate Systems

A brief digression is necessary at this point in order to define the jovigraphic longitude systems used throughout this article. To define the longitude completely, it is necessary to specify (*a*) a polar axis and period of rotation, (*b*) a zero meridian, and (*c*) the sense in which the longitude angle increases, e.g. whether positive angles correspond to clockwise or counterclockwise rotations as viewed from above the planet.

The polar axis used in the definitions of the spacecraft and radio longitude systems coincides with the spin axis of Jupiter. Hence both coordinate systems are fundamentally related to the optical Systems I and II. The coordinate system generally used for the potential function derived from the spacecraft data is a (right-handed) spherical coordinate system (r, θ, ϕ) with positive longitudes defined as being counterclockwise when viewed from the north as shown in Figure 1. In contrast, the convention among astronomers has been to use a longitude system in which longitudes increase in a clockwise direction as viewed from above Jupiter's north pole. In this system, the central meridian longitude of Jupiter as seen from Earth increases in time.

In the analysis of the spacecraft data it is generally assumed that interior field sources rotate at a uniform rate that is determined from the radio astronomical observations of Jupiter. The radio observations give a rotation rate that appears to be constant and is currently known to within about one part in 10^6. A period of $09^h55^m25\overset{s}{.}711 \pm 0\overset{s}{.}04$ is consistent with the most recent determinations from decimetric and decametric data.

Traditionally, the period used by radio astronomers to describe their results is $09^h55^m29\overset{s}{.}37$ which is approximately 0.36 seconds shorter than the modern rotation period. This period, together with the assumption that the radio system coincides with the optical System II on January 1, 1957 ($00^h00^m00\overset{s}{.}0$ U.T.), leads to the System III (1957.0) longitudes which are widely used to present the radio astronomy data. In this longitude system, the magnetic poles and other radio features drift to higher longitudes at a rate of ~ +3° per year due to the slightly inaccurate period. This drift rate is particularly confusing when attempts are made to compare data taken years apart. As an example, Figure 2 shows the System III (1957.0) longitude of the north magnetic pole as determined by a number of different investigators. A new radio system designated $\lambda_{III}(1965)$, proposed by Riddle & Warwick (1976) to eliminate this drift (see also Seidelmann & Divine 1977), is currently coming into use. Transformation from a longitude in System III (1957.0), $\ell_{III}(1957.0)$, to a longitude in System III (1965), $\ell_{III}(1965)$, is accomplished using $\ell_{III}(1965) = \ell_{III}(1957.0) + 0.007 - (0.0083169)(t - 2438761.5)$ where t is the time at Jupiter in Julian Ephemeris days and ℓ is in degrees.

In order to compare the Pioneer models with radio astronomy observations, it is, of course, desirable to express the longitude in a common system. The Pioneer investigators have made this comparison easier by presenting their models in a coordinate system which is nearly identical to the System III (1957.0) coordinate system on a particular date. Acuña & Ness (1976a, b) use the equations given by Mead (1974) to accomplish this comparison. Smith et al (1976) define a coordinate system which coincides with System III (1957.0) on December 3, 1974 ($00^h00^m00^s.0$ U.T.) and thereafter Jupiter is assumed to rotate with a period of $09^h55^m29^s.7$ as recommended by Mead (1974).

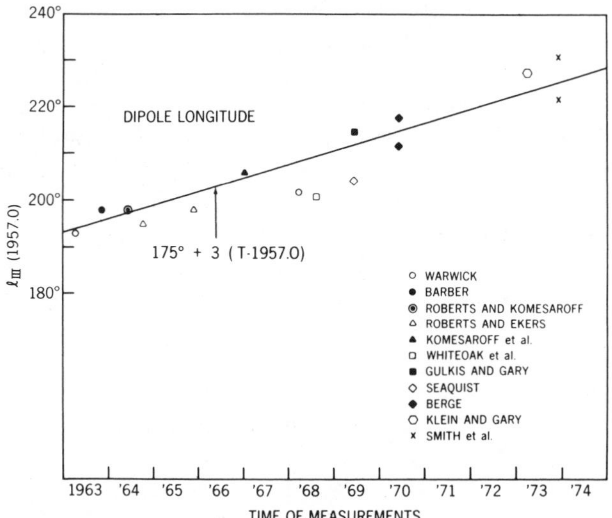

Figure 2 System III longitude of the north pole of the magnetic dipole as determined by a number of different investigators. The drift (3°/yr) is due to the inaccurate rotation rate used in the definition of System III longitude (after Mead 1974).

DIPOLE FIELD PARAMETERS

Dipole Parameters from Radio Astronomy Observations

The nonthermal decimeter component of radiation from Jupiter originates as a result of relativistic electrons moving in Jupiter's magnetic field, thereby producing synchrotron emission. An ultrarelativistic electron spiraling in a magnetic field emits the bulk of its radiation into a narrow cone about the direction of the instantaneous velocity of the electron. The radiation is in general elliptically polarized, with its sense determined by the direction of the magnetic field. Provided that the geometry of the magnetic field is that of a dipole, and that the ultrarelativistic electrons are predominantly in flat helix orbits, then the net radiation produced is beamed into the plane of the magnetic equator. The radiation is predominantly linearly polarized with the E-vector also in the magnetic equator. These characteristics of synchrotron radiation allow one to infer the dipole moment provided that the emission can be viewed from a number of different angles.

Detailed discussions of synchrotron emission of electrons trapped in a dipole field have been given by Chang & Davis (1962), Chang (1962), Thorne (1963), Legg & Westfold (1968), and others. These studies reveal that if the dipole axis is inclined with respect to the rotational axis and the synchrotron emission is viewed along a line perpendicular to the rotational axis, then the intensity of the radiation as seen by an observer will undergo a maximum twice each rotation, whenever the magnetic equator is viewed edge on. The two minima will occur when the two magnetic poles are tipped directly toward and away from the observer. The plane of linear polarization must rock back and forth relative to the rotational equator as the planet rotates. One or the other of the magnetic poles must lie on the central meridian whenever the electric vector of the polarized component is perpendicular to the axis of rotation. Also, a small amount of circular polarization should be present in the synchrotron emission and the sign of the circular polarization should vary as the planet rotates.

The beaming of the radiation, the rocking of the plane of polarization, and the variability of the circular polarization are all clearly visible in the ground-based data thereby supporting the dipole model. An historical account of these observations is given in Berge & Gulkis (1976). In Figure 3, we show the recent data of Neidhöfer et al (1977) which illustrate these effects. The observations were taken in August 1975 with the Effelsburg 100-m telescope operating at 11-cm wavelength.

Along the top of Figure 3 is a series of views of a body-centered dipole shown at successive values of the central meridian longitude (CML) of Jupiter. The arrows drawn under each view indicate the linear polarization directions. Twice during each rotation of Jupiter the magnetic equator is viewed edge on. The variation of the position angle of the electric vector between these two positions is a measure of the angle of the tilt of the dipole with respect to the rotation axis. The CML of the magnetic poles is indicated by the two vertical dashed lines. It can easily be deduced which of the two central meridian longitudes contains the magnetic pole in the northern (or southern) hemisphere by following the direction of polarization with rotation.

Roberts & Komesaroff (1965) give a good discussion of the beaming and polarization data in terms of a centered dipole model. They assume that the direction of linear polarization is perpendicular to the dipole

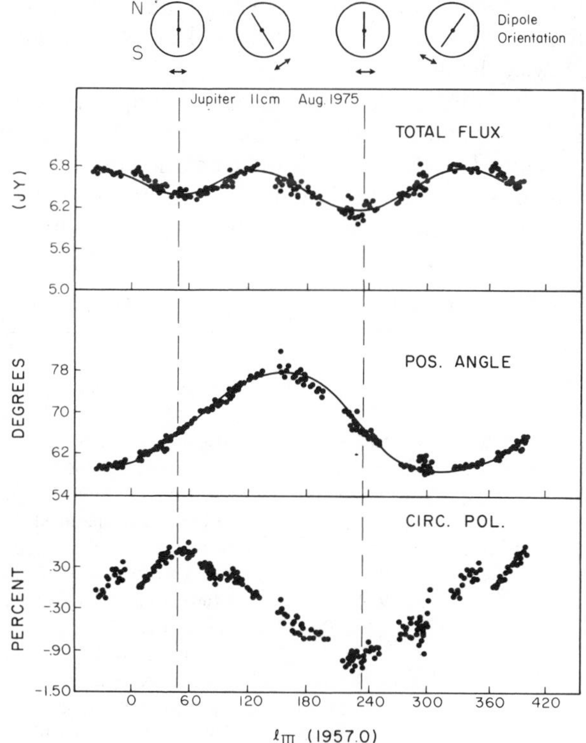

Figure 3 Variations in the flux density, the position angle, and the circular polarization as a function of System III central meridian longitude. The figures along the top of the page suggest the orientation of a body centered dipole which corresponds to the data (after Neidhöfer et al 1977).

axis and that the dipole axis is inclined at an angle, β, to the axis of rotation. Then the position angle, p, of the received polarization is given by

$$\tan(p - p_N) = \frac{-\tan\beta\,\sin(\lambda_{\text{III}} - \lambda_{\text{III}}^{\circ})\sec D_{\text{E}}}{1 - \tan\beta\,\tan D_{\text{E}}\cos(\lambda_{\text{III}} - \lambda_{\text{III}}^{\circ})}, \tag{8}$$

where p_N is the position angle of the normal to the axis of rotation, λ_{III} and $\lambda_{\text{III}}^{\circ}$ are the longitudes of the central meridian and the magnetic pole and D_{E} is the declination of the Earth as seen from Jupiter. The position angle of the electric vector is measured from celestial north in a counterclockwise sense. A fit of the observed position angle data to Equation (8) yields an estimate of β and $\lambda_{\text{III}}^{\circ}$. Provided that β is small and noting that $D_{\text{E}} \lesssim 3°$, Equation (8) can be written as

$$p - p_N = -\beta\sin(\lambda_{\text{III}} - \lambda_{\text{III}}^{\circ}). \tag{9}$$

In this form, the relationship between the amplitude of the position angle variations and the tilt is clear. The peak-to-peak amplitude of the position angle data is twice the dipole tilt.

In general, the position angle variation is not a simple sinusoid (Roberts & Komesoroff 1965) and it is not possible to define uniquely the dipole tilt without a detailed model. A simple interpretation of the position angle data is that the peak-to-peak variation of the electric vector is twice the dipole tilt. Another interpretation is that the fundamental harmonic term

Table 1 Estimates of the dipole tilt from position angle-longitude variations[a]

Wavelength (cm)	D_{E} (Deg)	β (Deg)	Reference
6	+0.8	8.8 ±1.0	Morris et al (1968)
	−0.8	9.16±0.27	Whiteoak et al (1969)
	−3.1	9.22±0.15	Gardner & Whiteoak (1977)
11	+2.8	9.9 ±0.26	Roberts & Komesaroff (1965)
	+2.8	9.8[b]	Komesaroff & McCulloch (1967)
	+0.8	9.6	Komesaroff & McCulloch (1967)
	+1.6	9.5	Komesaroff & McCulloch (1975)
	−2.5	9.52±0.11	Gardner & Whiteoak (1977)
	−2.1	9.47±0.12	Gardner & Whiteoak (1977)
	−3.1	9.44±0.09	Gardner & Whiteoak (1977)
	−2.8	9.43±0.13	Gardner & Whiteoak (1977)
21	+2.9	10.2 ±0.24	Roberts & Komesaroff (1965)
	−2.9	9.78±0.24	Gardner & Whiteoak (1977)
	−2.8	10.04±0.25	Gardner & Whiteoak (1977)

[a] Gardner & Whiteoak (1977).
[b] A revised estimate for the observations by Roberts & Komesaroff (1965).

of the Fourier series representation of the position angle data represents the dipole term. The amplitude yields the tilt angle, β, while the phase yields the longitude of the pole. Higher order terms are then assumed to represent departures from a centered dipole. As an example, Komesaroff & McCulloch (1975) give the following expression for the position angle variation referred to the epoch of Pioneer 10 encounter:

$$\begin{aligned} P(\lambda_{\mathrm{III}}) - P^\circ = & -9^\circ\!.5 \sin(\lambda_{\mathrm{III}} - 224^\circ) + 1^\circ\!.0 \sin 2(\lambda_{\mathrm{III}} - 118^\circ) \\ & -0.5 \sin 3(\lambda_{\mathrm{III}} - 221^\circ), \end{aligned} \tag{10}$$

from which we would deduce a dipole tilt of 9°.5. Half the peak-to-peak variation for the same function is 9°.3.

In Table 1, we give the data compiled by Gardner & Whiteoak (1977) for the dipole tilt based on the Fourier expansion method. The longitude of the pole in System III (1957.0) depends on the epoch of the measurement as shown in Figure 2.

The direction of the magnetic field can be deduced from measurements of the circular polarization. Berge (1965) noted that when the pole in the northern hemisphere was tipped toward the Earth, the circularly polarized component was in the lefthand sense. Since the emitting electrons appeared from the Earth to be moving clockwise, Berge deduced that the direction of the field at the equator must be southward.

Roberts & Komesaroff (1965) were the first to show how the magnetic field strength in the radiation belt can in principle be deduced from a measurement of the circular polarization. Berge (1965) originally estimated the average field intensity in the radiating region to be 0.17 gauss $< \bar{B} <$ 17 gauss. More recently, Komesaroff et al (1970) used the circular polarization measurements to limit the range to 0.4 gauss $< \bar{B} <$ 1.0 gauss. Assuming the models are valid for $r_J = 2$, where the bulk of the emission arises, then the equatorial field strength is between 3 and 15 gauss. The large uncertainty arises because of the dependence of the radiated power on both the field strength and the energy and pitch angle distributions of the relativistic electrons.

Dipole Parameters from Spacecraft Measurements

One of the first results obtained from the Pioneer 10 encounter with Jupiter in December 1973 was an unambiguous measurement which proved that the jovian field was not simply a centered dipole (Smith et al 1974a, b). The preliminary data analysis assumed that the source was a centered dipole and the vector field measurements were used to derive the equivalent magnetic moment that would produce the observed field (Equation 3). When successive field measurements were converted to the three dipole components and then to the equivalent dipole moment and the latitude and longitude of the pole, the position of the magnetic pole was found to move around systematically in jovigraphic coordinates as the planet rotated (see Figure 4). A consequence of this finding is that a more complicated model is needed in order to infer the dipole parameters.

The dipole parameters derived from various models and based on Pioneer magnetometer data are contained in Table 2. The parameters listed are the dipole moment, $|\mathbf{M}|$, the tilt angle, β (or equivalently the co-latitude, θ), and the longitude of the dipole axis. The latter is equivalent to the longitude of the magnetic pole located in the northern hemisphere of Jupiter. In Table 2, all longitudes have been converted to System III (1957.0), epoch 1974.9 (the Pioneer 11 encounter), which we designate as $\lambda_{\rm III}(1974.9)$. By convention the sense of the dipole is defined such that it corresponds to the direction from the negative to positive magnetic pole or alternatively lies along the normal to the plane of the equivalent current loop and has the same direction as the field lines interior to the current. Thus, at Jupiter, the interior dipole points nearly northward (except for the tilt angle) and gives rise to an external field which is southward at the equator.

The parameters in Table 2 have been obtained from a variety of publications which are identified in the table. Basically three different

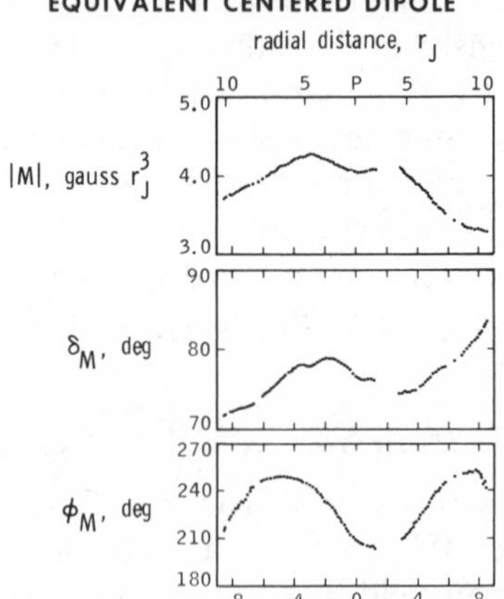

Figure 4 The magnetic moment (**M**), the latitude (δ_M), and the System III longitude (ϕ_M) of the centered dipole source that would give rise to the field observed by Pioneer 10 near periapsis. Significant deviations are present throughout the interval, showing that the field cannot be represented accurately by such a dipole (after Smith et al 1974b).

Table 2 Dipole moment

Model	M (Gauss r_J^3)	β (Deg)	λ_{III}(1974.9) (Deg)	Reference
P11(3,2)S	4.192	10.0	227.2	Smith et al (1975)
P10(3,2)	4.050	11.2	232.0	Davis et al (1975)[a]
P11(3,2)	4.165	10.1	227.9	Davis et al (1975)[a]
P10,11(3,2)	4.124	10.1	227.2	Davis et al (1975)[a]
P11(3,2)A	4.208	10.0	228.9	Davis et al (1975)[a]
P11 O_4	4.28	9.6	231.8	Acuña & Ness (1976a,b)
P10 D_2	4.00	10.6	225.0	Smith et al (1974b)
P11 D_4	4.225	10.8	230.9	Smith et al (1976)
P11 OTD	4.35	9.5	238.9	Acuña & Ness (1976a,b)
P10 DD	4.37	9.7	228.1	Melville (1976)

[a] Note that the publication date (1975) of the models in rows 2 through 5 precedes that of the models published in the proceedings of the Tucson Meeting (Smith et al 1976). However, the models indentified as Davis et al (1975) were actually derived subsequent to the Tucson Meeting and are believed by the investigators to be more accurate than the earlier published results.

types of models have yielded information about the dipole: spherical harmonic analyses, eccentric or offset dipoles, and dual dipoles.

Table 2 contains the results of several different spherical harmonic analyses. There are several reasons for these different models apart from their being based on Pioneer 10 or Pioneer 11 data alone or on both data sets taken together. In some analyses, the data were multiplied by weighting factors proportional to the radial distance raised to a prescribed power (Davis & Smith 1976). This weighting was introduced in order to cope with large differences in the observed field magnitude which could lead to the analysis being dominated by a relatively few measurements near Jupiter. Some analyses included spherical harmonic terms representing the contribution from currents external to the region of observation such as currents beyond $\sim 10r_J$ attributable to trapped particles. The generalized equation for the magnetic potential then contains an additional double series like that in Equation (1) but one which involves ascending, rather than descending powers of r (Smith et al 1976).

The dipole parameters in Table 2 are obtained as the lowest order terms of the spherical harmonic analysis as explained previously. The model designated O_4 is one such analysis that includes terms through octupole ($n = 3$) and is based on Pioneer 11 measurements made with the fluxgate magnetometer (Acuña & Ness 1976a,b). The other spherical harmonic analyses are based on the vector helium magnetometer (VHM) measurements (Smith et al 1976), including data obtained with both Pioneers 10 and 11 and are designated by the orders of the internal and external terms, e.g. P11(3, 2) means Pioneer 11 data were used to obtain three internal orders (dipole, quadrupole, and octupole) and two external orders (dipole and quadrupole). In addition to adding the external source terms to the field potential function, the VHM data analysis included a factor of the form $r^{-3/2}$ to weight the data. On the other hand, the O_4 analysis did not use weights. While weights do not lead, in general, to major differences, readers interested in their effect on the analysis are referred to Table 1 of Davis & Smith (1976).

The VHM model designated P11(3, 2)A includes data obtained during occultation of the spacecraft by Jupiter. These data were stored in an on-board memory but ambiguities have always been present with regard to the roll orientation of the spacecraft about the spin axis. The standard spherical harmonic analysis was therefore extended to include the occultation measurements of the axial field component parallel to the spin axis of Pioneer which do not depend on knowledge of the spacecraft roll attitude. Thus, this model makes use of data acquired nearest the planet during the occultation interval of 40 minutes.

The results based on VHM data yield $M = 4.13 \pm 0.08$ (2%) Gr_J^3, $\beta = 10.^\circ6 \pm 0.^\circ7$, and $\lambda_{III}(1974.9) = 229.^\circ6 \pm 2.^\circ4$. The values written in this form are the means of the maximum and minimum values and the corresponding deviations. The average and standard deviation were not used because the differences are not statistical but reflect the differing nature of the models. The maximum deviation for the dipole moment is also given in parentheses as a percentage of the moment. The result based on the fluxgate magnetometer O_4 model is $M = 4.28$ Gr_J^3, $\beta = 9.^\circ6$, and $\lambda_{III}(1974.9) = 235^\circ$.

Table 2 also contains the results of three offset dipole models, D_2 being the model derived from Pioneer 10 data, D_4 the model based on Pioneer 11 VHM results and OTD (for Offset Tilted Dipole) being obtained from the Pioneer 11 fluxgate magnetometer. The means and maximum deviations for these models are $M = 4.8 \pm 0.18$ (4.3%) Gr_J^3, $\beta = 10.^\circ2 \pm 0.^\circ7$, and $\lambda_{III}(1974.9) = 233.^\circ4 \pm 5.^\circ5$.

The final model in Table 2 is the main dipole inferred from a dual dipole model (DD). A 15-parameter version of this model derived by D. E. Jones and J. G. Melville was used (Melville 1976). The three additional parameters beyond the twelve needed to describe the dipoles are the components of a uniform field attributed to an external source. This model was based on Pioneer 10 data and yields $M = 4.37$ Gr_J^3, $\beta = 9.^\circ7$, and $\lambda_{III}(1974.9) = 228.^\circ1$.

The main conclusion to be drawn from the various models is that there is substantial agreement regarding the dipole parameters. The dipole

Table 3 Comparison of dipole parameters

Parameter	Decimetric radio emission	Pioneer 10 and 11 spacecraft
Inclination to rotation axis—(average)	9°6	10.2
—(range)	(8.8–10.2)	(9.5–11.2)
Polarity of magnetic pole in northern hemisphere	North	North
Longitude of magnetic pole in northern hemisphere epoch 1974.9	227–230	227.2–238.9
Dipole moment gauss r_J^3	3–15	4.0–4.4

moment appears to be known to within a few percent, the tilt angle to within at least 1°, and the longitude of the magnetic pole to within a few degrees.

Comparison of Radio Astronomy and Spacecraft Dipole Parameters

Table 3 shows a comparison between the spacecraft and radio astronomy determinations of the dipole parameters. It can be seen that there is good general agreement between the two methods of measurement. The spacecraft measurements when averaged together (a procedure which is not strictly correct since the data are not independent) yield a tilt and longitude which are nearly coincident with the radio astronomy determinations. The strength of the dipole moment was poorly known from the radio astronomy data and this large uncertainty has been eliminated by the spacecraft measurements.

HIGHER ORDER MULTIPOLES AND MAGNETIC ANOMALIES

Earth-based and spacecraft measurements have both revealed that Jupiter's magnetic field departs from a simple dipole. In the spacecraft data, the departures are revealed by the existence of nonzero spherical harmonic coefficients of a degree higher than one in the potential function for the field. The radio astronomy data, on the other hand, reveal the departures from a dipole field through various asymmetric variations in brightness, position angle, etc which are observed as Jupiter rotates. In this section, we discuss the asymmetries observed in the radio astronomy data and give the higher order multipoles inferred from the spacecraft measurements.

Asymmetries from Radio Astronomy

BRIGHTNESS DISTRIBUTION Branson (1968) measured the angular distribution of emission across Jupiter at 21 cm for three different ranges of central meridian longitude. The maps show a small but significant asymmetry in the form of a hot spot near longitude 221° [λ_{III}(1957.0), epoch 1974.9]. Maps and strip scans at 21 cm made by dePater & Dames (1978) during the Pioneer 10 flyby in December 1973 similarly showed the existence of a hot region near the longitude of $285 \pm 10°$ [λ_{III}(1957.0), epoch 1974.9]. Although Branson's maps were averaged over 120° of longitude, this is unlikely to account for the longitude difference between these two sets of observations. An enhanced volume emissitivity of ~ 1.6 in the vicinity of the hot spot was estimated by dePater & Dames (1978). Such a change could be due to local changes in the relativistic electron population, the magnetic field direction, or the magnetic field strength.

A topic related to the brightness distribution is the location of the nonthermal emission relative to the mass center of Jupiter. Since an asymmetry in the brightness distribution (or a dipole offset) can cause the apparent position of the centroid of the decimeter emission to vary with planetary rotation, a number of measurements of the centroid have been carried out. Table 4 gives the results of these observations. Taken individually none of the results represents a significant displacement of the radio centroid relative to the visible disk. However, they are all mutually consistent and, taken together, they indicate a displacement of about $0.10 r_J$ in the polar direction and $\sim 0.08 r_J$ in the equatorial direction in a longitude plane near 240° (epoch 1974.9).

POSITION ANGLE OF LINEAR POLARIZATION One of the best-determined parameters of the decimetric radiation is the variation of position angle (of linear polarization) with central meridian longitude. This curve departs

Table 4 Location of the centroid of the decimetric emission[a]

Wavelength (cm)	Polar displacement (r_J)	Equatorial displacement distance (r_J)	C.M. longitude at epoch 1974.9 (Deg)	Reference
11	$+0.10 \pm 0.20$	0.07 ± 0.05	251	Roberts & Ekers (1966)
11	$+0.20 \pm 0.23$	0.02 ± 0.05	232	McCulloch & Komesaroff (1973)
11	$+0.02 \pm 0.14$	0.11 ± 0.06	235 ± 1	Stannard & Conway (1976)
21	$+0.05 \pm 0.05$	0.08 ± 0.05	242 ± 25	Berge (1974)

[a] After Stannard & Conway (1976).

significantly from the simple sinusoidal variation predicted for a dipole field. When the declination of Earth is nonzero, the variation anticipated for a dipole is no longer a pure sinusoid. However, the second and higher order harmonic coefficients are less than 0.5% of the first harmonic in the extreme case $D_{\mathrm{E}} = \pm 3°$. The observations, on the other hand, typically lead to an amplitude of the second harmonic which is $\sim 10\%$ of the fundamental while the third harmonic is about half as large. The results of a recent set of measurements by Neidhöfer et al (1977) at 11 cm yielded the following result for the first three Fourier terms: (exclusive of the constant term)

$$P(\lambda_{\mathrm{III}}) \propto -9.^{\circ}40 \sin(\lambda_{\mathrm{III}} - 225.^{\circ}3) + 1.^{\circ}09 \sin 2(\lambda_{\mathrm{III}} - 105.^{\circ}3) - 0.^{\circ}05 \sin 3(\lambda_{\mathrm{III}} - 177.^{\circ}5) \tag{11}$$

for the epoch of the Pioneer 10 encounter.

Conway & Stannard (1972) have interpreted the departures from a simple sinusoid in terms of a magnetic anomaly which they locate near 220° longitude, close to the longitude of the north magnetic pole. This interpretation is based primarily on the suggestion that the "enhanced emission" observed by Branson (1968) is related to a magnetic anomaly. In view of this anomaly, these authors suggest that the position angle data may be fitted to a simple sinusoid if the hemisphere centered on the

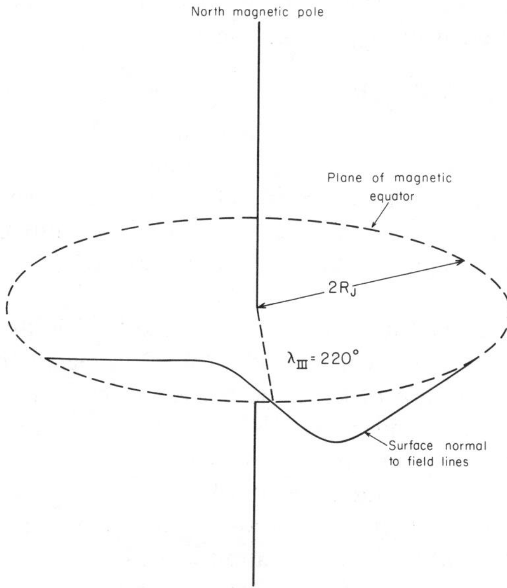

Figure 5 Sketch showing a suggested distortion of the surface normal to magnetic field lines at a distance of 2 r_{J} (after Conway & Stannard 1972).

north magnetic pole is excluded from the data. Figure 5 shows the suggested distortion of the surface normal to the magnetic field at a distance of two planetary radii. This suggestion, while interesting, needs to be viewed with caution. The recent observations by dePater & Dames (1978) show the hot spot has shifted approximately $+60°$ between 1967 and 1972. On the other hand, the position angle data have remained nearly constant. These results appear to contradict the original hypothesis.

BEAMING If Jupiter's magnetic field were a perfect dipole, its synchrotron emission would depend only on the magnetic latitude of the observer. The fact that the emissions at northern and southern magnetic latitudes are different is evidence of a field distortion (e.g. Roberts & Komesaroff 1965, Roberts & Ekers 1968). Warwick (1967) has noted that if the flux density measurements of Roberts and Ekers are plotted versus $\phi_m^1 = |\phi_m - 1.2°|$ where ϕ_m is the jovicentric magnetic latitude of the Earth, then the beaming appears to be symmetric. Warwick suggests that the distortion is due to a magnetic quadrupole so arranged that the field at the pole in the northern hemisphere is slightly weaker than that in the southern hemisphere.

DEGREE OF POLARIZATION The variation of the degree of linear polarization and circular polarization with rotation both show asymmetrical behavior similar to the beaming data; however, these effects have not been as widely discussed as those previously mentioned.

Spacecraft Measurements

DIPOLE OFFSET Table 5 contains the vector displacement of Jupiter's dipole, which has been inferred from various analyses of the Pioneer data. The offset is given in spherical coordinates with the displacement expressed in jovian radii, the latitude, δ, measured from Jupiter's rotational equator, and the longitude expressed in System III (1957.0) for epoch 1974.9.

The results given in Table 5 can be combined to yield the mean values and maximum deviations corresponding to the different methods. The offsets inferred from eccentric dipole models lead to $R = 0.089 \pm 0.021$ (24%), $\delta = 1.7 \pm 14.3°$, and $\lambda_{\text{III}} = 191.3 \pm 12.8°$. The offsets inferred from the various spherical harmonic analysis models lead to $R = 0.12 \pm 0.012$ (10%), $\delta = 12.3° \pm 20.3$, and $\lambda_{\text{III}} = 186.9 \pm 13.7°$. These values may be compared with those obtained from the 15 parameter dual dipole model which implies $R = 0.141$, $\delta = 20.4°$, and $\lambda_{\text{III}} = 233.4$.

Table 5 shows there is reasonable agreement among the various estimates of the offset, although the accuracy is evidently not as great

Table 5 Dipole offsets

Model[a]	R (r_J)	δ (Deg)	λ_{III}(1974.9) (Deg)
P11(3,2)S	0.112	5.3	176.1
P10(3,2)	0.132	32.6	200.5
P11(3,2)	0.128	4.0	175.8
P10,11(3,2)	0.132	6.3	176.5
P11(3,2)A	0.108	4.8	173.2
P11 O_4	0.131	−8.0	178.7
P10 D_2	0.110	15.9	178.5
P11 D_4	0.101	5.1	185.7
P11 OTD	0.068	−12.6	204.3
P10 DD	0.141	20.4	233.4

[a] The references associated with these models are the same as in Table 2.

as for the dipole parameters. Overall, the magnitude of the offset appears to be approximately $0.1 r_J$ and to be known to within at least 30%. The results indicate quite clearly that Jupiter's dipole is displaced from the planet's center by a distance which is well outside the limits of uncertainty. The Pioneer results also show that the offset is primarily equatorial with a mean latitude near 10°. If all the models in Table 5 are considered, the latitude is $10° \pm 23°$ which allows the offset to lie on either side of the equator and is consistent with zero. The System III (1957.0) longitude of the offset is near 205° within 29°.

MAGNETIC MULTIPOLES The fifteen spherical harmonic coefficients corresponding to the dipole, quadrupole, and octupole are given in Table 6 for the various Pioneer models. The coefficients (in gauss) are given for the same six models that are included in Table 2. Inspection of the table row-by-row shows good agreement between the quadrupole coefficients but occasional large discrepancies between corresponding octupole coefficients.

Basic limitations associated with the Pioneer observations, such as their being restricted to the two flight paths rather than being distributed over a sphere, raised the question as to how many terms could be accurately determined by harmonic analysis. One way to obtain an answer is to study how the coefficients are changed when different subsets of the data are analyzed. An analytical means of studying the stability of the higher order coefficients is provided by the so-called "condition numbers" (Lawson & Hanson 1974, Acuña & Ness 1976c). Although coefficients through hexadecapole ($n = 4$) have been calculated, the

Pioneer investigators agree that the spacecraft observations can only be extended with reasonable accuracy to the octupole moment. For that reason, the spherical harmonic results in Tables 6 and 7 are limited to the coefficients through order three.

Table 6 Spherical harmonic coefficients

Harmonic coefficients (gauss)	Model[a]					
	P11(3,2)S	P10(3,2)	P11(3,2)	P10,11(3,2)	P11(3,2)A	O_4
g_1^0	4.129	3.972	4.101	4.061	4.144	4.22
g_1^1	−0.492	−0.486	−0.488	−0.490	−0.481	−0.442
h_1^1	0.531	0.622	0.541	0.528	0.550	0.562
g_2^0	0.042	0.492	0.024	0.066	0.036	−0.203
g_2^1	−0.738	−0.696	−0.845	−0.871	−0.717	−0.871
h_2^1	−0.050	0.381	−0.072	−0.060	−0.078	−0.037
g_2^2	0.324	0.437	0.297	0.276	0.336	0.331
h_2^2	−0.381	−0.340	−0.435	−0.431	−0.317	−0.402
g_3^0	0.092	0.826	0.006	0.037	−0.047	−0.233
g_3^1	−0.413	−0.090	−0.684	−0.507	−0.606	−0.357
h_3^1	−0.084	−0.293	−0.056	−0.183	−0.044	−0.463
g_3^2	0.335	0.752	0.249	0.338	0.432	0.506
h_3^2	0.002	−0.174	−0.146	0.117	−0.158	0.096
g_3^3	−0.239	−0.191	−0.205	−0.487	−0.040	−0.292
h_3^3	0.118	−0.098	−0.300	0.092	0.136	0.233

[a] The references associated with each of these models are the same as in Table 2. These coefficients are valid in the r, θ, ϕ system at the epoch of Pioneer 11.

Table 7 Multipoles

Model[a]	Dipole moment ($G\ r_J^3$)	Quadrupole moment ($G\ r_J^4$)	Octupole moment ($G\ r_J^5$)
P11(3,2)S	4.19	0.89 (21%)	0.61 (15%)
P10(3,2)	4.05	1.09 (27)	1.19 (29)
P11(3,2)	4.17	1.00 (24)	0.83 (20)
P10,11(3,2)	4.12	1.01 (25)	0.82 (20)
P11(3,2)A	4.21	0.86 (20)	0.78 (19)
O_4	4.28	1.04 (24)	0.90 (21)

[a] For the references associated with each of these models, see Table 2.

Figure 6 is an isointensity contour map of the main field of Jupiter at the surface of the planet (assuming 1/15.4 flattening) and at $2r_J$ for the O_4 model (Acuña & Ness 1976a, b). In Figure 7, we show a comparison between the magnetic equator at $2r_J$ derived from the GSFC O_4 model and the VHM P11(3, 2) model. For these curves the magnetic equator is defined to be the locus of points where the field intensity is minimum along the field line. It is clear from this figure that the apparently minor differences in the spherical harmonic expansions can lead to significant differences in the location of the magnetic equator.

Information regarding the relative strength of the higher order terms in the spherical harmonic expansions is contained in Table 7. For this table, the coefficients corresponding to a given order have been combined by computing the square root of the sums of the square of the coefficients

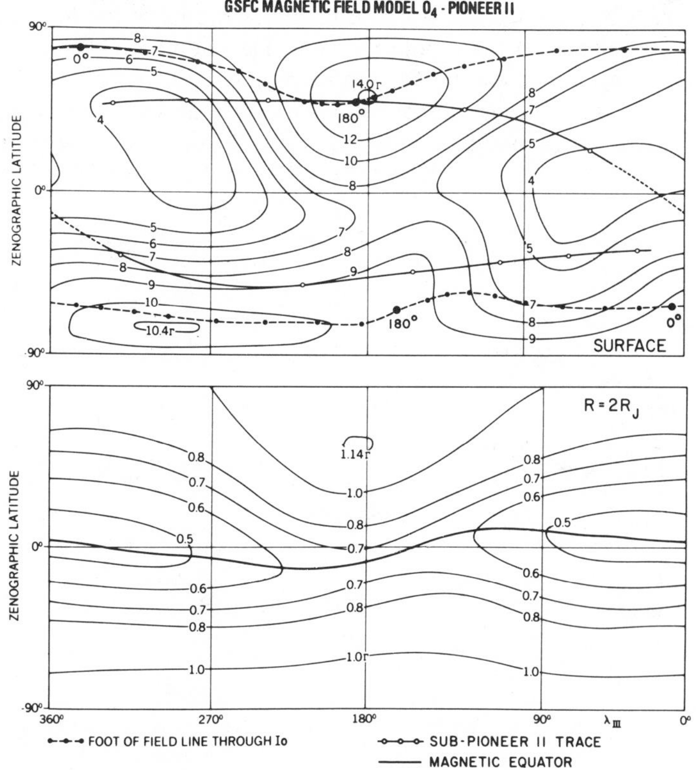

Figure 6 Isointensity contour maps of the main field of Jupiter at the surface of the planet (assuming 1/15.4 flattening) and at 2 r_J for the O_4 model. The trace of the Pioneer 11 trajectory is shown in the upper panel, as is the trace of the footprint of the flux tube associated with Io (after Acuña & Ness 1976a,b).

to obtain the dipole, quadrupole, and octupole moments. For example the quadrupole moment is given by

$$Q = [(g_2^0)^2 + (g_2^1)^2 + (h_2^1)^2 + (g_2^2)^2 + (h_2^2)^2]^{1/2}. \qquad (12)$$

The octupole moment is derived in a similar manner from the g_3^m and h_3^m terms. In addition to the magnitude, the ratio relative to the dipole moment (in percent) is given in parentheses.

As for the offset dipole, the results from the various models may be combined to yield mean values and the corresponding maximum deviations. Accordingly, the dipole moment is 4.17 ± 0.12 (3%) Gr_J^3, the quadrupole moment is 0.97 ± 0.12 (12%) Gr_J^4, and the octupole moment is 0.90 ± 0.29 (32%) Gr_J^5. The relative deviations associated with each moment rise from 3 to 12 to 32%, a clear indication of the increasing uncertainty to be attributed to the higher order terms.

There is also a substantial range in the strengths of the higher order moments relative to the dipole moment. Depending on the model, the ratio of the quadrupole to the dipole moment varies between 20 and 27%, while the octupole-dipole ratio is as small as 15% and as large as 29%. These ratios may be compared with the corresponding values for the geomagnetic field for which $Q/D = 0.13$ and $O/D = 0.09$. The larger

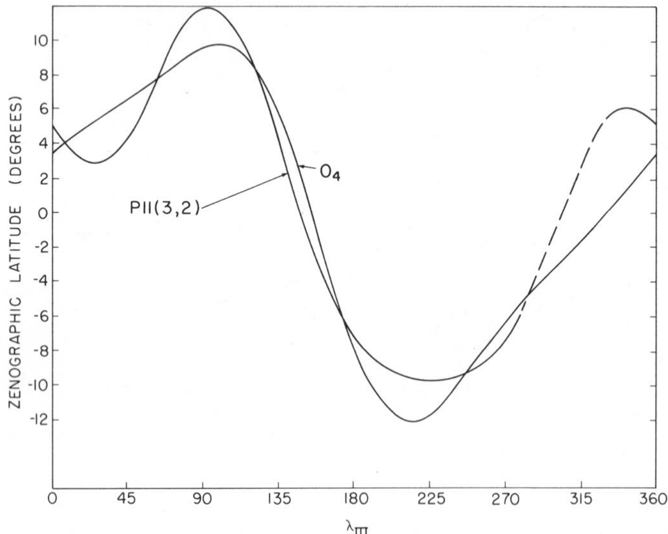

Figure 7 Magnetic equator at 2 r_J derived from the O_4 model (Acuña & Ness 1976a,b) and the P11(3,2) model (Davis et al 1975). The magnetic equator is defined to be the point where the magnetic field intensity is minimum along the field line.

ratios of the jovian multipoles have been interpreted as being qualitatively consistent with Jupiter's having a somewhat larger core than the Earth when each is compared to the diameter of the planet.

Comparison of Higher Order Moments and Magnetic Anomalies

The correlation of asymmetries in the Earth-based data with spacecraft data is an important problem on which there has been little work. It is difficult to make quantitative comparisons without a detailed model of the synchrotron emission and, except for the dipole model, this has not yet been achieved. Consequently, it is only possible at this time to make qualitative comparisons of the phenomenology of the Earth-based data with the in situ data. This procedure has serious limitations since the integrated properties of the synchrotron radiation produced in a complex magnetic field are not obvious a priori.

We may start out by comparing the radio centroid position with the offset dipole model derived from the spacecraft measurements. The latter indicates an offset of approximately $0.1r_J$, near the equatorial plane, and near the longitude 189°. The Earth-based measurements, which contain large uncertainties, yield a polar displacement of $0.1r_J$, an equatorial displacement of $0.08r_J$, and a longitude of 240°. We believe these two data sets are consistent, both implying a rather well-centered magnetic field. If any inconsistency exists between the data sets it is the polar displacement seen in the radio astronomy data which is absent from the spacecraft data.

The asymmetric variation of position angle with central meridian longitude is difficult to compare with the in situ data without a detailed model. Nevertheless, we have tried to make a comparison by using a number of simplifying assumptions. We assume that the position angle of the electric vector is determined solely by the electrons on the magnetic equator, that beaming is unimportant, that a thin shell model near $2r_J$ is representative of the belts, and that blockage of the radiation by the planet needs to be accounted for. Using these assumptions, we have calculated the net linear polarization as a function of central meridian longitude for the O_4 and P11(3,2) models. The results are shown in Figure 8 along with the experimental position angle curve given by Komesaroff & McCulloch (1976). The theoretical curves are seen to be in good agreement with the experimental data. They lie on both sides of the experiment data thereby suggesting that a good model for the synchrotron emission might help sharpen the in situ models.

Warwick's (1967) suggestion that the magnetic field might be weaker in

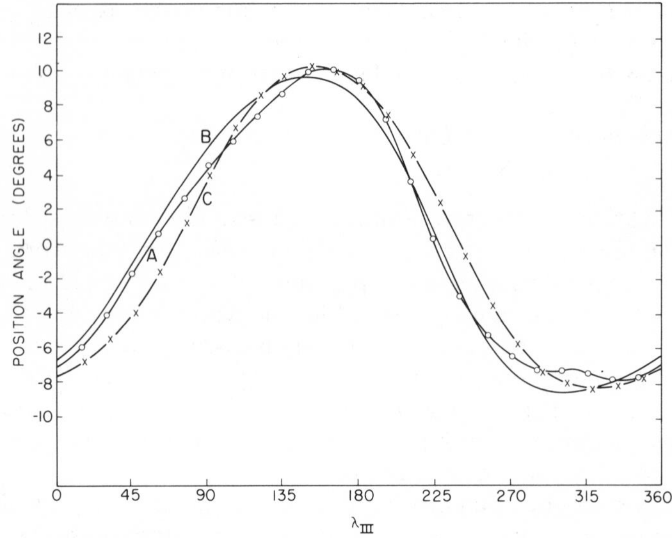

Figure 8 Theoretical and experimental curves of the position angle of the electric vector versus central meridian longitude. The curve labeled *B* is an experimental curve given by Komesaroff & McCulloch (1975) corrected to epoch 1974.9. Curves labeled *A* and *C* are derived from magnetic equators of Davis et al (1975) and Acuña & Ness (1976a,b), respectively, as shown in Figure 7.

the northern hemisphere than in the southern hemisphere in order to explain the asymmetrical beaming does not appear to be borne out by the isointensity contour plots of Figure 6. These show stronger fields in the northern hemisphere. A detailed model taking into account the vector direction of the magnetic field needs to be developed in order to investigate this effect more fully.

SECULAR CHANGES

Hide & Stannard (1976) (see also Runcorn 1967, Hide 1967) point out that, by analogy with the geomagnetic field which undergoes slow changes on the time scale of decades or longer, Jupiter's magnetic field may also undergo secular variations. Since observations of secular variations might provide otherwise unobtainable information on the deep interior of Jupiter, this subject is important and will be discussed here even though no unambiguous variations have been detected at this time.

In principal, secular variations could be detected by comparing in situ data spaced in time or by analyzing ground-based astronomy data accumulated over years. Ground-based data may reflect secular variations through (*a*) changes in rotation period, (*b*) changes in shape or orientation of the magnetic field, or (*c*) changes in the magnetic field strength.

To date, the only in situ measurements of Jupiter's magnetic field which can be used to search for secular variations were those made one year apart by Pioneers 10 and 11. By analogy with the geomagnetic field, whose moment has been observed to decrease roughly 5% per century and to drift westward approximately 0.2 degrees per year, this interval is too short to expect to see variations. Nevertheless, the Pioneer 10 and Pioneer 11 measurements yielded dipole moments which differed by 6% while the tilt angle and offset for the two models were in close agreement. The difference in the dipole moments is thought to represent an improvement in the measurement because of the more favorable Pioneer 11 trajectory. However, the possibility of secular change cannot be ruled out (Smith et al 1975).

Earth-based measurements, on the other hand, have been obtained for over two decades and a number of authors have examined these data for evidence of temporal variations (e.g. Hide & Stannard 1976). As ground-based measurement may provide the most practical means of searching for secular variations, we summarize, in the following section, the studies carried out to date.

Rate of Rotation of the Magnetic Field

Jovian decametric radio sources and the jovian decimetric synchrotron emissions are presumably tied to Jupiter's magnetic field. Hence, any changes in the rate of rotation of the magnetic field should be reflected in a longitude change of the various radio features. At decimetric wavelengths, rotation periods have been determined by comparing the position angle vs longitude or flux density vs longitude data taken years apart. A simple tracking of the position of the magnetic poles has also been used. At decametric wavelengths, rotation periods have been determined by measurements of source locations, storm commencement times, characteristic dynamic spectral features, and various statistical correlations. These are reviewed by Carr & Desch (1976).

Rotation rates obtained from observations of the decametric sources and the synchrotron emission have yielded periods which are in good agreement with each other to nearly 1 part in 10^6. Tables 8 and 9 (see Carr & Desch 1976, Berge & Gulkis 1976, Hide & Stannard 1976) give some recent determinations of the period.

While no unambiguous variations in the period have been measured, Carr & Desch (1976) mention that there is a suggestion of a decrease

Table 8 Decimetric determinations of Jupiter's rotation period

λ	Variation	Period	Reference
10 cm	Comb.	$9^h55^m29\overset{s}{.}70 \pm 0\overset{s}{.}05$	Bash et al (1964)
21	PA	29.37 ± 0.5	Roberts & Komesaroff (1965), Berge (1966)
21	S	29.50 ± 0.29	Davies & Williams (1966)
11	PA	29.83 ± 0.26	Komesaroff & McCulloch (1967)
6	PA	29.69 ± 0.05	Whiteoak, Gardner & Morris (1969)
13	S	29.72 ± 0.11	Gulkis & Gary (1971)
13	S	29.75 ± 0.05	Gulkis et al (1973)
21	S	29.72 ± 0.07	Berge (1974)
11	PA	29.76 ± 0.02	Komesaroff & McCulloch (1976)
6,11,21	PA	29.76 ± 0.02	Gardner & Whiteoak (1977)

Table 9 Decametric determinations of Jupiter's rotation period

Method	Period	Reference
Source A alignment 1966–1970	$09^h55^m29.72 \pm 0.17$	Carr & Desch (1976)
Commencement time	$29.70 + 0\overset{s}{.}5$	Duncan (1967, 1971)
Power spectrum	29.70 ± 0.02	Kaiser & Alexander (1972)
Occurrence probability	29.67 ± 0.01	Lecacheux (1974)

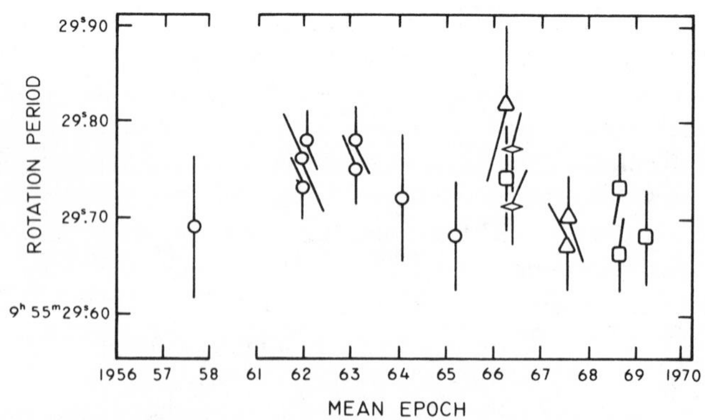

Figure 9 Mean rotation periods over intervals of approximately 12 years as a function of mean epoch of the interval. Frequencies are from 18 to 26.3 MHz (after Carr & Desch 1976).

in the measured rotation period from 1962 to 1969. The decrease, suggested by the data in Figure 9, amounts to 0.03 s. If this decrease is real, radio source positions will appear to move 0.26 degrees per year (in a longitude frame corrected for a modern period).

Position Angle Variations

The variation of polarization position angle with longitude of the central meridian reflects the geometry of the jovian magnetic field in a complex manner. While it has not been possible to work backward from the position angle data and derive more than the inclination of the dipole, it is nevertheless possible in principle to examine these data for possible signs of changes in the magnetic field shape. McCulloch (1975) has examined the amplitudes and phases of the Fourier components over an eleven year interval. He finds no significant phase variation with epoch for any harmonic, and the amplitudes of the first (dipole tilt) and third harmonic are essentially constant. The amplitude of the second harmonic may depend on the viewing angle (D_E).

Hide & Stannard (1976) used half the peak-to-peak variation in position angle to search for evidence of temporal variations in the inclination of the dipole. They find that within the error of their data, there is no marked change in the inclination of the dipole, although a least squares fit to the data suggests a secular decrease of 0.07 ± 0.03 degrees per year (standard error). Recent measurements by Neidhöfer et al (1977) do not support this trend.

Magnetic Field Strength

Long-term variations in the magnetic field strength may be reflected in the degree of circular polarization or total intensity of decimetric emission or possibly by variations of the upper decametric cutoff frequency near 40 MHZ (Hide & Stannard 1976). We are not aware of any systematic monitoring of the decametric cutoff frequency. Long-term variations of the decimetric flux density are now well established (e.g. Klein 1976), however, the underlying cause of these variations is not known. Klein (1976) argues that the variation is not caused simply by a change in the magnetic field intensity in the emission region because the observed variations at different wavelengths do not correspond to a simple change in the magnetic intensity.

Stannard & Conway (1976) have compared the degree of circular polarization measured at 21 cm and 49 cm over a three year interval. To within the errors of measurement, the circular polarization was found not to vary, thereby suggesting an upper limit of about 20% to any variations in the field strength between epoch 1967 and 1970.

Linear Polarization

Klein (1976) has examined the degree of linear polarization as a function of time for 21-cm, 13-cm, and 10-cm data. He finds that, with the exception of two observations in one year, the peak degree of polarization has not changed by more than 1% from the mean value during a twelve-year interval. Gardner & Whiteoak (1977) report a similar finding at 11 cm and 21 cm; however, between 1968 and 1971 they found the percentage polarization decreased by 17%. The reason for this decrease is not understood. However, if the 6-cm radiation originates closer to the planet than the longer wavelength radiation, then the variations could be due to secular variations of the higher order moments of the magnetic field which are expected to be more significant close to the planet.

CONCLUSION AND DISCUSSION

In this paper, we have attempted to bring together a description of the inner jovian magnetic field as determined by the Pioneer 10 and Pioneer 11 spacecraft and the Earth-based decimetric radio observations having a bearing on the magnetic field configuration. We have demonstrated that the dipole parameters determined from the two data sets are in good agreement. However, we also point out that little progress has been made to date in reconciling the asymmetries seen in the Earth-based data and the spacecraft data. Clearly, we are only in the first stages of bringing these synergistic measurements together.

It appears unlikely that there will be significant improvements to the in situ observations within the foreseeable future. Present plans for future space missions to Jupiter, namely the Voyager mission which is currently underway and the Galileo atmospheric probe and orbiter, will not provide magnetic field measurements any nearer the planet than $\simeq 4\ r_J$, i.e. at distances significantly larger than the closest approach of Pioneers 10 and 11. This limitation provides additional impetus to studies of the combined spacecraft and radio astronomy observations for the purpose of developing improved models of the jovian field.

Several steps need to be taken in order to make real progress toward improving the magnetic field models. On the theoretical side, models of the synchrotron emission for an arbitrary magnetic field configuration must be developed and studies carried out. A more accurate calculation of the linear polarization than the one presented in Figure 7 should allow one to close the gap between the VHM and the fluxgate magnetometer measurements. On the experimental side, we need high-resolution maps of the radiation belts in all polarizations. These measurements will show local distortions of the magnetic field directly, thereby allowing better representation of the spacecraft measurements. Also we believe that more observational studies of Jupiter should be carried out at wavelengths between 3 and 6 cm. These measurements may be more sensitive to the higher order magnetic moments than the longer wavelength measurements. There is a suggestion in Table 1 that the dipole tilt becomes progressively smaller as the wavelength decreases.

Finally, the decametric observations can provide strong constraints on the magnetic field configuration. If the sources of the decametric emission are close to the visible surface of Jupiter, these emissions might provide information on the highest order moments of Jupiter's field. However, a more fully developed theory of the decametric emission will need to be developed before such constraints can be applied.

ACKNOWLEDGMENTS

We are grateful to M. H. Acuña, A. Frandsen, D. E. Jones, M. J. Klein, J. Mannan, and N. F. Ness for important discussions relating to this work. We are especially grateful to L. Davis, Jr., for helpful comments and analyses affecting several sections of the article. We also thank M. Janssen, R. Newburn, and E. T. Olsen for reading the manuscript and making a number of suggestions for its improvement. This work presents the results of one phase of the research carried out at the Jet Propulsion Laboratory, California Institute of Technology, under contract NAS7-100, sponsored by the National Aeronautics and Space Administration.

Literature Cited

Acuña, M. H., Ness, N. F. 1976a. Results from the GSFC fluxgate magnetometer on Pioneer 11. See Gehrels 1976, pp. 830–47

Acuña, M. H., Ness, N. F. 1976b. The main magnetic field of Jupiter. *J. Geophys. Res.* 81:2917–22

Acuña, M. H., Ness, N. F. 1976c. The magnetic field of Jupiter. In *Magnetospheric Particles and Fields*, ed. B. M. McCormac, pp. 311–23. Dordrecht, Holland: Reidel

Bartels, J. 1936. Eccentric dipole approximating the Earth's magnetic field. *Terr. Magn.* 41:225–50

Bash, F. N., Drake, F. D., Gundermann, E., Helis, C. E. 1964. 10-cm observations of Jupiter, 1961–1963. *Astrophys. J.* 139: 975–85

Berge, G. L. 1965. Circular polarization of Jupiter's decimetric radiation. *Astrophys. J.* 142:1688–93

Berge, G. L. 1966. An interferometric study of Jupiter's decimeter radio emission. *Astrophys. J.* 146:767–98

Berge, G. L. 1974. The position and Stokes parameters of the integrated 21 cm radio emission of Jupiter and their variation with epoch and central meridian longitude. *Astrophys. J.* 191:775–84

Berge, G. L., Gulkis, S. 1976. Earth-based radio observations of Jupiter: Millimeter to meter wavelengths. See Gehrels 1976, pp. 621–92

Branson, B. F. B. A. 1968. High resolution radio observations of the planet Jupiter. *Mon. Not. R. Astron. Soc.* 139:155–62

Carr, T. D., Desch, M. D. 1976. Recent decametric and hectometric observations of Jupiter. See Gehrels 1976, pp. 693–773

Chang, D. B. 1962. Synchrotron radiation as the source of the polarized decimeter radiation from Jupiter. PhD thesis, California Institute of Technology. *Boeing*

Scientific Research Laboratories Document D1 82-0129

Chang, D. B., Davis, L., Jr. 1962. Synchrotron radiation as the source of Jupiter's polarized decimeter radiation. *Astrophys. J.* 136:567–81

Chapman, S., Bartels, J. 1940. *Geomagnetism*, pp. 639–68. London: Oxford Univ. Press

Conway, R. G., Stannard, D. 1972. Nondipole terms in the magnetic fields of Jupiter and the Earth. *Nature Phys. Sci.* 239:142–43

Davies, R. D., Williams, D. 1966. Observations of the continuum emission from Venus, Mars, Jupiter, and Saturn at 21.2 cm wavelength. *Planet. Space Sci.* 14:15–31

Davis, L., Jr., Jones, D. E., Smith, E. J. 1975. *The magnetic field of Jupiter*. Presented at American Geophysical Meeting, San Francisco, CA, 12 Dec.

Davis, L., Jr., Smith, E. J. 1976. The Jovian magnetosphere and magnetopause. In *Magnetospheric Particles and Fields*, ed. B. M. McCormac, pp. 301–10. Dordrecht, Holland: Reidel

dePater, I., Dames, H. A. C. 1978. Jupiter's radiation belts and atmosphere. *Astron Astrophys.* In press

Duncan, R. A. 1967. Jupiter's rotation period. *Planet. Space Sci.* 15:1687–94

Duncan, R. A. 1971. Jupiter's rotation. *Planet. Space Sci.* 19:391–98

Gardner, F. F., Whiteoak, J. B. 1977. Linear polarization observations of Jupiter at 6, 11 and 21 cm wavelengths. *Astron. Astrophys.* 60:369–75

Gehrels, T., ed. 1976. *Jupiter*. Tucson: Univ. Arizona Press. 1254 pp.

Gulkis, S., Gary, B. 1971. Circular polarization and total-flux measurements of Jupiter at 13.1 cm wavelength. *Astron. J.* 76:12–16

Gulkis, S., Gary, B., Klein, M., Stelzried, C. 1973. Observations of Jupiter at 13 cm wavelength during 1969 and 1971. *Icarus* 18:181–91

Hide, R. 1967. On the dynamics of Jupiter's interior and the origin of his magnetic field. In *Magnetism and the Cosmos*, ed. W. R. Hindmarsh, F. J. Lowes, P. H. Roberts, S. K. Runcorn, pp. 378–95. New York: Elsevier. 436 pp.

Hide, R., Stannard, D. 1976. Jupiter's magnetism: Observations and theory. See Gehrels 1976, pp. 767–87

Kaiser, M. L. Alexander, J. K. 1972. The Jovian decametric rotation period. *Astrophys. Lett.* 12:215–17

Klein, M. J. 1976. The variability of the total flux density and polarization of Jupiter's decimetric radio emission. *J. Geophys. Res.* 81:3380–82

Komesaroff, M. M., McCulloch, P. M. 1967. The radio rotation period of Jupiter. *Astrophys. Lett.* 1:39–41

Komesaroff, M. M., McCulloch, P. M. 1975. Asymmetries of Jupiter's magnetosphere. *Mon. Not. R. Astron. Soc.* 172:91–95

Komesaroff, M. M., McCulloch, P. M. 1976. Evidence for the unexpected time-stable symmetry of the Jovian magnetosphere. *J. Geophys. Res.* 81:3407–11

Komesaroff, M. M., Morris, D., Roberts, J. A. 1970. Circular polarization of Jupiter's decimetric emission and the Jovian magnetic field strength. *Astrophys. Lett.* 7:31–36

Lawson, C. L., Hanson, R. J. 1974. *Solving Least Square Problems*. New York: Prentice Hall

Lecacheux, A. 1974. Periodic variations of the position of Jovian decameter sources in longitude (System III) and phase of Io. *Astron. Astrophys.* 37:301–4

Legg, M. P. C., Westfold, K. C. 1968. Elliptic polarization of synchrotron radiation. *Astrophys. J.* 154:499–514

McCulloch, P. M. 1975. Long term variations in Jupiter's 11 cm radio emission. Proc. ASA2:340–42

McCulloch, P. M., Komesaroff, M. M. 1973. Location of the Jovian magnetic dipole. *Icarus* 19:83–86

Mead, G. D. 1974. Magnetic coordinates for the Pioneer 10 Jupiter encounter. *J. Geophys. Res.* 79:3514–21

Melville, J. G. 1976. Discussion. See Gehrels 1976, pp. 826–27

Morris, D., Whiteoak, J. B., Tonking, F. 1968. The linear polarization of radiation from Jupiter at 6 cm wavelength. *Aust. J. Phys.* 21:337

Neidhöfer, J., Booth, R. S., Morris, D., Wilson, W., Biraud, F., Ribes, J. C. 1977. New measurements of the Stokes parameters of Jupiter's 11 cm radiation. *Astron. Astrophys.* 61:321–28

Riddle, A. C., Warwick, J. W. 1976. Redefinition of System III longitude. *Icarus* 27:457–59

Roberts, J. A., Ekers, R. D. 1966. The position of Jupiter's Van Allen Belt. *Icarus* 5:149–459

Roberts, J. A., Ekers, R. D. 1968. Observations of the beaming of Jupiter's radio emission at 620 and 2650 Mc/sec. *Icarus* 8:160–65

Roberts, J. A., Komesaroff, M. M. 1965. Observations of Jupiter's radio spectrum and polarization in the range from 6 to 100 cm. *Icarus* 4:127–56

Runcorn, S. K. 1967. On the Rotation of Jupiter. In *Magnetism and the Cosmos*, ed. W. R. Hindmarsh, F. J. Lowes, P. H. Roberts, S. K. Runcorn, pp. 365–77. New York: Elsevier. 436 pp.

Seidelmann, P. K., Divine, N. 1977. Evaluation of Jupiter longitudes in System III (1965). *Geophys. Res. Lett.* 4:65–68

Smith, E. J., Davis, L., Jr., Jones, D. E. 1976. Jupiter's magnetic field and magnetosphere. See Gehrels 1976, pp. 788–829

Smith, E. J., Davis, L., Jr., Jones, D. E., Coleman, P. J., Jr., Colburn, D. S., Dyal, P., Sonett, C. P. 1975. Jupiter's magnetic field, magnetosphere, and interaction with the solar wind: Pioneer 11. *Science* 188: 451–54

Smith, E. J., Davis, L., Jr., Jones, D. E., Colburn, D. S., Coleman, P. J., Jr., Dyal, P., Sonett, C. P. 1974a. Magnetic field of Jupiter and its interaction with the solar wind. *Science* 183:305–6

Smith, E. J., Davis, L., Jr., Jones, D. E., Coleman, P. J., Jr., Colburn, D. S., Dyal, P., Sonett, C. P., Frandsen, A. M. A. 1974b. The planetary magnetic field and magnetosphere of Jupiter: Pioneer 10. *J. Geophys. Res.* 79:3501–13

Sommerfeld, A. 1964. *Partial Differential Equations in Physics*, pp. 123–35, 149–52. New York: Academic

Stannard, D., Conway, R. G. 1976. Recent observations of the decimetric radio emission from Jupiter. *Icarus* 27:447–52

Thorne, K. S. 1963. The theory of synchrotron radiation from stars with dipole magnetic fields. *Astrophys. J. Supp.* 8:1–30

Warwick, J. W. 1967. Radiophysics of Jupiter. *Space Sci. Rev.* 6:841–91

Whiteoak, J. B., Gardner, F. F., Morris, D. 1969. Jovian linear polarization at 6 cm wavelength. *Astrophys. Lett.* 3:81

Ann. Rev. Earth Planet. Sci. 1979. 7:417–42

SYNTHETIC SEISMOGRAMS

D. V. Helmberger and L. J. Burdick[1]

Seismological Laboratory, California Institute of Technology, Pasadena, California 91125

INTRODUCTION

In the past few years, our understanding of earth structure and of earthquakes has increased dramatically because of our greatly increased ability to interpret seismograms. Most of this progress can be attributed to the development of the necessary techniques for numerically synthesizing seismograms on a computer. Each of the processes that affect the waveform such as the seismic source or propagation of the wave through the earth is modeled by a linear time dependent operator. These linear operators are then convolved together to reproduce an analog waveform which can be compared directly to the data. In this arrangement, it is very simple to determine the nature and the magnitude of the effect that the various operators have on the waveform. The direct P and S body phases have proved to be the most useful types of seismic wave for modeling to date. A series of source studies based on body wave data has recently been completed. It has led to a completely new outlook on the relative importance of the rupture process versus the interaction of the seismic wave field with the earth's free surface in determining the body waveform. We shall review these studies in some detail. These source studies produced reliable source models which were then used to study the effects of the velocity structure of the upper mantle and of the shallow velocity structure near the source. We shall also discuss these results. Analogous progress which will not be covered in this review has also been made on modeling surface waves with time domain synthetics. Because of the much greater wavelengths involved, this line of research has not led to the type of detailed information about earthquakes and the earth which has come from body waves. It has led, nonetheless, to important evidence concerning the amount of slow deformation that accompanies some earthquakes and it appears to be much more useful for studying very large earthquakes. Important work in this field has been presented by Kanamori & Stewart (1976) and Gilbert & Dziewonski (1975).

[1] Now at Lamont-Doherty Geological Observatory, Columbia University, Palisades, New York 10964.

0084-6597/79/0515-0417$01.00

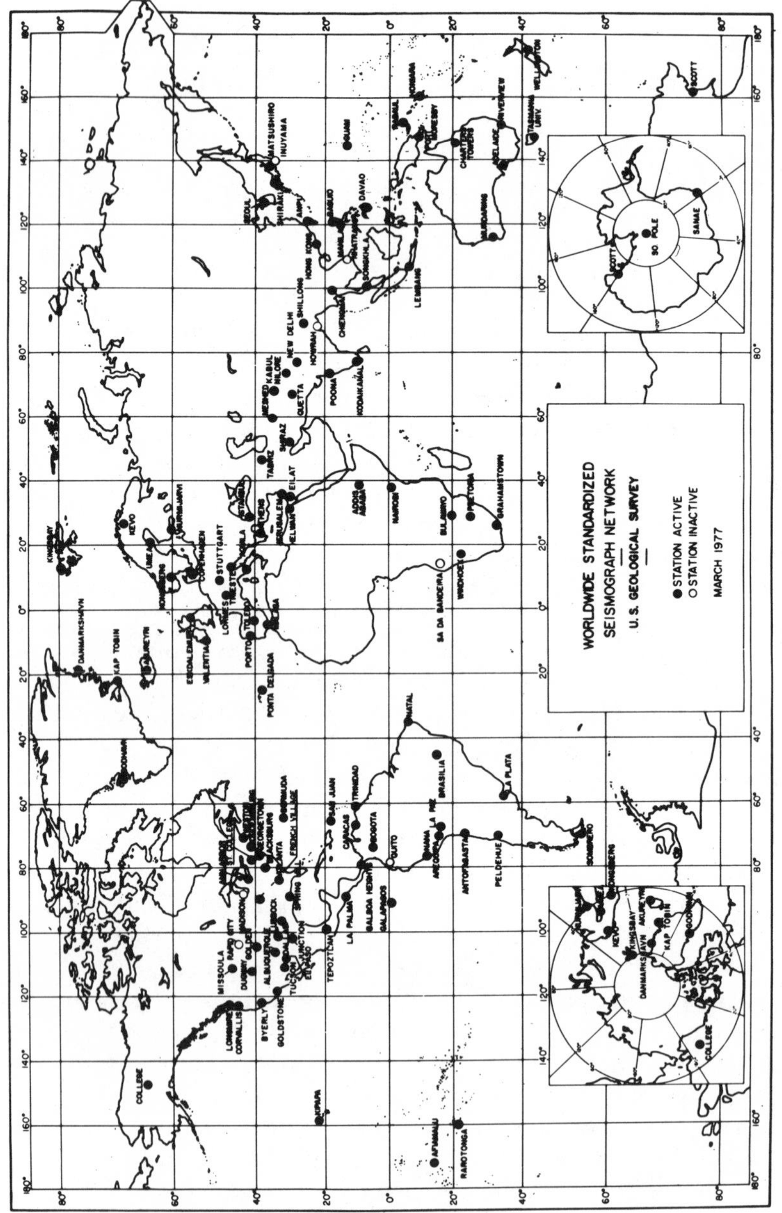

Figure 1 WWSS global network (after USGS).

Because we will be dealing with body waves we will be primarily concerned with events in the 5.0 to 6.5 magnitude class. These events are sufficiently large to be well recorded on the world wide seismic system network (WWSSN) but not so large as to drive the instruments off scale. The WWSSN global network is displayed in Figure 1. Note that the coverage is uneven with dense coverage occurring in affluent nations. This means that earthquakes in some regions like southern California provide much more complete data sets. For this reason, many of the most complete synthetic seismogram studies have been performed on California earthquakes. One such event was the 4/9/68 Borrego Mountain earthquake. A typical body wave recording of the event is shown for reference in Figure 2. The WWSSN stations provide six records: three components of motion from a long period instrument and three components of motion from a short period instrument. We shall show that most of the features of the first 20 seconds or so of the records in Figure 2 can be explained by a relatively simple source model which radiates a simple one-sided pulse. The complications in the waveforms are caused by the interaction of this pulse with the free surface near the source and the recording instrument.

There are four basic types of linear operators which are generally included in the computation of a synthetic seismogram. These represent the seismic instrument, $I(t)$, the attenuation operator, $A(t)$, the source operator, $E(t)$, and a wave propagation operator, $M(t)$. The instrument operator is of course well known. The attenuation operator does not introduce any structure into the waveforms in most problems of interest, but only has the effect of smoothing them. It must be included in the

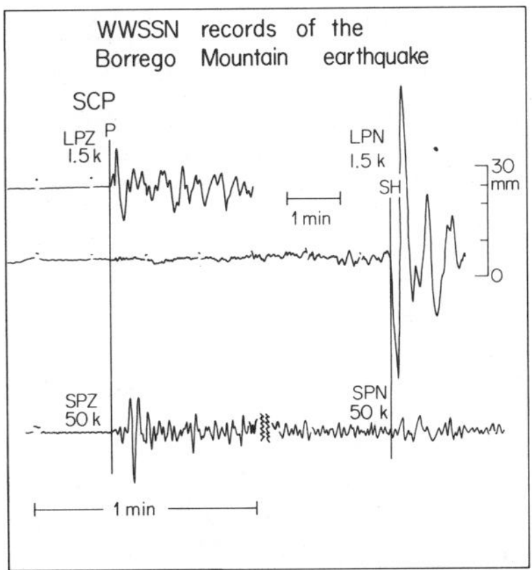

Figure 2 Example observations of the Borrego Mountain earthquake as recorded by SCP (State College, Penn.) which shows the long and short period P waves on the vertical component and the SH waves on the horizontal component.

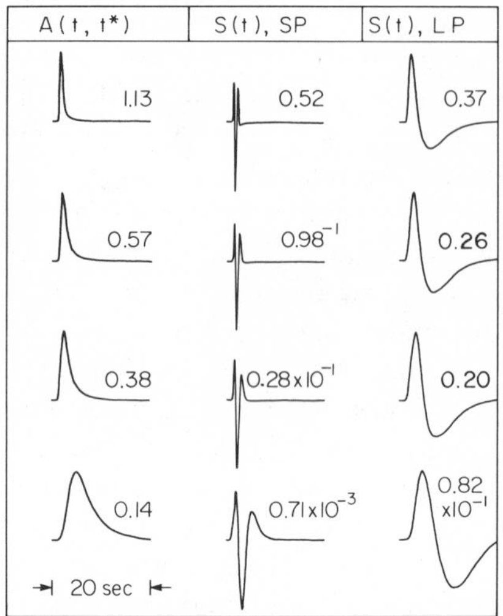

Figure 3 The first column displays $A(t,t^*)$ as a function of t^* assuming 0.5, 1.0, 1.5, and 4, from top to bottom. The middle column displays the convolution of $A(t,t^*)$ with the short period instrumental response, and similarly for the long period on the right. The relative amplitudes are indicated by the numbers above each trace.

calculations, nonetheless, because the relative amount of smoothing does vary with range and wave type and the effect on amplitudes of short period records is substantial. This is illustrated in Figure 2. The long period shear waves have clearly been more strongly smoothed than the long period P waves. The short period P waves are much larger than the short period S waves. The reason, as is well known, is that S waves are attenuated much more strongly than P waves. Because the attenuation operator is a very smooth function which does not vary significantly, it is often convenient to combine the instrument and Q operators into a single operator, $S(t)$, given by

$$S(t) = I(t) * A(t),$$

where * is the convolution operator. The attenuation operator which is most commonly used could be more properly written as $A(t,t^*)$ where t^* is a measure of the average attenuation along the ray path. The attenuation model is due to Futterman (1962) and Carpenter (1967). Most recent estimates of t^* are near 1 for compressional waves and 4 or larger for body waves (Anderson & Hart 1977, Burdick 1978). Figure 3 further illustrates the profound effect of the relative attenuation rates on absolute amplitudes of body waves. Using the composite operator $S(t)$ we can now write the complete expression for the synthetic seismogram $SS(t)$ as

$$SS(t) = S(t) * M(t) * E(t).$$

We will characterize earthquakes as distributed shear dislocations (see Haskell 1964 and Savage 1966) since this source description is particularly amenable to analysis. Following this kinematic approach, we do not necessarily have to understand the detailed mechanics taking place in the fault zone but merely state the amount of slip that occurs in the small elements of the faulting surface and its time history. Once the dislocation history is established by modeling all the data, we can then set about interpreting the slip in terms of the physical processes where the various assumptions about stress and frictional properties can be tested. We will limit this paper to the first step, namely an attempt at inferring the dislocation history in the presence of $M(t)$.

The techniques for calculating $M(t)$ can be quite complex or quite simple depending on the particular epicentral range. In Figure 4 we have divided the earth into different regions where different approaches to calculating $M(t)$ are required. At teleseismic ranges, Δ_T, from about 28 to 90° the effects of wave propagation through the earth are negligible,

$$M(t) = \delta(t).$$

The process of constructing synthetic seismograms beyond 90° where the earth's core produces a shadow zone is quite complicated and will not be discussed in this presentation. We refer the interested reader to Richards (1973). At ranges from 12° to 28°, Δ_{UPM}, the direct waves are reflected by the upper mantle structure, and a technique such as generalized ray theory which can account for triplications and low velocity zones must be used to compute $M(t)$. At ranges less than 12°, Δ_C, the crustal waveguide begins to play a major role and calculations become much more difficult. This is also the range in which lateral variations become more pronounced.

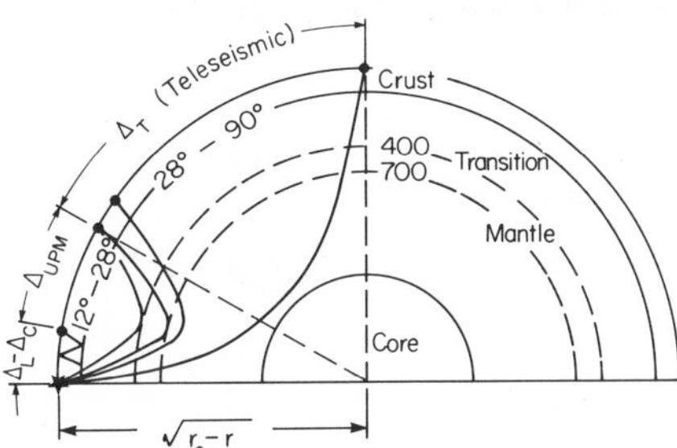

Figure 4 For convenience we have divided the epicentral distances from 0 to 90 degrees into Δ_T, Δ_{UPM}, Δ_C, and Δ_L where some example ray paths are shown schematically. Note that for distances less than Δ_T one normally receives several arrivals returning from various velocity jumps.

At still smaller ranges we get into the local field, Δ_L, where the source near-field terms and the very fine details of faulting become important. In brief, the closer the particular field is to the source, the more difficult the calculations of $M(t)$ become.

As might be expected, the first significant progress in body waveform modeling was made using records from Δ_T ranges. Because $M(t)$ was negligible, it was only necessary to model the source and attenuation operators. Since the source varies rapidly with azimuth and distance and attenuation does not, it was also possible to independently model these operators. After good source models were determined for several earthquakes, data from closer fields were modeled using the known source models with sophisticated methods for computing the earth response. We shall review this work in the same order beginning with the teleseismic field source studies, then moving to studies in the upper mantle, crustal, and local fields.

SOURCE DESCRIPTION DETERMINED BY TELESEISMIC WAVEFORMS

Suppose that the earth is homogeneous and that an earthquake can be idealized as a point shear dislocation situated at some depth, then one would expect to see a relatively simple waveform at all stations (see Figure 5). If the event is sufficiently deep, one would see the direct *P*, *pP*, the reflected arrival, and *sP*, the converted shear reflected arrival, all separated in time; but if it is shallow, these phases arrive together to make a relatively complicated signal. A similar set of arrivals, *S*, *sS*, and *pS*, comprise the SV waveform. The horizontally polarized shear component, S_{SH}, will have just a direct *S* and a reflected *sS*. If the source were a simple explosion emitting a pulse, $f(t)$ with strength ϕ_0, then the displacement potential for the direct ray at some distance *R* would be

$$\phi(R,t) = -\phi_0 \, f(t-R/\alpha)/\gamma R$$

where α is the compressional velocity and the constant, $\gamma = f(\infty)$, is simply

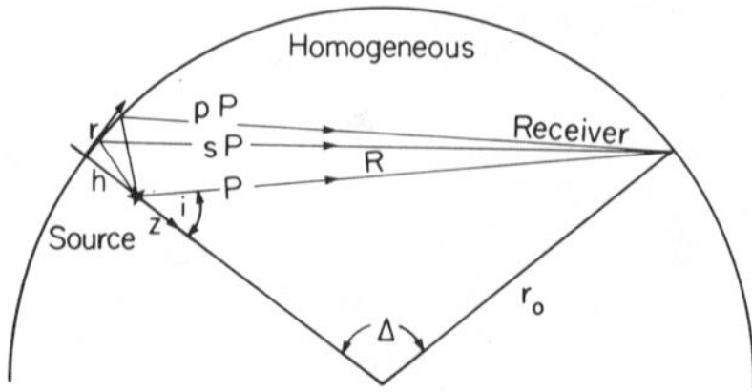

Figure 5 Schematic representation of the raypaths associated with the free surface interaction for a homogeneous earth.

used to normalize the pulse to unit area. Inside the earth the displacement is defined by the gradient of the potential or

$$W(x,z,t) = -\phi_0 (Z/R)\, f(t-R/\alpha)/\gamma R^2$$
$$-\phi_0 (Z/R) \frac{\mathrm{d}f\,(t-R/\alpha)}{\mathrm{d}t} /\alpha\gamma R.$$

Keeping the far-field term, or assuming $R \gg 1$, we obtain

$$W = -\phi_0 (\eta_\alpha) \frac{\mathrm{d}f\,(t-R/\alpha)}{\mathrm{d}t} /\gamma R$$

where $\eta_v = \cos i/v$. Similarly, the radial component becomes

$$Q = \phi_0 (p) \frac{\mathrm{d}f\,(t-R/\alpha)}{\mathrm{d}t} /\gamma R$$

where $p = \sin i/\alpha$, the ray parameter.

If we account for the free surface at the receiver, the expression becomes slightly more complicated with η_α and p replaced by functions R_{PZ} and R_{PR} as given by Helmberger (1974). The $(1/R)$ factor accounts for the geometric spreading of a spherical wave in a homogeneous medium. Because the earth has a velocity gradient, a more complex correction is required. Defining an effective $1/R$ correction by ψ we obtain

$$\psi^2 = \left(\alpha \tan i\right) \Big/ \left(r_0^3 \cos i \sin \Delta \left|\frac{\mathrm{d}^2 T}{\mathrm{d}\Delta^2}\right|\right)$$

where we have assumed $h \ll r_0$ and where Δ is the separation between the source and receiver in radians, Υ is the travel time, h is the source depth, and r_0 is the radius of the earth. Thus the geometric spreading can be determined by simply measuring the travel time along the surface with sufficient accuracy to obtain the second derivative. This result is given in Bullen (1965) and is accurate for the earth at teleseismic distances.

For a point shear dislocation these expressions become more complicated because of the radiation patterns (see Figure 6 and Langston &

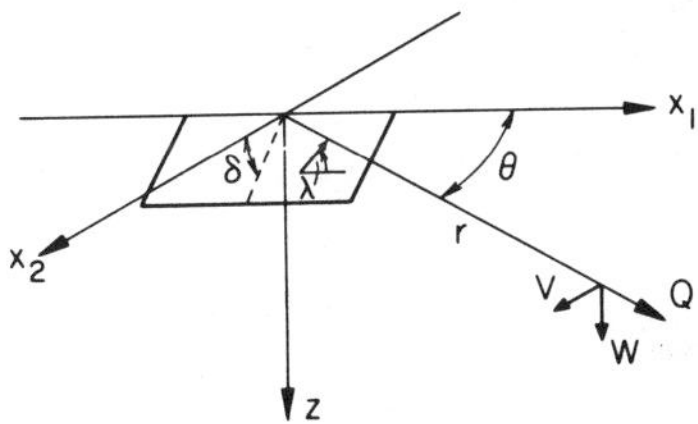

Figure 6 Coordinate system for the dislocation formulation. Z is positive downwards (after Langston & Helmberger 1975).

Helmberger 1975). These expressions are

$$\phi = \frac{M_0}{4\pi\rho_0} \sum_{j=1}^{3} A_j C_j f(t-t_\alpha)/\gamma R$$

$$\Omega = \frac{M_0}{4\pi\rho_0} \sum_{j=1}^{3} A_j SV_j f(t-t_\beta)/\gamma R$$

$$X = \frac{M_0}{4\pi\rho_0} \sum_{j=1}^{2} A_{j+3} SH_j f(t-t_\beta)/\gamma R$$

where $f(t)$ describes the dislocation or slip history and $(M_0/4\pi\rho_0)$ describes the strength with

$M_0 = D_0\mu(\pi r^2)$, moment

r = fault dimension

D_0 = average dislocation

μ = local shear modulus

ρ_0 = local density

α = compressional velocity

β = shear velocity

t_α, t_β = delay times

$(1/R)$ = geometric spreading.

The A's define the horizontal pattern

$A_1(\theta,\lambda,\delta) = \sin 2\theta \cos \lambda \sin \delta + 1/2 \cos 2\theta \sin \lambda \sin 2\delta$

$A_2(\theta,\lambda,\delta) = \cos \theta \cos \lambda \cos \delta - \sin \theta \sin \lambda \cos 2\delta$

$A_3(\theta,\lambda,\delta) = 1/2 \sin \lambda \sin 2\delta$

$A_4(\theta,\lambda,\delta) = \cos 2\theta \cos \lambda \sin \delta - 1/2 \sin 2\theta \sin \lambda \sin 2\delta$

$A_5(\theta,\lambda,\delta) = -\sin \theta \cos \lambda \cos \delta - \cos \theta \sin \lambda \cos 2\delta$,

where

θ = strike from the end of the fault plane

λ = rake angle

δ = dip angle.

If we assume a vertical fault ($\delta = 90°$) for $\lambda = 180°$, which is the case of a pure strike-slip fault, we obtain $A_1 = \pm \sin 2\theta$, $A_2 = A_3 = 0$. If $\lambda = \pm 90°$, we obtain $A_2 = \mp \sin \theta$, $A_1 = A_3 = 0$, which is the case of a pure dip-slip fault. The C's, SV's, and SH's describe the vertical radiation patterns.

$$C_1 = -p^2 \qquad SV_1 = -\varepsilon p \eta_\beta \qquad SH_1 = \frac{1}{\beta^2}$$

$$C_2 = 2\varepsilon p \eta_\alpha \qquad SV_2 = (\eta_\beta^2 - p^2) \qquad SH_2 = \frac{\varepsilon}{\beta^2}\frac{\eta\beta}{p}$$

$$C_3 = (p^2 - 2\eta_\alpha^2) \qquad SV_3 = 3\varepsilon p \eta_\beta$$

where

$$\varepsilon = \begin{matrix} +1 & z > h \\ -1 & z < h. \end{matrix}$$

The indices correspond to pure strike-slip (1), pure dip-slip (2), and a 45° dip-slip viewed at $\theta = 45°$ (3). Any other orientation can be simply obtained by a linear combination of the three fundamental faults.

Slip History

At this stage we need to make some statement about the expected dislocation behavior. Does the dislocation or slip across the fault occur instantaneously making $f(t)$ a step function, or does the slip have some characteristic duration depending on its length? Brune (1970) supposes the latter and, furthermore, that $f(t)$ can be parameterized by

$$\mathrm{d}f/\mathrm{d}t = \alpha^2 t \exp(-\alpha t),$$

where α is inversely proportional to the fault length. Thus, faults idealized to point sources have overall durations directly related to their length or dimension. A slightly more complicated slip history has been used in most modeling attempts since Brune's one-parameter model, while being intuitively pleasing, is not flexible enough. The far-field source function, $\mathrm{d}f/\mathrm{d}t$, used here will have a trapezoid shape described in general by three time parameters, δt's. The length of the positive, zero, and negative slope segments is given by δt_1, δt_2, and δt_3 with an estimate of the overall duration given by

$$\Upsilon_0 = \delta t_1/2 + \delta t_2 + \delta t_3/2$$

Note that the strain drop during an earthquake is (D_0/r) and, since stress is directly proportional to strain, one obtains stress drop, $\Delta\sigma$, by accurate measurements of (D_0/r) or moment and duration. Assuming a fixed moment, we obtain a high $\Delta\sigma$ for a short duration and a low $\Delta\sigma$ for a long duration. Thus, we can obtain information about the stress conditions in the earth by studying the waveforms or spectra of body phases as discussed by Brune (1970).

Synthetic Seismograms for Shallow Dislocation Sources

Synthetic seismograms can be constructed at Δ_T distances by simply putting together the various sub-operations discussed in the previous sections. We will assume $t_\alpha^* = 1$, see Figure 3, and the $(1/R)$ geometric spreading term is accurately given by Langston & Helmberger (1975) or the (ψ) computed from travel times. With the ray parameter given by 0.05 sec/km (or $\Delta = 80°$) we generated some example synthetics displayed in Figure 7. The stick diagram given in the first column displays the strength, polarity, and timing of the three interacting phases, *P*, *pP*, and *sP*, for the three fundamental faults. The reflected phases are computed by using the appropriate plane wave reflection coefficient. Crustal layering can be incorporated by applying ray summations as is also discussed by Langston & Helmberger (1975); however, these three phases give adequate results in most situations. The waveshapes are seriously distorted by the interferences with each fault orientation having its own characteristics. A similar set of waveforms involving the shear sources can be computed following the same procedure but with different waveshape characteristics. Thus, we can compare these synthetics with observed waveshapes such as

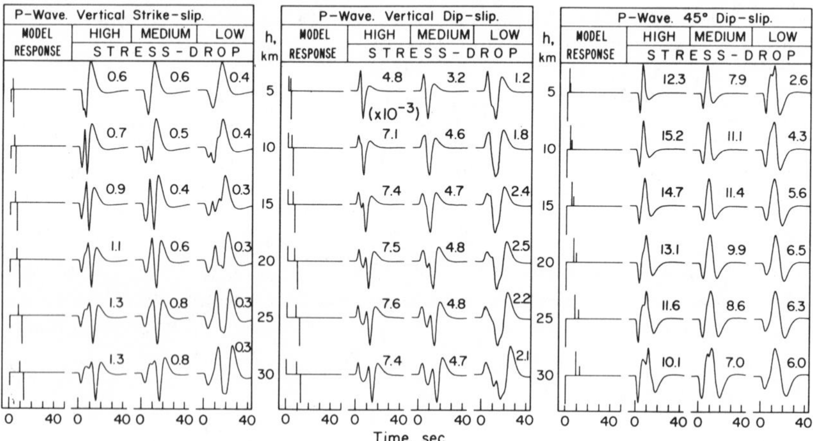

Figure 7 P-wave synthetic seismograms for the three potential terms with varying depth and time functions. The numbers to the upper right are actual potential amplitudes without the $(M_0/4\pi\rho_0)$, $(1/R)$ decay, and receiver functions included. The source time parameters, δt's, are high stress-drop (0.5, 1.0, 0.5), medium stress-drop (1.0, 3.0, 1.0) and low stress-drop (2.0, 6.0, 2.0). After Langston & Helmberger (1975).

those given in Figure 2 to determine the fault parameters, that is, strike, dip, slip vector, fault depth, moment, and δt's. This procedure has been conducted on about twenty events with considerable success. We will briefly review some of these studies starting with the event referred to earlier, namely the Borrego earthquake.

In Figure 8 we display the comparison of observations and synthetics from the Borrego earthquake study by Burdick & Mellman (1976). Note the coherence between neighboring stations such as the three east coast stations (WES, OGD, SCP) and the South American stations (NNA, ARE, LPB) which, of course, is necessary for the modeling technique to be meaningful. The faulting had a strike-slip orientation along a line 69° west of north with the northeastern side moving south relative to the southwestern side, producing the usual plot of dilations versus compression. Note that for points near the node such as BOG and BHP one finds difficulty in determining the polarity of the direct P since it is nearly zero. However, the remaining portion of the waveform is quite strong because of the dip-slip (sP) contribution. In Figure 9 we present the comparisons of the SH waves ($S+sS$) with the idealized point source model, and also the results of assuming a finite fault. The finite fault model can be pictured

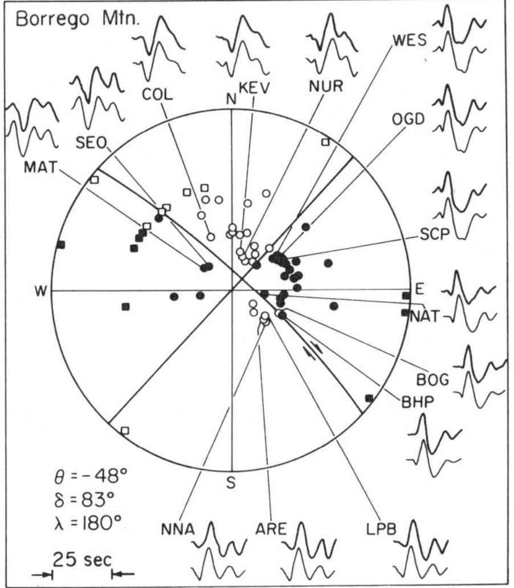

Figure 8 Observed (*top*) and synthetic (*bottom*) long period P waveforms at 14 WWSS stations. The P-first motion plot is represented by the equal area stereographic projection of the lower half of the focal sphere. Black dots indicate compression (upward breaking P) and open circles indicate dilatation (downward breaking P). The heavy solid lines denote the nodal planes used in determining the fault orientation, θ (strike), δ (dip), λ (slip direction). Modified after Burdick & Mellman (1976).

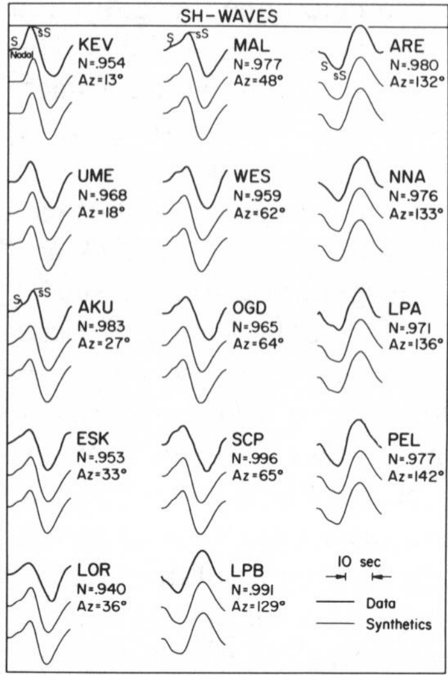

Figure 9 The observed SH waves (*top*) predicted by the model (*middle*) and the synthetics predicted by the same model with the main shock replaced by a finite source model. The stations are shown in order of increasing azimuthal deviation from the NW extension of the fault trace. After Burdick & Mellman (1976).

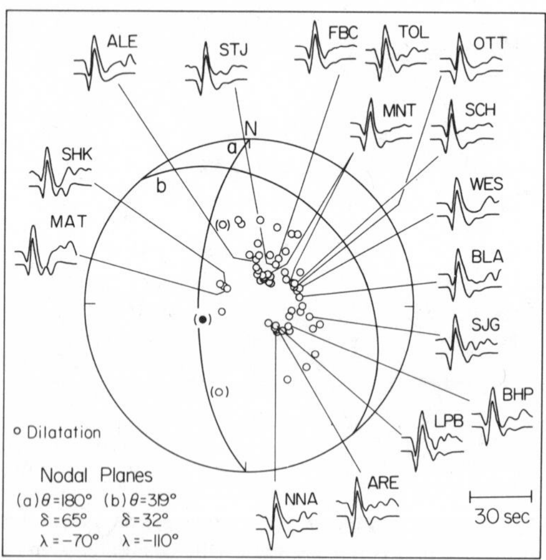

Figure 10 Focal mechanism plot showing the P-first motions for the Oroville earthquake, August 1, 1975. The observed P waveforms are given on the top and synthetic results on the bottom. Modified after Langston & Butler (1976).

as a collection of point sources in space which turn on at the appropriate times to simulate rupturing. For the case shown in Figure 9 we started the faulting at a depth of 9 km and allowed circular growth to a radius of 8 km. The directivity, the term applied to the geometrical dependence of the time function, is apparently too small to see under these circumstances. The best position to observe directivity is in the local field ($\Delta = \Delta_L$) near the ends of the fault which we will discuss later.

The P waveforms for shallow strike-slip faults appear to be dominated by the phase *sP* as displayed synthetically in Figure 7 and observed in Figure 8. Normal faults on the other hand generally dip at about 60° and are less affected by *sP* as is the case in Figure 10. Note that for this type of orientation the phases *pP* and *sP* have the same polarity (see Figure 7) and the summation of the two pulses remains nearly constant with azimuth. Thus, the nature of the seismograms at Δ_T is rather monotonous from a source mechanism point of view. Essentially, the same situation exists for thrust events such as the San Fernando earthquake; results are displayed in Figure 11. In this particular case the event was a double earthquake involving two different fault planes producing heavy interference effects. We will return to this interesting earthquake later when we discuss modeling local earthquakes. An example of a still more complicated

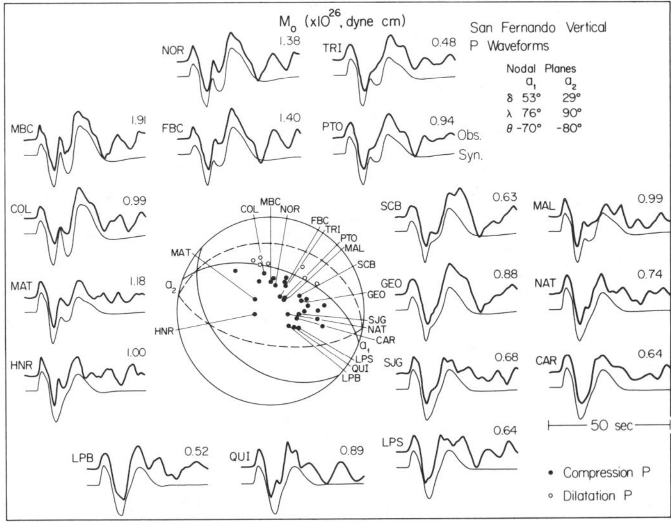

Figure 11 First motion plot showing the focal mechanism for the bottom section of the fault that broke first with nodal plane (a_1). The upper section has nodal plane (a_2) given by the dashed lines. The parameters for the top section of the fault were determined by synthetic modeling with the comparison displayed. After Langston (1978).

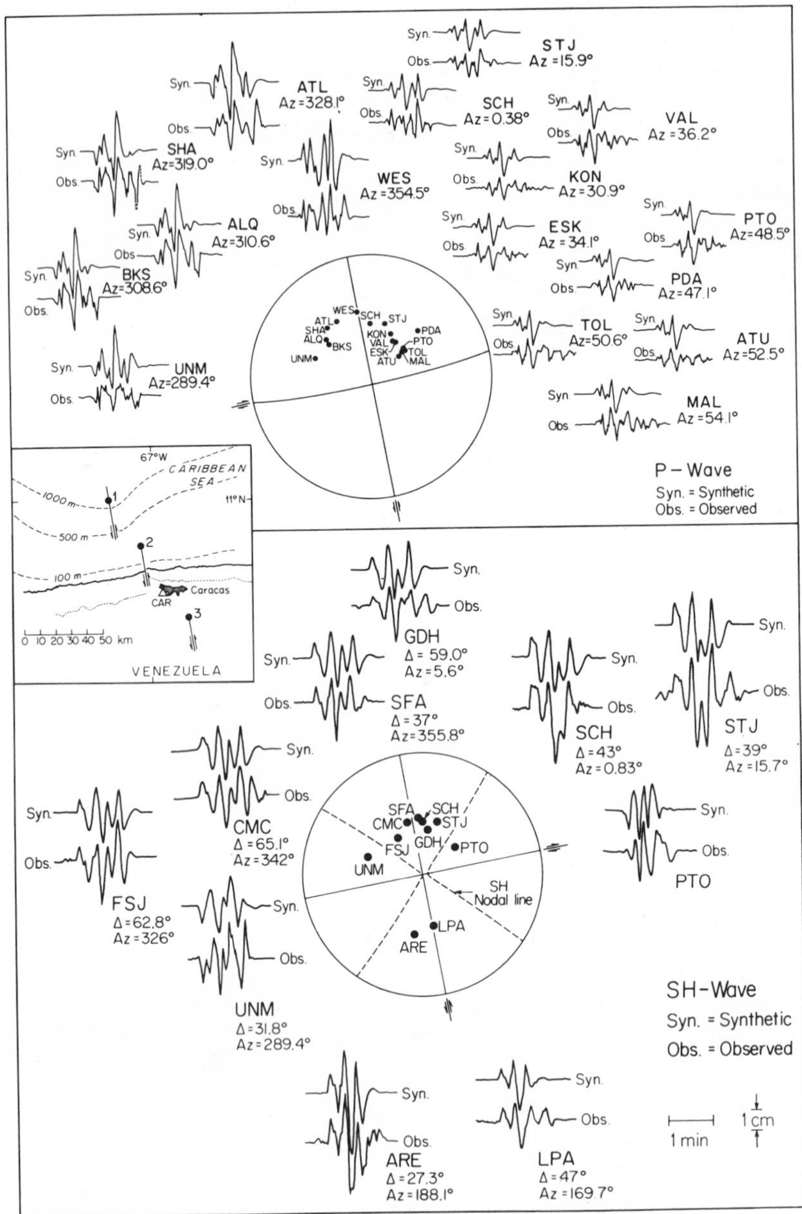

Figure 12 Comparison of observations (*bottom*) with synthetics (*top*) for the Caracas, Venezuela earthquake of July, 1967 (after Rial 1978).

modeling achievement is presented in Figure 12. This event is a composite of three strike-slip earthquakes located at 1, 2, and 3. The process of rupture progressed southwards starting at a depth of 8 km and proceeding to a depth of 28 km with an average velocity of 3 km/sec. The moments are in the ratios of 1, 2, and 2, respectively.

For our last teleseismic example, we consider a simple event with its epicenter located in the upper mantle, see Figure 13. Note that direct P is small in the northern and southern azimuths near the P-node and that the phase *pP* occurring about 30 seconds after the onset is relatively large. Thus, the long period behavior over the first 20 seconds is strongly controlled by the reflectivity of the crustal structure. The model presented here has a substantial low velocity zone at the base of the crust which is in agreement with the receiver structure determinations for stations in this region, see Langston (1977). The receiver structure is obtained by studying the waveshapes of P as recorded on the horizontal components and the SV waveforms, or essentially using all those components of motion ignored in the above examples. These other components are particularly

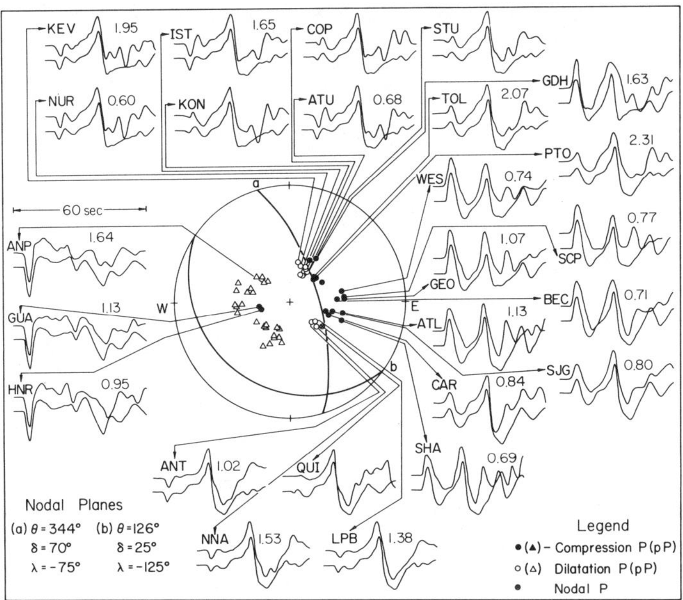

Figure 13 Focal mechanism plot for the Puget Sound earthquake April 29, 1965, indicating a nearly vertical normal fault. A comparison of observed (*top*) and synthetic (*bottom*) is quite good and required a model containing a substantial low velocity zone at the base of the crust. After Langston & Blum (1977).

susceptible to waveform distortions caused by the crustal structure beneath the receiver (see for example Burdick & Langston 1977), and, thus, these waveforms provide an excellent geophysical tool for studying the lower crust.

UPPER MANTLE STUDIES

The structure of the upper mantle has been the object of relatively active study over the last decade because of the importance of understanding the physical properties of the earth at depths where plates interact. Some general features such as the existence of major transitions near 400 and 650 km have been known for many years, but the details of the vertical structure as well as lateral variations have proven difficult to obtain. The reason is the lack of resolving power of the conventional seismological tools, namely travel times and ($dT/d\Delta$) measurements at a few observatories. Given these circumstances one can easily see the motivation to use more of the seismogram than just the travel time. The use of short period synthetic seismograms in upper mantle studies started with Helmberger & Wiggins (1971) and was used by Wiggins & Helmberger (1973) and Dey-Sarker & Wiggins (1976) to produce a number of models for several different regions. The synthetics in the above studies were used as a guide in the interpretation of data. It was seldom possible to match the seismic waveforms to the degree displayed in the teleseismic modeling discussed in the last section. Some possible reasons were the lack of knowledge

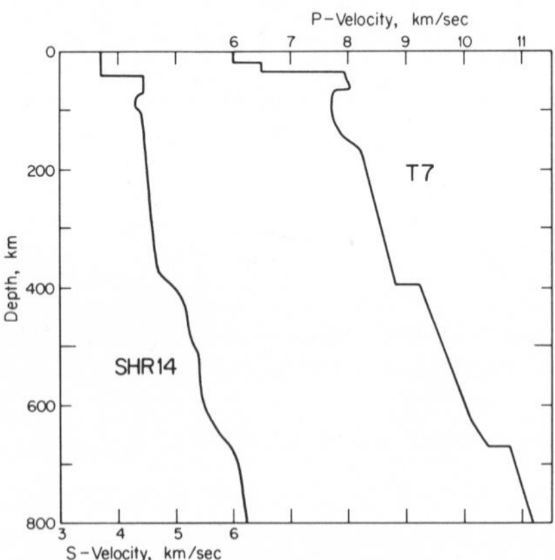

Figure 14 Velocity depth functions for T7 (Burdick & Helmberger 1978) and SHR14 (Helmberger & Engen 1974). The transition zones are probably sharp for both velocity models but cannot be determined by the S waves alone because of the absence of short period information.

about the source function at short periods and crustal distortions. Thus, we were led to consider the long period data as a supplementary measurement of the properties of the upper mantle since it is much more stable.

A model which fits travel times, $dT/d\Delta$, and much of the short and long period data for western United States is displayed in Figure 14. A reduced travel time plot displaying P-wave first arrivals along with the triplication branches for the model is given in Figure 15. Constructing synthetics for models containing triplications is quite complicated in that the waveform changes with range. In Figure 16 we display the smoothed delta function responses of Model T7 generated by two commonly used methods, namely generalized ray theory and reflectivity (see Burdick & Orcutt 1978). The implicit assumption made earlier when modeling Δ_T data was that the mantle did not distort the time function or that the earth behaved like a delta function once attenuation effects were removed. This means that the large spike occurring at the start of $\Delta = 28°$ will be the only response at greater ranges. To make a synthetic seismogram at ranges less than Δ_T we perform convolution of the waveforms given in Figure 16 with the effective waveform generated by the techniques discussed in the previous section. The procedure is presented in Figure 17 along with some example comparisons. Synthetic waveforms containing the effects of instrument, attenuation, source time function, idealized direct *P*, *pP*, and *sP*, etc. are given in the left hand column. The upper mantle delta function response appropriate for that range is plotted in the middle and the final convolved response on the right. Note that the final response at BLA ($\Delta = 29.4°$)

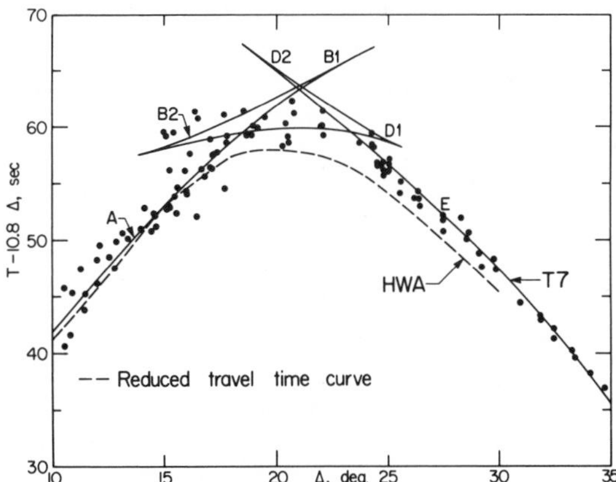

Figure 15 This figure compares the theoretical reduced first arrival travel time curves for HWA (Wiggins & Helmberger 1973) and T7 to the observed data from nuclear explosions. The triplication branches of T7 are included to aid in our discussion, After Burdick & Helmberger (1978).

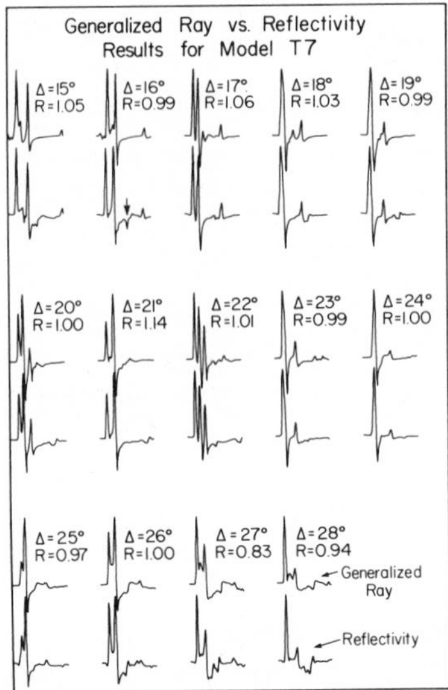

Figure 16 The responses from a reflectivity calculation (*bottom*) are compared to those from a generalized ray (*top*) for model T7: (*R*) is the actual amplitude ratio of the generalized ray response over the reflectivity. The source function was a narrow Gaussian filter (smoothed delta function). After Burdick & Orcutt (1978).

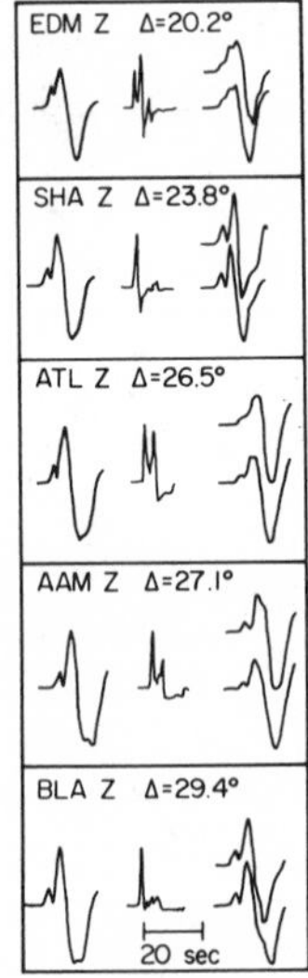

Figure 17 Triplication fits: first column contains the predicted waveshape due to the source only; second column is the delta function response of the model T7; third column is the convolution of the first two columns. The observed waveforms from the Borrego Mountain earthquake are on the top for comparison.

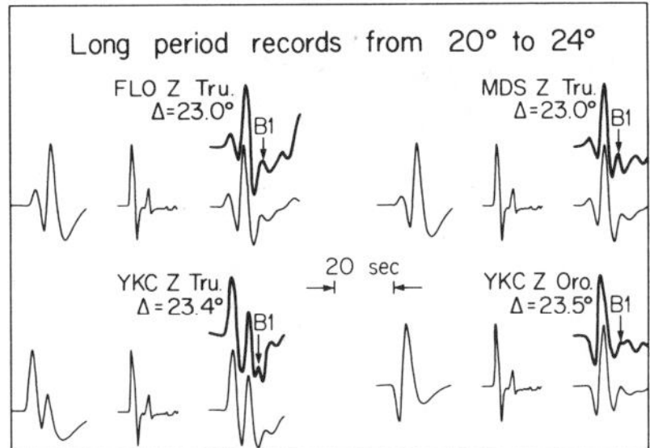

Figure 18 The long period records obtained from the Truckee (Tru.) and Oroville (Oro.) earthquake shown in this figure contain three arrivals. The arrival marked *B1* is a reflection from the 400-km discontinuity (see Figure 15). Arrivals from the 670-km discontinuity and the direct arrival (*D1*) occur at the start of the record. After Burdick & Helmberger (1978).

looks very much like the three observations at SCP, OGD, WES displayed earlier in Figure 8, whereas at smaller ranges the waveforms vary rapidly. After examining Figure 17 for a time one might suppose that the distortions at, say, EDM or SHA could have some other cause such as source complications or a substantial receiver interaction. One possible check against this situation is to examine a number of events and station pairs. In Figure 18 we show four observations at roughly the same range where we use the same upper mantle model but vary the source. Note that the Truckee event which was modeled teleseismically (Burdick 1977) looks considerably different in the northern direction (YKC) versus eastern direction (FLO). Nevertheless, the same upper mantle function works for both which gives us some confidence in our analysis. The steps involved in obtaining model T7 by fitting all this information are well documented by Burdick & Helmberger (1978). We used a tedious trial and error procedure. The next step is to implement a formal inversion on a computer which uses the cross-correlation of the observed waveforms with the synthetics in a direct inversion scheme. This method is now being developed and preliminary results indicate a high degree of resolution if the source function is well known. Spurred by these recent successes we have expanded our earthquake modeling efforts into other geographical regions in an attempt to gain a clearer picture of lateral variation in upper mantle structure.

SOURCE DESCRIPTION DETERMINED BY LOCAL OBSERVATIONS

Modeling the teleseismic long period waveforms tells us about the strength, orientation, depth, and overall duration, but does not tell us about the details of rupture. We would at least like to find its direction and velocity. These quantities can only be determined by strong motion seismology or by modeling the local field. However, to accomplish this we must model or account for propagational distortions caused by local crustal structure. In this section, we will apply recently developed synthetic modeling techniques to the interpretation of the faulting characteristics for a relatively simple event, namely the Borrego Mountain 1968, and a complex event, San Fernando 1971.

Point Dislocations in Layered Models

The basic technique used in constructing strong motion synthetics is to assume that an arbitrary distribution of dislocations representing a fault can be modeled by a summation of a large number of point shear dislocations distributed properly in space and time. The actual number of points required is related directly to the wavelength of interest. Next, one computes the Green's function which represents the response of the local structural model, usually assumed to a layered halfspace to a point shear dislocation with a delta function slip. An example calculation using the generalized ray method (Helmberger & Malone 1975) is displayed in Figure 19. In general, the delta function response is somewhat difficult to handle numerically and we use the temporal integral of the Green's func-

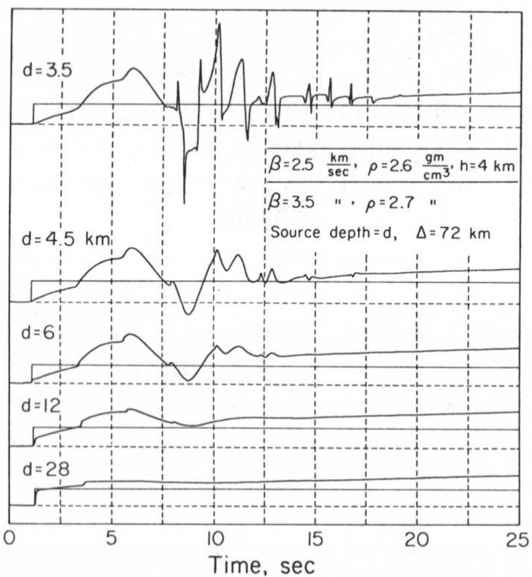

Figure 19 Step function responses at the surface assuming a point strike-slip dislocation situated at various depths. The amplitudes are scaled in relation to the top trace with the step response for a homogeneous half space (bottom properties) displayed for comparison. After Helmberger & Malone (1975).

tion, called the step response. The difference in response between the half-space and the layer over the halfspace is especially noticeable for the shallow source. The classical type of Love wave dispersion becomes well developed when the layer contains the source and is well understood in terms of the interference of rays. Comparing the responses with the source situated just above and below the layer boundary (see $d = 3.5$ and $d = 4.5$), one finds similar long period behavior as expected from physical considerations. Producing a theoretical displacement for a given slip history, $S(t)$, requires a convolution of $S(t)$ with the derivative of the step response. The ground motion resulting from realistic faulting motions at least at the longer periods is now modeled by superposition of those simple point source elements.

Modeling Strong Motions

We will begin this discussion by showing some observations from the Borrego Mountain earthquake that can be largely explained by applying the above techniques.

Unfortunately, not many local recordings are available for the Borrego Mountain event because of its remote location. El Centro was the nearest site ($\Delta = 60$ km) where the event was recorded by Carder displacement meters and the standard accelerographs. An example of the strongest component of motion is displayed in Figure 20 showing the motion in

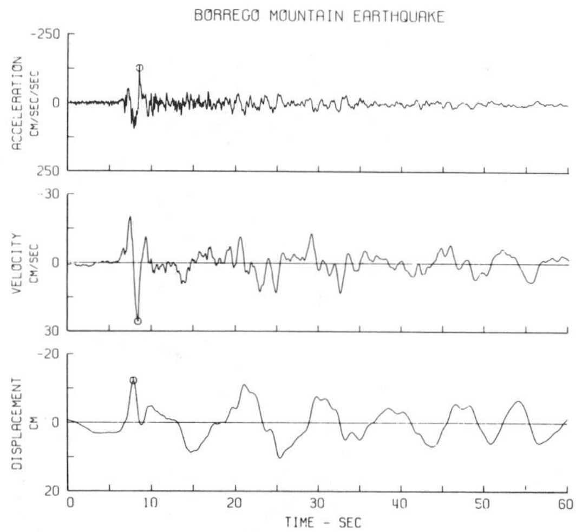

Figure 20 Acceleration, velocity, and displacement motions for the north-south component recorded at El Centro of the April 8 Borrego Mountain earthquake 1968. US Coast and Geodetic Survey et al (1968).

terms of acceleration, and processed velocity and displacement. Since the earthquake was essentially a strike-slip event and the station is only 8° from the strike of the fault, we expected predominantly *SH* motion with small motions on the vertical and radial components; all of these features are well documented by Heaton & Helmberger (1977). The observed *SH* displacement is displayed at the top of Figure 21 with a highly idealized synthetic model on the bottom. Several models have been constructed to fit the first 40 sec of motion. A 2.9-km thick layer with shear velocity of 1.5 km/sec overlying a halfspace with shear velocity of 3.3 km/sec gives a good overall fit to the Love wave portion of the record for a variety of distributed sources at depths ranging from 4 to 10 km. The source distribution in this example is particularly simple. It is just two point sources with slip histories chosen to be compatible with the teleseismic information. The interpretation is that massive faulting (high $\Delta\sigma$) occurred at a depth of 9 km producing the powerful direct arrival followed by rupturing in the upward direction, thus producing the observed Love waves. The detailed or fine-scale properties of the faulting are still not resolved due to lack of near-in data. Some synthetics displaying various plausible assumptions about the rupture parameters are given in Figure 22. In this exercise we assumed a rectangular fault with different epicenter locations, Δ_E, and allowed the elements to activate in sequence, simulating a rupture velocity. The moment was adjusted to obtain record strengths comparable to the data. Note that the horizontal rupture direction and slip history are particularly important with respect to the short period or high frequency

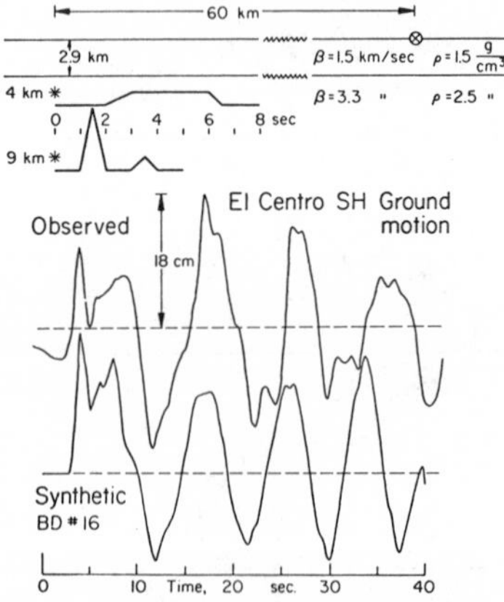

Figure 21 Comparison of the observed strong ground motion (transverse component) of the Borrego Mountain earthquake (April 9, 1968; $M_L = 6.4$) at El Centro, California, with ground motion computed on the basis of the simple source model schematically illustrated in the upper left embedded in the crustal structure parameterized in the upper right.

information. Thus, the strong acceleration and velocity spike at the beginning of the record in Figure 20 could be caused by a point source or substantial faulting motion directed towards the station.

For our second example, we considered the San Fernando earthquake which is well enough recorded to actually determine the rupture parameters independently. The data set obtained from this event is huge and has been discussed at length by Hanks (1975) where he shows the applicability of seismological techniques to the interpretation of strong motion records. He demonstrates the existence of surface waves and displays numerous examples of the coherency of signals from neighboring stations. The near-in stations show particularly large accelerations and have been discussed by Mikumo (1973), Hanks (1975), and Trifunac (1974). The latter author obtained relatively good fits for the three components of motion obtained from the Pacoima Dam site. In fact, the data from this station allows relatively tight bounds to be put on the rupture process.

We approached this data set following the same strategy used in the Borrego Mountain study—namely, we initially constrained the fault description to fit the teleseismic waveform data. Unfortunately, the waveforms produced by San Fernando, while being as coherent from station to station as those displayed in Figure 2, are quite complicated with rupturing

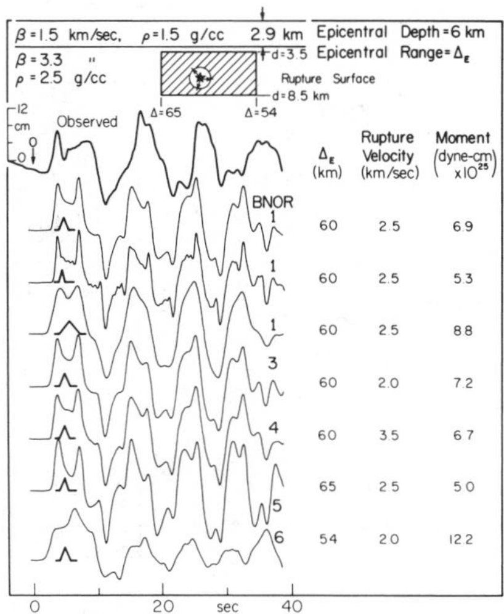

Figure 22 Comparison of observed with synthetic SH ground motion for models consisting of a rectangular fault with uniform offset which initiates at a point and propagates radially to the edge of the rectangle. In models BNOR5 and BNOR6 the rupture propagates unilaterally toward and away from El Centro, respectively. Rupture propagation is bilateral in the other models shown. The far-field time function for each point on the fault is displayed directly beneath the beginning of each synthetic. After Heaton & Helmberger (1977).

occurring on two fault planes, starting at depth and propagating towards the surface (see Langston 1978). A rectangular model similar to the one he proposes is displayed in Figure 23. We assumed a halfspace in modeling the nearest stations and computed the Green's functions on a 0.5 km spacing. After a diligent search (see Heaton 1978), we arrived at the slip distribution given by the contours. It explains many of the observed properties including the static offsets. Although it is necessary to present numerous observations and possible slip distributions to support this proposed model, the comparison between the synthetics and data for PAC and LKH are the most indicative; that is, to produce the required change in amplitude between PAC and LKH requires very strong focusing to the south. This requires a relatively narrow fault at depth, probably less than 6 km wide with about 2 m of displacement. The rupture velocity is 2.8 km/sec for the bottom segment and 1.8 km/sec for the top section. Note that this model indicates rather small offsets beneath PAC and massive faulting towards the south probably within a km of the surface. Because of the high apparent accelerations in the region just to the north of the surface break and because of the large stress drop implied (approaching the breaking strength of rocks), this portion of the faulting surface is being studied in detail.

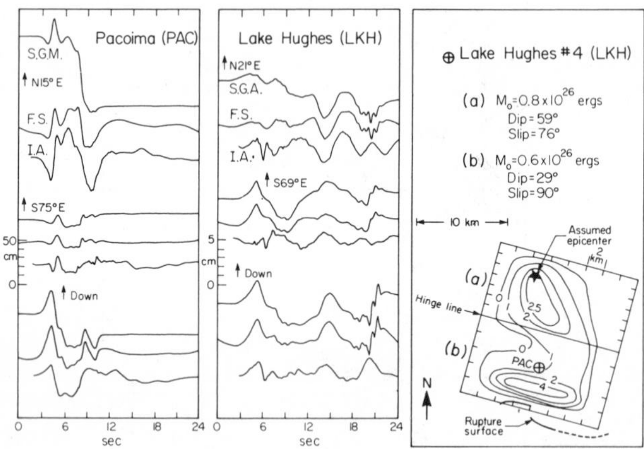

Figure 23 Relative geometry showing the locations of PAC (Pacoima Dam) and LKH (Lake Hughes) relative to the San Fernando fault structure and a comparison of displacements (*top*), synthetic with instrumental response (*middle*), and observed (*bottom*). Note the factor of 10 difference in amplitude scale between PAC and LKH which indicates the strong dependence on rupture direction. The contours of displacement on the fault are expressed in meters. Modified after Heaton (1978).

DISCUSSION

In this review we have restricted ourselves to the uses of synthetic seismograms in the interpretation of observed body phases from earthquakes. For convenience, we divided the epicentral distances into Δ_T, Δ_{UPM}, Δ_C, and Δ_L corresponding to the part of the earth influencing the observation. At teleseismic distances we showed that long period body phases from shallow earthquakes are coherent at neighboring stations and that the observed waveform could be decomposed in a manner that allows determination of faulting parameters. The corresponding comparison between short period records is, in general, much less coherent. This may mean that the rupturing process is smooth when viewed at long wavelengths but rough on a shorter scale. Another possibility is that heterogeneities along the ray path scatter the shorter wavelengths or, perhaps, both processes are involved. We showed how once the long periods were modeled and the faulting parameters determined, the upper mantle transition zones could be studied by using the triplication data at Δ_{UPM} distances. This line of research will probably proceed rapidly in the near future because of the recently developed fast asymptotic methods of Wiggins (1976), Chapman (1976), and Mellman & Helmberger (1978). These methods yield synthetics very much like those displayed in Figure 16 at much lower cost (Burdick & Orcutt 1978). At shorter distances, the crustal waveguide plays a major role in that a relatively large amount of energy is trapped between the free surface and the crust-mantle transition (see Helmberger 1973). At still shorter distances, Δ_L, the details of the rupturing processes become more important and the simulation process more complex. This subject, called strong motion seismology, has become a very important branch of the field. It is providing a wealth of new information concerning the details of the rupture process both in attempts at prediction and in judging the hazards to crucial structures near active faults.

Literature Cited

Anderson, D. L., Hart, R. S. 1977. Attenuation models for the earth. *Phys. Earth Planet. Inter.* In press

Brune, J. N. 1970. Tectonic stress and the spectra of seismic waves from earthquakes. *J. Geophys. Res.* 75:4997–5009

Bullen, K. E. 1965. *An Introduction to the Theory of Seismology*, Ch. 8. Third Edition. Cambridge: University Press

Burdick, L. J. 1977. *Broad-band seismic studies of body waves.* PhD thesis. California Institute of Technology, Pasadena

Burdick, L. J. 1978. t* for S waves with a continental raypath. *Bull. Seismol. Soc. Am.* In press

Burdick, L. J., Helmberger, D. V. 1978. The upper mantle P velocity structure of the western United States. *J. Geophys. Res.* 83:1699–712

Burdick, L. J., Langston, C. A. 1977. Modeling crustal structures through the use of converted phases in teleseismic body-wave forms. *Bull. Seismol. Soc. Am.* 67:677–91

Burdick, L. J., Mellman, G. R. 1976. Inversion of the body waves of the Borrego Mountain earthquake to the source mechanism. *Bull. Seismol. Soc. Am.* 66: 1485–99

Burdick, L. J., Orcutt, J. A. 1978. A comparison of the generalized ray and reflectivity methods of waveform synthesis. *Geophys. J.* In press

Carpenter, E. W. 1967. Teleseismic signals calculated for underground, underwater, and atmospheric explosions. *Geophysics* 32:17–32

Chapman, C. H. 1976. A first motion alternative to geometrical ray theory. *Geophys. Res. Lett.* 3:153–56

Dey-Sarker, S. K., Wiggins, R. A. 1976. Upper mantle structure in western Canada. *J. Geophys. Res.* 81:3619–32

Futterman, W. I. 1962. Dispersive body waves. *J. Geophys. Res.* 67:5279–91

Gilbert, F., Dziewonski, A. M. 1975. An application of normal mode theory to the retrieval of structural parameters and source mechanisms from seismic spectra. *Phil. Trans. R. Soc. London* 278:187–269

Hanks, T. C. 1975. Strong ground motion of the San Fernando, California earthquake: ground displacements. *Bull. Seismol. Soc. Am.* 65:193–226

Haskell, N. A. 1964. Total energy and energy spectral density of elastic wave radiation from propagating faults. *Bull. Seismol. Soc. Am.* 54:1811–31

Heaton, T. H. 1978. *Generalized ray models of strong ground motion.* PhD thesis. California Institute of Technology, Pasadena

Heaton, T. H., Helmberger, D. V. 1977. A study of the strong ground motion of the Borrego Mountain, California earthquake. *Bull. Seismol. Soc. Am.* 67:315–30

Helmberger, D. V. 1973. On the structure of the low-velocity zone. *Geophys. J. R. Astron. Soc.* 34:251–63

Helmberger, D. V. 1974. Generalized ray theory of shear dislocations. *Bull. Seismol. Soc. Am.* 64:45–64

Helmberger, D. V., Engen, G. R. 1974. Upper mantle shear structure. *J. Geophys. Res.* 79:4017–28

Helmberger, D. V., Malone, S. D. 1975. Modeling local earthquakes as shear dislocations in a layered half-space. *J. Geophys. Res.* 80:4881–88

Helmberger, D. V., Wiggins, R. A. 1971. Upper mantle structure of midwestern United States. *J. Geophys. Res.* 76:3229–45

Kanamori, H., Stewart, G. S. 1976. Mode of strain release along the Gibbs fracture zone, mid-Atlantic Ridge. *Phys. Earth Planet. Inter.* 11:312–32

Langston, C. A. 1977. Corvallis, Oregon, crustal and upper mantle receiver structures from teleseismic P and S waves. *Bull. Seismol. Soc. Am.* 67:713–24

Langston, C. A. 1978. The February 9, 1971 San Fernando earthquake. *Bull. Seismol. Soc. Am.* 68:1–30

Langston, C. A., Blum, D. E. 1977. The April 29, 1965, Puget Sound earthquake and the crustal and upper mantle structure of western Washington. *Bull. Seismol. Soc. Am.* 67:693–711

Langston, C. A., Butler, R. 1976. Focal mechanism of the August 1, 1975 Oroville earthquake. *Bull. Seismol. Soc. Am.* 66: 1111–20

Langston, C. A., Helmberger, D. V. 1975. A procedure for modeling shallow dislocation sources. *Geophys. J. R. Astron. Soc.* 42:117–30

Mellman, G. R., Helmberger, D. V. 1978. A modified first motion approximation for the synthesis of body wave seismograms. *Geophys. J. R. Astron. Soc.* In press

Mikumo, T. 1973. Faulting process of the San Fernando earthquake of February 9, 1971, inferred from static and dynamic nearfield displacements. *Bull. Seismol. Soc. Am.* 63:249–64

Rial, J. A. 1978. The Caracas, Venezuela earthquake of July, 1967: a multiple source event. *J. Geophys. Res.* In press

Richards, P. G. 1973. Calculation of body waves, for caustics and tunnelling in core phases. *Geophys. J. R. Astron. Soc.* 35:243–64

Savage, J. C. 1966. Radiation from a realistic model of faulting. *Bull. Seismol. Soc. Am.* 56:577–92

Trifunac, M. D. 1974. A three-dimensional dislocation model for the San Fernando, California earthquake of February 9, 1971. *Bull. Seismol. Soc. Am.* 64:149–72

US Coast and Geodetic Survey, Seismological Field Survey and California Institute of Technology, Earthquake Engineering Research Laboratory. 1968. Strong-motion instrumental data on the Borrego Mountain earthquake of 9 April 1965. No. 119

Wiggins, R. A. 1976. Body wave amplitude calculations—II. *Geophys. J. R. Astron. Soc.* 46:1–10

Wiggins, R. A., Helmberger, D. 1973. Upper mantle structure of Western United States. *J. Geophys. Res.* 78:1870–80

Ann. Rev. Earth Planet. Sci. 1979. 7: 443–72

THE ROLE OF NO AND NO_2 IN THE CHEMISTRY OF THE TROPOSPHERE AND STRATOSPHERE

Paul J. Crutzen

National Center for Atmospheric Research, P.O. Box 3000, Boulder, Colorado 80307

INTRODUCTION

The main importance of the oxides of nitrogen (NO, NO_2) in atmospheric chemistry is their role in the determination of the earth's ozone distribution. In the stratosphere ($\approx$ 10–50 km) nitric oxide (NO) is formed mainly by the oxidation of nitrous oxide (N_2O). Nitric oxide and its oxidation product nitrogen dioxide (NO_2) then participate in an important set of catalytic reactions which transfer ozone (O_3) to molecular oxygen and which are effective especially above about 24 km. Surprisingly, however, at lower altitudes nitric oxide acts catalytically to produce ozone by natural "smog" reactions. Although the effects of such reactions are clearly observable during photochemical smog episodes near urban centers (see, for example, the discussion by Altshuller & Bufalini 1971), there is no reason to assume that they are restricted to such environments. Analogous reactions should occur also outside the urban boundary layer during the oxidation of carbon monoxide and natural hydrocarbons (especially methane, isoprene, and the terpenes) in the presence of nitric oxide (Crutzen 1973, Zimmerman et al 1978).

Like water vapor and carbon dioxide, ozone plays an essential role in the radiation budget of the earth's atmosphere below about 80 km. This is because of its radiative properties in the ultraviolet and in the infrared. The absorption of ultraviolet radiation protects the biosphere from harmful radiation. In this process electronically excited $O(^1D)$ atoms are generated:

$$O_3 + h\nu \rightarrow O(^1D) + O_2, \qquad \lambda \lesssim 310 \text{ nm}.$$

0084-6597/79/0515-0443$01.00

In contrast to ground state atomic oxygen, excited oxygen atoms react rapidly with several minor gaseous constituents in the atmosphere, such as nitrous oxide and water vapor (H_2O), leading to the formation of nitric oxide and hydroxyl (OH):

$$O(^1D) + N_2O \rightarrow 2\ NO,$$

$$O(^1D) + H_2O \rightarrow 2\ OH.$$

The hydroxyl radical reacts with many gases which otherwise would be inert in the troposphere, such as carbon monoxide (CO), hydrocarbons (Levy 1971, 1974), and chlorinated hydrocarbons. If no OH were present in the troposphere, the atmospheric abundances of these gases would be much larger than observed in the atmosphere, in many cases by orders of magnitude. One of the most important tasks in atmospheric chemistry is to establish the origin of tropospheric ozone, because ozone affects the chemical composition and thermal characteristics of the atmosphere in decisive ways.

According to classical views, tropospheric ozone emanates from the stratosphere and is destroyed at the earth's surface (Junge 1962, 1963, Mohnen et al 1977, Danielsen & Mohnen 1977). It is clear, however, that the production of ozone in the troposphere under the influence of man's activities and involving catalysis by NO and NO_2 should be substantial. Consequently, this is one of the issues that we review.

The effects of additions of nitric oxide to the stratosphere under the influence of a number of human activities and the effect on stratospheric ozone have been a subject of considerable concern, stimulating intense research activities during the last decade on the chemistry and distribution of the oxides of nitrogen in the stratosphere. Such additions may occur due to emissions in the exhaust gases of high-flying aircraft, especially large fleets of supersonic aircraft cruising between 16 and 20 km (Johnston 1971, Crutzen 1971). Since nitrous oxide is produced in the soils and waters by microbiological processes (Pratt et al 1977), increased agricultural activities, in particular the use of nitrogen fertilizer, may well lead to an accumulation of N_2O in the atmosphere. In addition, nitrous oxide is produced by combustion (Weiss & Craig 1976, Pierotti & Rasmussen 1976). Because the oxidation of N_2O by $O(^1D)$ leads to production of NO in the stratosphere, the concentrations of ozone will be affected (Crutzen 1974a, McElroy 1976). In this article we only briefly review the potential consequences of an increase in N_2O mixing ratios in the atmosphere. Discussions of the factors influencing increased N_2O release from soils and the different views on this subject have been presented elsewhere (Crutzen 1976, McElroy et al 1976, 1977, Crutzen & Ehhalt 1977, Liu et al 1976a, 1977, Sze & Rice 1976, Pratt et al 1977, Hahn &

Junge 1977). A thorough study of current knowledge on the effects of nitrates in the environment, including a review of the nitrogen fertilizer-ozone problem, has just been published (NAS 1978).

The importance of ozone, and therefore of the oxides of nitrogen, for the terrestrial radiation budget and average vertical temperature profile is now well recognized. Although reductions in the total abundance of atmospheric ozone allow more solar radiation to penetrate to ground level, and tend to produce higher surface temperatures, the decrease in the downward longwave radiation emitted by CO_2, O_3, and H_2O from a cooler lower stratosphere containing less ozone would actually cause a decrease in surface temperatures. Using radiative-convective models Manabe & Wetherald (1967) and Ramanathan et al (1976) have calculated lower surface temperatures for atmospheres containing less than standard ozone concentrations throughout the atmosphere. Global surface temperatures would be lower by about 1.5–3°C in an atmosphere containing no ozone at all. For situations in which ozone is removed at higher levels ($\approx$24 km) and increased at lower levels (as will occur for NO_x injections), the effects are probably opposite. The climatological aspects of stratospheric composition changes are discussed by Liu et al (1976b) and by Ramanathan & Coakley (1978). Since nitrous oxide itself is a significant absorber of infrared radiation of atmospheric and terrestrial origin, the direct effects of atmospheric nitrous oxide increases on the terrestrial radiation budget are also of interest (Wang et al 1976). Recent research in theoretical dynamic meteorology indicates the important role of stratospheric thermal stability in controlling the vertical transfer of wave energy between the troposphere and stratosphere (Bates 1977, Lindzen & Tung 1978). According to Bates (1977) substantial reductions in upper stratospheric temperatures by 10 K, as a consequence of industrial activities, could indirectly lead to marked changes in the horizontal heat flux by ultra-long planetary waves in the troposphere.

Nitric oxide is produced in very large amounts in the upper atmosphere, mostly in polar regions, from the action of ultraviolet and X-ray radiation, from precipitating solar electrons in auroras (above 100 km), from solar protons in the upper stratosphere and mesosphere, and from galactic cosmic rays in the lower stratosphere. All these emissions are in some way dependent on solar activity, especially in the upper layers of the atmosphere. Correlations between solar activity and climate have been shown to exist on long time scales (Eddy 1976, 1977) and may be due to variations in the solar energy output. However, solar activity could also influence climate and the dynamics of the atmosphere through the previously discussed chain of effects ($NO_x \rightarrow O_3 \rightarrow$ static stability $\rightarrow$ planetary wave reflections).

Table 1 Abbreviations of groups of chemical constituents of importance, especially to the ozone budget of the stratosphere. The symbols within parentheses denote densities in molecules per cubic centimeter.

$O_x = O + O_3$ (odd oxygen)
$NO_x = N + NO + NO_2$ (active nitrogen)
$HNO_x = HNO + HONO + HONO_2 + HO_2NO_2$
$NOX = NO_3 + NO_2NO_3 + HNO_x + ClONO_2 + PAN + \cdots$
$(NOX) = (NO_3) + 2(N_2O_5) + (HNO_x) + (ClONO_2) + (PAN) + \cdots$
$NX = NO_x + NOX$ (odd nitrogen oxides, odd nitrogen)
$ClO_x = Cl + ClO$ (active chlorine)
$ClX = ClO_x + HCl + ClOH + ClONO_2$ (odd chlorine)
$HO_x = H + OH + HO_2$ (active hydrogen)
$HX = HO_x + HNO_x + HOOH$ (odd hydrogen)
$(HX) = (HO_x) + (HNO_x) + 2(H_2O_2)$

This article concentrates on the important role played by the oxides of nitrogen in the chemistry of the global atmosphere. Among the nitrogen oxides, NO and NO_2 are by far the most important because of their ability to affect ozone through catalytic reactions. Chemical interactions with several other atmospheric gases produce a great number of species of oxides of nitrogen. The terminology adopted in this discussion is presented in Table 1. Unfortunately, a generally accepted terminology does not exist and occasionally one encounters terms like Cl_x, NO_x, and NO_y defined differently by different authors. The most recent compilation of reaction rate coefficients can be found in the review by Hampson & Garvin (1977).

THE TROPOSPHERE

Chemistry of the Oxides of Nitrogen NO and NO_2

Production of ozone in polluted environments by photochemical oxidation of the unburned hydrocarbons in automobile exhaust gases requires the presence of nitric oxide and nitrogen dioxide as catalysts. The set of reactions responsible for the formation of ozone is basically the following:

$$R + O_2 + M \rightarrow RO_2 + M \qquad R2$$
$$RO_2 + NO \rightarrow RO + NO_2 \qquad R3$$
$$NO_2 + h\nu \rightarrow NO + O \qquad \lambda \lesssim 400 \text{ nm} \qquad R4$$
$$O + O_2 + M \rightarrow O_3 + M \qquad R5$$

$$N1: R + 2\,O_2 \rightarrow RO + O_3$$

In this set of reactions the symbol R denotes a radical species such as H,

CH_3, and $CH_3C(O)$, which are intermediate products of hydrocarbon and carbon monoxide oxidation in the atmosphere (Altshuller & Bufalini 1971, Demerjian et al 1974). These reactions cause a stepwise dissociation of molecular oxygen, which is a biradical and attaches easily to other radicals. Direct dissociation of O_2 cannot take place in the troposphere because the solar ultraviolet radiation that reaches the troposphere is insufficiently energetic to split the molecule.

There is no reason to believe that the production of ozone in the troposphere is restricted to polluted urban environments. Ozone should also be produced in the non-urban troposphere, if the necessary photochemical ingredients are present. The oxidation of carbon monoxide (CO) in the presence of nitric oxide provides the simplest mechanism for ozone formation in "clean" air (Crutzen 1973):

$$CO + OH \rightarrow H + CO_2 \qquad R1$$
$$H + O_2 + M \rightarrow HO_2 + M \qquad R2a$$
$$HO_2 + NO \rightarrow HO + NO_2 \qquad R3a$$
$$NO_2 + h\nu \rightarrow NO + O \qquad R4$$
$$O + O_2 + M \rightarrow O_3 + M \qquad R5$$

$$N2: CO + 2O_2 \rightarrow CO_2 + O_3$$

It follows, therefore, that the oxidation of CO in the atmosphere could yield one ozone molecule for each carbon monoxide molecule oxidized in the atmosphere. This set of reactions requires the presence of sufficient NO to insure its occurrence since other reactants compete with NO for the HO_2 radical. Without sufficient NO, one of the following reaction sequences, which do not produce ozone, may occur:

$$CO + OH \rightarrow H + CO_2(2x) \qquad R1$$
$$H + O_2 + M \rightarrow HO_2 + M(2x) \qquad R2a$$
$$HO_2 + HO_2 \rightarrow H_2O_2 + O_2 \qquad R15a$$
$$H_2O_2 + h\nu \rightarrow 2OH \qquad R16$$

$$N3: 2CO + O_2 \rightarrow 2CO_2$$

or:

$$CO + OH \rightarrow H + CO_2 \qquad R1$$
$$H + O_2 + M \rightarrow HO_2 + M \qquad R2a$$
$$HO_2 + O_3 \rightarrow OH + 2O_2 \qquad R10$$

$$N4: CO + O_3 \rightarrow CO_2 + O_2$$

Only recently have the rate coefficients for reaction R3a (Howard & Evenson 1977) and R10 (Zahniser & Howard 1978) been determined satisfactorily, and reaction R3a was shown to be very fast. It is easy to estimate the concentration of nitric oxide required in order for reaction sequence N2 to dominate over N4. Assuming a typical clean air ground level ozone volume mixing ratio of about 20 ppbv (2×10^{-8}), and adopting the rate constants $k_{10} = 1.4 \times 10^{-14} \exp(-580/T)$ and $k_{3a} = 3.3 \times 10^{-12} \exp(254/T)$ cm^3 mol^{-1} s^{-1}, from the work of Howard and co-workers we arrive at a critical volume mixing ratio for NO of 0.005 ppbv. For NO mixing ratios above this very small value, ozone production via the reaction set N2 is larger than its destruction by the reaction set N4. We review the few available measurements of "clean air" NO_x concentrations in the following section, but note here that from available observations, it seems most likely that background concentrations of NO in the troposphere are indeed substantially larger than 0.005 ppbv over the industrialized continental areas in the Northern Hemisphere (N.H.). In the Southern Hemisphere (S.H.) and in unindustrialized areas of the N.H. the situation should be quite different. There is, therefore, a clear possibility that significant in situ photochemical ozone production takes place in the N.H. troposphere by reaction cycle N2, while in the S.H. ozone loss by reaction set N4 may be more important (Fishman & Crutzen 1978). Additional loss of ozone occurs in both hemispheres as reaction R6a is partially ($\lesssim 10\%$) followed by reaction R17

$$O_3 + h\nu \rightarrow O(^1D) + O_2 \qquad \text{R6a}$$

$$O(^1D) + H_2O \rightarrow 2OH \qquad \text{R17}$$

The dominant fate of $O(^1D)$ atoms in the atmosphere is otherwise deactivation by N_2 and O_2 molecules to yield ground state $O(^3P)$, which immediately undergoes reaction R5 yielding no net effect on ozone concentrations.

Considerable amounts of tropospheric ozone may also be formed during the photochemical oxidation of industrial and natural hydrocarbons in the presence of NO. Among the natural emissions, isoprene (C_5H_8) and terpenes ($C_{10}H_{16}$) are emitted by tree foliage (Zimmerman et al 1978), and methane (CH_4) (Ehhalt 1974) is formed in highly anaerobic environments, especially swamps and marshes, by the decay of organic matter. Carbon monoxide is in all cases an intermediate product of the oxidation of these hydrocarbon gases and the natural atmospheric CO source is probably larger than that provided by the fossil fuel combustion processes (Zimmerman et al 1978). Among possible oxidation sequences leading to the production of carbon monoxide and molecular hydrogen from methane (CH_4) oxidation is (Levy 1971, 1974):

Reaction	
$CH_4 + OH \rightarrow CH_3 + H_2O$	
$CH_3 + O_2 + M \rightarrow CH_3O_2 + M$	R2b
$CH_3O_2 + NO \rightarrow CH_3O + NO_2$	R3b
$CH_3O + O_2 \rightarrow CH_2O + HO_2$	
$CH_2O + h\nu \rightarrow CO + H_2$	
$HO_2 + NO \rightarrow OH + NO_2$	R3a
$NO_2 + h\nu \rightarrow NO + O$ (2x)	R4
$O + O_2 + M \rightarrow O_3 + M$ (2x)	R5

N5: $CH_4 + 4O_2 \rightarrow H_2O + CO + H_2 + 2O_3$

The efficiency of this set of reactions in producing ozone again depends on the abundance of NO in the atmosphere. A competitive reaction of HO_2 with CH_3O_2 introduces a reaction chain that does not lead to ozone production:

Reaction	
$CH_4 + OH \rightarrow CH_3 + H_2O$	
$CH_3 + O_2 + M \rightarrow CH_3O_2 + M$	
$CH_3O_2 + HO_2 \rightarrow CH_3O_2H + O_2$	R15b
$CH_3O_2H + h\nu \rightarrow CH_3O + OH$	
$CH_3O + O_2 \rightarrow CH_2O + HO_2$	
$CH_2O + h\nu \rightarrow CO + H_2$	

N6: $CH_4 + O_2 \rightarrow CO + H_2 + H_2O$

Because the reaction $HO_2 + NO \rightarrow OH + NO_2$ is very fast, nitric oxide plays an important role in determining the concentration of perhydroxyl radical (HO_2). Lower concentrations of NO, therefore, allow higher concentrations of HO_2. If we adopt similar rate coefficients for reactions R3b and R15b, as for R3a and R15a respectively, we can again estimate that reaction with nitric oxide and subsequent ozone formation (chain N5) will be more important than the chain which does not produce ozone (N6) for nitric oxide volume mixing ratio larger than about 0.005 ppbv. For much lower volume mixing ratios we do not expect significant ozone production from methane oxidation in the troposphere. We note also that species like CH_3O_2H (methyl-hydroperoxide), other organic hydroperoxides, as well as H_2O_2 (hydrogen peroxide), are quite soluble in water and may be removed from the atmosphere in rainwater before photodissociation occurs, so that production of CO in reaction cycle N6 is in no way guaranteed. The photodissociation probability of CH_3O_2H is not known, but is normally assumed to be close to that of H_2O_2, which has a tropospheric lifetime against photolysis of a few days. The oxidative power of the peroxides in rainwater may strongly influence the oxidation of SO_2 to H_2SO_4 as hypothesized by Crutzen (1975) and convincingly shown in the laboratory by Penkett et al (1978). Through the combined effect of reactions R3 and R15, the probability of photochemical gas phase formation of the peroxides (RO_2H) is inversely proportional to the square of the nitric oxide concentrations. Consequently, the conversion of SO_2 to H_2SO_4 in cloud droplets by H_2O_2 could strongly depend on the concentrations of NO_x in the air. Thus the contribution of HNO_3 to precipitation acidification via the reaction $OH + NO_2(+M) \rightarrow HNO_3(+M)$ may be favored over that of H_2SO_4 in regions close to pollution sources, where NO concentrations are high. If this hypothesis is indeed correct, it would follow that increasing additions of NO_x to the atmosphere would have the effect of pushing the oxidation zone of SO_2 to H_2SO_4 further away from the pollution source areas.

The oxidation schemes of isoprene and the terpenes in the atmosphere are more complicated than those of CH_4, but ozone production occurs much faster than during the oxidation of methane, if NO is present (Zimmerman et al 1978). In these cases, the abundance of nitric oxide is again of great importance in defining the oxidation paths and products, and the possibility of a substantial removal of the peroxides and other oxygenated hydrocarbon intermediates should be considered, if insufficient nitric oxide ($\lesssim 0.005$ ppbv) is present. Little information is presently available to allow estimations of the importance of homogeneous versus heterogeneous reactions in the troposphere and much of the present discussion is, therefore, rather descriptive.

The exact magnitude of carbon monoxide and ozone production in the troposphere for the moment must remain quite uncertain. Some rough estimates of maximum production rates of carbon monoxide in the troposphere can be made, however, by assuming that each carbon atom emitted into the atmosphere as isoprene and methane will yield one molecule of carbon monoxide.

Adopting the estimated worldwide production rates of methane of 400–620 $\times 10^{12}$ g C/yr (Ehhalt 1974) and 830 $\times 10^{12}$ g C/yr for terpenes and isoprene (Zimmerman et al 1978), we arrive at a maximum average global CO column production rate of about 4×10^{11} mol cm^{-2} s^{-1} due to the oxidation of methane and isoprene/terpenes, if all oxidations lead to CO. In comparison, the industrial input of carbon monoxide from incomplete combustion amounts on the average to about 9×10^{10} mol cm^{-2} s^{-1} (Seiler 1974). NO and CO are emitted together in industrial and automotive emissions, so that these CO emissions are most likely to yield ozone by reaction sequence N2. As we will see, the atmospheric residence time of NO_x must be very short, so that the presence of enough NO_x to produce ozone during natural hydrocarbon oxidation is not certain.

We have seen that during the oxidation of CH_4 to CO, there is a production of two ozone molecules (mechanism N5). Similar results apply to the oxidation of other hydrocarbons (Zimmerman et al 1978). If the oxidation of all organic compounds (including that of CO) were to yield only two ozone molecules per carbon atom emitted, then the average global yield of ozone would already be about 10^{12} mol cm^{-2} s^{-1}, compared to an estimated average flux of ozone out of the stratosphere into the troposphere in the N.H. of less than 10^{11} mol cm^{-2} s^{-1} (Mohnen et al 1977, Danielsen & Mohnen 1977) and a ground level average global ozone destruction rate of 4–7.6 $\times$ 10^{10} mol cm^{-2} s^{-1} (Fabian & Junge 1970). Such an ozone column production rate is clearly too large; the sink of ozone provided by the sequence of reaction R6a followed by R17 is an order of magnitude smaller (Fishman & Crutzen 1978). Consequently, there are probably large portions in the troposphere where the NO volume mixing ratio is less than 0.005 ppbv. This would also imply a substantial removal of organics from the atmosphere in rainfall and the existence of quite large concentrations of peroxides, alcohols, and aldehydes in air with low NO concentrations.

A substantial fraction of the ozone molecules in the troposphere has probably been produced in situ (especially in the N.H.) and is not of stratospheric origin. We should note that the stratospheric flux of ozone into the troposphere occurs in bursts over restricted geographical areas in the vicinity of tropopause breaks. Such meteorological processes are clearly much more easily observable than the in situ ozone formation and destruction processes taking place virtually everywhere on a time scale of weeks. In view of the important chemical role played by tropospheric ozone, a careful examination of the origin of tropospheric ozone is necessary. This requires a much better and more extensive data base on the worldwide tropospheric and lower-stratospheric ozone distribution than is presently available, especially in the S.H. The issue is clearly of substantial importance for the prediction of future changes in the distribution of some photochemically and radiatively active minor constituents in the earth's atmosphere, which are affected by OH radical attack. The observed ozone distributions in the N.H. and S.H., taking into account differences in stratosphere-troposphere exchange and the much larger destruction rates of ozone on soil than on sea water, require a four-fold larger annual flux of ozone from the stratosphere to the troposphere in the N.H. than in the S.H. (Fishman & Crutzen 1978). Meteorological parameters which, indirectly, should be indicative of stratosphere-troposphere exchange do not suggest such a large difference in meteorological behavior between hemispheres. However, insufficient meteorological data in the S.H. make this study rather tentative. Nevertheless, we should seriously consider the hypothesis of substantial ozone production in the N.H. and ozone destruction in the S.H. The concentration of NO is one of the most important, and least known, photochemical parameters in this regard. Some rate coefficients of essential reactions may still not be well enough known for tropospheric applications. For instance, the reaction coefficients determined for some reactions involving HO_2 at low pressure probably need to be adjusted for applications at tropospheric conditions. In addition, an increase in the rates of the reactions $HO_2 + OH \rightarrow H_2O + O_2$ and $2HO_2 \rightarrow H_2O_2 + O_2$ has been observed following water vapor additions in laboratory systems (Hamilton 1975, Hamilton & Naleway 1976, Cox 1978). Similar complications could prove to be the case with other reactions involving HO_2.

Nitrogen Oxide Observations

There are few reliable observations of the concentrations of the oxides of nitrogen NO, NO_2, and HNO_3 in the non-urban troposphere. The optical measurements of Noxon (1978, 1979) clearly show that most of the vertical column mass of NO_2 (and therefore NO) is located in the stratosphere, when observations are made in relatively unpolluted air masses. During such conditions, the total NO_2 abundance in the troposphere is mostly well below 10^{15} cm^{-2}, even in the industrial eastern United States. This normally applies if measurements are made at a distance of more than 50 km from urban centers outside the urban plume. Assuming a scale height of 2 km for the NO_2 mixing ratio, the average volume mixing ratio of NO_2 is less than 0.25 ppbv. Near cities, higher tropospheric NO_2 concentrations on the order of 100 ppbv or more are typically found (Trijonis 1978). Noxon's measurements of background, continental, NO_x concentrations, which are much below 1 ppbv, cast considerable doubt on the value of the often quoted literature estimates of Robinson & Robbins (1971), which were based on older and presumably less reliable techniques. Subsequent measurements include those of Lodge et al (1974) in the humid tropics in Panama (0.1–0.3 ppbv), Drummond (1977) near Laramie, Wyoming (0.1–0.4 ppbv of NO_x), and Moore (1974), who measured 0.1–0.3 ppbv of NO_2 near Boulder, Colorado. Ritter et al (1978) report NO_x volume mixing ratios between 0.3 and 0.5 ppbv during "clean air" conditions in northern Michigan and an average of 0.2 ppbv at a station in the Rocky Mountains at a 3-km altitude 30 km northeast of

Denver. Measurements over Colorado and Wyoming in December by Kley et al (1978) give ground level mixing ratios of about 1 ppbv for NO_x and a fall-off to 0.15 ppbv above 6 km. Ground level observations by the same researchers at the same Rocky Mountain station give NO_x volume mixing ratios between 0.02 and 0.2 ppbv. Considering Noxon's low background tropospheric NO_2 concentration determinations ($\lesssim 0.25$ ppbv) and the high urban concentrations of about 100 ppbv (Trijonis 1978), clearly the residence time of NO_x in the troposphere must be remarkably short, maybe less than one day.

The most extensive tropospheric measurement of nitric acid (HNO_3) in the "clean" troposphere has been conducted by Huebert & Lazrus (1978, 1979) during the worldwide chemistry expeditions "GAMETAG" (an acronym for Global Atmospheric Measurement Experiment of Tropospheric Aerosols and Gases). These measurements were made on board the NCAR Electra aircraft in 1977 and 1978. These observations show average nitric acid volume ratios between about 0.2 and 0.8 ppbv over continental mid-latitudes (US and Canada), while at all other locations (mostly over the Pacific) measured mixing ratios were lower, <0.03–0.15 ppbv in the lowest two kilometers and <0.03–0.3 ppbv in the middle troposphere. From these data we can, for the sake of discussion, make some rough estimates of the background volume mixing ratios of NO_x. To a first approximation the reactions affecting the concentrations of NO_x in the unpolluted troposphere are

$$HNO_3 + h\nu \rightarrow OH + NO_2 \qquad R18$$
$$OH + HNO_3 \rightarrow H_2O + NO_3 \qquad R19$$
$$NO_3 + NO \rightarrow 2NO_2$$
$$NO_2 + h\nu \rightarrow NO + O \qquad R4$$
$$O + O_2 + M \rightarrow O_3 + M \qquad R5$$
$$NO + O_3 \rightarrow NO_2 + O_2 \qquad R7$$
$$NO_2 \rightarrow \text{nitrates (heterogeneous, e.g. wet removal)}$$
$$OH + NO_2(+M) \rightarrow HNO_3(+M) \qquad R20$$

Note that in this set of reactions the photolysis of nitric acid provides a source of NO_2 and, therefore, of NO_x as reactions R3, R4, and R7 will immediately establish a steady state ratio between the concentrations of NO and NO_2. Neglecting heterogeneous removal and assuming that there are no other atmospheric sources of NO_x than HNO_3 decomposition, and stationary state conditions for NO_x, we can derive the upper limit to the concentration of NO_x from the equation

$$(NO_2) < \frac{[J_{18} + k_{19}(OH)](HNO_3)}{k_{20}(OH)}$$

Assuming tropospheric global daytime concentrations of OH to be near 10^6 mol cm^{-3} (Perner et al 1976, Davis et al 1976) and adopting appropriate values for the reaction coefficients and the dissociation probability of nitric acid, we arrive at (NO) $\lesssim 4 \times 10^{-2}$ (HNO_3), since (NO) $\approx$ (NO_2) in the lower "clean" troposphere. This leads to NO volume mixing ratios over continental mid-latitudes of $8–32 \times 10^{-12}$ to less than 6×10^{-12} over non-continental areas. The calculated concentrations over continental areas are generally smaller than those obtained by direct measurements of NO. Therefore in the following section of this paper we explore possible sources of NO_x in the free troposphere other than from photolysis of nitric acid.

Sources and Sinks of Tropospheric NO_x

The input of nitric oxide into the troposphere consists of an anthropogenic source of about 20×10^{12} g N/yr from combustion engines (Pratt et al 1977), a contribution by lightning of $8–40 \times 10^{12}$ g N/yr (Noxon 1978, Chameides et al 1977), and a possible production of 10^{13} g N/yr in soils (Galbally & Roy 1978). Adopting the known rate constants for the reaction $NH_3 + OH \rightarrow NH_2 + H_2O$ and assuming an average OH concentration of 10^6 cm^{-3}, we estimate a tropospheric lifetime of NH_3 against gas phase destruction of about 3 months. Because this is so long, it seems likely that most ammonia emitted into the atmosphere is removed by processes other than gas phase destruction. Furthermore, since it is not certain that NH_2 will oxidize to NO_x, the production of NO_x from NH_3 oxidation should be less than about 8 Tg N/yr. This estimate is based on a total ammonia source of 73–100 Tg N/yr (Söderlund & Svensson 1976, Dawson 1977) and an average tropospheric lifetime of less than one week. Soils were previously thought to release NO into the atmosphere (see e.g. Robinson & Robbins 1971) in amounts much larger than the industrial input. This view is unacceptable in view of the global atmospheric NO_x observations and estimated residence times.

For the reaction $NO_2 + OH(+M) \rightarrow HNO_3(+M)$ to provide an NO_x residence time of one day, average daytime OH concentrations must be about 2.5×10^6 mol cm^{-3}. From observations, average global mid-latitude molecular densities of hydroxyl are probably about 10^6 cm^{-3} or larger (Perner et al 1976, Davis et al 1976, Crutzen & Fishman 1977). Most of the industrial input of NO_x, together with copious amounts of other pollutants, such as reactive hydrocarbons, occurs, however, in or near urban centers. In such environments the hydroxyl concentrations in the planetary boundary layer are expected to be substantially higher than in the "clean" troposphere, so that reaction with OH may provide the required efficient sink for NO_x. The nitric acid formed in this conversion should then be removed on a time scale of less than one month to explain the measurements of Huebert & Lazrus (1978, 1979). Some of the nitric acid formed in the polluted boundary layer may, however, be converted back to NO_x after dispersion into the background troposphere as the equilibrium value of the ratio $(HNO_3)/(NO_2)$ decreases from 100 to 25 to 3 for OH concentrations of respectively 10^7, 10^6, and 10^5 cm^{-3}. It is possible that gaseous organic nitrates play a substantial role as a sink for urban NO_x and a global source of NO_x for outside industrial areas, and provide another source of NO_x in the "clean" troposphere.

Organic peroxy-nitrates are produced by recombination reactions $RO_2 + NO_2(+M) \rightarrow RO_2NO_2(+M)$. In smog chamber experiments (see Demerjian et al 1974) the observed peroxy-nitrates originated only from the peroxy-acyl radicals $CH_3C(=O)O_2$, $CH_3CH_2C(=O)O_2$, and their higher homologues, and not from the peroxy-alkyl radicals, such as CH_3O_2, $C_2H_5O_2$, etc, indicating that the peroxy-alkyl nitrates are less stable than the peroxy-acyl nitrates. This experimental finding has now been confirmed by a review of kinetic and thermochemical studies on the thermal stability and kinetics of $CH_3C(=O)O_2NO_2$ (PAN), $CH_3O_2NO_2$, and HO_2NO_2 (Cox & Roffey 1977, Hendry & Kenley 1977, 1979, Cox 1978, Graham et al 1977).

Among the peroxy-acyl nitrates the most important species, both in smog chamber experiments and in the atmosphere, seems to be peroxy-acetyl nitrate (PAN), which is formed from the peroxy-acetyl (PA) radical, $CH_3C(=O)O_2$, by the reaction

$$PA + NO_2(+M) \rightarrow PAN(+M).$$

PAN is not stable in the atmosphere, but is decomposed according to the following chain

$$PAN \rightleftarrows PA + NO_2 \qquad R21a\rightarrow, R21b\leftarrow$$

$$PA + NO \rightarrow CH_3 + CO_2 + NO_2 \qquad R22.$$

These reactions yield the following formula for the photochemical lifetime of PAN during daytime when NO is present:

$$\tau_{PAN} = \frac{1}{k_{21a}}\left\{1 + \frac{k_{21b}(NO_2)}{k_{22}(NO)}\right\}$$

with

$$k_{21a} = (7.9 \times 10^{14}) \exp\left(\frac{-12540}{T}\right)$$

and with $k_{21b} = 1.4 \times 10^{-12}$ cm^3 mol^{-1} s^{-1} and $k_{22} = 2.4 \times 10^{-12}$ cm^3 mol^{-1} s^{-1} (Cox & Roffey 1977).

With $(NO_2) \approx (NO)$ during daytime and no PAN loss at night, because (NO) disappears, we compute the following values for the atmospheric residence time of PAN against photochemical destruction as a function of altitude, adopting temperature tabulations in the US standard atmosphere: at $z = 0$ km, $T = 288$ K, $\tau(\text{PAN}) = 0.4$ days; $z = 2$ km, $T = 275$ K, $\tau(\text{PAN}) = 2.9$ days; $z = 4$ km, $T = 262$ K, $\tau(\text{PAN}) = 28$ days; $z = 6$ km, $T = 249$ K, $\tau(\text{PAN}) \approx 1$ yr; $z = 8$ km, $T = 235$ K, $\tau(\text{PAN}) = 15$ yr.

We note that the lifetime of PAN against photochemical destruction increases rapidly with altitude, becoming longer than one year above 5 km. Photolysis of PAN below this altitude is probably negligible (E. R. Stephens, personal communication). Above 5 km PAN is probably slowly decomposed by OH radical attack and maybe by photolysis. No information on either of these processes is currently available. It is clear that PAN formation will be favored during the cold winter months (Hendry & Kenley 1979). Concentrations of PAN similar to or larger than those of nitric acid have been measured in contaminated air in West Covina near Los Angeles. According to Spicer et al (1976) and Spicer (1977), on the average about 80% of the NO_x removed from the polluted air ended up as PAN (average daily volume mixing ratio ≈ 12 ppbv against nitric acid ≈ 3 ppbv). Only a little particulate nitrate (≈ 1 ppbv) was formed. From an extensive series of measurements in St. Louis, Missouri, and West Covina, California, the ratio $(\text{PAN}+HNO_3)/(NO+NO_2+\text{PAN}+HNO_3)$ in the afternoon was shown to range between 0 and 76% in St. Louis and between 2 and 54% in West Covina, with an average value of about 11% in both cities (Spicer et al 1976). As these measurements were conducted at sites continuously supplied with fresh automotive NO_x and hydrocarbon emissions, the observations were clearly made in air masses in which transformations of NO_x to PAN and HNO_3 were still taking place. In fact, both smog chamber experiments and theoretical calculations (Spicer et al 1976, Hendry & Kenley 1979) indicate a buildup of PAN to concentrations larger than NO_2 in the later phases of photochemical smog formation. The hydrocarbons which operate in the transformation of NO_x to PAN in St. Louis and West Covina no doubt come overwhelmingly from automotive exhaust. In rural environments, especially near forests, hydrocarbon emissions consist mainly of isoprene and terpenes from tree foliage. Clearly then, it is not only important to consider nitric acid, but also and possibly more importantly PAN and other vapor phase organic nitrates as precursors to NO_x in the global atmosphere. This is important as PAN is surprisingly slowly lost from the atmosphere by wet removal. A laboratory study by Garland & Penkett (1976) gave a deposition rate of PAN on water surfaces slower than that of ozone. Global transport of NO_x may, therefore, occur in the manner shown in

Figure 1. Whether or not PAN can indeed be a "hiding place" for NO_x has to be determined from measurements of NO_x and PAN in the unpolluted troposphere far from urban centers. Fortunately, good techniques to make low concentration measurements of these gases are now becoming available.

In considering non-photochemical removal rates of NO, NO_2, and their oxidation products from the atmosphere, we must consider the role of rainout and washout (wet removal) in the atmosphere. According to Rodhe & Grandell (1973) the efficiency of this process is to an important degree determined by the frequency of precipitation, and can at the fastest lead to a removal time of a few days (2–4 days for Stockholm, Sweden). This time is about equal to the estimated residence time of nitric acid, but seems to be too long to explain the residence time of NO_x. Gas phase reactions must, therefore, be more important.

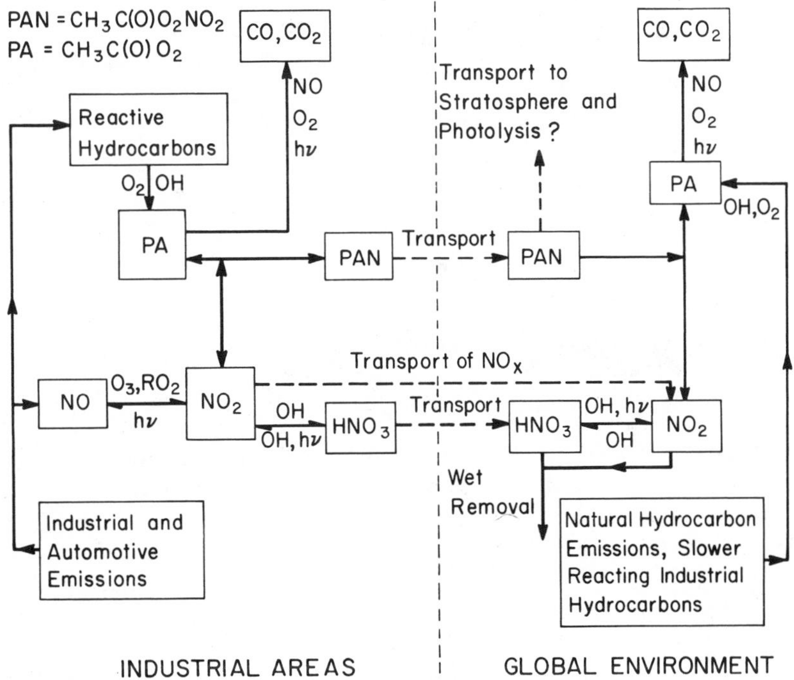

Figure 1 Transport of NO_x, HNO_3, and $CH_3C(O)O_2NO_2$ (= PAN) from urban centers to the "clean" troposphere. PAN has a long lifetime in the middle or upper troposphere and may deliver NO_x in the warmer lower troposphere at large distances from the urban centers in which PAN was initially formed. Likewise, nitric acid formed in polluted environments may supply NO_x to the "clean" troposphere.

Finally, we must also consider the effects of "dry" deposition. Hill (1971) observed a deposition velocity of 2 cm/s of NO_2 on an alfalfa canopy with an NO_2 volume mixing ratio of 50 ppbv (the deposition velocity of NO was only 0.1 cm/s). Similar removal rates were determined by Tingey (1968) on alfalfa and oats. With a deposition velocity of the order of 2 cm/s, soils, vegetation, and water surfaces may provide a significant sink for tropospheric NO_x (and presumably nitric acid) yielding a residence time in the lower atmosphere of a few days.

OXIDES OF NITROGEN IN THE STRATOSPHERE

At the end of the last decade our knowledge of stratospheric photochemistry was truly rudimentary. Considering the supreme importance of ozone to the atmosphere and the biosphere, the slow evolution of knowledge up to that time seems most remarkable. It was then still commonly accepted that the photochemical processes affecting ozone in the stratosphere, as proposed by Chapman (1930), were adequately explained by

(*a*) formation through photolysis of O_2:

$$O_2 + h\nu \rightarrow 2O \qquad \lambda \lesssim 240 \text{ nm}$$
$$O + O_2 + M \rightarrow O_3 + M \ (2x)$$

$$\text{net:} \quad 3O_2 \rightarrow 2O_3$$

(*b*) removal through the reactions:

$$O_3 + h\nu \rightarrow O + O_2 \qquad \lambda \lesssim 1140 \text{ nm}$$
$$O + O_3 \rightarrow 2O_2$$

$$2O_3 \rightarrow 3O_2$$

Regarding the troposphere, it was believed that ozone was essentially inert when transported below about 25 km and only destroyed by contact with the earth's surface. During the last decade, the progress in our knowledge of the photochemistry of atmospheric ozone has been truly revolutionary. Indeed, comparing the present literature with that written before 1970 one wonders whether it describes the chemistry of the same planet.

Now it is clear that the oxides of nitrogen play an important role in defining the abundance and distribution of stratospheric ozone. Interest in the stratospheric role of the oxides of nitrogen started with the recognition that NO and NO_2 could catalyze the destruction of ozone in the stratosphere and that human activities could lead to significant increases in the stratospheric abundance of NO_x (Johnston 1971, Crutzen 1970). Ozone destruction was postulated to occur, with X = NO, via the cycle of reactions:

$$O_3 + h\nu \rightarrow O + O_2 \qquad R6$$
$$O + XO \rightarrow O_2 + X \qquad R8$$
$$O_3 + X \rightarrow O_2 + NO_2 \qquad R7$$

$$N7: \quad 2O_3 \rightarrow 3O_2$$

The existence of an effect from the catalytic cycle involving nitric oxide and leading to ozone destruction in the stratosphere below 45 km was clearly demonstrated by the observations of a sudden decrease in ozone concentrations lasting for several weeks immediately following the huge solar proton event of August, 1972 (Heath et al 1977). The detected decrease in ozone was very close to that earlier calculated through use of a two-dimensional, photochemical model of the atmosphere (Heath et al 1977).

We may infer from the preceding discussion that direct transfer of NO_x and HNO_3 from the troposphere to the stratosphere must be essentially forbidden by the very short tropospheric residence time of NO_x of the order of one day. Observational support of this contention is provided by the global distribution measurements of nitric acid by Huebert & Lazrus (1978, 1979) in which nitric acid was shown to be present in the middle troposphere at volume mixing ratios ranging from less than 0.03 to 0.3 ppbv. In contrast, stratospheric nitric acid volume mixing ratios are, on the average, close to 5 ppbv (Lazrus & Gandrud 1974). In addition, in the previous section we referred to the observations of Noxon (1978), which show that most of the atmospheric NO_2 is likewise located in the stratosphere.

Less reactive and less water soluble oxides of nitrogen do, however, penetrate the tropopause. There may possibly be some upward transfer of organic nitrate molecules (in particular PAN). There is, however, little doubt about the importance of nitrous oxide (N_2O). This gas seems very inert in the troposphere and is photochemically destroyed in the stratosphere (Bates & Hays 1967), mainly through the photolysis reaction

$$N_2O + h\nu \rightarrow N_2 + O$$

with smaller contributions coming from the reactions

$$N_2O + O(^1D) \rightarrow N_2 + O_2$$
$$\rightarrow 2NO.$$

Although only about 10% of the nitrous oxide that enters the stratosphere is removed by the latter reaction (Johnston et al 1978), this reaction is of substantial importance because it provides an important, and probably a dominant, source of nitric oxide in the ozone layer (Nicolet & Vergison 1971, Crutzen 1971, McElroy & McConnell 1971).

Other natural sources of stratospheric nitric oxide are provided by the action of galactic cosmic rays at high geomagnetic latitudes especially in the lower stratosphere (Nicolet 1975) and by sporadic outbursts of solar protons following some solar flares, the so-called solar proton events (SPE) or polar cap absorption events (PCA) (Crutzen et al 1975). Very large amounts of NO_x are produced above 100 km by EUV radiation and in auroras by the ionizing and dissociative action of energetic electrons of solar origin. Using a one-dimensional photochemical model, Strobel et al (1970) and Strobel (1971) estimated that only a negligible fraction of the nitric oxide produced at high altitudes could reach the stratosphere. This is due to the efficient reactions (Cieslik & Nicolet 1974), $(NO+h\nu \rightarrow N+O) + (N+NO \rightarrow N_2+O) = 2NO \rightarrow N_2+O_2$, which occur, however, only in the sunlit atmosphere above 45 km (Nicolet 1975). In darkness, no nitrogen atoms are formed and those present are transformed into nitric oxide by reactions with molecular oxygen and ozone. During the long polar night conditions NO may be transported downwards into the stratosphere by the general circulation of the upper atmosphere (Murgatroyd & Singleton 1961, Geisler & Dickinson 1968, Crutzen 1972). This idea is supported by some modern developments in the theory of the large scale, Lagrangian-formulated, dynamics of the upper atmosphere (McIntyre 1977, Matsuno & Nakamura 1978, Dunkerton 1978). According to Offerman (personal communication, 1978) downward transport of NO near the mesopause (≈ 80 km) does apparently also occur at middle and high latitudes during periods of enhanced D-region radio wave absorption (D-region winter anomalies). Such downward transport, if it occurred to a significant degree, would be of special interest, as NO production in the high atmosphere is large and related to solar activity (Strobel et al 1970, Crutzen 1970).

Table 2 shows estimates of the gross NO_x production rates in the atmosphere above 100 km. Stolarski et al (1975) show that the absorbed solar EUV energy ($\lambda < 1025$ Å) is about equally split between photoelectrons and ion pair production that becomes stored as chemical energy. Photoelectron impact dissociates N_2 into $N(^2D)$ and $N(^4S)$ and in the global mean about 2.5×10^{10} cm^{-2} s^{-1} dissociations occur above 100 km. The gross global NO_x production rate is, thus, 2.7×10^{29} s^{-1} above 100 km. This source can vary by a factor of two as the solar EUV output varies over the solar cycle. $N(^4S)$ and $N(^2D)$ are produced not only by

Table 2 Sources of atmospheric NO_x (1 Tg = 10^{12} g)

Troposphere	
Lightning:	8–40 Tg N/yr (Noxon 1978, Chameides et al 1977)
Combustion:	20 Tg N/yr (Söderlund & Svensson 1976)
Soil exhalations:	10 Tg N/yr (Galbally & Roy 1978)
Ammonia oxidation:	<8 Tg N/yr (this review)
Stratosphere	
N_2O oxidation:	1 Tg N/yr (Johnston et al 1978)
Galactic cosmic rays:	0.024–0.036 Tg N/yr (Nicolet 1975); solar maximum-minimum
Solar proton event:	0.12 Tg N (1972 maximum, Crutzen et al 1975)
Nuclear bombs:	0.024 Tg N per megaton TNT (Johnston et al 1973)
Thermosphere	
Meteoroids:	0.17 Tg N/yr (Park & Menees 1978)
Solar EUV:	200–400 Tg N/yr; solar minimum-maximum

photoelectron dissociation but also by ion chemistry as discussed by Strobel et al (1970), Strobel (1971), Rusch et al (1975), and Roble et al (1978).

At high latitudes, the aurora is a source of NO through electron dissociation by N_2 into $N(^4S)$ and $N(^2D)$ and ion chemistry. The global production rate by auroral processes is highly variable and difficult to estimate. Akasofu (1976) using satellite photographs of auroral structure estimated the total area in the polar cap that was subjected to auroral particle bombardment to vary between 4×10^5 and 1.8×10^7 km^2 for geomagnetic quiet and intense auroral substorm conditions. Using the particle data measured by the Atmospheric Explorer satellite, Torr & Torr (1976) determined the latitude distribution of the auroral energy deposition and mean particle spectrum. Above about 60° geomagnetic latitude the energy input was about 2 ergs cm^{-2} s^{-1} with an electron particle spectrum that peaked about 1.5 keV. The total dissociation rate of N_2 by auroral electrons for this energy input and spectrum is about 1.7×10^{10} cm^{-2} s^{-1} above 100 km with peak production at an altitude of 115 km. These values were calculated using the aurora model of Roble & Rees (1977). This is a lower limit of the production rate and the actual rate could be a factor of four larger, since one must also consider the complex ion chemistry. If the production rate is multiplied by the total area of auroral emissions as evaluated by Akasofu then the global auroral NO_x production rate (considering magnetically conjugate auroral zones) exceeds 6×10^{26} s^{-1} for geomagnetic quiet periods and 2.5×10^{28} s^{-1} for intense auroral substorm conditions. Although these rates are smaller than the global EUV production rates, they may be important because of the probability of downward transport at high latitudes during winter. It follows clearly from a comparison of the NO_x production rates in Table 2 that a very small leakage of thermospheric NO_x into the stratosphere may have important consequences for the photochemistry of the stratosphere.

Increased levels of nitric oxide in the stratosphere can be caused in particular by three human activities:

(*a*) direct injection of nitric oxide from the exhaust gases of aircraft flying in the stratosphere (Johnston 1971, Crutzen 1971);
(*b*) injection of nitric oxide in the fireballs of nuclear explosions (Johnston et al 1973, Goldsmith et al 1973); and
(*c*) man's agricultural manipulations, e.g. the increased use of nitrogen fertilizer, leading to the release of nitrous oxide from soils and waters to the atmosphere and thereby increasing the stratospheric levels of nitric oxide (Crutzen 1974a, McElroy 1976).

Many detailed assessment reports have been published over the past years on the stratospheric effects of human activities (NAS 1975, NASA 1977, NAS 1978). We therefore devote our attention only to a brief account of present knowledge of stratospheric photochemistry, an extremely complex subject because of the strong interactions of the oxides of nitrogen with other gases. Photochemically derived species from water vapor and chlorine compounds affect ozone, as does NO_x, by catalytic reactions according to the generalized scheme N7 with X = H and OH (Bates & Nicolet 1950, Hampson 1964) and X = Cl (Stolarski & Cicerone 1974, Crutzen 1974b, Wofsy & McElroy 1974).

None of these authors foresaw a major role of ClO_x in the stratosphere. A dramatic change took place, however, when Molina & Rowland (1974) discovered the growing production in the stratosphere of the very powerful catalysts Cl and ClO as a result of the photolysis of the chlorofluoromethane gases $CFCl_3$ and CF_2Cl_2, by then widely and increasingly used in refrigerators and as propellants in spray cans. With chlorine chemistry taking on great importance in stratospheric photochemistry, the chemical interactions between the ClX, HX, and NX reactant groups became of substantial interest (see Table 1 and Figure 2). If not for these interactions, NO_x, HO_x, and ClO_x would each independently destroy ozone catalytically. It is the interactions between the groups themselves that make the effects of these substances on ozone and on one another much more difficult to predict.

Cl atoms react mainly with O_3 through

$$Cl + O_3 \rightarrow ClO + O_2 \qquad \text{(see N7).}$$

Coupling of the NX and ClX groups occurs as follows:

$$NO + ClO \rightarrow Cl + NO_2 \qquad R23.$$

The net effect of these two reactions is to shift the balance from NO to NO_2, thereby enhancing NX catalyzed ozone destruction, when NO_2 reacts with O in reaction R8, but reducing ClX catalyzed destruction, when it dissociates to yield O by reaction R4. The effects of NO_x additions to the stratosphere must, therefore, depend on the ClX concentrations, which at any time are present in the stratosphere. Reaction 23 also shifts the equilibrium between Cl and ClO towards Cl. When more Cl is present, more ClO_x is converted to non-reactive HCl, through the

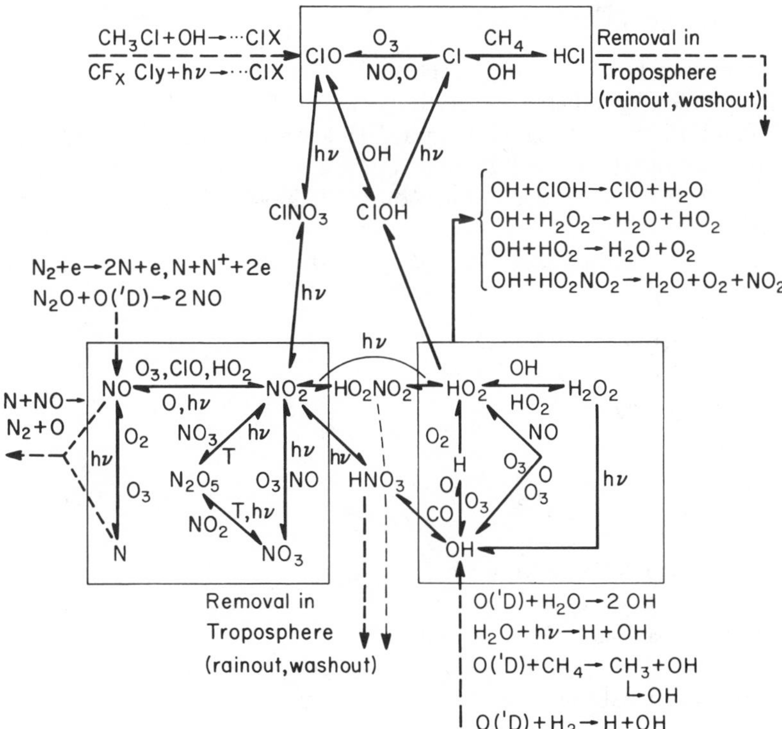

Figure 2 Interactions between the NX, HX, and ClX families of reactants. The transformation reactions within each family of reactants occur generally at much faster rates than the net rates of formation (and destruction) of NX, ClX, and HX. This forms the basis for the "quasi steady state assumption" (QSSA), which is often used with great advantage in time-dependent models. The importance of ClOH and HO_2NO_2 is doubtful.

reactions

$$Cl + CH_4 \rightarrow HCl + CH_3$$

$$Cl + HO_2 \rightarrow HCl + O_2$$

and thus the ClX-catalyzed destruction of ozone is repressed. However, the HX group is coupled to the NX group via

$$OH + NO_2(+M) \rightarrow HNO_3(+M) \qquad R2a$$

$$HO_2 + NO \rightarrow OH + NO_2 \qquad R3a.$$

R2a mainly stores active NO_x as HNO_3 (compare HCl, above), thereby protecting ozone from NO_x catalyzed destruction. The reaction that reforms NO_x from HNO_3,

$$HNO_3 + h\nu \rightarrow OH + NO_2,$$

is rather slow below about 30 km.

Reaction R3a is therefore very important below 35 km, since it shifts the equilibrium between HO_2 and OH towards OH, thus enhancing production of HNO_3 via R2a. Reaction R3a has recently been shown to be extremely fast (Howard & Evenson 1977), having the effect of making HNO_3 the predominant NX compound below 30 km.

Reaction R3a also diverts the catalytic, ozone destroying HX reaction set

$$\begin{array}{l} OH + O_3 \rightarrow HO_2 + O_2 \\ HO_2 + O_3 \rightarrow OH + 2O_2 \\ \hline \text{net:} \quad 2O_3 \rightarrow 3O_2 \end{array}$$

into a "do-nothing" cycle,

$$\begin{array}{c} OH + O_3 \rightarrow HO_2 + O_2 \\ HO_2 + NO \rightarrow OH + NO_2 \\ NO_2 + h\nu \rightarrow NO + O \\ O + O_2 + M \rightarrow O_3 + M \\ \hline \text{no net chemical effect,} \end{array}$$

thereby decreasing ozone loss. This will be true under present atmospheric conditions with moderate stratospheric NO_x additions.

Finally, the enhancement of OH through R3a tends to push HCl back to the ozone destroying compounds ClO and Cl via the reaction

$$HCl + OH \rightarrow Cl + H_2O.$$

Other possible reactions coupling the ClX, NX, and HX groups are the following recombination reactions:

$$ClO + NO_2(+M) \rightarrow ClONO_2(+M)$$

$$HO_2 + NO_2(+M) \rightarrow HO_2NO_2(+M)$$

$$ClO + HO_2 \rightarrow ClOH + O_2.$$

The mixed molecules $ClONO_2$, $HONO_2$, HO_2NO_2, and ClOH are all photolyzed effectively at greater heights. The photolysis probabilities of ClOH and HNO_4 and their reaction kinetics are not well known, but it seems most probable that they are sufficiently large that the importance of these compounds in stratospheric chemistry is not great. Stratospheric observations show an upper limit for HO_2NO_2 of less than 10% of the abundance of HNO_3 (Murcray et al 1978), indicating that HO_2NO_2 is probably efficiently removed by photolysis at longer wavelengths than 300 nm. Spectral studies on the absorption cross sections of ClOH also show appreciable absorption at wavelengths longer than 300 nm (Molina & Molina 1978). These experiments are difficult to conduct, however, and the results do not agree with the theoretical calculations by Jaffe & Langhoff (1977). If ClOH photolysis were fast, the impact of ClOH formation would be minor (there is only a slight catalysis of the reaction $HO_2 + O_3 \rightarrow OH + 2O_2$). A slow photolysis rate would, however, allow more ClX to be stored as ClOH (which does not react with ozone) and furthermore give rise to a catalytic cycle $(ClO + HO_2 \rightarrow ClOH + O_2) + (ClOH + OH \rightarrow ClO + H_2O) = (OH + HO_2 \rightarrow H_2O + O_2)$, which could influence the odd hydrogen balance of the stratosphere.

The mixed molecules would exist mainly below about 30 km, providing stratospheric ozone protection from otherwise substantially larger destruction. As this complex chemistry occurs in a region of the stratosphere much affected by still badly known transport processes, modeling of the chemistry of the lower stratosphere has become rather complicated. There is also considerable uncertainty about the abundance of gaseous chlorine compounds in the stratosphere. Given the known abundances of natural and anthropogenic organic chlorine compounds (CH_3Cl, $CFCl_3$, CF_2Cl_2, CH_3CCl_3, CCl_4) in the upper troposphere, the maximum possible volume mixing ratio of ClX in the upper stratosphere should be about 2 ppbv. Available measurements of stratospheric ClO leave the possibility open for far larger concentrations of stratospheric ClX: Anderson et al (1977, 1978) reported mixing ratios larger than 2 ppbv of ClO alone in the 35–50 km altitude region (on one occasion 7 ppbv).

The measurements by Anderson et al (1977, 1978) seem to have been carried out with great care and considerable ingenuity. Nevertheless, substantial difficulties are connected with such a high concentration of ClO. First, a large enough ClX source to explain these measurements is not presently known. Second, with so much powerful catalyst around, the concentration of ozone should be exceptionally low, or alternatively the large abundance of ClO should be compensated by a very low abundance of NO_2 (ClO reacts about six times faster with O than NO_2 does). The measured volume mixing ratios of ClO at 35–40 km vary widely between 0.6 and 7 ppbv from experiment to experiment, and ozone does not vary that extremely in this altitude range. It would, therefore, be a remarkable coincidence if the ClO and NO_2 volume mixing ratios were exactly anti-correlated such that six times the mixing ratio of ClO and one time that of NO_2 would always add up to approximately the same total number. In favor of the prevalence of lower volume mixing ratios of ClO are the ground and aircraft millimeter-wavelength emission measurements of Waters et al (1978), which show that profile peak ClO volume mixing ratios during their observation period (May–August 1977) cannot have exceeded 1 ppbv.

In general, it now seems from detailed photochemical model calculations that ozone is very sensitive to anthropogenic ClX additions to the stratosphere (NASA 1977). To estimate the influence of anthropogenic additions of NO_x to the stratosphere, knowledge of the background concentration of ClX in the stratosphere is of considerable importance. With a large atmospheric ClX concentration, additions of NO_x tend to reduce ClX destruction of ozone below 35 km by more conversion of active chlorine ClO_x to HCl and by enhanced production of the mixed ClX-NX molecules. NO_x thus affects stratospheric ozone in complex ways and it now seems that the total ozone column density is only slightly dependent on NO_x additions to the stratosphere (Duewer et al 1977, Crutzen & Howard 1978). Specifically, we now expect that NO_x additions to the present stratosphere will lead to smaller ozone concentrations above about 24 km and larger values below 24 km (Crutzen & Howard 1978). It is difficult to predict the effects of stratospheric aircraft and of increased use of nitrogen fertilizer on the stratospheric ozone column, and it is doubtful that a one-dimensional model can be used to estimate these effects. With ozone increases occurring below 24 km, it seems quite possible that the total ozone column will increase at high latitudes (where there is a lot of ozone below 24 km) and decrease at low latitudes (where most ozone is located above 24 km). Because the lower stratosphere is important for climate, the increased heating of this region caused by larger solar ultraviolet radiation penetration into a layer with more ozone can lead to higher tropopause temperatures and more water vapor in the stratosphere (Liu et al 1976b).

It is clear that many years of research are still required to establish the consequences of a number of human activities. It is essential to obtain more measurements of the global distribution of photochemically reactive compounds in the stratosphere. The ClX content of the stratosphere must be measured and more independent observations are clearly necessary. There are now many measurements available of NX compounds (e.g., Ackerman et al 1975, Chaloner et al 1975, Drummond et al 1977, Goldman et al 1970, 1978, Harries 1978, Lazrus & Gandrud 1974, Loewenstein et al 1975, Mason & Horvath 1976, Murcray et al 1973, 1978, Patel et al 1974, Ridley et al 1973, 1975, Toth et al 1973). Although no dramatic disagreement exists here between observations and theoretical model results, agreement is also not sufficiently good (Harries 1978, Crutzen & Howard 1978). Extensive reviews of measurements of stratospheric minor constituents have been given in various assessment studies (NAS 1975, 1976). An extensive study comparing observations with model calculations was performed by Wofsy (1978). The substantial influence of transport processes on the distribution of minor stratospheric constituents was again shown very convincingly by the observational studies of Noxon (1978, 1979), who reported large day-to-day variations in the total column abundance of NO_2, which correlated with the passage of meteorological disturbances in the lower stratosphere. Similar correlations between total ozone recordings and meteorological variations in the lower stratosphere have, of course, been known to exist for many years (Dütsch 1974). A remarkable additional finding of Noxon (1979) is a persistent sharp decline of the total vertical NO_2 column abundance by about a factor of two polewards of 50°N and 50°S during the winter season, when a well-defined polar vortex patterns exists. This "cliff" seems to reflect a sharp gradient in the temperatures below about 30 km and may be explained by the temperature dependence of reaction R7 and stratospheric circulation characteristics. At high latitudes during winter we must further consider the formation of N_2O_5 (Crutzen 1971).

CONCLUSIONS

The oxides of nitrogen, NO and NO_2, are extremely important in atmospheric chemistry, especially because of their role in the radical chemistry and the production and destruction of tropospheric and stratospheric ozone. The role of ozone in tropospheric photochemistry is explained by its role in the formation of the extremely important OH radical, which attacks many otherwise inert gases in the troposphere. In addition, reactions involving OH (and its byproduct H_2O_2) lead to the formation of the acid gases HNO_3 and H_2SO_4. These gases are important in determining the acidity of precipitation. Because of catalytic ozone-forming reactions, involving NO and NO_2, tropospheric ozone concentrations, especially in the N.H., may be influenced strongly by photochemical in situ production. Anthropogenic activities contribute substantially to the total NO_x source in the atmosphere.

The oxides of nitrogen, NO and NO_2, are removed very efficiently from the lower troposphere. The processes responsible for their fast removal are, however, not well known. Several gaseous organic nitrates, especially PAN, formed in polluted boundary layers, are very stable above about 4 km in the troposphere and may, therefore, carry "captured" NO_x over substantial distances on the globe. After return to the lower, warmer layers of the troposphere PAN will break up, thereby releasing NO_2. This hypothesis of a possible mechanism for long distance transport of NO_x has yet to be tested by observations in "unpolluted" air.

In the stratosphere, NO_x participates in many reactions influencing the equilibria in the ClX and HX photochemical systems in such a way that the total ozone abundance may be relatively insensitive to NO_x additions to the stratosphere. The vertical profile of ozone may, however, be significantly altered with higher ozone concentrations below 24 km and lower concentrations above. Profile changes of this kind may play a role in climatic and dynamical processes.

The large production of NO_x above about 100 km is dependent on solar activity. A small leakage of this NO_x to the stratosphere may have very interesting consequences.

Acknowledgments

The National Center for Atmospheric Research (NCAR) is sponsored by the National Science Foundation. Drs. Raymond Roble and Manfred Rees generously provided me with their estimates of odd nitrogen production rates in the thermosphere. I thank Dr. Jack Fishman and Ms. Susan Solomon for their assistance in the preparation of this review.

Literature Cited

Ackerman, M., Fontanella, J. C., Frimout, D., Girard, A., Louisnard, N., Muller, C. 1975. Simultaneous measurements of NO and NO_2 in the stratosphere. *Planet. Space Sci.* 23:651–60

Akasofu, S.-I. 1976. Recent progress in studies of DMSP auroral photographs. *Space Sci. Rev.* 19:169–215

Altshuller, A. P., Bufalini, J. J. 1971. Photochemical aspects of air pollution: a review. *Environ. Sci. Technol.* 5:39–64

Anderson, J. G., Grassl, H. J., Shetter, R. E., Margitan, J. J. 1978. Stratospheric free chlorine measured by balloon borne in situ resonance fluorescence. *J. Geophys. Res.* Submitted

Anderson, J. G., Margitan, J. J., Stedman, D. H. 1977. Atomic chlorine and the chlorine monoxide radical in the stratosphere: three in situ observations. *Science* 198:501–4

Bates, D. R., Hays, P. B. 1967. Atmospheric nitrous oxide. *Planet. Space Sci.* 15:189–97

Bates, D. R., Nicolet, M. 1950. The photochemistry of water vapor. *J. Geophys. Res.* 55:301–27

Bates, J. R. 1977. Dynamics of stationary ultra-long waves in middle latitudes. *Q. J. R. Meteorol. Soc.* 103:397–430

Chaloner, C. P., Drummond, J. R., Houghton, J. T., Jarnot, R. F., Roscoe, H. K. 1975. Stratospheric measurements of H_2O and the diurnal change of NO and NO_2. *Nature* 258:696–97

Chameides, W., Stedman, D., Dickerson, R., Rusch, D., Cicerone, R. 1977. NO_x production in lightning. *J. Atmos. Sci.* 34:143–49

Chapman, S. 1930. A theory of upper atmospheric ozone. *Mem. R. Meteorol. Soc.* 3:103–25

Cieslik, S., Nicolet, M. 1974. The aeronomic dissociation of nitric oxide. *Planet. Space Sci.* 21:925–38

Cox, R. A. 1978. Kinetics of HO_2 radical reactions of atmospheric interest. Paper 15. WMO Symposium on the Geophysical Aspects and Consequences of Changes in the Composition of the Stratosphere. *WMO-No. 511.* World Meteorological Organization, Geneva, Switzerland

Cox, R. A., Roffey, M. J. 1977. Thermal decomposition of peroxyacetyl-nitrate in the presence of nitric oxide. *Environ. Sci. Technol.* 11:900–6

Crutzen, P. J. 1970. The influence of nitrogen oxides on the atmospheric ozone content. *Q. J. R. Meteorol. Soc.* 96:320–25

Crutzen, P. J. 1971. Ozone production rates in an oxygen-hydrogen-nitrogen oxide atmosphere. *J. Geophys. Res.* 76: 7311–27

Crutzen, P. J. 1972. SST's—a threat to the earth's ozone shield. *Ambio* 1:41–51

Crutzen, P. J. 1973. A discussion of the chemistry of some minor constituents in the stratosphere and troposphere. *Pageoph* 106–108:1385–99

Crutzen, P. J. 1974a. Estimates of possible variations in total ozone due to natural causes and human activities. *Ambio* 3: 201–10

Crutzen, P. J. 1974b. A review of upper atmospheric photochemistry. *Can. J. Chem.* 52:1569–81

Crutzen, P. J. 1975. A two-dimensional photochemical model of the atmosphere below 55 km: estimates of natural and man-caused perturbations due to NO_x. Fourth Conference on CIAP, U.S. Department of Transportation, p. 276

Crutzen, P. J. 1976. Upper limits on atmospheric ozone reductions following increased application of fixed nitrogen to the soil. *Geophys. Res. Lett.* 3:169–72

Crutzen, P. J., Ehhalt, D. H. 1977. Effects of nitrogen fertilizers and combustion on the stratospheric ozone layer. *Ambio* 6: 112–17

Crutzen, P. J., Fishman, J. 1977. Average concentrations of OH in the troposphere, and the budgets of CH_4, CO, H_2, and CH_3CCl_3. *Geophys. Res. Lett.* 4:321–24

Crutzen, P. J., Howard, C. J. 1978. The effect of the $HO_2 + NO$ reaction rate constant on one-dimensional model calculations of stratospheric ozone perturbations. *Pageoph* 116:497–510

Crutzen, P. J., Isaksen, I. S. A., Reid, G. C. 1975. Solar proton events: stratospheric sources of nitric oxide. *Science* 189:457–59

Danielsen, E. F., Mohnen, V. A. 1977. Project Dustorm Report: Ozone Transport, In situ measurements, and meteorological analyses of tropopause folding. *J. Geophys. Res.* 82:5867–77

Davis, D. D., Heaps, W., McGee, T. 1976. Direct measurements of natural tropospheric levels of OH via an aircraft borne tunable dye laser. *Geophys. Res. Lett.* 3:331–34

Dawson, G. A. 1977. Atmospheric ammonia flux from undisturbed land. *J. Geophys. Res.* 82:3125–33

Demerjian, K. L., Kerr, J. A., Calvert, J. G. 1974. The mechanisms of photochemical smog formation. *Adv. Environ. Sci. Technol.* 4:1–262

Drummond, J. 1977. *Atmospheric measurement of nitric oxide using chemiluminescence.* PhD thesis. University of Wyoming, Laramie

Drummond, J. W., Rosen, J. M., Hofmann, D. J. 1977. Balloon-borne chemiluminescent measurement of NO to 45 km. *Nature* 265:319–20

Duewer, W. H., Wuebbles, D. J., Ellsaesser, H. W., Chang, J. S. 1977. NO_x catalytic ozone destruction: sensitivity to rate coefficients. *J. Geophys. Res.* 82:935–42

Dunkerton, T. 1978. On the mean meridional mass motions of the stratosphere and mesosphere. *J. Atmos. Sci.* 35:2325–33

Dütsch, H. U. 1974. The ozone distribution in the atmosphere. *Can. J. Chem.* 52:1491–1504

Eddy, J. A. 1976. The Maunder Minimum. *Science* 192:1189–1202

Eddy, J. A. 1977. Climate and the changing sun. *Climatic Change* 1:173–90

Ehhalt, D. H. 1974. The atmospheric cycle of methane. *Tellus* 26:58–70

Fabian, P., Junge, C. E. 1970. Global rate of ozone destruction at the earth's surface. *Arch. Meteorol. Geophys. Bioklimatol.* A19:161–72

Fishman, J., Crutzen, P. J. 1978. The origin

of ozone in the troposphere. *Nature* 274: 855–58

Galbally, I. E., Roy, C. R. 1978. Loss of fixed nitrogen from soils by nitric oxide exhalation. *Nature* 275:734–35

Garland, J. A., Penkett, S. A. 1976. Absorption of peroxy acetyl nitrate and ozone by natural surfaces. *Atmos. Environ.* 10: 1127–31

Geisler, J. E., Dickinson, R. E. 1968. Vertical motions and nitric oxide in the upper mesosphere. *J. Atmos. Terr. Phys.* 30:1505–21

Goldman, A., Fernald, F. G., Williams, W. J., Murcray, D. G. 1978. Vertical distribution of NO_2 in the stratosphere as determined from balloon measurements of solar spectra in the 4500 Å region. *Geophys. Res. Lett.* 5:257–56

Goldman, A., Murcray, D. G., Murcray, F. H., Williams, W. J. 1970. Identification of the ν_3 NO_2 band in the solar spectrum observed from a balloon-borne spectrometer. *Nature* 225:443–44

Goldsmith, P., Tuck, A. F., Fost, J. S., Simmons, E. L., Newson, R. L. 1973. Nitrogen oxides, nuclear weapon testing, Concorde and stratospheric ozone. *Nature* 244:545–49

Graham, R. A., Winer, A. M., Pitts, J. N., Jr. 1977. Temperature dependence of the unimolecular decomposition of pernitric acid and its atmospheric implications. *Chem. Phys. Lett.* 51:215–20

Hahn, J., Junge, C. 1977. Atmospheric nitrous oxide: a critical review. *Z. Naturforsch. A* 32:190–214

Hamilton, E. J., Jr. 1975. Water vapor dependence of the kinetics of the self-reaction of HO_2 in the gas phase. *J. Chem. Phys.* 63:3682–83

Hamilton, E. J., Jr., Naleway, C. A. 1976. Theoretical calculation of strong complex formation by the HO_2 radical: $HO_2 \cdot H_2O$ and $HO_2 \cdot NH_3$. *J. Phys. Chem.* 80:2037–40

Hampson, J. 1964. *Photochemical behavior of the ozone layer.* Quebec: CARDE. 280 pp.

Hampson, R. F., Garvin, D., eds. 1977. Reaction rate and photochemical data for atmospheric chemistry—1977. *NBS Special Publication 513.* US Dept. Commerce. 106 pp.

Harries, J. E. 1978. Ratio of HNO_3 to NO_2 concentrations in daytime stratosphere. *Nature* 274:235

Heath, D. F., Krueger, A. J., Crutzen, P. J. 1977. Solar proton event: influence on stratospheric ozone. *Science* 197:886–89

Hendry, D. G., Kenley, R. A. 1977. Generation of peroxy radicals from peroxy nitrates (RO_2NO_2). Decomposition of peroxyacyl nitrates. *J. Am. Chem. Soc.* 99:3198–99

Hendry, D. G., Kenley, R. A. 1979. Atmospheric chemistry of peroxynitrates. *Nitrogenous Air Pollutants: Chemical and Biological Implications*, ed. D. Grosjean. Ann Arbor Science Publishers. In press

Hill, A. C. 1971. Vegetation: a sink for atmospheric pollutants. *J. Air Poll. Cont. Assoc.* 21:341–46

Howard, C. J., Evenson, D. K. 1977. Kinetics of the reaction of HO_2 with NO. *Geophys. Res. Lett.* 4:437–40

Huebert, B. J., Lazrus, A. 1978. Global tropospheric measurements of nitric acid vapor and particulate nitrate. *Geophys. Res. Lett.* 5:577–80

Huebert, B. J., Lazrus, A. L. 1979. Tropospheric measurements of nitric acid vapor and particulate nitrate. See Hendry & Kenley 1979

Jaffe, R. L., Langhoff, S. R. 1977. Theoretical study of the photodissociation of HOCl. *J. Chem. Phys.* 68:1638–48

Johnston, H. 1971. Reduction of stratospheric ozone by nitrogen oxide catalysis from SST exhaust. *Science* 173:517–22

Johnston, H. S., Serang, O., Podolske, J. 1978. Instantaneous global nitrous oxide photochemical rates. *Rev. Geophys. Space Phys.* In press

Johnston, H. S., Whitten, G., Birks, J. 1973. Effect of nuclear explosions on stratospheric nitric oxide and ozone. *J. Geophys. Res.* 78:6107–28

Junge, C. E. 1962. Global ozone budget and exchange between stratosphere and troposphere. *Tellus* 14:363–77

Junge, C. E. 1963. *Air Chemistry and Radioactivity.* New York: Academic. 382 pp.

Kley, D., McFarland, M., Drummond, J. W., Schmeltekopf, A. L. 1978. Altitude profile and surface measurements of tropospheric NO and NO_2. *J. Geophys. Res.* To be submitted

Lazrus, A. L., Gandrud, B. W. 1974. Distribution of stratospheric nitric acid vapor. *J. Atmos. Sci.* 31:1102–8

Levy, H. II. 1971. Normal atmosphere: large radical and formaldehyde concentrations predicted. *Science* 173:141–43

Levy, H. II. 1974. Photochemistry of the troposphere. *Adv. Photochem.* 9:5325–32

Lindzen, R. S., Tung, K. K. 1978. Wave overreflection and shear instability. *J. Atmos. Sci.* 35:1626–32

Liu, S. C., Cicerone, R. J., Donahue, T. M., Chameides, W. L. 1976a. Limitations of fertilizer induced ozone reduction by the long lifetime of the reservoir of fixed nitrogen. *Geophys. Res. Lett.* 3:157–60

Liu, S. C., Cicerone, R. J., Donahue, T. M., Chameides, W. L. 1977. Sources and sinks of atmospheric N_2O and the possible ozone reduction due to industrial fixed nitrogen fertilizers. *Tellus* 29:251–63

Liu, S. C., Donahue, T. M., Cicerone, R. J., Chameides, W. L. 1976b. Effect of water vapor on the destruction of ozone in the stratosphere perturbed by Cl_x or NO_x pollutants. *J. Geophys. Res.* 18:3111–18

Lodge, J. P., Machado, P. A., Pate, J. B., Sheesley, D. C., Wartburg, A. F. 1974. Atmospheric trace chemistry in the American humid tropics. *Tellus* 26:250–53

Loewenstein, M., Savage, H. F., Whitten, R. C. 1975. Seasonal variations of NO and O_3 at altitudes of 18.3 and 21.3 km. *J. Atmos. Sci.* 32:2185–90

Manabe, S., Wetherald, R. T. 1967. Thermal equilibrium of the atmosphere with a given distribution of relative humidity. *J. Atmos. Sci.* 24:241–59

Mason, C. J., Horvath, J. J. 1976. The direct measurement of nitric oxide concentrations in the upper atmosphere by a rocket-borne chemiluminescent detector. *Geophys. Res. Lett.* 3:391–94

Matsuno, T., Nakamura, K. 1978. The Eulerian and Lagrangian mean meridional circulations in the stratosphere at the time of a sudden warming. *J. Atmos. Sci.* In press

McElroy, M. B. 1976. Chemical processes in the solar system. In *Chemical Kinetics*, ed. D. R. Herschback. *Int. Rev. Sci.* 9:127–204. London: Butterworth

McElroy, M. B., Elkins, J. W., Wofsy, S. C., Yung, Y. L. 1976. Sources and sinks for atmospheric N_2O. *Rev. Geophys. Space Phys.* 14:143–50

McElroy, M. B., McConnell, J. C. 1971. Nitrous oxide: a natural source of stratospheric NO. *J. Atmos. Sci.* 28:1095–98

McElroy, M. B., Wofsy, S. C., Yung, Y. L. 1977. The nitrogen cycle: perturbations due to man and their impact on atmospheric N_2O and O_3. *Philos. Trans. R. Soc. Ser. B.* 277:159–81

McIntyre, M. 1977. Wave transport in stratified, rotating fluids. *IAU Colloquium No. 28. Problems in Stellar Convection. Lecture Notes in Physics*. Berlin: Springer

Mohnen, V. A., Hogan, A., Coffey, P. 1977. Ozone measurements in rural areas. *J. Geophys. Res.* 82:5889–95

Molina, L. T., Molina, M. J. 1978. The ultraviolet spectrum of HOCl. *J. Phys. Chem.* 82:2410–14

Molina, M. J., Rowland, F. S. 1974. Stratospheric sink for chlorofluoromethanes: chlorine atom-catalysed destruction of ozone. *Nature* 249:810–12

Moore, H. 1974. Isotopic measurement of atmospheric nitrogen compounds. *Tellus* 26:169–74

Murcray, D. G., Goldman, A., Csoeke-Poeckh, A., Murcray, F. H., William, W. J., Stocker, R. N. 1973. Nitric acid distribution in the stratosphere. *J. Geophys. Res.* 78:7033–38

Murcray, D. G., Williams, W. J., Barker, D. B., Goldman, A., Bradford, C., Cook, G. 1978. Measurements of constituents of interest in the photochemistry of the ozone layer using infrared techniques. WMO Symposium on the Geophysical Aspects and Consequences of Changes in the Composition of the Stratosphere. *WMO-No. 511*, pp. 61–68

Murgatroyd, R. J., Singleton, F. 1961. Possible meridional circulations in the stratosphere and mesosphere. *Q. J. R. Meteorol. Soc.* 87:125–35

NASA. 1977. Chlorofluoromethanes and the Stratosphere. *NASA Reference Publication 1010*, ed. R. D. Hudson, p. 266

NAS. 1975. Environmental Impact of Stratospheric Flight. National Academy of Sciences

NAS. 1976. Halocarbons: Effects on Stratospheric Ozone. National Academy of Sciences

NAS. 1978. Nitrates: an environmental assessment. National Academy of Sciences

Nicolet, M. 1975. On the production of nitric oxide by cosmic rays in the mesosphere and stratosphere. *Planet. Space Sci.* 23:637–49

Nicolet, M., Vergison, E. 1971. L'oxyde azoteux dans la stratosphère. *Aeronomica Acta* A-91

Noxon, J. F. 1978. Tropospheric NO_2. *J. Geophys. Res.* 83:3051–57

Noxon, J. F. 1979. Stratospheric NO_2. II. Global behavior. *J. Geophys. Res.* Submitted

Park, C., Menees, G. P. 1978. Odd nitrogen production by meteoroids. *J. Geophys. Res.* 83:4029–35

Patel, C. K. N., Burkhardt, E. G., Lambert, C. A. 1974. Spectroscopic measurements of stratospheric nitric oxide and water vapor. *Science* 184:1173–76

Penkett, S. A., Jones, R. M. R., Brice, K. A., Eggleton, A. E. J. 1978. The importance of atmospheric ozone and hydrogen peroxide in oxidising sulphur dioxide in cloud and rainwater. *Atmos. Environ.* In press

Perner, D., Ehhalt, D. H., Paetz, H. W., Roth, E. P., Volz, A. 1976. OH radicals in the lower troposphere. *Geophys. Res. Lett.* 3:466–68

Pierotti, D., Rasmussen, R. 1976. Combustion as a source of nitrous oxide in the atmosphere. *Geophys. Res. Lett.* 3:265–67

Pratt, P. F., Barber, J. C., Corrin, M. L., Goering, J., Hauck, R. D., Johnston, H. S., Klute, A., Knowles, R., Nelson, D. W.,

Pickett, R. C., Stephens, E. R. 1977. *Climatic Change* 1:109–35

Ramanathan, V., Callis, L. B., Boughner, R. E. 1976. Sensitivity of surface temperature and atmospheric temperature to perturbations in the stratospheric concentration of ozone and nitrogen-dioxide. *J. Atmos. Sci.* 33:1092–1112

Ramanathan, V., Coakley, J. A. Jr. 1978. Climate modelling through radiative-convective models. *Rev. Geophys. Space Phys.* In press

Ridley, B. A., Bruin, J. T., Schiff, H. I., McConnell, J. C. 1975. In situ measurements of NO using a balloon-borne chemiluminescent instrument. *J. Geophys. Res.* 80:1925–29

Ridley, B. A., Schiff, H. I., Shaw, A. W., Bates, L., Howlett, L. C., Levaux, H., Megill, L. R., Ashenfelter, T. E. 1973. In situ measurements of nitric oxide in the stratosphere. *Nature* 245:310–11

Ritter, J. A., Stedman, D. H., Kelly, T. J. 1978. Ground level measurements of NO, NO_2, and O_3 in rural air. See Hendry & Kenley 1978

Robinson, E., Robbins, R. C. 1971. Sources, abundance and fate of gaseous atmospheric pollutants—supplement. *American Petroleum Institute Publication No. 4015*

Roble, R. G., Rees, M. H. 1977. Time-dependent studies of the aurora: Effects of particle precipitation on the dynamic morphology of ionospheric and atmospheric properties. *Planet. Space Sci.* 25: 991–1010

Roble, R. G., Stewart, A. I., Torr, M. R., Rusch, D. W., Wand, R. H. 1978. The calculated and observed ionospheric properties during Atmospheric Explorer-C satellite crossings over Millstone Hill. *J. Atmos. Terr. Phys.* 40:21–33

Rodhe, H., Grandell, J. 1973. On the removal time of aerosol particles from the atmosphere by precipitation scavenging. *Tellus* 24:442–54

Rusch, D. W., Stewart, A. I., Hays, P. B., Hoffman, J. H., 1975. The NI (5200 Å) dayglow. *J. Geophys. Res.* 80:2300–5

Seiler, W. 1974. The cycle of atmospheric CO. *Tellus* 26:116–35

Söderlund, R., Svensson, B. H. 1976. The global nitrogen cycle. *Ecol. Bull.* 22:23–77. Stockholm

Spicer, C. W. 1977. The fate of nitrogen oxides in the atmosphere. *Adv. Environ. Sci. Technol.* 7:163–261

Spicer, C. W., Gemma, J. L., Schumacher, P. M., Ward, G. F. 1976. The fate of nitrogen oxides in the atmosphere. 2nd year report to Coordinating Research Council, Batelle, Columbus Laboratories. 111 pp.

Stolarski, R. S., Cicerone, R. J. 1974. Stratospheric chlorine: a possible sink for ozone. *Can. J. Chem.* 52:1610–15

Stolarski, R. S., Hays, P. B., Roble, R. G. 1975. Atmospheric heating by solar EUV radiation. *J. Geophys. Res.* 80:2266–76

Strobel, D. F. 1971. Odd nitrogen in the mesosphere. *J. Geophys. Res.* 76:8384–93

Strobel, D. F., Hunten, D. M., McElroy, M. B. 1970. Production and diffusion of nitric oxide. *J. Geophys. Res.* 75:4307–21

Sze, N. D., Rice, H. 1976. Nitrogen cycle factors contributing to N_2O production from fertilizers. *Geophys. Res. Lett.* 3:343–46

Tingey, D. T. 1968. *Foliar absorption of nitrogen dioxide.* MA thesis. Dept. Botany, Univ. Utah

Torr, D. G., Torr, M. R. 1976. Global characteristics of 0.2 to 26 keV charged particles at F region altitudes. *Geophys. Res. Lett.* 3:305–8

Toth, R. A., Farmer, C. B., Schindler, R. A., Raper, O. F., Schaper, P. W. 1973. Detection of nitric oxide in the lower atmosphere. *Nature* 244:7–8

Trijonis, J. 1978. Empirical relationships between atmospheric nitrogen dioxide and its precursors. *EPA-600/3-78-018.* Environmental Sciences Research Laboratory, US Environ. Protect. Agency

Wang, W. C., Yung, Y. L., Lacis, A. A., Mo, T., Hansen, J. E. 1976. Greenhouse effects due to man-made perturbations of trace gases. *Science* 194:685–90

Waters, J. W., Gustincic, J. J., Kakar, R. K., Roscoe, H. K., Swanson, P. N., Phillips, T. G., de Graauw, T., Kerry, A. R. 1978. Aircraft search for millimeter-wavelength emissions by stratospheric ClO. *J. Geophys. Res.* Submitted

Weiss, R. F., Craig, H. 1976. Production of atmospheric N_2O by combustion. *Geophys. Res. Lett.* 3:751–53

Wofsy, S. C. 1978. Temporal and latitudinal variation of stratospheric trace gases: a critical comparison between theory and experiment. *J. Geophys. Res.* 83:364–78

Wofsy, S. C., McElroy, M. B. 1974. HO_xNO_x and ClO_x: their role in atmospheric photochemistry. *Can. J. Chem.* 52:1544–59

Zahniser, M., Howard, C. J. 1978. The reaction of HO_2 with O_3. *J. Chem. Phys.* To be submitted

Zimmerman, P. R., Chatfield, R. B., Fishman, J., Crutzen, P. J., Hanst, P. L. 1978. Estimates on the production of CO and H_2 from the oxidation of hydrocarbon emissions from vegetation. *Geophys. Res. Lett.* 5:679–82

Ann. Rev. Earth Planet. Sci. 1979. 7: 473–502

PALEOZOIC PALEOGEOGRAPHY

A. M. Ziegler and C. R. Scotese
Department of Geophysical Sciences, University of Chicago, Chicago, Illinois 60637

W. S. McKerrow
Department of Geology, Oxford University, Oxford, England

M. E. Johnson
Department of Geology, Williams College, Williamstown, Massachusetts 02167

R. K. Bambach
Department of Geological Sciences, Virginia Polytechnic Institute and State University, Blacksburg, Virginia 24060

INTRODUCTION

In this paper we present a set of seven new paleographic reconstructions showing lithofacies distributions and land-sea relationships for selected intervals from the Cambrian Period to the Permian Period. Our reconstructions are "non-Pangaean", as it is now well established that the plate tectonics process was operating during the Paleozoic, and probably in some form during the Proterozoic as well (Burke & Dewey 1973). Certainly the hallmark of plate convergence—the paired andesitic volcanic chains and oceanic ophiolite sequences—are well developed in older fold belts (Zonenshayn 1973). The plate tectonic history of the Paleozoic resulted in a world in which continents, initially disseminated in low latitudes, gradually moved together to form a continuous barrier from pole to pole. Thus, geographic relationships existed in the Paleozoic that have not been subsequently attained, and this has interesting and important consequences for the biogeographer, paleoclimatologist, and tectonicist alike.

In the present paper we first discuss the advantages and limitations of all the kinds of evidence—paleomagnetic, biogeographic, tectonic, and

0084-6597/79/0515-0473$01.00

climatic—that have been used to determine the orientations and interrelationships of the continents in the Paleozoic. Secondly, we describe the methods used to "flesh-out" the base maps with the regional topographic and bathymetric details. Finally, we discuss the individual paleogeographic reconstructions with special reference to plate tectonic events and to the degree of confirmation of the paleomagnetic predictions by the distribution of the climatically sensitive sediments.

CONTINENTAL RELATIONSHIPS

A paper on "Phanerozoic World Maps" (Smith, Briden & Drewry 1973), presenting outline maps based on paleomagnetic predictions, proved to be a milestone in understanding continental distributions during the Paleozoic. Recently there have been several attempts to make more refined maps by adding new paleomagnetic data and by incorporating information from the fields of paleoclimatology (Zonenshayn & Gorodnitskiy 1977a,b), biogeography (Ziegler et al 1977b), and tectonics (Kanasewich, Havskov & Evans 1978, Morel & Irving 1978). These reconstructions differ in detail, but the basic outline of continental motion in the Paleozoic is emerging. The prime difficulty is that there is no direct way to measure longitudinal separation of continental pairs. This is because the older ocean floor has either been subducted or so deformed against the continental margins that it cannot be used to determine relative longitude in the way that it has been used in the Cenozoic and Mesozoic Eras (see Laughton 1975).

Paleomagnetic Evidence

Paleomagnetic research is the only source of quantitative data on the past orientations of continents. Extensive catalogues of polar determinations have been published in recent years (McElhinny 1973, Irving, Tanczyk & Hastie 1976), although there still are gaps in our knowledge of the Paleozoic apparent polar wander paths of some of the major paleocontinents, such as China and Gondwana.

A major problem for the paleogeographer is that the paleomagnetic data are often inconsistent. Secondary magnetizations may be acquired long after deposition through the effects of metamorphism or weathering. Moreover, areas marginal to continents, such as the Baja Peninsula, may move significant distances relative to the continental nucleus, along faults that may later become obscure. In selecting paleomagnetic data for our use, we have, where possible, used poles which 1. are self-consistent and lie on a well defined polar wander path, 2. have been thermally, magnetically, or chemically "cleaned", 3. were obtained from rocks that do not bear a metamorphic overprint, and 4. were all obtained from a single platform area.

Because the Earth's magnetic field is axisymmetric, it is not possible to determine the relative longitude of continents using paleomagnetic data, unless it can be assumed that the continents involved were traveling on the same lithospheric plate. A modern example of this is the Indian Ocean plate which carries three separate continents, India, Australia, and Arabia. An apparent example of this situation in the past is the association of Baltica (northern Europe) with the African portion of Gondwana during the early Paleozoic. The rotation of this pair is such that Baltica is east of Africa in the Cambrian but north of Africa by the Silurian, so that the relative distance can be established paleomagnetically when there is latitudinal separation and then applied to the earlier time when there is a longitudinal relationship.

Biogeographic Evidence

The distribution of faunal provinces provides a useful check on the latitudes predicted by paleomagnetism and, in addition, can put some constraints on the longitudinal separation of continents. It is well known that biogeographic barriers are due both to climate and to geographic distance. However, severe problems arise because plants and animals have remarkable distributional abilities. A case in point is the present Indo-Pacific Province which spans about 180° of longitude. Conversely, sharp environmental gradients, such as those along shelf margins, may persist in the same region for long periods and result in faunal changes that might be falsely interpreted as indicating geographic separation. Nonetheless, biogeographic patterns have proven useful in determining the east-to-west order of the continents during the Lower Paleozoic, a time when several continents occupied the same latitudinal belt. For instance, in the Ordovician, northern Europe contains shelf faunas which differ from those in North American and Siberia (Ziegler et al 1977b), yet Zonenshayn & Gorodnitskiy (1977a) show northern Europe between North America and Siberia. In the present paper we place northern Europe adjacent to Africa, and North America and Siberia together, giving a better account of the biogeographic patterns.

Much work remains to be done on Paleozoic biogeography. Several symposia on the subject exist (Middlemiss, Rawson & Newall 1971, Hughes 1973, Hallam 1973, Ross 1974); however, few authors have treated more than one taxonomic group for more than one or two geologic periods. Most of these studies recognize but three or four provinces. A preliminary compilation of described Paleozoic biogeographic provinces using our reconstructions (Ziegler et al 1977b) indicates that most of the ancient provinces actually are on the scale of realms or regions in the sense used by biogeographers studying the Recent (Neill 1969, Briggs 1974). The large biogeographic subdivisions recognized in the Paleozoic often have a latitudinal distribution which can be interpreted as reflecting climatic influence. Future work will doubtless be wider in scope, and should provide a useful test for current reconstructions.

Floral and faunal diversity gradients may also provide information on paleolatitude. Recent studies of Permian faunas (Vine 1973, Waterhouse & Bonham-Carter 1975) indicate that China, with its relatively diverse faunas, was probably lower in latitude than shown in Pangaean reconstructions, and indeed was probably not part of Eurasia until the Mesozoic.

Tectonic Evidence

The timing of continental splits and collisions can be accurately determined using tectonic, geochronologic, sedimentologic, and biogeographic evidence (McKerrow & Ziegler 1972). This information has been very useful in establishing relative positions of the continents for certain intervals of time. For instance, areas in Iran and Tibet that are north of the Cenozoic sutures of the Zagros and Indus share with areas to the south the same Lower and Middle Paleozoic floras (Norin 1946, p. 78), stratigraphy (Stocklin 1974), and paleomagnetic orientations (Becker, Forster & Soffel 1973). Apparently they were part of the Arabian-Indian-Australian portion of Gondwana during this interval. There is evidence for rifting along this margin of Gondwana in the late Paleozoic (Gealey 1977, p. 1187, Veevers & Cotterill 1978, p. 344). As Norin (1946, p. 4) put it, "In Kashmir, the Tethyan development was introduced by enormous outpourings of basaltic lavas, the Panjal Trap, which began in the Lower Permian, culminated in the Middle Permian, and lasted, locally, into the Triassic." By the Jurassic, Tibet shared the same floras with Siberia (Gansser 1964, p. 34), indicating that the ocean between them had closed. These data suggest that Iran and Tibet rifted from Gondwana and moved north to become part of Asia, much in the same way that India and Australia did in subsequent periods.

In addition to tectonic events, tectonic patterns are proving useful in orienting the continents of the Paleozoic world. One might expect that subduction zones of the past would have had continuity from continent to continent, as they do at present in the "ring of fire" around the Pacific Ocean. At some time, all the major Paleozoic continents have had active andesitic volcanic chains along at least one of their margins. We have arranged the reconstructions on the assumption that most of these compressive margins were in continuous belts. It has been possible to do this within the constraints provided by the paleomagnetic and biogeographic data.

Climatic Evidence

Wegener (1929) used the distribution of climatically sensitive sedimentary rock types as support for his theory of continental drift; with the advent of paleomagnetic studies in the past three decades, there have been attempts to expand this study (Briden 1968, Drewry, Ramsay & Smith 1974, Volkheimer 1969). In 1973, Robinson published a very thoughtful paper titled "Paleoclimatology and Continental Drift" in which she provided models of continental climates, as well as Permian and Triassic paleogeographic maps with inferred surface-wind circulation patterns. She emphasized the effect of precipitation on sedimentary rocks, particularly those in which chemical or biologic processes play a major role, and pointed out that precipitation patterns, being influenced by wind directions, create climatic gradients with an east-west component. This she contrasted with temperature, which is more closely tied to latitude and also has some influence on sedimentary rock types.

An important question must be raised at this point regarding the adequacy of the present Earth as a climatic model for periods with vastly different continental arrangements and land-sea proportions. Why do surface winds blow from the east in low and high latitudes, but from the west in mid latitudes? Would this always have been the case? If the Earth were not rotating, upper-level winds would blow poleward and surface winds equatorward as a result of the differential heating between equator and pole. However, these cells are deformed and, in effect, latitudinally compressed by the Earth's rotation. The dry descending air associated with these cells occurs at about 20° to 30° north and south of the equator in the major desert belts. It then flows poleward and equatorward along the surface but is deflected by the Coriolis effect to become the mid-latitude westerlies and low-latitude easterlies, respectively. Air also descends in the polar regions and is deflected to become the polar easterlies. Thus wind directions are ultimately controlled by physical processes related to differential heating and the Earth's rotation. It has been established that, as a result of tidal friction, the Earth's rotation has slowed with time (Rosenberg & Runcorn 1975). One would consequently predict that the dry belts would have been closer to the equator in earlier geologic periods because the stronger Coriolis forces would result in greater latitudinal compression of circulation cells. The magnitude of the change, however, has probably not exceeded twenty percent since the beginning of the Paleozoic Era. We conclude that the present Earth is a satisfactory model for climatic reconstructions of the past.

We have expanded Robinson's (1973) precipitation model by adding idealized continents with marginal mountain ranges on the west and on the east (Figure 1). It is important to note that precipitation belts are *not*

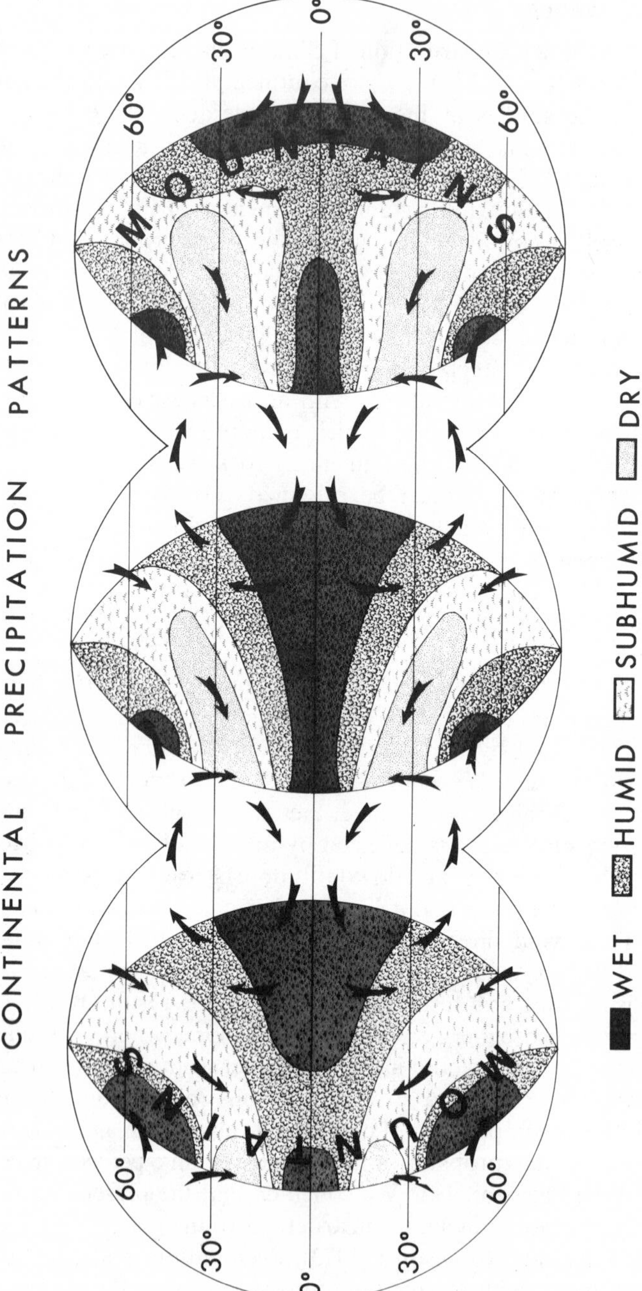

Figure 1 Idealized models of continental precipitation patterns (based on Robinson 1973, Figure 4, with mountains and surface winds added).

parallel with latitude. Only at the equator are similar conditions found on opposite coasts, and this is because of the low pressure associated with the intertropical convergence zone. Poleward of this, conditions tend to be very wet on windward coasts, that is, on eastern coasts in low latitudes and western coasts in mid latitudes, and very dry along the leeward sides of continents. The general effect is for the boundaries of precipitation belts to increase in latitude from west to east. This tendency is accentuated by mountain ranges.

The effects of high precipitation are represented in the rock record by coal swamps, thick clastic sequences, and glacial tillites, while dry climates are indicated by the occurrence of evaporites. There is a negative association of precipitation and the formation of carbonates, and one might expect carbonates to be best developed along west coasts in association with the subtropical highs. However, water temperature seems to be a factor of overwhelming importance in carbonate formation, and the highest temperatures are found along east coasts, as in the Caribbean and Australian carbonate seas. Here, easterly winds transport water heated beneath the tropical sun to western coasts where it is then carried to middle latitudes by western boundary currents such as the Gulf Stream.

As Robinson has pointed out, "Palaeoclimatology involves the study of the motion of all three complex systems of hydrosphere, atmosphere, and lithosphere, and hence it is the most difficult of Earth sciences." As in the case of biogeographic evidence, the paleoclimatic information must be interpreted very cautiously when reconstructions based on paleomagnetic data are evaluated. Yet all lines of information must be resolved before satisfactory reconstructions can be produced. We cannot claim to have arrived at this happy state in our Paleozoic reconstructions, and we present them as working models which will doubtless be altered as further information becomes known to us. As one of the great paleogeographers, L. J. Wills (1951), put it, "A paleogeographic map is not a Delphic Oracle to be blindly reverenced and believed, but an Aunt Sally at which to hurl one's own and other people's field observations. I, at any rate, shall not be surprised if Aunt Sally is knocked over completely or loses an arm or a leg. Such a success, however, on the part of the reader, carries with it the obligation to set up a more truthful dummy, capable of standing until it, in turn, is overthrown."

PALEOGEOGRAPHIC METHODS

In the paleogeographic maps presented in the following section we make a clearcut distinction between data and interpretation. Data are shown on the maps as discrete symbols, each representing the summation of gross lithologic types at a particular section which were deposited during a defined length of time. Interpretation is represented by shading of mountains, lowlands, shelves, and deeper seas. Ideally, the information on each map should represent an instant in time, but, practically speaking, it is possible only to correlate rocks, on a worldwide basis, to the stage level in the Paleozoic. Thus, the information on each map shows average conditions during an interval of about ten million years. This interval is short in comparison with the movement of plates and attendant climatic shifts, but long in relation to sea level fluctuations, particularly during periods like the Carboniferous. So, the distribution of the climatically sensitive sediments provides a relatively accurate test of the paleomagnetic predictions, while the position of the shorelines probably changed constantly during the intervals represented by the maps.

Lithologic Data

The lithologic information plotted on our maps came from numerous sources (see Appendix for references) and has been subdivided into nine major facies groups: 1. clastic sections containing a coarse clastic component, 2. fine clastics, 3. carbonates, 4. evaporites, 5. coal, 6. tillite, 7. ophiolite sequences, 8. the andesite suite, and 9. plateau basalts and dolerite dikes. The first three are the dominant lithologies and their symbols have been appropriately combined where mixed clastic and carbonate sequences occur. For the sake of simplicity in drafting the maps, the clastic and carbonate components of the remaining six lithotypes have been omitted. The data portrayal has been designed to emphasize the most paleogeographically significant lithologic components rather than to give a statistical account of their proportions. Thus we show any clastic sequence with a coarse fraction by the coarse clastic symbol. This indicates that traction or gravity currents operated during some portion of the interval represented by the map, and in turn requires an explanation in terms of a suitable source region for these sediments. We pay particular attention to purely carbonate sections as these are the most likely to be of Bahamian type, and therefore confirmatory of paleomagnetic predictions of low latitude, east coast situations.

The most climatically significant of sediments are the evaporites, coals, and tillites. The Paleozoic evaporites we have noted are probably all quasi-marine and consist of gypsum, anhydrite, and occasionally halite. Such minerals are forming in the dry valleys of Antarctica today (Torii 1975) and might conceivably have formed in marginal marine areas of the polar dry zones in the past. However, the majority of Paleozoic evaporites conform to the west coast subtropical zones, that is, to the coastlines that today are both dry and hot. Coal and tillite indicate high precipitation and are often found interbedded in the late Paleozoic, as they are in the Pleistocene. This climatic association is enhanced by the fact that glaciers upset drainage patterns, thereby creating swamps. Not all Paleozoic coals were associated with the temperate zone, however, and the equatorial rainy belt is especially well represented by coal swamps in the Carboniferous. This type of occurrence seems to be unique in time (Briden 1968, p. 184) and these swamps must have resulted from tectonic events associated with the collision of North America and Africa.

The volcanic rocks form in one of three tectonic environments, mid-ocean ridges, compressive plate boundaries, and continental rift zones. It is relatively easy to deduce these situations from the stratigraphic context. The mid-ocean spreading centers yield the ophiolite sequences, which often, however, are represented only by their upper layers, the pillow basalts. We use the word ophiolite, therefore, in a very loose sense, for spilites and associated pelagic cherts and shales. We point out that all such occurrences are surely allochthonous and were formed hundreds of kilometers from the symbols that represent them on the maps. The compressive plate boundaries are represented by the andesite suite of calc-alkaline volcanics and intrusives. This association is characterized by a variety of rock types ranging in composition from rhyolite to basalt and these are usually found interbedded with shallow water sediments. Finally, rifting usually leaves a record of plateau basalts and dolerite dikes, and these rocks are clearly associated with older continental crust.

Environmental Interpretation

Our subdivision of the world environment into mountains, lowlands, shelves, and oceans is the minimum needed for climatic and biogeographic interpretations. There is a natural bias toward data representing shelf environments because land deposits are vulnerable to erosion and the older deep sea deposits have been largely obliterated by tectonic processes. We show mountains in the vicinity of andesitic volcanism and in areas where continental collision was in progress. The shorelines and shelf areas have been interpreted on the basis of benthic community and lithofacies patterns. The shelves, as mapped, may include deeply subsided areas of continental crust, while the deep sea areas are interpreted as being crust of oceanic type. We show as marine some regions that have been traditionally regarded as land on the basis of unconformities. Many marine areas today are not receiving sediments, so the presence of an unconformity cannot be taken as evidence that a land area existed during the time span represented by the unconformity. At best, paleogeographic interpretation is subjective, and our hope is that, by plotting lithologic data as distinct from environmental interpretation, others will be induced to take the data, with its inherent patchiness, and to suggest alternative interpretations where desirable.

PALEOZOIC RECONSTRUCTIONS

The seven maps presented in the following pages are preliminary versions of a more extensive set being prepared at the University of Chicago for *An Atlas of Paleogeographic Maps*. The maps will be refined as more data are collected and critiques are received. Approximately thirty time intervals will be covered, ranging from the late Precambrian to the present, and paleogeographic, biogeographic, climatologic, and tectonic maps will be included.

The Mollweide Projection is used in the present paper. Like the Mercator Projection, lines of latitude are shown parallel, but the Mollweide has the additional advantages of being an equal area projection and of showing the whole Earth including the polar regions. We are preparing a companion paper for the *Journal of Geology* which will have alternative projections, and which will list the paleomagnetic determinations used and give more geographic base information.

The outlines of the Paleozoic continents used in the following reconstructions are from our world suture map (Ziegler et al 1977a, Figure 1). Laurentia consisted of nuclear North America, plus Greenland, Scotland, Spitzbergen, and parts of eastern Siberia, but not Florida or Avalonia (Nova Scotia and eastern Newfoundland). Apparently Alaska is a composite of several continental fragments that have traveled significant distances since the Paleozoic. Recent paleomagnetic evidence indicates that the North Slope of Alaska and presumably contiguous parts of Siberia were rotated counterclockwise from Arctic Canada in the Mesozoic (Newman, Mull & Watkins 1977). Southern Alaska, southwest of the Denali fault, apparently has come from the southeast as indicated by paleomagnetic studies (Packer & Stone 1974) and by offsets along the numerous dextral faults in Alaska and the Canadian Cordillera. The Kolyma Platform and certain other small areas in the Arctic region have been left off the maps because, as yet, we have no information on where they were in the Paleozoic.

Baltica consisted of the Russian Platform which is bounded on the south by the Hercynian Suture and on the east by the Uralian Suture. It is often assumed that England and Avalonia also were part of Baltica. However, we treat them separately on both paleomagnetic and faunal evidence. All these areas collided with Laurentia in the mid-Paleozoic, and all may have been part of Gondwana until this time. The paleomagnetic information is suggestive but not definitive on this point. The name

Laurussia is applied to the larger Laurentia-Baltica continent in the late Paleozoic.

Gondwana originally included South America, Africa, Arabia, India, Australia, and Antarctica as well as Florida, southern Europe, Turkey, Iran, Afghanistan, Tibet, and Malaya. As has been mentioned, rifting along the northern margin of Gondwana must have begun by the late Paleozoic along lines that subsequently became the Zagros and Indus Sutures.

Asia east of the Urals is usually portrayed in reconstructions as a single block but in fact was traversed by several Paleozoic oceans (Ziegler et al 1977a). Kazakhstania was separated from Siberia along the Irtysch Suture. China remains a paleogeographic enigma because there is so little paleomagnetic data for this area. It does seem certain that China was not adjacent to Siberia in the Paleozoic, as a prominent suture, the Mongolian geosyncline of Kobayashi (1971), can be traced across southern Mongolia, through Manchuria, and into the Sikhote Alin area of the Soviet Union. This geosyncline contains Triassic flysch, and Paleozoic areas to the north yield Siberian floras and "boreal" faunas, while areas to the south have Cathaysian floras and "austral" faunas. Tibet and Malaya are also separated from China by a Triassic "Indosinian geosyncline" (Jen & Chu 1970), and indeed, Tibet may consist of several island arcs or continental fragments successively accreted to Asia (Chang & Cheng 1973). In our reconstructions we have kept North and South China together despite the fact that they are separated by a major Mesozoic fold belt, the Tsin-Ling axis. Lower Paleozoic faunas (Lu et al 1974, Palmer 1972, Whittington 1973) and Upper Paleozoic floras (Chaloner & Meyen 1973) are shared by North and South China indicating that the two halves, if separated, were not far apart. Finally, China is traversed by major east-west transcurrent faults which may have displaced South China as much as 1000 km to the east with respect to some parts of North China (Molnar & Tapponier 1975). Much palinspastic work thus remains to be done in this large area before the paleogeography can be fully understood.

Cambrian Paleogeography

The Cambrian evidently began during a phase of continental rifting. The margins of three of the Paleozoic continents, southeastern Laurentia (North America), northwestern Baltica (northern Europe), and northern Gondwana (Australia to Pakistan) show evidence of rifting; this consists of aulacogen formation in Oklahoma (Hoffman, Dewey & Burke 1974, p. 41), southern Quebec (Bailey 1977), southern Norway (Stormer 1967, p. 210), and of extensive plateau basalts in northern Australia and Pakistan (Veevers 1976, p. 186). The seas gradually advanced from the

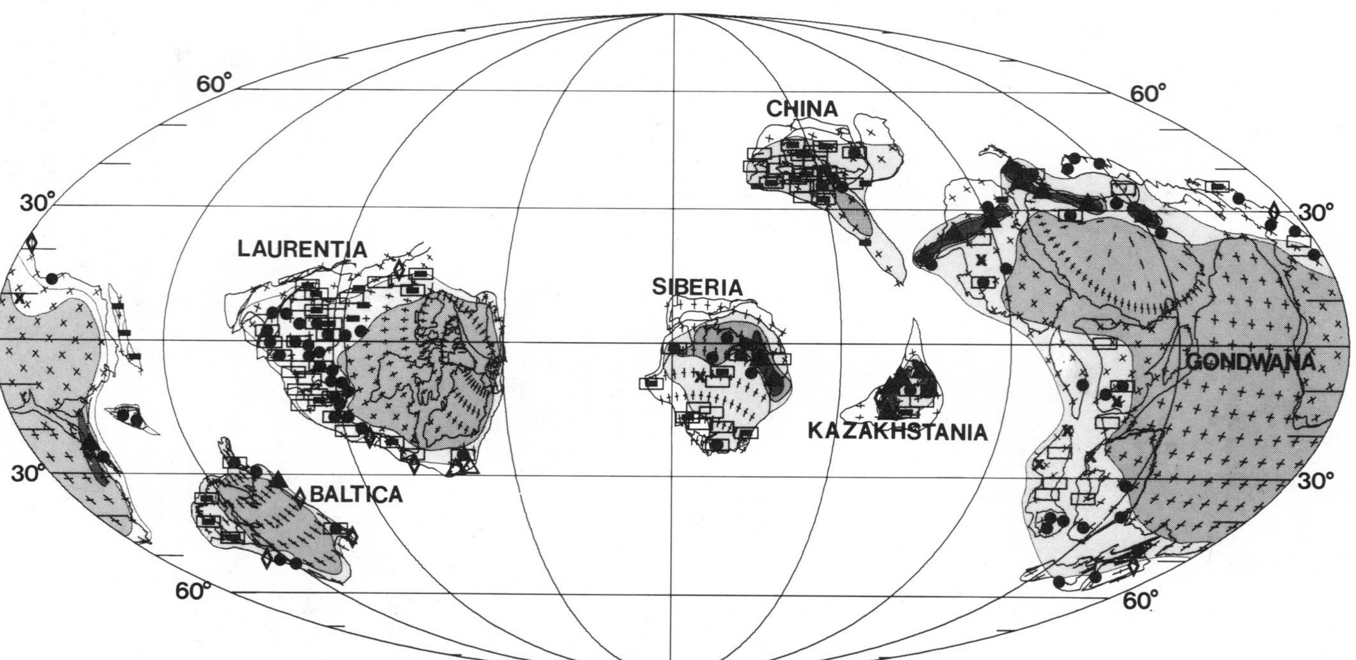

Figure 2 Middle late Cambrian (Franconian) paleogeographic reconstruction. Shading and symbols for Figures 2–8 as follows: dark grey = mountains, medium grey = lowlands, light grey = shallow shelves, white = deeper shelves and oceanic areas, solid triangles = andesitic volcanics, open triangles = plateau basalts, diamonds = ophiolites, dots = coarse clastics, dashes = fine clastics, boxes = carbonates, X's = evaporites, C's = coals, T's = tillites.

margins onto these Cambrian continents and by the middle late Cambrian, the interval shown on Figure 2, achieved a maximum coverage for the period.

A striking feature of the late Cambrian is the relatively low-latitude position of the continents, leaving polar oceans of wider extent than at any subsequent time. We assume that the only high mountains were those associated with the andesitic chains along the margins of Siberia, Kazakhstania, and the Malayan-Australian-Antarctic portion of Gondwana. All three belts had a compressive history that began in the Precambrian and extended through the whole Paleozoic.

Good agreement is found between the continental orientations predicted by paleomagnetic data and the distribution of climatically sensitive rock types. The equatorial rainy belt is well represented by coarse clastics, especially in Laurentia where the Cambrian rocks are widely exposed. Evaporites are found in the northern subtropical belt, which crossed central South America and Australia, and in the southern subtropical belt which crossed northern Siberia as well as India, and adjacent parts of Iran and Arabia. Pure carbonates of late Cambrian age are developed in outer shelf regions of those continents and do not extend poleward of 40° latitude, as is the case today.

Ordovician Paleogeography

By the Middle Ordovician, Gondwana was moving toward the south pole (Figure 3) resulting in the glaciation of both North and South Africa at the very end of the period (Beuf et al 1971). Subduction began along the margins of the Iapetus Ocean in New England and Scotland on the North American side, and in Wales and Norway on the European side. In fact this latter zone may have been continuous with the Tasman zone of Australia by way of western South America and Antarctica. Though there is radiometric evidence of Ordovician plutonic activity in Argentina (Acenolaza 1976), the presence of an island arc or Andean margin has yet to be substantiated by the discovery of volcanic equivalents.

The distribution of climatically sensitive sediments in the Middle Ordovician can be satisfactorily explained using this reconstruction. From Upper Cambrian through the Ordovician, Laurentia underwent a counterclockwise rotation but remained in low latitudes. Coarse clastics continued to be eroded from the interior. Evaporites associated with this continent are symmetrically disposed at about 20° north and south latitude.

In Siberia, evaporites plot to within a few degrees of the equator, suggesting that this position may be inaccurate and that the continent may have been slightly north of the position shown. Carbonates, in general, occur within the latitudinal limits of their present distribution, except in

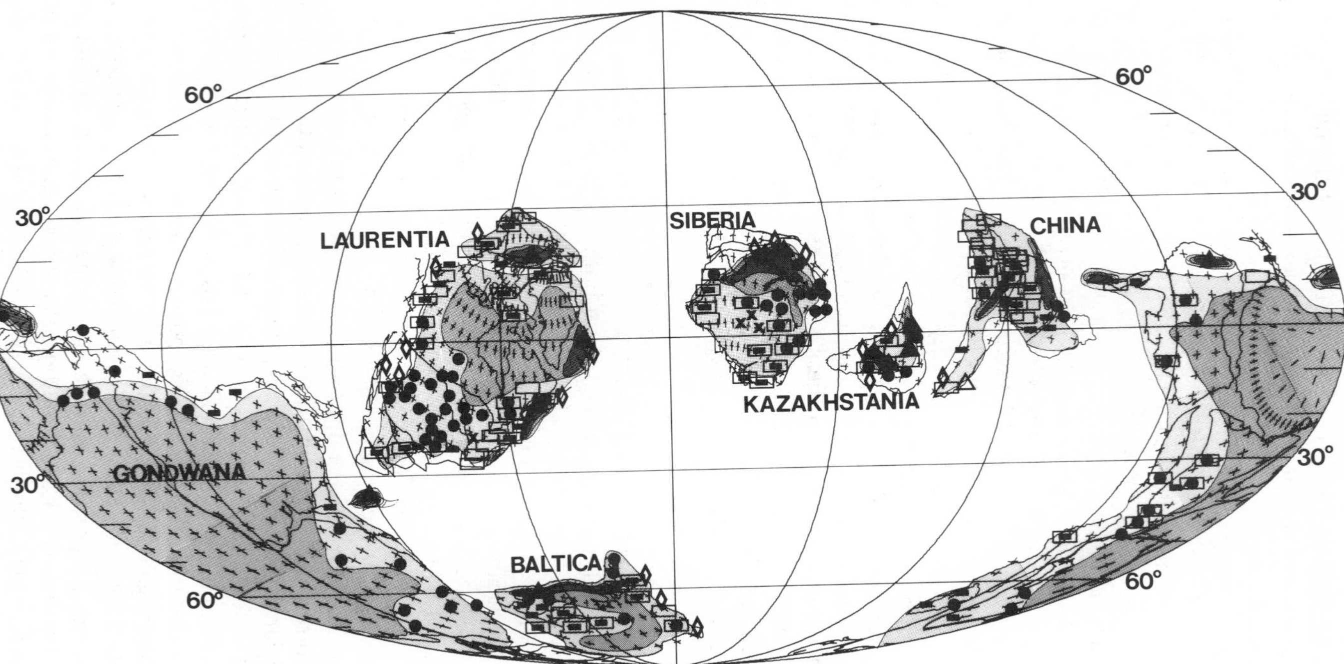

Figure 3 Middle Ordovician (Llandeilo-Caradoc) paleogeographic reconstruction (see Figure 2 for shading and symbols).

Baltica where they extend up to 70° south. Again, the pole chosen may be in error, though it must be pointed out that in such a configuration the warm western boundary currents would have likely been deflected southward toward Baltica.

Silurian Paleogeography

The Silurian began with a sea-level rise that has been related to the melting of continental ice sheets in the African and South American portions of Gondwana (Sheehan 1973). A paleogeographic map for the Lower Silurian has already been published (Ziegler et al 1977a). The interval chosen for this paper is the Middle Silurian (Figure 4), a time when shallow seas covered perhaps more of the continents than at any other time. One would expect that at times such as these, temperatures would have been mild, worldwide, and climatic zones more nearly parallel with latitude. Silurian evaporites of the Michigan Basin of North America, the Pechora Basin of northern Europe, and the Canning and Caernarvon Basins of western Australia were confined 10° to 30° north and south of the equator. Carbonates were deposited in low latitudes and were especially developed in North America. The Cambrian and Ordovician clastic source area associated with the mid-continental high and Canadian Shield had long since disappeared.

During the Middle Silurian compressive plate boundaries were well developed around southern Gondwana extending from Australia across eastern Antarctica and up along the western margin of South America. They may have been connected with subduction zones in Malaya and South China to the northeast and Baltica and Siberia to the west. This would place Laurentia, which possessed its own subduction zone along the eastern seaboard, outside this ancient "ring of fire".

Devonian Paleogeography

By the Devonian (Figure 5), the Iapetus Ocean which separated Laurentia and Baltica closed, resulting in the Caledonian and Acadian Orogenies of northern Europe and eastern North America. This collision, which resulted in the formation of the "Old Red Continent", was the first of several Paleozoic continental collisions. It is usually assumed that the relative position of Baltica and Laurentia at the time of the opening of the present Atlantic was also their Devonian configuration. However, recent paleomagnetic studies (Kent & Updyke 1978, Van der Voo, French & French 1979) indicate that Baltica collided in a more southerly position and then sinistral transcurrent faulting, in association with the collision of Gondwana in the Carboniferous, reactivated this boundary, Baltica being shifted to the north.

Figure 4 Middle Silurian (Wenlock) paleogeographic reconstruction (see Figure 2 for shading and symbols).

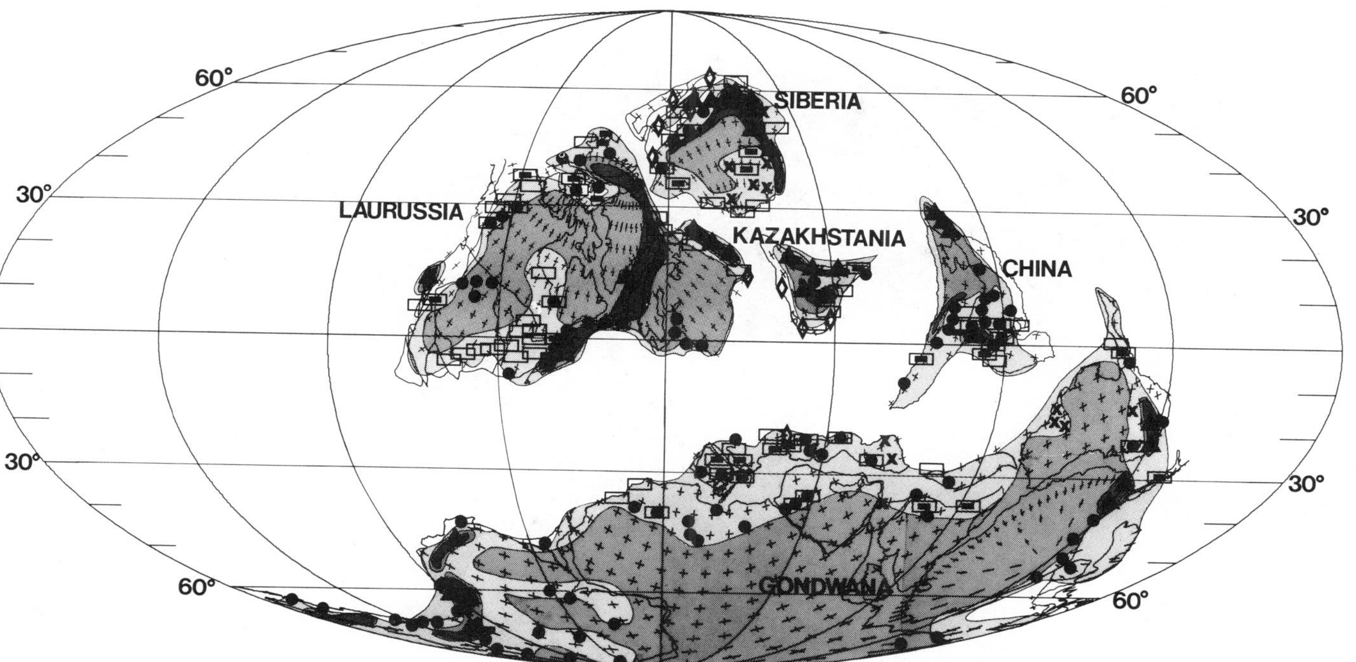

Figure 5 Late early Devonian (Emsian) paleogeographic reconstruction (see Figure 2 for shading and symbols).

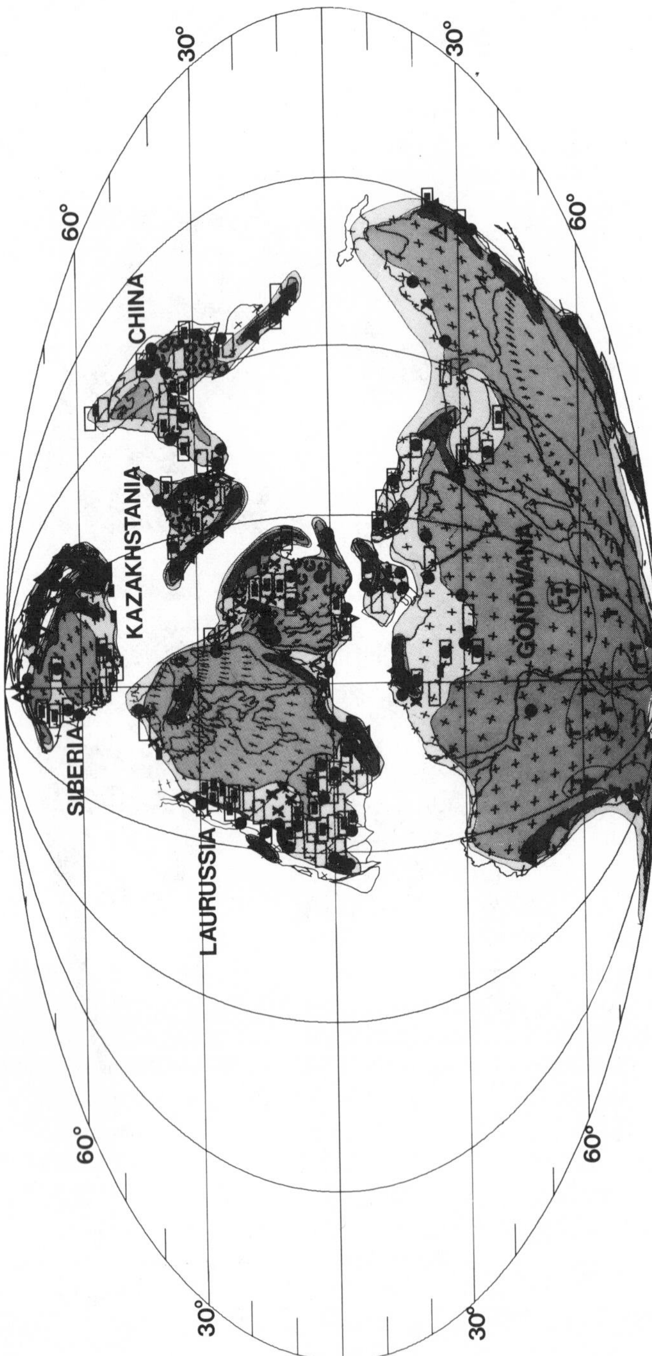

Figure 6 Late early Carboniferous (Visean) paleogeographic reconstruction (see Figure 2 for shading and symbols).

The tectonic events which occurred along other margins of Laurentia and Baltica at about this time may represent readjustments related to their collision. Along the western and northern sides of Laurentia, the Antler, Cariboo, and Ellesmere Orogenies, which were initially non-magmatic in character, began in the Devonian (Burchfiel & Davis 1975, p. 369, Douglas et al 1970, p. 415, Thorsteinsson & Tozer 1970, p. 566). Also, the Uralian margin of Baltica was transformed from a passive margin to a compressive margin in the late Silurian and Devonian (Ivanov et al 1975, p. 107). Similarly, andesitic volcanism began along the southern margin of Baltica in the Greater Caucasus about the Middle Devonian (Khain 1975, p. 141), and in the late Devonian the Donets Trough formed (Pistrak & Pashova 1972, p. 212) possibly as a result of back-arc spreading.

The climatic indicators conform to expectations except for evaporites in Siberia which occur too far north. Possibly the Devonian pole bears an overprint from a later period when Siberia was farther north.

Early Carboniferous Paleogeography

By this time, coal swamps and glaciers were well developed and, together with the evaporites and carbonates, provide abundant criteria for testing the accuracy of the reconstruction (Figure 6). The coals are all relatively low-latitude in occurrence and are limited to the windward sides of Laurussia, Kazakhstania, and China, in the easterly trade wind belts. The tillites are associated with the southern African and South American portions of Gondwana from 50° to 80° south and are concentrated on the western side of the supercontinent, where westerlies would have provided the necessary moisture for glaciation. Evaporites are widely distributed between 5° and 30° north and south and tend to occur most frequently on the western sides of large land masses or high mountain ranges. Carbonates are confined to low latitudes except in some of the northern continents, particularly Siberia, and we are not sure whether this represents an error in the reconstruction, or whether conditions were abnormally warm in this ocean-dominated hemisphere.

The consuming plate margins of the Devonian were maintained in the Carboniferous and new ones were added to the facing margins of Laurussia and Gondwana, heralding the collision of these continents in the late Carboniferous. Areas of active volcanism during the early Carboniferous included Morocco, central Europe, the Greater Caucasus, and the lands extending from Turkey to Tibet.

Late Carboniferous Paleogeography

During the Carboniferous, Gondwana rotated clockwise, colliding with Laurussia in the Upper Carboniferous (Figure 7). The rotation of these

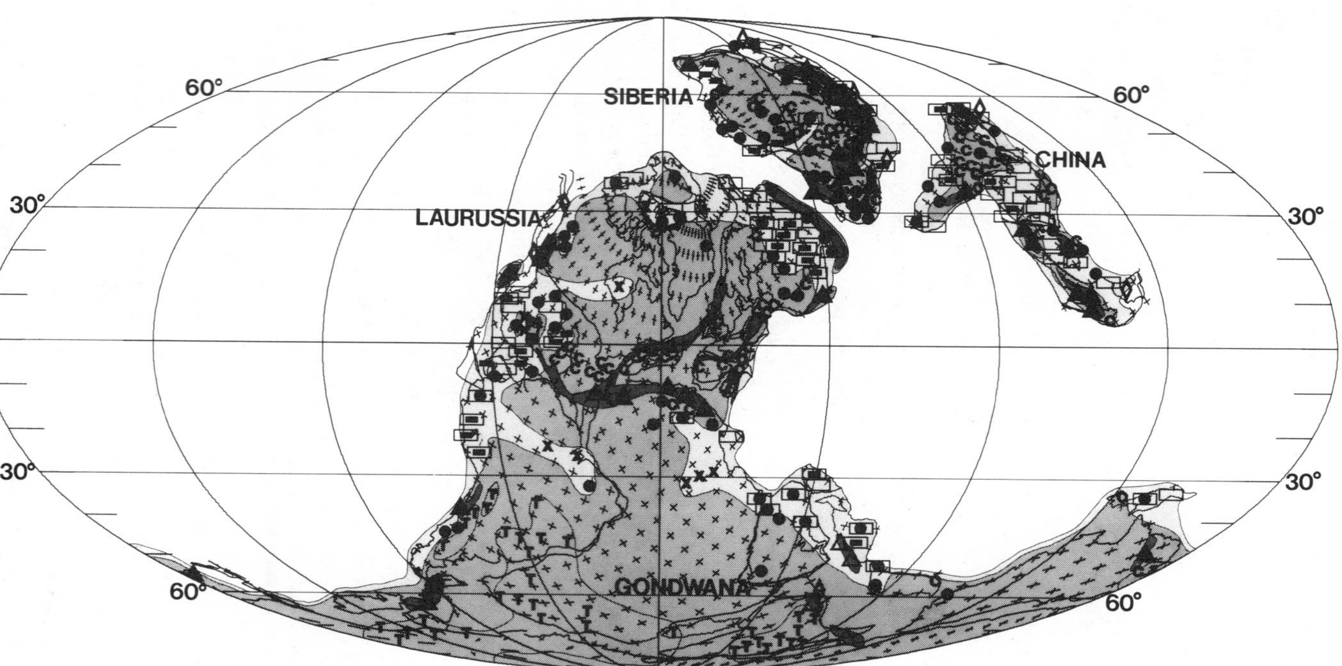

Figure 7 Middle late Carboniferous (Westphalian CD) paleogeographic reconstruction (see Figure 2 for shading and symbols).

two continents widened the Tethyan Seaway, and the collision resulted in the formation of the Ouachita, Appalachian, Mauritanide, and Hercynian foldbelts. Readjustments along the former Baltica-Laurentia collision zone took the form of transcurrent faulting seen in Great Britain and Maritime Canada, and the opening of graben in the North Sea and Gulf of St. Lawrence, as Europe was driven northward with respect to North America. Also the block faulting which produced the Ancestral Rockies can be traced, in the subsurface, to the Ouachita foldbelt and was probably related to this great event (Coney 1973). Evidently Siberia and Kazakhstania collided about the same time, as marine conditions between the two ceased. The supercontinent of Pangaea had begun to form.

The late Carboniferous is rich in climatically sensitive sediments, and their distribution conforms closely to predictions based on the climatic models (Figure 1). The tropical rainy zone is represented by the belt of coal swamps extending from the midwestern United States through Europe, and is widest on the windward coast, that is, from the Donets Basin of southern Russia to Morocco. The study of cellularly well-preserved wood of many taxa reveals no or only faint growth rings demonstrating that these coal floras were not subjected to seasonal interruptions in growth (Chaloner & Creber 1973). The subtropical dry belts are represented by the evaporites and rise in latitude from west to east, as do modern desert belts. The temperate precipitation belts are indicated by the Siberian and Gondwanian coals, in which growth rings preserved in plant fossils indicate strong seasonal interruption of growth (Chaloner & Meyen 1973), and the cold south polar region is shown by the tillites. Warm currents would have been deflected north and south along the western margin of Tethys and this would account for the otherwise surprisingly high-latitude carbonates of India and China, and the thermophilic vegetation of China.

Permian Paleogeography

The continental orientations of the Permian (Figure 8) do not differ significantly from the Upper Carboniferous, but the climates do, particularly in low-latitude areas. Evidently, the rise of the mountains along the contact of Gondwana and Laurussia completely blocked the moist equatorial easterlies, resulting in very dry conditions in North America and Europe. In contrast with the coal swamps of the Carboniferous, Permian coal swamps were limited to the temperate belts. The last remnants of the late Paleozoic ice sheets were restricted to the mountainous regions of eastern Australia during late Permian.

A prominent tectonic event in the Permian was the rifting of Tibet-Iran-Turkey from the northern margin of Gondwana as discussed earlier in this paper. These fragments, together with China and Malaya, did not

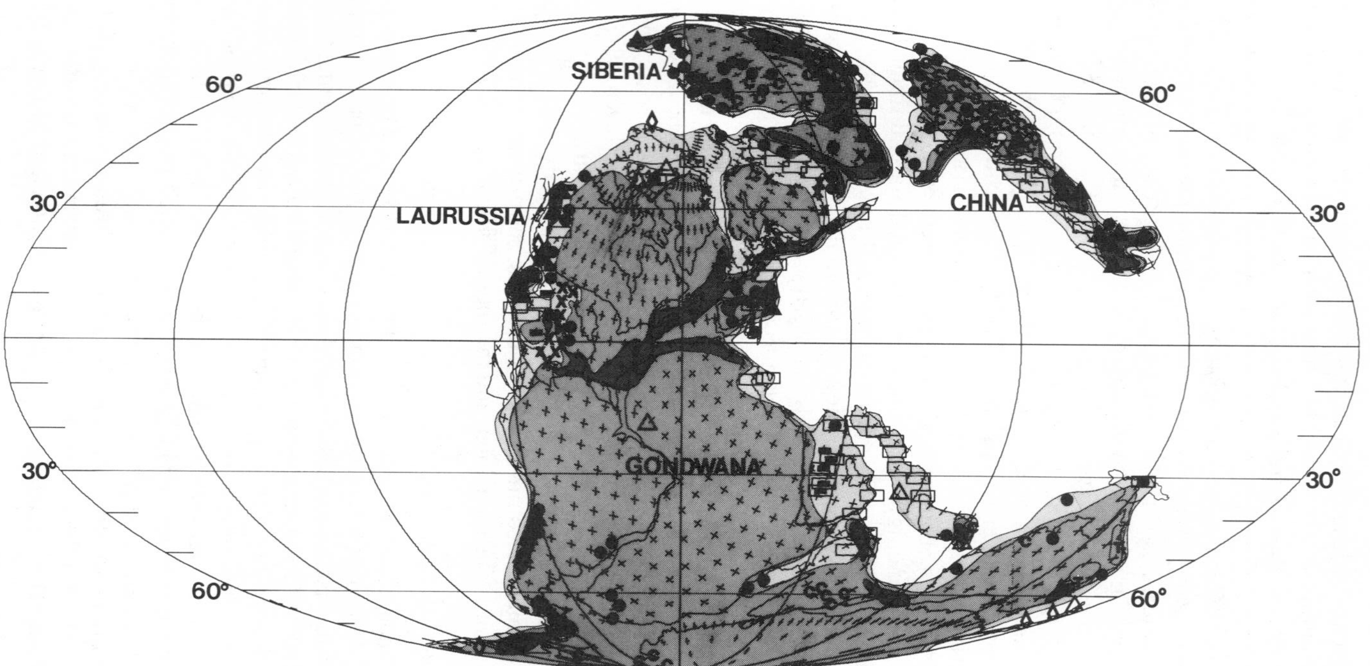

Figure 8 Early late Permian (Kazanian) paleogeographic reconstruction (see Figure 2 for shading and symbols).

become part of Asia until the Indosinian Orogeny of the late Triassic. Confusion has resulted in the literature because of the dual affinities of many southern Asian areas to both Gondwana and northern Asia (Meyerhoff & Meyerhoff 1978). We believe the explanation lies in a long continued pattern of rifting along northern Gondwana and accretion of the derived fragments to Asia. Pangaea was not completely assembled until the late Triassic and then it began to fragment in the early Jurassic with the opening of the central Atlantic Ocean.

CONCLUSIONS

The distribution of climatically sensitive sediments shown on our reconstructions for the Paleozoic is in good agreement with expectations based on the model of the Earth's present atmospheric and oceanic circulation patterns. This indicates that the available paleomagnetic data is reasonably accurate. We have not tried to adjust any of the orientations within the confidence intervals calculated for the poles, but feel that the few cases in which mismatches occur could be resolved by doing this. We do not mean to imply that the paleomagnetic data is completely adequate yet, nor that the climatic information is diagnostic in all cases. There is, however, mutual confirmation for paleomagnetically well-studied continents which have a good stratigraphic record and are large enough to cover several climatic zones at any given time.

Although the Earth's present circulation patterns and their general climatic consequences can be detected in the past, we certainly do not mean to imply that climate has been constant through time. The proportion of land, and its latitudinal array, must have been very important in controlling world temperature and precipitation. The heat derived from solar radiation is absorbed and redistributed in the oceans, and by contrast, lost over land areas during the nights and the winters. From this, one would expect that the world climate of periods like the Recent, the Permo-Carboniferous, and the late Precambrian, with much land in high latitudes, would be generally cool and this is confirmed by glaciations of these times. At the other extreme were times like the early Paleozoic and the late Mesozoic with large expanses of shelf seas associated with relatively low latitude continents. The occurrence during such times of carbonates in higher latitudes than present may be evidence of more uniform temperature conditions, rather than errors in the reconstructions.

There is a pervasive tectonic pattern in the collisional history of the northern continents through the Phanerozoic. Suture zones in general are progressively younger from northwest to southeast: 1. the Siluro-Devonian Caledonian/Acadian Orogenies involved North America and northern Europe, 2. the Carboniferous Appalachian/Hercynian/Uralian Orogenies added Africa, central Europe, Kazakhstan and Siberia to this piece, 3. the Triassic Indosinian Orogeny added China, 4. the Tertiary Alpine/Himalayan Orogenies added southern Europe, Arabia, and India, and 5. the future "Javan" Orogeny will add Australia. Most, and possibly all, of these continents were originally part of Gondwana. Much remains to be done to sort out the paleogeographic details, particularly in the early Paleozoic, but this will come with refined paleomagnetic, biogeographic, tectonic, and climatic studies.

ACKNOWLEDGMENTS

This work is supported by the Shell Development Company, the National Science Foundation (Grant EAR 76-23180 to A.M.Z.), and the Amoco International Oil Company. The computer graphics were done by Colleen Barton, and technical assistance was provided by Beth Gierlowski and Roger Humphreville. Kirk Hansen helped with the climatic interpretations.

Literature Cited

Acenolaza, F. G. 1976. The Ordovician System in Argentina and Bolivia. In *The Ordovician System: Proceedings of a Palaeontological Association Symposium, Birmingham, 1974*, ed. M. G. Bassett, pp. 479–87. Univ. of Wales Press and National Museum of Wales. 696 pp.

Bailey, D. K. 1977. Lithosphere control of continental rift magnetism. *J. Geol. Soc. London* 133: 103–6

Becker, H., Forster, H., Soffel, H. 1973. Central Iran, a former part of Goldwanaland?—palaeomagnetic evidence from Infra Cambrian rocks and iron ores of the Bafq area, central Iran. *Z. Geophys.* 39: 953–63

Beuf, S., Biju-Duval, B., deCharpel, O., Rognon, P., Gariel, O., Bennacef, A. 1971. Les grès du Paléozoique Inférieur au Sahara. *Publications de l'Institut Francais du Pétrole Collection "Science et Technique du Pétrole" no. 18.* Paris: Technip 465 pp.

Briden, J. C. 1968. Paleoclimatic evidence of a geocentric axial dipole field. In *The History of the Earth's Crust, a Symposium*, ed. R. A. Phinney, pp. 178–94. New Jersey: Princeton Univ. Press. 244 pp.

Briggs, J. C. 1974. *Marine Biogeography*. New York: McGraw-Hill. 475 pp.

Burchfiel, B. C., Davis, G. A. 1975. Nature and controls of Cordilleran orogenesis, western United States: extensions of an earlier synthesis. *Am. J. Sci.* 275–A: 363–96

Burke, K., Dewey, J. F. 1973. An outline of Precambrian plate development. In *Implications of Continental Drift to the Earth Sciences*, ed. D. H. Tarling, S. K. Runcorn, 2: 1035–45. New York: Academic. 1184 pp.

Chaloner, W. G., Creber, G. T. 1973. Growth rings in fossil woods as evidence of past climates. In *Implications of Continental Drift to the Earth Sciences*, ed. D. H. Tarling, S. K. Runcorn, 1: 425–37. New York: Academic. 622 pp.

Chaloner, W. G., Meyen, S. V. 1973. Carboniferous and Permian floras of the northern continents. In *Atlas of Palaeobiogeography*, ed. A. Hallam, pp. 169–86. Amsterdam, London, New York: Elsevier. 531 pp.

Chang, C. F., Cheng, H. L. 1973. Some tectonic features of the Mt. Jolmo Lungma area, southern Tibet, China. *Scientia Sinica* 14: 247–65

Coney, P. J. 1973. Non-collision tectogenesis in western North America. See Burke & Dewey 1973, pp. 713–27

Douglas, R. J. W., Gabrielse, H., Wheeler, J. O., Stott, D. F., Belyea, H. R. 1970. Geology of western Canada—Chapter VIII. In *Geology and Economic Minerals of Canada*, ed. R. J. W. Douglas, pp. 365–488. *Econ. Geol. Rep. No. 1.* Geol. Surv. Canada. 838 pp.

Drewry, G. E., Ramsay, A. T. S., Smith, A. G. 1974. Climatically controlled sediments, the geomagnetic field, and trade wind belts in Phanerozoic time. *J. Geol.* 82: 82:531–53

Gansser, A. 1964. *Geology of the Himalayas.* London, New York, Sydney: Interscience. 289 pp.

Gealey, W. K. 1977. Ophiolite obduction and geologic evolution of the Oman Mountains and adjacent areas. *Geol. Soc. Am. Bull* 88:1183–91

Hallam, A., ed. 1973. *Atlas of Paleobiogeography.* Amsterdam, New York, London: Elsevier. 532 pp.

Hoffman, P., Dewey, J. F., Burke, K. 1974. Aulacogens and their genetic relation to geosynclines, with a Proterozoic example from Great Slave Lake, Canada. In *Modern and Ancient Geosynclinal Sedimentation,* ed. R. H. Dott, Jr., R. H. Shaver, pp. 38–55. *Soc. Econ. Paleotol. Mineral., Spec. Publ. 19.* 380 pp.

Hughes, N. F., ed. 1973. Organisms and Continents through Time. *Palaeontol. Assoc. Spec. Pap. No. 12.* 334 pp.

Irving, E., Tanczyk, E., Hastie, J. 1976. Catalogue of Paleomagnetic directions and poles—Paleozoic results 1949–1975. *Geomagnetic Ser. No. 5.* Energy, Mines and Resources, Canada. 99. pp.

Ivanov, S. N., Perfiliev. A. A., Efimov, A. A., Smirnov, G. A., Necheukhin, V. M., Fershtater, G. B. 1975. Fundamental features in the structure and evolution of the Urals. *Am. J. Sci.* 275-A: 107–30

Jen, C-S., Chu, C-C. 1970. Geosynclinal Indosinian Structure in the Lanp'ing-Weisi Region, Western Yunnan. *Inst. Geol. Rev.* 12:447–63

Kanasewich, E. R., Havskov, J., Evans, M. E. 1978. Plate tectonics in the Phanerozoic. *Can. J. Earth Sci.* 15:919–55.

Kent, D. V., Updyke, N. D. 1978. Paleomagnetism of the Devonian Catskill Red Beds: evidence for motion of the coastal New England-Canadian maritime region relative to cratonic North America. *J. Geophys. Res.* 83:4441–50

Khain, V. E. 1975. Structure and main stages in the techtonomagmatic development of the Caucasus: an attempt at geodynamic interpretation. *Am. J. Sci.* 275-A: 131–56

Kobayaski, T. 1971. The Manmo Group of the Mongolian Geosyncline in Manchuria and adjacent areas. *Geol. Min. Res. Far East* 3:3–69

Laughton, A. S. 1975. Tectonic evolution of the northeast Atlantic Ocean: A review. *Nor. Geol. Unders. Pub.* 29:169–93

Lu, Y. H., Chu, C. H., Chien, Y. Y., Lin, H. L., Chow, T. Y., Yuan, K. X. 1974. Bio-environmental control hypothesis and its application to Cambrian biostratigraphy and paleozoogeography. *Nanking Inst. Geol. Paleontol. Mem., Academic Sinica* 5:27–110

McElhinny, M. W. 1973. *Palaeomagnetism and Plate Tectonics.* Cambridge: Cambridge Univ. Press. 358 pp.

McKerrow, W. S., Ziegler, A. M. 1972. Palaeozoic oceans. *Nature* 240:92–94

Meyerhoff, H. A., Meyerhoff, A. A. 1978. Spreading history of the eastern Indian Ocean and India's northward flight from Antarctica and Australia: discussion and reply. *Geol. Soc. Am. Bull.* 89:637–40

Middlemiss, F. A., Rawson, P. F., Newall, G., eds. 1971. *Faunal Provinces in Space and Time. Geol. J. Spec. Issue No. 4.* Liverpool: Seel House. 236 pp.

Molnar, P., Tapponnier, P. 1975. Cenozoic tectonics of Asia: Effects of a continental collision. *Science* 189:419–26

Morel, P., Irving, E. 1978. Tentative paleocontinental maps for the early Phanerozoic and Proterozoic. *J. Geol.* 86:535–61

Neill, W. T. 1969. *The Geography of Life.* New York: Columbia Univ. Press. 480 pp.

Newman, G. W., Mull, C. G., Watkins, N. D. 1977. Northern Alaska paleomagnetism, plate rotation, and tectonics. *Alaska Geol. Soc. Symp., Indications of Alaska Plate Tectonics,* pp. 16–19

Norin, E. 1946. *Geological Explorations in Western Tibet.* Rept. Sino-Swedish Exped., Publ. 29. Stockholm: Aktiebolaget Thule. 214 pp.

Packer, D. R., Stone, D. B. 1974. Paleomagnetism of Jurassic rocks from southern Alaska, and the tectonic implications. *Can. J. Earth Sci.* 11:976–97

Palmer, A. R. 1972. Problems in Cambrian biogeography. *24th Int. Geol. Congr.,* Sec. 7, Paleontology, pp. 310–15

Pistrak, R. M., Pashova, N. T. 1972. Structural position of the Centers of Volcanism in the Pripyat-Donets Aulacogen. *Geotectonics* 6:212–14

Robinson, P. L. 1973. Palaeoclimatology and continental drift. See Chaloner & Creber 1973, pp. 451–76

Rosenberg, G. D., Runcorn, S. K., eds. 1975. *Growth Rhythms and the History of the Earth's Rotation.* New York: Wiley. 559 pp.

Ross, C. A., ed. 1974. *Paleogeographic Provinces and Provinciality. Soc. Econ. Paleontol. Mineral., Spec. Publ. No. 21.* 233 pp.

Sheehan, P. 1973. The relation of late Ordovician glaciation to the Ordovician-Silurian changeover in North American brachiopod faunas. *Lethaia* 6:147–54

Smith, A. G., Briden, J. C., Drewry, G. E. 1973. Phanerozoic world maps. In

Organisms and Continents through Time, ed. N. F. Hughes. *Palaeontol. Assoc. Spec. Pap. No. 12*: 1–42
Stocklin, J. 1974. Possible ancient continental margins in Iran. In *The Geology of Continental Margins*, ed. C. A. Burk, C. L. Drake, pp. 873–88. New York: Springer. 1009 pp.
Stormer, L. 1967. Some aspects of the Caledonian Geosyncline and foreland west of the Baltic Shelf. *Q. J. Geol. Soc. London* 123: 183–214
Thorsteinsson, R., Tozer, E. T. 1970. Geology of the Arctic Archipelago—Chapter X. In *Geology and Economic Minerals of Canada*, ed. R. J. W. Douglas, pp. 547–90. *Econ. Geol. Rep. No. 1*. Geol. Surv. Canada. 838 pp.
Torii, T. 1975. Geochemical and geophysical studies of dry valleys, Victoria Land in Antarctica. *Mem. Natl. Inst. Polar Res. Tokyo, Japan. Spec. Issue No. 4*. 89 pp.
Van der Voo, R., French, A. N., French, R. B. 1979. A paleomagnetic pole position from the folded late Devonian Catskill Redbeds, and its megatectonic implications. *Geology* 7. In press
Veevers, J. J. 1976. Early Phanerozoic events alongside the Australasian-Antarctic Platform. *Geol. Soc. Aust. J.* 23: 183–206
Veevers, J. J., Cotterill, D. 1978. Western margin of Australia: evolution of a rifted arch system. *Geol. Soc. Am. Bull.* 89: 337–55
Vine, F. J. 1973. Organic diversity, palaeomagnetism, and Permian palaeogeography. See Smith et al 1973, pp. 61–77
Volkheimer, W. 1969. Palaeoclimatic evolution in Argentina and relations with other regions of Gondwana. In *Gondwana Stratigraphy. Int. Union Geol. Sci. Symp.*, pp. 551–88. Paris: UNESCO, 1173 pp.
Waterhouse, J. B., Bonham-Carter, G. F. 1975. Global distribution and character of Permian biomes based on brachiopod assemblages. *Can. J. Earth Sci.* 12: 1085–1146
Wegener, A. 1929. *The Origin of Continents and Oceans*. Transl. John Biram, 1966 New York: Dover. 246 pp. (from German)
Whittington, H. B. 1973. Ordovician Trilobites. See Hallam 1973, pp. 13–18
Wills, L. J. 1951. *Palaeogeographic Atlas of the British Isles and Adjacent Parts of Europe*. London: Blackie. 64 pp.
Ziegler, A. M., Hansen, K. S., Johnson, M. E., Kelly, M. A., Scotese, C. R., Van der Voo, R. 1977a. Silurian continental distributions, paleogeography, climatology, and biogeography. *Tectonophysics* 40: 13–51
Ziegler, A. M., Scotese, C. R., McKerrow, W. S., Johnson, M. E., Bambach, R. K. 1977b. Paleozoic biogeography of continents bordering the Iapetus (Pre-Caledonian) and Rheic (Pre-Hercynian) Oceans. In *Paleontology and Plate Tectonics*, ed. R. M. West, *Milwaukee Public Museum, Spec. Publ. Biol. Geol.* 2: 1–34
Zonenshayn, L. P. 1973. The evolution of central Asiatic geosynclines through seafloor spreading. *Tectonophysics* 19: 213–32
Zonenshayn, L. P., Gorodnitskiy, A. M. 1977a. Paleozoic and Mesozoic reconstructions of the continents and oceans. Article 1—Early and middle Paleozoic reconstructions. *Geotectonics* 11: 83–94
Zonenshayn, L. P., Gorodnitskiy, A. M. 1977b. Paleozoic and Mesozoic reconstructions of the continents and oceans. Article 2—Late Paleozoic and Mesozoic reconstructions. *Geotectonics* 11: 159–72

Appendix: References for Paleogeographic Data Points

General

Bassett, M. G., ed. 1976. *The Ordovician System: Proceedings of a Palaeontological Association Symposium*, Birmingham, September 1974. Cardiff: Univ. of Wales Press and National Museum of Wales. 696 pp.
Congrès Géologique International, 1956–1966. *Lexique Stratigraphique International*. Paris: Centre National de la Recherche Scientifique
Fairbridge, R. W., ed. 1975. *The Encyclopedia of World Regional Geology, Part I: Western Hemisphere (Including Antarctica and Australia)*. Encyclopedia of Earth Sciences, Vol. 8. Stroudsburg, Pennsylvania: Dowden, Hutchinson, & Ross. 704 pp.
Holland, C. H. 1971. *Cambrian of the New World*. New York: Interscience. 456 pp.
International Union of Geological Sciences. 1969. *Gondwana Stratigraphy*. France: UNESCO. 1173 pp.
International Union of Geological Sciences, Commission on Stratigraphy. 1970. *Second Gondwana Symposium: Proceedings and Papers*. Pretoria, South Africa: Council for Scientific and Industrial Research. 689 pp.
Oswald, D. H., ed. 1967. *International Symposium on the Devonian System.*

Calgary, Alberta: Alberta Society of Petroleum Geologists. Vol. 1, 1055 pp.; Vol. 2, 1377 pp.

North America

Armstrong, A. K. 1974. Carboniferous carbonate depositional models, preliminary lithofacies and paleotectonic maps, Arctic Alaska. *Am. Assoc. Petrol. Geol. Bull.* 58:621–45

Berry, W. B. N., Boucot, A. J. 1970. Correlation of the North American Silurian rocks. *Geol. Soc. Am. Spec. Pap. 102.* 289 pp.

Bird, J. M., Dewey, J. F. 1970. Lithosphere plate—continental margin tectonics and the evolution of the Appalachian Orogen. *Geol. Soc. Am. Bull.* 81:1031–60

Briggs, G., ed. 1974. Carboniferous of the Southeastern United States. *Geol. Soc. Am. Spec. Pap. 148.* 361 pp.

Brosge, W. P., Dutro, J. T. 1973. Paleozoic rocks of northern and central Alaska. *Am. Assoc. Petrol. Geol. Mem.* 19:361–75

Churkin, M. Jr. 1969. Paleozoic tectonic history of the Arctic Basin north of Alaska. *Science* 165:549–55

Churkin, M. Jr., Eberlein, G. D. 1977. Ancient borderland terranes of the North American Cordillera: Correlation and micro-plate tectonics. *Geol. Soc. Am. Bull.* 88:769–86

Cook, T. D., Bally, A. W., eds. 1975. *Stratigraphic Atlas of North and Central America.* Princeton, New Jersey: University Press. 272 pp.

Dewey, J. 1969. Evolution of the Appalachian/Caledonian Orogen. *Nature* 222:124–29

Douglas, R. J. W. 1970. *Geology and Economic Minerals of Canada. Geol. Surv. Canada, Econ. Geol. Rep. No. 1.* Ottawa. 838 pp.

Fullagar, P. 1971. Age and origin of plutonic intrusions in the Piedmont of the southeastern Appalachians. *Geol. Soc. Am. Bull.* 82:2845–62

Haller, J. 1971. *Geology of the East Greenland Caledonides.* Switzerland: Interscience. 413 pp.

Hatcher, R. D. Mr. 1972. Developmental model for the Southern Appalachians. *Geol. Soc. Am. Bull.* 83:2735–60

Henriksen, N. 1973. Caledonian geology of Scoresby Sund Region, central east Greenland. *Am. Assoc. Petrol. Geol. Mem.* 19:440–52

Lenz, A. C. 1972. Ordovician to Devonian history of northern Yukon and adjacent District of Mackenzie. *Bull. Can. Petrol. Geol.* 20:321–361

Lopez-Ramos, E. 1969. Marine Paleozoic rocks of Mexico. *Am. Assoc. Petrol. Geol. Bull.* 53:2399–2422

Mallory, W. W. 1972. *Geologic Atlas of the Rocky Mountain Region.* Rocky Mountain Assoc. Geol., Denver, CO. Denver: A. B. Hirschfield. 331 pp.

McCrossan, R. G., Glaister, R. P., eds. 1964. *Geological History of Western Canada.* Calgary: Alberta Soc. Petrol. Geol. 232 pp.

McKerrow, W. S., Ziegler, A. M. 1971. The Lower Silurian paleogeography of New Brunswick and adjacent areas. *J. Geol.* 79:635–46

Mesolella, K. J., McCormick, L. M., Ormiston, A. R., Robinson, J. D. 1974. Cyclic deposition of Silurian carbonates and evaporites in Michigan Basin. *Am. Assoc. Petrol. Geol. Bull.* 58:34–62

Pojeta, J., Kriz, J., Berdan, J. M. 1976. Silurian-Devonian pelecypods and Paleozoic stratigraphy of subsurface rocks in Florida and Georgia and related Silurian pelecypods from Bolivia and Turkey. *US Geol. Surv. Prof. Pap. No. 879.* 32 pp.

Rankin, D. W. 1976. Appalachian salients and recesses: Late Precambrian continental breakup and the opening of the Iapetus Ocean. *J. Geophys. Res.* 81:5605–19.

Shaver, R. H. 1977. Silurian reef geometry—new dimensions to explore. *J. Sediment. Petrol.* 47:1409–24

Sheehan, P. M. 1975. Evaluation of brachiopod faunas from Solis Limestone (Ordovician, Silurian, and Lower Devonian), Chihuahua, Mexico. *Am. Assoc. Petrol. Geol. Bull.* 59:1010–14

Tillement, B. A., Peniguel, G., Guillemin, J. P. 1976. Marine Pennsylvanian rocks in Hudson Bay. *Bull. Can. Petrol. Geol.* 24:418–39

Wanless, H. R., Tubb, J. B. Jr., Gednetz, D. E., Weiner, J. L. 1963. Mapping sedimentary environments of Pennsylvanian cycles. *Geol. Soc. Am. Bull.* 74:437–86

Williams, H. 1969. Pre-Carboniferous development of Newfoundland Appalachians. *Am. Assoc. Petrol. Geol. Mem.* 12:32–58

Zen, E-an, White, W. S., Hadley, J. B., Thompson, J. B. Jr. 1968. *Studies of Appalachian Geology: Northern and Maritime.* New York: Interscience. 475 pp.

Europe

Anderson, T. A. 1975. Carboniferous subduction complex in the Harz Mountains, Germany. *Geol. Soc. Am. Bull.* 86:77–82

Aubouin, J. 1965. *Geosynclines.* Amsterdam: Elsevier. 335 pp.

Bennison, G. M., Wright, A. E. 1969.

The Geological History of the British Isles. New York: St. Martin's Press. 406 pp.
Brunn, J. H., Mercier, J. 1971. Esquisse de la structure et de l'évolution géologique de la Grèce. *Tectonique de l'Afrique, Sci. Terre* 6:103–11 UNESCO
Burchfiel, B. C. 1976. Geology of Romania. *Geol. Soc. Am. Spec. Pap. 158.* 82 pp.
Charlesworth, J. K. 1966. *The Geology of Ireland.* Edinburgh, London: Oliver & Boyd. 276 pp.
Cielinski, S., Czerminske, J., Kozlowski, S., Pajchlowa, M., Ryka, W., Slacka, A. 1971. *Lithostratigraphical table of geological deposits in Poland.* Warsawa, Poland: Instytut Geologiczny. 2 p.
Cordarecea, A., Bercia, I., Boldur, C., Constantinof, D., Maier, O., Marinescu, F., Mercus, D., Nastaseanu, S. 1968. Geological structure of the southwestern Carpathians. *Inst. Geol. Congr., 23rd, Guide to Excursion 49 AC-Romania.* 49 pp.
Dessila-Codarcea, M. 1968. Évolution préalpine des massifs crystallaphylliens des Carpates roumaines. *Int. Geol. Congr., 23rd.* 3:121–27
Falke, H., ed. 1976. *The Continental Permian in Central, West, and South Europe.* Dordrecht, Holland: Reidel. 352 pp.
Flugel, H. W. 1975. Einige probleme des Variszikums von Neo-Europe. *Geol. Runds.* 64:1–62
Flugel, H. W., Schonlaub, H. P. 1972. Geleitworte zur stratigraphischen tabelle des Palaozoikums von Osterreich. *Verh. Geol.* B—A 2:187–98
Foose, R. M., Manheim, R. 1975. Geology of Bulgaria: a review. *Am. Assoc. Petrol. Geol. Bull.* 58:303–35
Fulop, J. 1968. Geology of the Transdanubian Central Mountains. *Int. Geol. Congr., 23rd Guide to Excursion 39C, Hungary.* 50 pp.
Garetskiy, R. G. 1970. On the Basement of the Moesice Plate. *Geotectonics* 4:272–77
Gardiner, P. R. R. 1975. Plate tectonics and the evolution of the southern Irish Caledonides. *Sci. Proc. Royal Dublin Soc., Ser. A, Pt. III, Geodynamics Prog.* 5:358–96
Gee, D. G. 1975. A tectonic model for the central part of the Scandinavian Caledonides. *Am. J. Sci.* 275-A:468–515
Gignoux, M. 1955. *Stratigraphic Geology.* San Francisco: Freeman. 682 pp.
Goncalves, F. 1974. Carta Geológica de Portugal, Folha 36-B, Estremoz. *Serv. Geol. Portugal.* Lisbon. 64 pp.
Harland, W. B. 1969. Contribution of Spitzbergen to understanding of evolution of North Atlantic region. *Am. Assoc. Petrol. Geol. Mem.* 12:817–51
Holland, C. H., ed. 1974. *Cambrian of the British Isles, Norden, and Spitsbergen.* London, New York, Sydney, Toronto: Wiley. 300 pp.
Jaeger, V. H. 1976. Das Silur und Unterdevon vom thuringischen typ in Sardinien und seine regionalgeologische bedeutung. *Nova Acta Lepoldina* 45:263–99
Julivert, M. 1971. Decollement tectonics in the Hercynian Cordillera of northwest Spain. *Am. J. Sci.* 270:1–29
Leggo, P. J., Tanner, P. W. G., Leake, B. E. 1969. Isochron study of Donegal granite and certain Dalradian rocks of Britain. *Am. Assoc. Petrol. Geol. Mem.* 12:354–62
Mannil, R. 1966. Evolution of the Baltic Basin during the Ordovician. *Inst. Geol. Akad. Nauk Estonia SSR.* 200 pp.
McKerrow, W. S. 1962. The chronology of Caledonian folding in the British Isles. *Proc. Nat. Acad. Sci.* 68:1905–13
Oxburgh, E. R. 1967. An outline of the geology of the central Eastern Alps. *Proc. Geol. Assoc.* 79:1–128
Pegrum, R. M., Rees, G., Naylor, D. 1975. *Geology of the north-west European continental shelf—volume 2—The North Sea.* London: Graham Trotman Dudley. 225 pp.
Robins, B., Gardner, P. M. 1975. The magmatic evolution of the Seiland Province, and Caledonian plate boundaries in northern Norway. *Earth Planet. Sci. Lett.* 26:167–78
Smith, D. B., Brunstrom, R. G. W., Manning, P. I., Simpson, S., Shotton, F. W. 1974. A correlation of Permian rocks in the British Isles. *J. Geol. Soc. London* 130:1–45
Sokolowski, S. 1970. *Geology of Poland.* Vol. 1, Stratigraphy, Part I, Pre-Cambrian and Paleozoic. Warsaw: Wydawnictwa Geologiczne. 651 pp.
Walter, R. 1972. Palaogeographie des Siluriums in Nord-, Mittel- und Westeuropa. *Geotektonische Forsch.* 41:1–180
Ziegler, A. M. 1970. Geosynclinal development of the British Isles during the Silurian Period. *J. Geol.* 78:445–79
Ziegler, P. A. 1975. The geological evolution of the North Sea area in the tectonic framework of north western Europe. *Norges Geol. Unders.* 316:1–27

Asia

Amantov, V. A., Blagonravov, V. A., Borzakovsky, Yu. A., Durante, M. V., Zonenshain, L. P., Luwsandarsan, B., Matrosov, P. S., Suetenko, O. D., Fulunnova, I. B., Khasin, R. A. 1970.

Stratigraphy and tectonics of the Mongolian Peoples Republic. *Trans. Academic Nauk.* 1:8–65. Joint Soviet Mongolian Scientific Research Geological Expedition

Belov, A. A. 1971. Paleozoic Tectonics of the western and central Taurus (Turkey). *Geotectonics* 5:31–38

Bender, F. 1968. *Geologie Von Jordanien.* Berlin: Gebrüder Borntraeger. 230 pp.

Berry, W. B. N., Boucot, A. J. 1972. Correlation of the Southeast Asian and Near Eastern Silurian rocks. *Geol. Soc. Am. Spec. Pap. 137.* 65 pp.

Brinkman, R. 1976. *Geology of Turkey.* Amsterdam, Oxford, New York: Elsevier. 158 pp.

Burton, C. K. 1972. Outline of the geological evolution of Malaya. *J. Geol.* 80:293–309

Frakes, L. A., Kemp, E. M., Crowell, J. C. 1975. Late Paleozoic glaciation: Part VI, Asia. *Geol. Soc. Am. Bull.* 86:454–64

Gansser, A. 1964. *Geology of the Himalayas.* London, New York, Sydney: Interscience. 289 pp.

Gobbett, D. J., Hutchison, C. S. 1973. *Geology of the Malay Peninsula.* New York, London, Sydney, Toronto: Interscience. 438 pp.

Gupta, H. K. 1973. *Seminar on geodynamics of the Himalayan region.* Hyderabad: Natl. Geophys. Res. Inst. 221 pp.

Hamada, T., Igo, H., Kobayashi, T., Kioke, T. 1975. Older and Middle Palaeozoic formations and fossils of Thailand and Malaysia. *Geol. Palaeontol. Southeast Asia* 15:1–38

Li, S. K., ed. 1956. Regional Stratigraphic Table of China (draft). Translated into English, 1958, by Joint Publication Research Service, U.S. Dept. of Commerce, Washington, D.C. *J.P.R.S.* 18:538. 511 pp.

Mehdiratta, R. C. 1962. *Geology of India, Pakistan, Ceylon, and Burma.* Delhi: Atma Ram. 203 pp. 3rd ed.

Minato, M., Gorai, M., Hunahashi, M., eds. 1965. *The Geologic Development of the Japanese Islands.* Tokyo: Tsukiji Shokan. 442 pp.

Nanking University, Department of Geology. 1974. Granitic rocks of different geological periods of southeastern China and their genetic relations to certain metallic mineral deposits. *Scientia Sinica* 17:55–72

Ohta, Y. 1973. *Geology of the Nepal Himalayas.* Japan: Saikon. 286 pp.

Powers, R. W., Ramirez, L. F., Redmond, C. D., Elberg, E. L. 1966. Geology of the Arabian Peninsula: Sedimentary geology of Saudi Arabia. *US Geol. Surv. Prof. Pap.* 560-D. 147 pp.

Stocklin, J. 1968. Structural history and tectonics of Iran: A review. *Am. Assoc. Petrol. Geol. Bull.* 52:1229–58

Stocklin, J. 1974. Possible Ancient Continental Margins in Iran. In *The Geology of Continental Margins,* ed. C. A. Burk, C. L. Drake, pp. 873–88. New York, Heidelberg, Berlin: Springer-Verlag. 1009 pp.

Takai, F., Matsumoto, T., Toriyama, R., eds. 1963. *Geology of Japan.* Berkeley, Los Angeles: Univ. Calif. Press. 279 pp.

Tateiwa, I. 1967. Synopsis of the geological systems of Korea. *Geol. Min. Res. Far East* 1:3–24

Thein, M., Haq, B. T. 1969. The Pre-Paleozoic and Paleozoic stratigraphy of Burma: a brief review. *Union Burma J. Sci. Tech.* 2:275–87

Toriyama, R., Hamada, T., Igo, H., Ingavat, R., Kanmera, K., Kobayashi, T., Koike, T., Ozawa, T., Pitakpaivan, K., Piyasin, S., Sakagami, S., Yanagica, J., Yin, E. H. 1975. The Carboniferous and Permian systems in Thailand and Malaysia. *Geol. Palaeontol. Southeast Asia* 15:39–76

Van Bemmelen, R. W. 1949. *The Geology of Indonesia.* Vol. 1A. The Hague: Govt. Print. Off. 732 pp.

Vinogradov, A. P., ed. 1968. *Atlas of the lithological-paleogeographical maps of the U.S.S.R.,* Vol. 1. Moscow: Min. Geol. U.S.S.R., Acad. Sci. U.S.S.R. 52 pp.

Vinogradov, A. P., ed. 1969. *Atlas of the lithological-paleogeographical maps of the U.S.S.R.,* Vol. 2. Moscow: Min. Geol. U.S.S.R., Acad. Sci. U.S.S.R. 65 pp.

Wittekindt, H. 1973. Erlauterungen zur geologischen karte von zentral- und sud-Afganistan 1:500000. *Bundenanst. Bodenforselrung.* Hannover. 109 pp.

Wolfart, V. R. 1967. Zur entwicklung der Palaozoischen Tethys in Vorderasien. *Erdoel Kohle Erdgas Petrochem.* 20:168–80

South America

Berry, W. B. N., Boucot, A. J. 1972. Correlation of the South American Silurian rocks. *Geol. Soc. Am. Spec. Pap. 133.* 59 pp.

Bigarella, J. J. 1973. Geology of the Amazon and Parnaiba Basins. In *The Ocean Basins and Margins,* ed. A. E. M. Nairn, F. G. Stehli. 1:25–86. The South Atlantic. New York, London: Plenum. 583 pp.

Frakes, L., Crowell, J. 1969. Late Paleozoic glaciation: I, South America. *Geol. Soc. Am. Bull.* 80:1007–42

Harrington, H. J. 1962. Paleogeographic development of South America. *Bull. Am. Assoc. Petrol. Geol.* 46:1773–1814

Jenks, W. F., ed. 1956. Handbook of South American Geology. *Geol. Soc. Am. Mem.* 65. 378 pp.

McBride, S. L., Caelles, J. C., Clark, A. H., Farrar, E. 1976. Palaeozoic radiometric age provinces in the Andean basement, latitudes 25°–30° S. *Earth Planet. Sci. Lett.* 29:373–83

Rocha-Campos, A. C. 1973. Upper Paleozoic and Lower Mesozoic paleogeography, and paleoclimatological and tectonic events in South America. *Can. Soc. Petrol. Geol.* 2:398–424

Stewart, J. W., Evernden, J. F., Snelling, N. J. 1974. Age determinations from Andean Peru: a reconnaissance survey. *Geol. Soc. Am. Bull.* 85:1107–116

Africa

Berry, W. B. N., Boucot, A. J. 1973. Correlation of the African Silurian rocks. *Geol. Soc. Am. Spec. Pap. 147.* 83 pp.

Black, R., Girod, M. 1970. Late Paleozoic to Recent igneous activity in West Africa and its relationship to basement structure. In *African magmatism and tectonics*, ed. T. N. Clifford, I. G. Glass, pp. 185–210. Edinburgh: Oliver & Boyd. 461 pp.

El-Nakkady, S. E. 1958. *Stratigraphic and Petroleum Geology of Egypt. Univ. Assiut. Monogr. Ser. No. 1.* Assiut. 215 pp.

Follot, J. 1952. Ahnet et Mouydir. *Int. Geol. Congr., 19th, Monogr. Reg. Ser. 1.* Algerie. 80 pp.

Frakes, L. A., Crowell, J. C. 1970. Late Paleozoic glaciation: II, Africa exclusive of the Karroo Basin. *Geol. Soc. Am. Bull.* 81:2261–86

Furon, R. 1963. *Geology of Africa.* New York: Hafner. 377 pp.

Furon, R. 1964. *Le Sahara: Géologie Ressources Minérales.* Paris: Payot. 313 pp. 2nd ed.

Haughton, S. H. 1969. *Geological History of Southern Africa.* Geol. Soc. South Africa. 535 pp.

Hollard, H., Jacquemont, P. 1970. Le Gothlandien, le Devonien et le Carbonifere des regions du Dra et du Zemoul. *Morocco Serv. Geol. Notes Mem.* 225: 7–33

Jaeger, V. H., Bonnefous, J., Massa, D. 1975. Le Silurien en Tunisie; ses relations avec le Silurian de Libye nord-occidentale. *Bull. Soc. Geol. France* 17:68–76

Menchikoff, N. 1957. Les grandes lignes de la Géologie Saharienne. *Rev. Geogr. Phys. Géol. Dynam.* 1:37–45

Rust, I. C. 1973. The evolution of the Paleozoic Cape Basin, southern margin of Africa. In *The Ocean Basins and Margins, The South Atlantic*, ed. A. E. M. Nairn, F. G. Stehli, 1:247–276. New York, London: Plenum. 583 pp.

Sougy, J. 1969. Grandes lignes structuales de la chaine des Mauritanides et de son avant-pays (socle Precambrian et sa couverture Infracambrienne et Paleozoique), Afrique de l'Ouest. *Bull. Soc. Geol. France* 11:133–49

Van Houten, F. B., Brown, R. H. 1977. Latest Paleozoic–early Mesozoic paleogeography, northwestern Africa. *J. Geol.* 85:143–56

Antarctica

Adie, R. J., ed. 1972. *Antarctic Geology and Geophysics.* Oslo: Universitelstorlaget. 876 pp.

Elliot, D. H. 1975. Tectonics of Antarctica: A review. *Am. J. Sci.* 275-A:45–106

McAlester, A., Doumani, G. 1966. Bivalve ecology in the Devonian of Antarctica. *J. Paleontol.* 40:752–55

Shergold, J. H., Cooper, R. A., Mackinnon, D. I., Yochelson, E. L. 1976. Late Cambrian brachiopoda, mollusca, and trilobita from Northern Victoria Land, Antarctica. *Palaeontology* 19:247–91

Australia

Audley-Charles, M. G. 1968. The geology of Portuguese Timor. *Geol. Soc. London, Mem. 4.* 76 pp.

Brown, D. A., Campbell, K. S. W., Crook, K. A. W. 1968. *The Geological Evolution of Australia and New Zealand.* Oxford: Pergamon. 409 pp.

Veevers, J. J. 1976. Early Phanerozoic events alongside the Australasian-Antarctic Platform. *Geol. Soc. Aust. J.* 23:183–206

AUTHOR INDEX

T

U

V

CUMULATIVE INDEXES

CONTRIBUTING AUTHORS VOLUMES 3–7

CHAPTER TITLES VOLUMES 3–7

ORDER FORM **ANNUAL REVIEWS INC.**

Please list on the order blank on the reverse side the volumes you wish to order and whether you wish a standing order (the latest volume sent to you automatically upon publication each year). Volumes not yet published will be shipped in month and year indicated. Prices subject to change without notice. Out of print volumes subject to special order.

NEW.... to be published in 1980

ANNUAL REVIEW OF PUBLIC HEALTH — $17.00 per copy ($17.50 outside USA)
Volume 1 available May 1980

SPECIAL PUBLICATIONS

ANNUAL REVIEW REPRINTS: CELL MEMBRANES, 1975-1977 (published 1978)
A collection of articles reprinted from recent Annual Review series.
Soft cover — $12.00 per copy ($12.50 outside USA)

THE EXCITEMENT AND FASCINATION OF SCIENCE (published 1965)
A collection of autobiographical and philosophical articles by leading scientists.
Clothbound — $6.50 per copy ($7.00 outside USA)

THE EXCITEMENT AND FASCINATION OF SCIENCE, VOLUME 2:
Reflections by Eminent Scientists (published 1978)
Hard cover — $12.00 per copy ($12.50 outside USA)
Soft cover — $10.00 per copy ($10.50 outside USA)

HISTORY OF ENTOMOLOGY (published 1973)
A special supplement to the ANNUAL REVIEW OF ENTOMOLOGY series.
Clothbound — $10.00 per copy ($10.50 outside USA)

ANNUAL REVIEW SERIES

Annual Review of ANTHROPOLOGY	$17.00 per copy ($17.50 outside USA)
Volumes 1-7 (1972-1978) currently available	Volume 8 available October 1979
Annual Review of ASTRONOMY AND ASTROPHYSICS	$17.00 per copy ($17.50 outside USA)
Volumes 1-16 (1963-1978) currently available	Volume 17 available September 1979
Annual Review of BIOCHEMISTRY	$18.00 per copy ($18.50 outside USA)
Volumes 28-47 (1959-1978) currently available	Volume 48 available July 1979
Annual Review of BIOPHYSICS AND BIOENGINEERING	$17.00 per copy ($17.50 outside USA)
Volumes 1-7 (1972-1978) currently available	Volume 8 available June 1979
Annual Review of EARTH AND PLANETARY SCIENCES	$17.00 per copy ($17.50 outside USA)
Volumes 1-6 (1973-1978) currently available	Volume 7 available May 1979
Annual Review of ECOLOGY AND SYSTEMATICS	$17.00 per copy ($17.50 outside USA)
Volumes 1-9 (1970-1978) currently available	Volume 10 available November 1979
Annual Review of ENERGY	$17.00 per copy ($17.50 outside USA)
Volumes 1-3 (1976-1978) currently available	Volume 4 available October 1979
Annual Review of ENTOMOLOGY	$17.00 per copy ($17.50 outside USA)
Volumes 7-23 (1962-1978) currently available	Volume 24 available January 1979
Annual Review of FLUID MECHANICS	$17.00 per copy ($17.50 outside USA)
Volumes 1-10 (1969-1978) currently available	Volume 11 available January 1979

(continued on reverse side)

Annual Review of GENETICS	$17.00 per copy ($17.50 outside USA)
Volumes 1-12 (1967-1978) currently available	Volume 13 available December 1979
Annual Review of MATERIALS SCIENCE	$17.00 per copy ($17.50 outside USA)
Volumes 1-8 (1971-1978) currently available	Volume 9 available August 1979
Annual Review of MEDICINE: Selected Topics in the Clinical Sciences	$17.00 per copy ($17.50 outside USA)
Volumes 1-3, 5-15, 17-29 (1950-1952, 1954-1964, 1966-1978) currently available	Volume 30 available April 1979
Annual Review of MICROBIOLOGY	$17.00 per copy ($17.50 outside USA)
Volumes 14-32 (1960-1978) currently available	Volume 33 available October 1979
Annual Review of NEUROSCIENCE	$17.00 per copy ($17.50 outside USA)
Volume 1 currently available	Volume 2 available March 1979
Annual Review of NUCLEAR AND PARTICLE SCIENCE	$19.50 per copy ($20.00 outside USA)
Volumes 9-28 (1959-1978) currently available	Volume 29 available December 1979
Annual Review of PHARMACOLOGY AND TOXICOLOGY	$17.00 per copy ($17.50 outside USA)
Volumes 1-3, 5-18 (1961-1963, 1965-1978) currently available	Volume 19 available April 1979
Annual Review of PHYSICAL CHEMISTRY	$17.00 per copy ($17.50 outside USA)
Volumes 9-29 (1958-1978) currently available	Volume 30 available November 1979
Annual Review of PHYSIOLOGY	$17.00 per copy ($17.50 outside USA)
Volumes 19-40 (1957-1978) currently available	Volume 41 available March 1979
Annual Review of PHYTOPATHOLOGY	$17.00 per copy ($17.50 outside USA)
Volumes 1-16 (1963-1978) currently available	Volume 17 available September 1979
Annual Review of PLANT PHYSIOLOGY	$17.00 per copy ($17.50 outside USA)
Volumes 10-29 (1959-1978) currently available	Volume 30 available June 1979
Annual Review of PSYCHOLOGY	$17.00 per copy ($17.50 outside USA)
Volumes 4, 5, 8, 10-29 (1953, 1954, 1957, 1959-1978) currently available	Volume 30 available February 1979
Annual Review of SOCIOLOGY	$17.00 per copy ($17.50 outside USA)
Volumes 1-4 (1975-1978) currently available	Volume 5 available August 1979

To ANNUAL REVIEWS INC., 4139 El Camino Way, Palo Alto, CA 94306 USA (415-493-4400)
Please enter my order for the following publications:
(Standing orders: indicate which volume you wish order to begin with)

________________, Vol(s). ____ Standing order ____
________________, Vol(s). ____ Standing order ____
________________, Vol(s). ____ Standing order ____
________________, Vol(s). ____ Standing order ____

Amount of remittance enclosed $______
Please bill me for the amount $______

California residents please add sales tax.
Prices subject to change without notice.

SHIP TO (Include institutional purchase order if billing address is different)

Name ________________
Address ________________
____
________________ Postal Code ________
Signed ________________ Date ________

___ Send free copy of annual Prospectus for current year
___ Send free back contents brochure for Annual Review(s) of ________________

PLACE
STAMP
HERE

ANNUAL REVIEWS INC.
4139 EL CAMINO WAY
PALO ALTO, CALIFORNIA 94306, USA